Square Matrices of Order 2

Vasile Pop • Ovidiu Furdui

Square Matrices of Order 2

Theory, Applications, and Problems

Foreword by Dennis S. Bernstein

 Springer

Vasile Pop
Department of Mathematics
Technical University of Cluj-Napoca
Cluj-Napoca, Romania

Ovidiu Furdui
Department of Mathematics
Technical University of Cluj-Napoca
Cluj-Napoca, Romania

ISBN 978-3-319-85526-4   ISBN 978-3-319-54939-2   (eBook)
DOI 10.1007/978-3-319-54939-2

Mathematics Subject Classification (2010): 15A03, 15A04, 15A09, 15A15, 15A16, 15A18, 15A21, 15A23, 15A24, 15A27, 15A60, 15A99, 15B10, 15B33, 15B36, 15B51, 15B57, 11B83, 40A05, 40A30, 97I30, 97I50

Printed on acid-free paper

This Springer imprint is published by Springer Nature
The registered company is Springer International Publishing AG
The registered company address is: Gewerbestrasse 11, 6330 Cham, Switzerland

Gold, when multiplied, conspire against his master; but books, when multiplied, make great use of those who have them.
—St. John Chrysostom (347–407)

He who neglects learning in his youth loses the past and is dead for the future.
—Euripides (480 B.C.–406 B.C.)

To Alina

O. F.

Foreword

Mathematics began with the positive integers. Next came fractions, the number zero, and negative numbers. Attempts to extract the roots of cubic polynomials led to the imaginary unit $i = \sqrt{-1}$, whose name suggests something fictitious and suspect. Beyond the complex numbers, quaternions enlist a trio of noncommuting units. Vectors arise from the need to keep track of multiple scalars, such as coordinates in a plane or three-dimensional space.

Matrices, which represent linear functions on vectors, are of immense practical importance in science and engineering, and the prosaic matrix-vector equation $Ax = b$ is possibly the most important equation in all of numerical computation. The solution of a partial differential equation may entail a matrix A whose order is in the thousands or millions, and the order of the Google page rank matrix, which represents the interdependence of web sites, is several billion. Large matrices represent real problems whose solution requires vast computing resources and innovative algorithms. Matrices are the bread and butter of modern computation.

In this unique and charming book, Vasile Pop and Ovidiu Furdui eschew large matrices and instead focus their attention on the simplest possible case, namely, matrices of size 2×2. This raises two questions: Why consider such a special case? and How much interesting mathematics can there be on 2×2 matrices? The second question has a quick answer: A surprising amount. The first question requires more discussion.

Since a 2×2 matrix has four entries, this book is devoted to the properties of a mathematical structure specified by a quadruple of real numbers. For example, the matrix $A = \begin{pmatrix} 1 & 2 \\ -2 & 1 \end{pmatrix}$ provides a representation of the complex number $1 + 2i$. Moreover, using $B = \begin{pmatrix} 3 & 4 \\ -4 & 3 \end{pmatrix}$ to represent $3 + 4i$, the sum $A + B$ represents $4 + 6i$. Likewise, AB represents $(1 + 2i)(3 + 4i)$, and A^{-1} represents $\frac{1}{1+2i}$. Consequently, 2×2 matrices can be used to represent and manipulate complex numbers, all without the need for the imaginary unit i.

As a physical application of 2×2 matrices, the vector differential equation $x' = Ax$ with $A = \begin{pmatrix} 0 & 1 \\ -\omega_n^2 & 0 \end{pmatrix}$ represents the simple harmonic oscillator with natural frequency ω_n. The extension $A = \begin{pmatrix} 0 & 1 \\ -\omega_n^2 & -2\zeta\omega_n \end{pmatrix}$ includes the effect of damping, where ζ is the damping ratio. The special case $A = \begin{pmatrix} 0 & 1 \\ 0 & 0 \end{pmatrix}$ models a mass without a spring or dashpot. Note that A is not zero but satisfies $A^2 = 0$; A is nilpotent. No nonzero real, complex, or quaternionic scalar can be nilpotent, and this observation shows that matrices have properties that are not shared by scalars.

This book is innovative in its concept and scope, and exciting in its content. It shows that a "simple" setting can illuminate and expose the beauty and intricacies of a rich and useful mathematical construction. The authors of this book, both of whom are world-renowned for the development and solution of mathematical problems, have gathered a truly splendid collection of problems, ideas, and techniques. I await the sequel on $n \times n$ matrices. After reading this wonderful book, dear reader, I am sure that you will share my anticipation.

Ann Arbor, MI, USA   Dennis S. Bernstein
August 2016

Preface

This book is the fruit of the authors' work in the last decade in teaching linear algebra classes and in preparing the students for university entrance examinations tests as well as for national and international student competitions like Traian Lalescu, a Romanian mathematical competition, Seemous, and IMC.

The goal of this book is to discuss *completely* and *in detail* all important topics related to the theory of square matrices of order two and the theory of vector spaces of dimension two with all of their implications in the study of plane geometry, the algebraic curves of degree 2, the conics, the extension of elementary functions from calculus to functions of matrices, and the applications of matrix calculus to mathematical analysis.

We believe this book, which treats exclusively the applications of the matrix calculus for square matrices of order 2, is necessary in the literature for the following reasons:

- this is, perhaps, the *first book in the literature* that collects, in a single volume, the theory, the applications, and the problems involving square matrices of order 2;
- the book is written in a way that is accessible to anyone with a modest background in mathematics;
- the topics and problems extend naturally to matrices of order n, and the techniques and the ideas in the book, which are highly original, can be used in linear algebra;
- the large number of problems of various degrees of difficulty, from easy to difficult and challenging, offers the reader a valuable source for learning the basics of linear algebra and matrix theory;

(continued)

- the ideas and problems in the book are original, many of the problems being proposed by the authors in journals from Romania (*Gazeta Matematică Seria B*, *Gazeta Matematică Seria A*) and abroad (*the American Mathematical Monthly*, *Mathematics Magazine*, *the College Mathematics Journal*) and others will see the light of publication for the first time in the literature.

To whom is this book addressed?

To you, the reader, who likes matrix theory and to those who have not had the chance of being introduced yet to the fascinating world of matrices. The book is geared towards high school students who wish to study the basics of matrix theory, as well as undergraduate students who want to learn the first steps of linear algebra and matrix theory, a fundamental topic that is taught nowadays in all universities of the world.

We also address this book to first- and second-year graduate students who wish to learn more about an application of a certain technique on matrices, to doctoral students who are preparing for their prelim exams in linear algebra as well as to anyone who is willing to explore the strategies in this book and savor a splendid collection of problems involving matrices of order 2.

We also address this book to professionals and nonprofessionals who can find, in a single volume, everything one needs to know, from the ABCs to the most advanced topics of matrices of order 2 and their connection to mathematical analysis and linear algebra.

The book is a must-have for students who take a linear algebra class and for those who prepare for mathematical competitions like Putnam, Seemous, and International Mathematical Competition for University Students.

The book can be used by our colleagues who teach elementary matrix theory in high school, by instructors who teach linear algebra in college or university, and by those who prepare students for mathematical competitions.

Why a book on matrices of order 2?

- First, because in any linear algebra class the students interact with matrices and the most simple of them all are those of order 2.
- Second, because there are lots of beautiful results in matrix theory, see the nice determinant formulae in Chapter 1, that hold only for square matrices of order 2, their extension to n-dimensional matrices loses the splendor, the simplicity, and the finesse of their proofs.
- Third, because in mathematics one needs to understand simple things first, and what is most simple in linear algebra is a matrix of order 2.

(continued)

- Fourth, because the authors wanted to gather together in a single volume, for the help of students and instructors, everything that should be known about matrices of order 2 and their applications. Although many excellent books on linear algebra have a chapter or special sections devoted to square matrices of order 2, in this text the reader has all the formulae one needs to know about the most basic topics of matrix theory.

The book offers an unusual collection of problems specializing on matrices of order 2 that are rarely seen. The problems vary in difficulty from the easiest ones involving the calculations of the nth power of a matrix to the most advanced like those in Chapter 4. Most of the problems in this book are new and original and see the light of publication for the first time. Others are inspired by several books that are not found in the western literature [47, 48, 50, 51]. Another important source of inspiration for some of the problems is the famous Romanian journal *Gazeta Matematică B*, the oldest mathematical publication in Romania, the first issue being published in 1895, and perhaps the first mathematical journal in the world with a problem column. We do not claim the originality of all the problems included in this volume, and we are aware that some exercises and results are either known or very old.

We solved most of the problems in detail, but there are a few exercises with no solutions. This is because we want to stimulate the readers not to follow our techniques for solving a problem but instead to develop their own methods of attack. A couple of problems are challenge problems. These could be viewed as more demanding problems that encourage the reader to be creative and stimulate research and discovery of original solutions for proving known results and establishing new ones.

The contents of the book

The book has six chapters and two appendices.

In Chapter 1, we go over the basic results and definitions and we give, among others, the structure of special matrices, such as idempotent, nilpotent, involutory, skew-involutory, and orthogonal. We also discuss the centralizer of a square matrix. This chapter contains special problems on the computation of determinants as well as a section of exercises on the classical algebraic structures such as groups, rings, fields of matrices and their properties.

Chapter 2 deals with the celebrated Cayley–Hamilton theorem, its reciprocal, the Jordan canonical form, the real and the rational canonical form of special matrices. Also, this chapter contains a section called quickies which is about problems that have an unexpected succinct solution.

In Chapter 3 we give formulae for the calculation of the nth power of a matrix, we study sequences defined by linear and homographic recurrence relations, we solve binomial matrix equations, and we review the famous equation of Pell. This chapter contains a special section devoted to the binomial equation $X^n = aI_2$, $a \in \mathbb{R}^*$, which we believe appears for the first time in the literature. The problems are new and original, see the problem about Viéte's formulae for a quadratic matrix equation, and challenge the reader to explore the tools discussed throughout the chapter.

Chapter 4, the **jewel of this book**, is a mixture of matrix theory and mathematical analysis. This chapter contains sequences and series of matrices, elementary functions of matrices, and introduces, we believe for the first time in the literature, the **Riemann zeta function** as well as the **Gamma function** of a square matrix. The problems vary in diversity, from the computations of an exponential function, the resolution of a system of differential equations to integrals of matrices, single or double, and the calculation of **Frullani matrix integrals**.

Chapter 5, **unique** in the literature, is about the study of special linear applications of the plane such as projections and reflections and their fundamental properties. Most of the problems in this chapter, which are gems of linear algebra, appear for the first time in the literature.

In Chapter 6 we use the Jordan canonical form to reduce the algebraic curves of degree 2, the conics, to their canonical form. The problems in this chapter are neither standard nor known: they range from the reduction of a conic to its canonical form to the study of an extremum problem or even to the calculation of a double integral over an elliptical domain.

Appendix A contains a bouquet of topics from linear algebra such as the computation of exponential and trigonometric functions of matrices to topics from classical analysis like the calculation of nonstandard series and the discussion of two Frullani integrals that are new in analysis.

In Appendix B, hopefully *new in linear algebra*, we solve the trigonometric matrix equations and invite the reader to explore further these topics.

The book is designed to fascinate the novice, puzzle the expert, and trigger the imaginations of all. It contains an unusual collection of problems in matrix theory which we think they are *splendid*. Whether the problems turn out to be splendid or not, and we believe they are!, that is for you the reader to decide. We hope you will enjoy both the problems and the theory.

Acknowledgments

We express our appreciation to professor Ioan Şerdean from Orăştie who provided us with reference materials for this book.

Many special thanks to Alina Sîntămărian who read large portions of this book and helped us to improve the presentation of this material. She has also provided reference materials as well as assisted the second author with her expertise in Latex to solve various problems that appeared during typesetting. Alina is also *responsible* for drawing all the figures in this book by using the GeoGebra package and she is to *thank* for valuable comments during the preparation of this book.

Thank you all and enjoy the **splendid** 2×2 **matrices**!!!

Cluj-Napoca, Romania   Vasile Pop
June 2016   Ovidiu Furdui

Contents

Foreword ix

Preface xi

Notations xxi

1   **Matrices of order 2** 1
    1.1   Definitions and notations 1
    1.2   Matrices and their special properties 7
    1.3   The set of complex numbers and matrices of order 2 17
    1.4   Problems 20
    1.5   A bouquet of group, ring, and field theory problems 28
    1.6   Solutions 35

2   **The Cayley–Hamilton Theorem** 63
    2.1   The Cayley–Hamilton Theorem 63
    2.2   The eigenvalues of symmetric matrices 73
    2.3   The reciprocal of the Cayley–Hamilton Theorem 74
    2.4   The characteristic polynomial of matrices XY and YX 75
    2.5   The Jordan canonical form 77
    2.6   Problems 81
    2.7   Solutions 89
    2.8   Quickies 103
    2.9   Solutions 104

3   **Applications of Cayley–Hamilton Theorem** 107
    3.1   The nth power of a square matrix of order 2 107
        3.1.1   Problems 110
        3.1.2   Solutions 116
    3.2   Sequences defined by systems of linear recurrence relations 122
        3.2.1   Problems 123
        3.2.2   Solutions 125

3.3   Sequences defined by homographic recurrence relations............. 129
       3.3.1   Problems.. 131
       3.3.2   Solutions... 133
3.4   Binomial matrix equations 137
       3.4.1   An artistry of binomial equations. The nth real
               roots of $aI_2, a \in \mathbb{R}^*$ 145
       3.4.2   Problems.. 149
       3.4.3   Solutions... 156
3.5   Pell's diophantine equation 173
       3.5.1   Problems.. 179
       3.5.2   Solutions... 180

4   Functions of matrices. Matrix calculus 183
    4.1   Sequences and series of matrices 183
    4.2   Elementary functions of matrices............................. 188
    4.3   A novel computation of function matrices 194
    4.4   Explicit expressions of e^A, $\sin A$, and $\cos A$ 199
    4.5   Systems of first order differential equations
           with constant coefficients 201
    4.6   The matrix Riemann zeta function............................. 205
    4.7   The matrix Gamma function 208
    4.8   Problems 211
    4.9   Solutions 237

5   Applications of matrices to plane geometry 281
    5.1   Linear transformations 281
    5.2   The matrix of special transformations........................ 283
    5.3   Projections and reflections of the plane 286
    5.4   Gems on projections and reflections.......................... 292
    5.5   The isometries of the plane 296
    5.6   Systems of coordinates on the plane.......................... 299
    5.7   Problems 300
    5.8   Solutions 303

6   Conics 315
    6.1   Conics 315
    6.2   The reduction of conics to their canonical form 319
    6.3   Problems 325
    6.4   Solutions 329

A   Gems of classical analysis and linear algebra......................... 347
    A.1   Series mirabilis 347
    A.2   Two quadratic Frullani integrals 352
    A.3   Computing e^{Ax} 354
    A.4   Computing $\sin Ax$ and $\cos Ax$ 356

B   Trigonometric matrix equations.. 359
   B.1   Four trigonometric equations 359

References... 365

Index.. 369

Notations

$\mathbb{N}$	the set of natural numbers ($\mathbb{N} = \{1, 2, 3, \ldots\}$)
$\mathbb{Z}$	the set of integers ($\mathbb{Z} = \{\ldots, -2, -1, 0, 1, 2, \ldots\}$)
$\mathbb{Q}$	the set of rational numbers
$\mathbb{Q}^*$	the set of nonzero rational numbers ($\mathbb{Q}^* = \mathbb{Q} \setminus \{0\}$)
$\mathbb{R}$	the set of real numbers
$\mathbb{R}^*$	the set of nonzero real numbers ($\mathbb{R}^* = \mathbb{R} \setminus \{0\}$)
$\overline{\mathbb{R}}$	the completed real line ($\overline{\mathbb{R}} = \mathbb{R} \cup \{-\infty, \infty\}$)
$\mathbb{C}$	the set of complex numbers
$\mathbb{C}^*$	the set of nonzero complex numbers ($\mathbb{C}^* = \mathbb{C} \setminus \{0\}$)
$\Re(z)$	the real part of the complex number z
$\Im(z)$	the imaginary part of the complex number z
$\binom{n}{k}$	the binomial coefficient indexed by n and k is the coefficient of x^k term in the polynomial expansion of the binomial power $(1 + x)^n$
$\mathcal{M}_2(\mathbb{F})$	the set of square matrices of order 2 with entries in $\mathbb{F} \in \{\mathbb{Z}, \mathbb{Q}, \mathbb{R}, \mathbb{C}\}$
$\mathrm{GL}_2(\mathbb{C})$	the set of invertible matrices
$\mathrm{SL}_2(\mathbb{C})$	the special linear group
A^T	the transpose of A
$\overline{A}$	the conjugate of A
$\mathrm{Tr}(A)$	the trace of A
$\det A$	the determinant of A
A^{-1}	the inverse of A
A_*	the adjugate of A, also denoted by $\mathrm{adj}(A)$
R_α	the rotation matrix of angle α, i.e., $R_\alpha = \begin{pmatrix} \cos\alpha & -\sin\alpha \\ \sin\alpha & \cos\alpha \end{pmatrix}$
J_A	the Jordan canonical form of the matrix A
$(F_n)_{n\geq 0}$	the Fibonacci sequence $F_0 = 0, F_1 = 1$ and $F_{n+1} = F_n + F_{n-1}$, $n \geq 1$
$(L_n)_{n\geq 0}$	the Lucas sequence $L_0 = 2, L_1 = 1$ and $L_{n+1} = L_n + L_{n-1}$, $n \geq 1$

$AX = \lambda X, \ X \neq 0$	the eigenvalue-eigenvector equation
$\rho(A)$	the spectral radius of A
$\mathrm{Spec}(A)$	the spectrum of $A \in \mathcal{M}_2(\mathbb{C})$
$V_1 \oplus V_2$	the direct sum of vector (sub)spaces V_1 and V_2
$\mathrm{Ker} f_A$	the kernel of the linear application f_A
$\mathrm{Im} f_A$	the image of the linear application f_A
H_n	the nth harmonic number
	$H_n = 1 + 1/2 + 1/3 + \cdots + 1/n$
$\zeta(3)$	Apéry's constant
	$\zeta(3) = \sum_{n=1}^{\infty} 1/n^3 = 1.2020569031\ldots$
ζ	the Riemann zeta function
	$\zeta(z) = \sum_{n=1}^{\infty} 1/n^z = 1 + 1/2^z + 1/3^z + \cdots + 1/n^z$
	$+ \cdots, \quad \Re(z) > 1$
Li_2	the Dilogarithm function
	$\mathrm{Li}_2(z) = \sum_{k=1}^{\infty} z^k/k^2 = -\int_0^z \frac{\ln(1-t)}{t} \mathrm{d}t, \quad |z| \leq 1$
Li_n	the Polylogarithm function
	$\mathrm{Li}_n(z) = \sum_{k=1}^{\infty} z^k/k^n = \int_0^z \frac{\mathrm{Li}_{n-1}(t)}{t} \mathrm{d}t, \quad |z| \leq 1 \text{ and } n \neq 1, 2$
Γ	the Gamma function (Euler's Gamma function)
	$\Gamma(z) = \int_0^{\infty} x^{z-1} \mathrm{e}^{-x} \mathrm{d}x, \quad \Re(z) > 0$

Chapter 1
Matrices of order 2

*Any work has mistakes. Mistakes are an incentive
to do better. There comes a day when the worker
dies but the world has used his work and the pain
that brought a new work.*

Nicolae Iorga (1871–1940)

1.1 Definitions and notations

Definition 1.1 Let F be a set of real or complex numbers. By a square matrix of order 2 with entries of F we understand an array having two rows and two columns

$$A = \begin{pmatrix} a_{11} & a_{12} \\ a_{21} & a_{22} \end{pmatrix},$$

where $a_{ij} \in F$, $i, j \in \{1, 2\}$, are called the elements of the matrix A.

The ordered pair (a_{11}, a_{22}) is called the *main diagonal* of A and the ordered pair (a_{12}, a_{21}) is called the *secondary diagonal* of A.

A matrix of order 2 is denoted by $A = (a_{ij})_{i,j=1,2}$ and the set of all matrices of order two with complex entries is denoted by $\mathcal{M}_2(\mathbb{C})$. In this set we distinguish the following subsets

$$\mathcal{M}_2(\mathbb{Z}) \subset \mathcal{M}_2(\mathbb{Q}) \subset \mathcal{M}_2(\mathbb{R}) \subset \mathcal{M}_2(\mathbb{C}).$$

If $A = (a_{i,j})_{i,j=1,2}$ and $B = (b_{i,j})_{i,j=1,2}$, then we say that $A = B$ if and only if $a_{i,j} = b_{i,j}$ for all $i, j \in \{1, 2\}$.

Definition 1.2 *Addition of matrices.* Let $A, B \in \mathcal{M}_2(\mathbb{C})$,

$$A = \begin{pmatrix} a_{11} & a_{12} \\ a_{21} & a_{22} \end{pmatrix} \quad \text{and} \quad B = \begin{pmatrix} b_{11} & b_{12} \\ b_{21} & b_{22} \end{pmatrix}.$$

© Springer International Publishing AG 2017
V. Pop, O. Furdui, *Square Matrices of Order 2*, DOI 10.1007/978-3-319-54939-2_1

The *sum $A + B$* of matrices A and B is the matrix

$$A + B = \begin{pmatrix} a_{11} + b_{11} & a_{12} + b_{12} \\ a_{21} + b_{21} & a_{22} + b_{22} \end{pmatrix}.$$

We give below the properties of the addition of two matrices which can be verified by direct computation.

Lemma 1.1 *The following equalities hold:*

(a) *(commutativity)* $A + B = B + A, \quad \forall A, B \in \mathcal{M}_2(\mathbb{C})$;
(b) *(associativity)* $(A + B) + C = A + (B + C), \quad \forall A, B, C \in \mathcal{M}_2(\mathbb{C})$;
(c) *(the zero element) the zero matrix*

$$O_2 = \begin{pmatrix} 0 & 0 \\ 0 & 0 \end{pmatrix}$$

  verifies the following equality $A + O_2 = O_2 + A = A, \quad \forall A \in \mathcal{M}_2(\mathbb{C})$;
(d) *(the opposite element) $\forall A \in \mathcal{M}_2(\mathbb{C})$ there is $-A \in \mathcal{M}_2(\mathbb{C})$ such that $A + (-A) = (-A) + A = O_2$. If $A = (a_{ij})_{i,j=1,2}$, then $-A = (-a_{ij})_{i,j=1,2}$.*

Remark 1.1 Lemma 1.1 shows that $(\mathcal{M}_2(\mathbb{Z}), +)$, $(\mathcal{M}_2(\mathbb{Q}), +)$, $(\mathcal{M}_2(\mathbb{R}), +)$, and $(\mathcal{M}_2(\mathbb{C}), +)$ are abelian groups.

Definition 1.3 *Multiplication of matrices.*
  Let $A, B \in \mathcal{M}_2(\mathbb{C})$,

$$A = \begin{pmatrix} a_{11} & a_{12} \\ a_{21} & a_{22} \end{pmatrix} \quad \text{and} \quad B = \begin{pmatrix} b_{11} & b_{12} \\ b_{21} & b_{22} \end{pmatrix}.$$

The *product AB* of matrices A and B is the matrix defined by

$$AB = \begin{pmatrix} a_{11}b_{11} + a_{12}b_{21} & a_{11}b_{12} + a_{12}b_{22} \\ a_{21}b_{11} + a_{22}b_{21} & a_{21}b_{12} + a_{22}b_{22} \end{pmatrix}.$$

In other words, the (i, j) entry of the matrix AB is obtained by adding the products of the corresponding entries of ith row of A with the corresponding entries of the jth column of B.

In general, the multiplication of matrices is *not commutative*, i.e., $AB \neq BA$. For example, if

$$A = \begin{pmatrix} 1 & 2 \\ -1 & 0 \end{pmatrix} \quad \text{and} \quad B = \begin{pmatrix} 2 & 1 \\ 1 & 2 \end{pmatrix},$$

then

$$AB = \begin{pmatrix} 4 & 5 \\ -2 & -1 \end{pmatrix} \neq \begin{pmatrix} 1 & 4 \\ -1 & 2 \end{pmatrix} = BA.$$

Next we give the properties of the multiplication of matrices which can be proved by direct computation.

Lemma 1.2 *The following equalities hold:*

(a) *(associativity)* $(AB)C = A(BC)$, $\forall A, B, C \in \mathcal{M}_2(\mathbb{C})$;
(b) *(distributivity to the left)* $A(B + C) = AB + AC$, $\forall A, B, C \in \mathcal{M}_2(\mathbb{C})$;
(c) *(distributivity to the right)* $(A + B)C = AC + BC$, $\forall A, B, C \in \mathcal{M}_2(\mathbb{C})$;
(d) *(the unit matrix) the unit matrix*

$$I_2 = \begin{pmatrix} 1 & 0 \\ 0 & 1 \end{pmatrix}$$

verifies the equality $AI_2 = I_2A = A$, $\forall A \in \mathcal{M}_2(\mathbb{C})$.

Remark 1.2 Lemma 1.2 shows that $(\mathcal{M}_2(\mathbb{Q}), \cdot)$, $(\mathcal{M}_2(\mathbb{R}), \cdot)$, and $(\mathcal{M}_2(\mathbb{C}), \cdot)$ are monoids.

Also, Lemma 1.1 and Lemma 1.2 show that $(\mathcal{M}_2(\mathbb{Q}), +, \cdot)$, $(\mathcal{M}_2(\mathbb{R}), +, \cdot)$, and $(\mathcal{M}_2(\mathbb{C}), +, \cdot)$ are noncommutative rings with the zero element O_2 and the unit element I_2.

Since the multiplication of matrices verifies part (a) of Lemma 1.2 we define the powers of a matrix $A \in \mathcal{M}_2(\mathbb{C})$ as follows: $A^0 = I_2$ (if $A \neq O_2$), $A^1 = A$, $A^2 = A \cdot A$, $A^3 = A^2 \cdot A, \ldots, A^n = A^{n-1} \cdot A, n \in \mathbb{N}$.

Definition 1.4 *Multiplication by scalars.*
  Let $\alpha \in \mathbb{C}$ and let $A \in \mathcal{M}_2(\mathbb{C})$,

$$A = \begin{pmatrix} a_{11} & a_{12} \\ a_{21} & a_{22} \end{pmatrix}.$$

The product of the complex number α and the matrix A is the matrix defined by

$$\alpha A = \begin{pmatrix} \alpha a_{11} & \alpha a_{12} \\ \alpha a_{21} & \alpha a_{22} \end{pmatrix}.$$

Lemma 1.3 *The following equalities hold:*

(a) $(\alpha + \beta)A = \alpha A + \beta A$, $\forall \alpha, \beta \in \mathbb{C}$, $\forall A \in \mathcal{M}_2(\mathbb{C})$;
(b) $\alpha(A + B) = \alpha A + \alpha B$, $\forall \alpha \in \mathbb{C}$, $\forall A, B \in \mathcal{M}_2(\mathbb{C})$;
(c) $\alpha(\beta A) = (\alpha\beta)A$, $\forall \alpha, \beta \in \mathbb{C}$, $\forall A \in \mathcal{M}_2(\mathbb{C})$;
(d) $1 \cdot A = A$, $\forall A \in \mathcal{M}_2(\mathbb{C})$.

Remark 1.3 The properties of Lemma 1.3 show that the groups $(\mathcal{M}_2(\mathbb{Q}), +)$, $(\mathcal{M}_2(\mathbb{R}), +)$, and $(\mathcal{M}_2(\mathbb{C}), +)$ are vector spaces over $\mathbb{Q}$, $\mathbb{R}$, and $\mathbb{C}$ respectively.

Definition 1.5 *Diagonal and triangular matrices.*

(a) A matrix of the form $\begin{pmatrix} a & 0 \\ 0 & d \end{pmatrix} \in \mathcal{M}_2(\mathbb{C})$ is called *diagonal.*

(b) Matrices of the form $\begin{pmatrix} a & b \\ 0 & d \end{pmatrix} \in \mathcal{M}_2(\mathbb{C})$ or $\begin{pmatrix} a & 0 \\ c & d \end{pmatrix} \in \mathcal{M}_2(\mathbb{C})$ are called *triangular.*

We use notations $\mathcal{M}_{2,1}(\mathbb{C})$ and $\mathcal{M}_{1,2}(\mathbb{C})$ for the set of columns and the set of rows with complex entries, respectively.

Operations with vectors

If $C_1 = \begin{pmatrix} x_1 \\ y_1 \end{pmatrix}$ and $C_2 = \begin{pmatrix} x_2 \\ y_2 \end{pmatrix}$, then

$$C_1 + C_2 = \begin{pmatrix} x_1 + x_2 \\ y_1 + y_2 \end{pmatrix} \quad \text{and} \quad \alpha C_1 = \begin{pmatrix} \alpha x_1 \\ \alpha y_1 \end{pmatrix}, \quad \alpha \in \mathbb{C}.$$

One can check that similar properties like those in Lemma 1.3 are verified by vectors and we have that $\mathcal{M}_{2,1}(\mathbb{C})$, $\mathcal{M}_{2,1}(\mathbb{R})$ and $\mathcal{M}_{2,1}(\mathbb{Q})$ are vector spaces over $\mathbb{C}$, $\mathbb{R}$, and $\mathbb{Q}$, respectively. Their *dimension* is 2 with *canonical basis* $\mathcal{B} = \{E_1, E_2\}$, where $E_1 = \begin{pmatrix} 1 \\ 0 \end{pmatrix}$ and $E_2 = \begin{pmatrix} 0 \\ 1 \end{pmatrix}$.

Definition 1.6 *Basis of vector spaces $\mathcal{M}_{2,1}(\mathbb{C})$, $\mathcal{M}_{2,1}(\mathbb{R})$, $\mathcal{M}_{2,1}(\mathbb{Q})$.*

The vectors $X_1 = \begin{pmatrix} x_1 \\ y_1 \end{pmatrix}$ and $X_2 = \begin{pmatrix} x_2 \\ y_2 \end{pmatrix}$ form a *basis* in $\mathcal{M}_{2,1}(\mathbb{C})$ if for any vector $X = \begin{pmatrix} x \\ y \end{pmatrix} \in \mathcal{M}_{2,1}(\mathbb{C})$ there exist and are unique $\alpha_1, \alpha_2 \in \mathbb{C}$ such that $X = \alpha_1 X_1 + \alpha_2 X_2$.

Nota bene. The vectors $E_1 = \begin{pmatrix} 1 \\ 0 \end{pmatrix}$ and $E_2 = \begin{pmatrix} 0 \\ 1 \end{pmatrix}$ form a basis in $\mathcal{M}_{2,1}(\mathbb{C})$ which is called the canonical basis. Any vector $X = \begin{pmatrix} x \\ y \end{pmatrix} \in \mathcal{M}_{2,1}(\mathbb{C})$ can be written uniquely as $X = xE_1 + yE_2$.

Lemma 1.4 *Two vectors $X_1 = \begin{pmatrix} x_1 \\ y_1 \end{pmatrix}$ and $X_2 = \begin{pmatrix} x_2 \\ y_2 \end{pmatrix}$ form a basis in $\mathcal{M}_{2,1}(\mathbb{C})$ if and only if the matrix $P = \begin{pmatrix} x_1 & x_2 \\ y_1 & y_2 \end{pmatrix} \in \mathcal{M}_2(\mathbb{C})$ is invertible ($\det P \neq 0$) and in this case P is called the matrix of passing from the canonical basis $\mathcal{B} = \{E_1, E_2\}$ to the basis $\mathcal{B}' = \{X_1, X_2\}$.*

Proof By definition, the vectors X_1 and X_2 form a basis in $\mathcal{M}_{2,1}\,(\mathbb{C})$ if and only if for any vector $X = \begin{pmatrix} x \\ y \end{pmatrix} \in \mathcal{M}_{2,1}\,(\mathbb{C})$ there exist and are unique the scalars $\alpha_1, \alpha_2 \in \mathbb{C}$ such that $\alpha_1 X_1 + \alpha_2 X_2 = X \Leftrightarrow$

$$\begin{cases} \alpha_1 x_1 + \alpha_2 x_2 = x \\ \alpha_1 y_1 + \alpha_2 y_2 = y \end{cases} \qquad \Leftrightarrow \qquad P \begin{pmatrix} \alpha_1 \\ \alpha_2 \end{pmatrix} = \begin{pmatrix} x \\ y \end{pmatrix}.$$

If P is invertible, i.e., $\det P \neq 0$, we get the unique solution $\begin{pmatrix} \alpha_1 \\ \alpha_2 \end{pmatrix} = P^{-1} \begin{pmatrix} x \\ y \end{pmatrix}$. If $\det P = 0$, then the system $P \begin{pmatrix} \alpha_1 \\ \alpha_2 \end{pmatrix} = \begin{pmatrix} 0 \\ 0 \end{pmatrix}$, in variables α_1, α_2, has infinitely many solutions. $\qquad\qquad \square$

Multiplication by vectors

■ *a row and a matrix*
   If

$$v = \begin{pmatrix} v_1 \\ v_2 \end{pmatrix} \quad \text{and} \quad A = \begin{pmatrix} a_{11} & a_{12} \\ a_{21} & a_{22} \end{pmatrix},$$

then

$$v^T A = (v_1 \ v_2) \begin{pmatrix} a_{11} & a_{12} \\ a_{21} & a_{22} \end{pmatrix} = (v_1 a_{11} + v_2 a_{21} \ \ v_1 a_{12} + v_2 a_{22});$$

■ *a matrix and a column*
   If

$$A = \begin{pmatrix} a_{11} & a_{12} \\ a_{21} & a_{22} \end{pmatrix} \quad \text{and} \quad C = \begin{pmatrix} c_1 \\ c_2 \end{pmatrix},$$

then

$$AC = \begin{pmatrix} a_{11} & a_{12} \\ a_{21} & a_{22} \end{pmatrix} \begin{pmatrix} c_1 \\ c_2 \end{pmatrix} = \begin{pmatrix} a_{11} c_1 + a_{12} c_2 \\ a_{21} c_1 + a_{22} c_2 \end{pmatrix}.$$

■ *a row and a column*
   If $L = (l_1 \ l_2)$ and $C = \begin{pmatrix} c_1 \\ c_2 \end{pmatrix}$, then

$$LC = (l_1 c_1 + l_2 c_2) \quad \text{and} \quad CL = \begin{pmatrix} c_1 l_1 & c_1 l_2 \\ c_2 l_1 & c_2 l_2 \end{pmatrix}.$$

Nota bene. We mention that if $A \in \mathscr{M}_2(\mathbb{C})$, $A \neq O_2$, then $\det A = 0$ if and only if $A = CL$, where $C = \begin{pmatrix} c_1 \\ c_2 \end{pmatrix} \neq \begin{pmatrix} 0 \\ 0 \end{pmatrix}$ and $L = (l_1 \; l_2) \neq (0 \; 0)$, i.e., a nonzero matrix A has *rank* 1 if and only if A can be written as the product of a nonzero column and a nonzero row. Thus, any matrix of rank 1 is of the following form

$$A = CL = \begin{pmatrix} c_1 l_1 & c_1 l_2 \\ c_2 l_1 & c_2 l_2 \end{pmatrix}.$$

We will be using the notation $(C_1 \mid C_2)$ for a 2×2 matrix having columns C_1 and C_2 and similarly $\begin{pmatrix} L_1 \\ L_2 \end{pmatrix}$ for a square matrix of order two with rows L_1 and L_2. We have the following formulae involving products of special matrices.

(a) If C_1 and C_2 are two columns, we have

$$A(C_1 \mid C_2) = (AC_1 \mid AC_2)$$

and, if L_1 and L_2 are two rows, then

$$\begin{pmatrix} L_1 \\ L_2 \end{pmatrix} A = \begin{pmatrix} L_1 A \\ L_2 A \end{pmatrix},$$

where $A \in \mathscr{M}_2(\mathbb{C})$ is a given matrix.

(b) We have

$$(C_1 \mid C_2) \begin{pmatrix} L_1 \\ L_2 \end{pmatrix} = C_1 L_1 + C_2 L_2$$

and

$$\begin{pmatrix} L_1 \\ L_2 \end{pmatrix} (C_1 \mid C_2) = \begin{pmatrix} L_1 C_1 & L_1 C_2 \\ L_2 C_1 & L_2 C_2 \end{pmatrix}.$$

(c) The product involving a diagonal matrix

$$\begin{pmatrix} \alpha_1 & 0 \\ 0 & \alpha_2 \end{pmatrix} \begin{pmatrix} L_1 \\ L_2 \end{pmatrix} = \begin{pmatrix} \alpha_1 L_1 \\ \alpha_2 L_2 \end{pmatrix}$$

and

$$(C_1 \mid C_2) \begin{pmatrix} \alpha_1 & 0 \\ 0 & \alpha_2 \end{pmatrix} = (\alpha_1 C_1 \mid \alpha_2 C_2).$$

1.2 Matrices and their special properties

Definition 1.7 *Elementary transformations.*
  Let $A \in \mathcal{M}_2(\mathbb{C})$. The following operations performed on the matrix A are called *elementary transformations*:

- changing two rows of A;
- multiplying a row of A by a nonzero complex number;
- adding a row to another row.

Similarly, one can define the elementary transformations by performing the corresponding operations on the columns of A.

Definition 1.8 *Elementary matrices.*
  A matrix $A \in \mathcal{M}_2(\mathbb{C})$ is called an *elementary matrix* if A is obtained from I_2 by applying an elementary transformation.

Let $a \in \mathbb{C}^*$ and let E_{1a} and E_{2a} be the following two special elementary matrices

$$E_{1a} = \begin{pmatrix} a & 0 \\ 0 & 1 \end{pmatrix} \quad \text{and} \quad E_{2a} = \begin{pmatrix} 1 & 0 \\ 0 & a \end{pmatrix}.$$

We observe that these two matrices have been obtained from the unit matrix I_2 by multiplication of the rows of I_2 by the complex number a.
  Operations with elementary matrices
  We have:

- the multiplication of a row of a matrix A by a complex number a is equivalent to multiplying A to the *left* by the corresponding elementary matrix E_{1a} or E_{2a}:

$$E_{1a}\begin{pmatrix} L_1 \\ L_2 \end{pmatrix} = \begin{pmatrix} aL_1 \\ L_2 \end{pmatrix} \quad \text{or} \quad E_{2a}\begin{pmatrix} L_1 \\ L_2 \end{pmatrix} = \begin{pmatrix} L_1 \\ aL_2 \end{pmatrix};$$

- the multiplication of a column of a matrix A by a complex number a is equivalent to multiplying A to the *right* by the corresponding elementary matrix E_{1a} or E_{2a}:

$$(C_1 \,|\, C_2)E_{1a} = (aC_1 \,|\, C_2) \quad \text{or} \quad (C_1 \,|\, C_2)E_{2a} = (C_1 \,|\, aC_2).$$

Let E_p be *the permutation matrix*

$$E_p = \begin{pmatrix} 0 & 1 \\ 1 & 0 \end{pmatrix}.$$

Then:

■ to change two rows of A means to multiply A to the left by E_p

$$E_p \left(\frac{L_1}{L_2} \right) = \left(\frac{L_2}{L_1} \right);$$

■ to change two columns of A means to multiply A to the right by E_p

$$(C_1 \,|\, C_2) E_p = (C_2 \,|\, C_1).$$

Let E_{12} and E_{21} be the elementary matrices corresponding to the addition of rows and columns respectively, i.e.

$$E_{12} = \begin{pmatrix} 1 & 0 \\ 1 & 1 \end{pmatrix} \quad \text{and} \quad E_{21} = \begin{pmatrix} 1 & 1 \\ 0 & 1 \end{pmatrix}.$$

Then:

■ to add the first row (the first column) of A to the second row (the second column) of A means to multiply to the left (the right) the matrix A by E_{12} (E_{21}):

$$E_{12} \left(\frac{L_1}{L_2} \right) = \left(\frac{L_1}{L_1 + L_2} \right) \quad \text{and} \quad (C_1 \,|\, C_2) E_{21} = (C_1 \,|\, C_1 + C_2);$$

■ to add the second row (the second column) of A to the first row (the first column) of A means to multiply to the left (the right) the matrix A by E_{21} (E_{12}):

$$E_{21} \left(\frac{L_1}{L_2} \right) = \left(\frac{L_1 + L_2}{L_2} \right) \quad \text{and} \quad (C_1 \,|\, C_2) E_{12} = (C_1 + C_2 \,|\, C_2).$$

Definition 1.9 The *transpose* of a matrix $A \in \mathcal{M}_2(\mathbb{C})$,

$$A = \begin{pmatrix} a_{11} & a_{12} \\ a_{21} & a_{22} \end{pmatrix},$$

is defined by

$$A^T = \begin{pmatrix} a_{11} & a_{21} \\ a_{12} & a_{22} \end{pmatrix}.$$

Thus, the transpose of a matrix A is obtained by taking the rows (respectively the columns) of A as the columns (respectively the rows) of A^T.

Property 1.1 If $A, B \in \mathcal{M}_2(\mathbb{C})$ and $\alpha \in \mathbb{C}$, then:

(a) $(A^T)^T = A$;
(b) $(A + B)^T = A^T + B^T$;
(c) $(\alpha A)^T = \alpha A^T$;
(d) $(AB)^T = B^T A^T$.

The next definition introduces various types of square matrices.

Definition 1.10 Let $A \in \mathcal{M}_2(\mathbb{C})$.

(a) A is *symmetric* if $A^T = A$. This implies $a_{12} = a_{21}$. Thus, a symmetric matrix is of the following form

$$A = \begin{pmatrix} a & b \\ b & c \end{pmatrix}.$$

(b) A is *antisymmetric or skew symmetric* if $A^T = -A$. This implies $a_{ij} = -a_{ji}$, $\forall i, j \in \{1, 2\}$. Thus, a symmetric matrix has the following form

$$A = \begin{pmatrix} 0 & b \\ -b & 0 \end{pmatrix}.$$

(c) The *conjugate* of $A = (a_{ij})_{i,j\in\{1,2\}}$ is the matrix $\overline{A} = (\overline{a_{ij}})_{i,j\in\{1,2\}}$, where $\overline{a_{ij}}$ is the complex conjugate of a_{ij}.
(d) The *conjugate transpose* (sometimes called the *adjoint* or *Hermitian adjoint*) of A is the matrix $A^* = (\overline{A})^T$. We note that

$$\left(A^*\right)^* = A, \quad \forall A \in \mathcal{M}_2(\mathbb{C}).$$

Remark 1.4 One can prove, see problem **1.41**, that any matrix $M \in \mathcal{M}_2(\mathbb{C})$ can be written uniquely as the sum of a symmetric matrix $S = \frac{1}{2}(M + M^T)$ (*the symmetric part* of M) and an antisymmetric matrix $A = \frac{1}{2}(M - M^T)$ (*the antisymmetric part* of M).

Definition 1.11 If $A = \begin{pmatrix} a_{11} & a_{12} \\ a_{21} & a_{22} \end{pmatrix} \in \mathcal{M}_2(\mathbb{C})$, then the *trace* of A is the complex number defined by $\mathrm{Tr}(A) = a_{11} + a_{22}$. In other words, the trace of a square matrix is the sum of the entries on the main diagonal.

Property 1.2 If $A, B \in \mathcal{M}_2(\mathbb{C})$ and $\alpha \in \mathbb{C}$ then:

(a) $\mathrm{Tr}(A + B) = \mathrm{Tr}(A) + \mathrm{Tr}(B)$;
(b) $\mathrm{Tr}(\alpha A) = \alpha \mathrm{Tr}(A)$;
(c) $\mathrm{Tr}(AB) = \mathrm{Tr}(BA)$;
(d) $\mathrm{Tr}(A) = \mathrm{Tr}(A^T)$.

Nota bene. In general $\text{Tr}(AB) \neq \text{Tr}(A)\text{Tr}(B)$.

Remark 1.5 If $\mathbb{F} \in \{\mathbb{Q}, \mathbb{R}, \mathbb{C}\}$, then the application $\text{Tr} : \mathscr{M}_2(\mathbb{F}) \to \mathbb{F}$ is a *linear functional* from the vector space $\mathscr{M}_2(\mathbb{F})$, over $\mathbb{F}$, to $\mathbb{F}$.

Definition 1.12 If $A = \begin{pmatrix} a_{11} & a_{12} \\ a_{21} & a_{22} \end{pmatrix} \in \mathscr{M}_2(\mathbb{C})$, then the *determinant* of A is defined by

$$\det A = \begin{vmatrix} a_{11} & a_{12} \\ a_{21} & a_{22} \end{vmatrix} = a_{11}a_{22} - a_{12}a_{21}.$$

Property 1.3 The following formulae hold:

(a) $\det(AB) = \det A \det B, \ \forall A, B \in \mathscr{M}_2(\mathbb{C})$;
(b) $\det(A_1 A_2 \cdots A_n) = \det A_1 \det A_2 \cdots \det A_n, \ \forall A_k \in \mathscr{M}_2(\mathbb{C}), k = \overline{1, n}, n \in \mathbb{N}$;
(c) $\det(A^n) = (\det A)^n, \ \forall A \in \mathscr{M}_2(\mathbb{C})$ and $n \in \mathbb{N}$;
(d) $\det(A^T) = \det A, \ \forall A \in \mathscr{M}_2(\mathbb{C})$;
(e) $\det(\alpha A) = \alpha^2 \det A, \ \forall A \in \mathscr{M}_2(\mathbb{C})$ and $\alpha \in \mathbb{C}$;
(f) $\det(-A) = \det A, \ \forall A \in \mathscr{M}_2(\mathbb{C})$;
(g) $\det(\overline{A}) = \overline{\det A}, \ \forall A \in \mathscr{M}_2(\mathbb{C})$.

Proposition 1.1 *If C_1, C_2 are the columns of a matrix A, i.e., $A = (C_1 \mid C_2)$, C' is a new column and $a \in \mathbb{C}$, then:*

■ $\det(C_1 \mid C_2) = -\det(C_2 \mid C_1)$;
■ $\det(aC_1 \mid C_2) = \det(C_1 \mid aC_2) = a \det(C_1 \mid C_2)$;
■ $\det(C_1 + C' \mid C_2) = \det(C_1 \mid C_2) + \det(C' \mid C_2)$.

If L_1, L_2 are the rows of a matrix A, i.e., $A = \begin{pmatrix} L_1 \\ L_2 \end{pmatrix}$, L' is a new row and $a \in \mathbb{C}$, then:

■ $\det \begin{pmatrix} L_1 \\ L_2 \end{pmatrix} = -\det \begin{pmatrix} L_2 \\ L_1 \end{pmatrix}$;
■ $\det \begin{pmatrix} aL_1 \\ L_2 \end{pmatrix} = \det \begin{pmatrix} L_1 \\ aL_2 \end{pmatrix} = a \det \begin{pmatrix} L_1 \\ L_2 \end{pmatrix}$;
■ $\det \begin{pmatrix} L_1 + L' \\ L_2 \end{pmatrix} = \det \begin{pmatrix} L_1 \\ L_2 \end{pmatrix} + \det \begin{pmatrix} L' \\ L_2 \end{pmatrix}$.

Remark 1.6 It is worth mentioning that the function $\det : \mathscr{M}_2(\mathbb{C}) \cong \mathbb{C}^2 \times \mathbb{C}^2 \to \mathbb{C}$ is an *alternating bilinear application*.

In general, $\det(A + B) \neq \det A + \det B$. However, the following formula for the determinant of the sum and the difference of two matrices holds true.

Lemma 1.5 A special determinant formula.

If $A, B \in \mathcal{M}_2(\mathbb{C})$, then $\det(A + B) + \det(A - B) = 2 \det A + 2 \det B$.

Proof The lemma can be proved by direct computation, see the solution of part (a) of problem **1.31**. □

Lemma 1.6 Famous determinant inequalities.

(a) If $A \in \mathcal{M}_2(\mathbb{R})$, then $\det(A^2 + I_2) \geq 0$.
(b) If a, b and c are real numbers such that $b^2 - 4ac \leq 0$, then

$$\det\left(aA^2 + bA + cI_2\right) \geq 0, \quad \forall A \in \mathcal{M}_2(\mathbb{R}).$$

Proof (a) We have $A^2 + I_2 = (A + iI_2)(A - iI_2)$ and it follows that $\det(A^2 + I_2) = \det(A + iI_2) \det(A - iI_2) = \det(A + iI_2)\overline{\det(A + iI_2)} = |\det(A + iI_2)|^2$.

(b) If $a = 0$ we have, since $b^2 - 4ac \leq 0$, that $b = 0$ and this implies that $\det(cI_2) = c^2 \geq 0$. If $a \neq 0$, then $aA^2 + bA + cI_2 = a\left[\left(A + \frac{b}{2a}I_2\right)^2 + \frac{4ac-b^2}{4a^2}I_2\right]$ and it follows that

$$\det(aA^2 + bA + cI_2) = a^2 \left| \det\left(A + \frac{b}{2a}I_2 + i\frac{\sqrt{4ac - b^2}}{2a}I_2\right) \right|^2.$$

The lemma is proved. □

Definition 1.13 If $A \in \mathcal{M}_2(\mathbb{C})$ and $\det A = 0$ we say that the matrix A is *singular* and if $\det A \neq 0$ we say that the matrix A is *nonsingular*.

We denote by $\mathrm{GL}_2(\mathbb{C})$ the set of all nonsingular matrices

$$\mathrm{GL}_2(\mathbb{C}) = \{A \in \mathcal{M}_2(\mathbb{C}) : \det A \neq 0\}.$$

A special subset of $\mathrm{GL}_2(\mathbb{C})$, denoted by $\mathrm{SL}_2(\mathbb{C})$, and called the *special linear group* is the set of all matrices having the determinant equal to 1, i.e.

$$\mathrm{SL}_2(\mathbb{C}) = \{A \in \mathcal{M}_2(\mathbb{C}) : \det A = 1\}.$$

Remark 1.7 The pairs $(\mathrm{GL}_2(\mathbb{Q}), \cdot)$, $(\mathrm{GL}_2(\mathbb{R}), \cdot)$, and $(\mathrm{GL}_2(\mathbb{C}), \cdot)$ are noncommutative groups called *linear groups* and $(\mathrm{SL}_2(\mathbb{Q}), \cdot)$, $(\mathrm{SL}_2(\mathbb{R}), \cdot)$, and $(\mathrm{SL}_2(\mathbb{C}), \cdot)$ are subgroups of them which are called *special linear groups*.

Definition 1.14 We say the matrix $A \in \mathcal{M}_2(\mathbb{C})$ is *invertible* if there exists $B \in \mathcal{M}_2(\mathbb{C})$ such that $AB = BA = I_2$. The matrix B is *unique* with this property, is called *the inverse* of A and is denoted by A^{-1}.

We have the following implications

$$A \in \mathcal{M}_2(\mathbb{C}) \text{ is invertible} \Leftrightarrow \det A \neq 0 \Leftrightarrow A \in \mathrm{GL}_2(\mathbb{C}).$$

One can prove, by direct calculations, that if $A \in \mathcal{M}_2(\mathbb{C})$,

$$A = \begin{pmatrix} a & b \\ c & d \end{pmatrix}$$

is invertible, then

$$A^{-1} = \frac{1}{\det A} A_*,$$

where

$$A_* = \begin{pmatrix} d & -b \\ -c & a \end{pmatrix},$$

is the *reciprocal* matrix also known as the *adjugate* of A. Sometimes this matrix is also denoted by $\mathrm{adj}(A)$.

Remark 1.8 We mention that, if $A \in \mathcal{M}_2(\mathbb{C})$ is invertible, another method for determining the inverse of A would be to use elementary transformations. More precisely, we consider the *matrix with blocks* $(A \mid I_2)$ and, by performing a sequence of elementary transformations we transform it to the matrix $(I_2 \mid B)$, in which case $B = A^{-1}$.

Property 1.4 If $A, B \in \mathcal{M}_2(\mathbb{C})$ are invertible and $\alpha \in \mathbb{C}^*$, then

(a) $(AB)^{-1} = B^{-1}A^{-1}$;
(b) $(\alpha A)^{-1} = \dfrac{1}{\alpha} A^{-1}$;
(c) $\left(A^T\right)^{-1} = (A^{-1})^T$;
(d) $(A^n)^{-1} = (A^{-1})^n$.

One can also prove, by mathematical induction, that

$$(A_1 A_2 \cdots A_n)^{-1} = A_n^{-1} A_{n-1}^{-1} \cdots A_1^{-1},$$

where $A_k \in \mathrm{GL}_2(\mathbb{C})$, $k = \overline{1, n}$, $n \in \mathbb{N}$.

The next definition introduces some special classes of matrices.

Definition 1.15 Let $A \in \mathscr{M}_2(\mathbb{C})$. Then:

(a) A is *involutory* if $A^2 = I_2$ (see problem 1.12);
(b) A is *skew involutory* if $A^2 = -I_2$ (see Example 1.2);
(c) A is *idempotent* if $A^2 = A$ (see problem 1.14);
(d) A is *nilpotent* if and only if $A^2 = O_2$ (see problem 1.8);
(e) A is *Hermitian* if $A^* = A$. If $A \in \mathscr{M}_2(\mathbb{R})$, then A is a symmetric matrix;
(f) A is *skew Hermitian* if $A^* = -A$. If $A \in \mathscr{M}_2(\mathbb{R})$, then A is an antisymmetric matrix;
(g) A is *normal* if $AA^* = A^*A$;
(h) A is *unitary* if $A^* = A^{-1}$. If $A \in \mathscr{M}_2(\mathbb{R})$, then A is an orthogonal matrix (see Example 1.1).

Lemma 1.7 *Let $A \in \mathscr{M}_2(\mathbb{C})$.*

(a) *A is Hermitian if and only if iA is skew Hermitian.*
(b) *A is involutory if and only if iA is skew involutory.*

Proof The proof is based on the definition of Hermitian, involutory, skew Hermitian, and skew involutory matrices respectively. $\qquad\square$

Example 1.1 **Special unitary matrices (real and complex).**

If $A \in \mathscr{M}_2(\mathbb{C})$ is a unitary matrix with $\det A = 1$, then

$$A = \begin{pmatrix} a & b \\ -\overline{b} & \overline{a} \end{pmatrix},$$

where $a, b \in \mathbb{C}$ with $|a|^2 + |b|^2 = 1$.

To see this, we let $A = \begin{pmatrix} a & b \\ c & d \end{pmatrix}$, where $a, b, c, d \in \mathbb{C}$. Since $\det A = 1$ we get that

$$A^{-1} = \begin{pmatrix} d & -b \\ -c & a \end{pmatrix} \quad \text{and} \quad A^* = (\overline{A})^T = \begin{pmatrix} \overline{a} & \overline{c} \\ \overline{b} & \overline{d} \end{pmatrix}.$$

We have $A^* = A^{-1}$ and this implies that $d = \overline{a}$, $c = -\overline{b}$ and since $ad - bc = 1$ we get that $|a|^2 + |b|^2 = 1$. Thus, $A = \begin{pmatrix} a & b \\ -\overline{b} & \overline{a} \end{pmatrix}$.

Similarly, if $A \in \mathcal{M}_2(\mathbb{C})$ is a unitary matrix with $\det A = -1$, then

$$A = \begin{pmatrix} a & b \\ \bar{b} & -\bar{a} \end{pmatrix},$$

where $a, b \in \mathbb{C}$ with $|a|^2 + |b|^2 = 1$.

In the particular case of real unitary (orthogonal) matrices we obtain

- *rotation matrices* $A = \begin{pmatrix} \cos\alpha & \sin\alpha \\ -\sin\alpha & \cos\alpha \end{pmatrix}$ with $\det A = 1$;

- *reflection matrices* $A = \begin{pmatrix} \cos\alpha & \sin\alpha \\ \sin\alpha & -\cos\alpha \end{pmatrix}$ with $\det A = -1$.

Nota bene. Let $\mathscr{L}$ be a line passing through the origin of the coordinate system which makes an angle α with the x-axis and let $M(x_M, y_M)$ be a point in the Cartesian plane. If $N(x_N, y_N)$ is the symmetric of M about the line $\mathscr{L}$, then one can check that

$$\begin{pmatrix} x_N \\ y_N \end{pmatrix} = \begin{pmatrix} \cos(2\alpha) & \sin(2\alpha) \\ \sin(2\alpha) & -\cos(2\alpha) \end{pmatrix} \begin{pmatrix} x_M \\ y_M \end{pmatrix}.$$

Because of this reason, the matrix $\begin{pmatrix} \cos(2\alpha) & \sin(2\alpha) \\ \sin(2\alpha) & -\cos(2\alpha) \end{pmatrix}$ is called *a reflection matrix*. For the definition of rotation matrices see problem **1.61**.

Definition 1.16 *Equivalent matrices.*
On the set $\mathcal{M}_2(\mathbb{C})$ we define the relation $A \approx B$ if and only if there exist $P, Q \in$ $GL_2(\mathbb{C})$ such that $B = QAP$. Two matrices A and B that satisfy this condition are called *equivalent*.

Definition 1.17 *Similar matrices.*
On the set $\mathcal{M}_2(\mathbb{C})$ we define the relation $A \sim B$ if and only if there exists $P \in$ $GL_2(\mathbb{C})$ such that $B = P^{-1}AP$. Two matrices A and B that verify this condition are called *similar*.

Remark 1.9 It can be proved that two matrices are *equivalent* if and only if they have the same *rank* and they are *similar* if and only if they have the same *Jordan canonical form*.

Definition 1.18 *Commuting matrices.*
We say the matrices $A, B \in \mathcal{M}_2(\mathbb{C})$ commute if $AB = BA$.

If $A \in \mathcal{M}_2(\mathbb{C})$, then we denote by $\mathscr{C}(A)$ the set of all matrices that commute with A, also known as the *centralizer of A* (see [38, p. 213])

$$\mathscr{C}(A) = \{X \in \mathscr{M}_2(\mathbb{C}) : AX = XA\}.$$

Now we give an important property of the centralizer of a square 2×2 matrix.

Theorem 1.1 The centralizer of a matrix A.

Let $A \in \mathscr{M}_2(\mathbb{C})$ and let $\mathscr{C}(A) = \{X \in \mathscr{M}_2(\mathbb{C}) : AX = XA\}$.

(a) If $A = kI_2$, $k \in \mathbb{C}$, then $\mathscr{C}(A) = \mathscr{M}_2(\mathbb{C})$;

(b) If $A \neq kI_2$, $k \in \mathbb{C}$, then $\mathscr{C}(A) = \{\alpha A + \beta I_2 : \alpha, \beta \in \mathbb{C}\}$.

Proof (a) If $A = kI_2$, $k \in \mathbb{C}$, we have nothing to prove since any square matrix commutes with kI_2.

(b) Let $A = \begin{pmatrix} a & b \\ c & d \end{pmatrix}$ and let $X = \begin{pmatrix} x & y \\ z & t \end{pmatrix}$. The equation $AX = XA$ implies that

$$\begin{cases} ax + bz = xa + yc \\ ay + bt = xb + yd \\ cx + dz = za + tc \\ cy + dt = zb + td \end{cases} \quad \text{or} \quad \begin{cases} bz = cy \\ y(a - d) = b(x - t) \\ z(a - d) = c(x - t). \end{cases}$$

If $a \neq d$, then $y = b\alpha$, $z = c\alpha$, where $\alpha = \dfrac{x - t}{a - d}$. We have $t = x - \alpha a + \alpha d = \beta + \alpha d$, where $\beta = x - \alpha a$. This implies that

$$X = \begin{pmatrix} \alpha a + \beta & b\alpha \\ c\alpha & \alpha d + \beta \end{pmatrix} = \alpha \begin{pmatrix} a & b \\ c & d \end{pmatrix} + \beta \begin{pmatrix} 1 & 0 \\ 0 & 1 \end{pmatrix}.$$

If $a = d$, then we distinguish between the cases when $x \neq t$ or $x = t$. If $x \neq t$ we get that $b = c = 0$, which contradicts the fact that $A \neq kI_2$. Thus, $x = t$ and we also have that $bz = cy$. Observe that b and c cannot be both 0 since this would contradict $A \neq kI_2$.

If $b = 0 \Rightarrow y = 0$ and we have

$$X = \begin{pmatrix} x & 0 \\ z & x \end{pmatrix} = \alpha \begin{pmatrix} a & 0 \\ c & a \end{pmatrix} + \beta \begin{pmatrix} 1 & 0 \\ 0 & 1 \end{pmatrix}, \quad \text{where } \alpha = \frac{z}{c}, \ \beta = x - \frac{az}{c}.$$

If $b \neq 0$ we have $z = \dfrac{cy}{b}$ and

$$X = \begin{pmatrix} x & y \\ \frac{cy}{b} & x \end{pmatrix} = \alpha \begin{pmatrix} a & b \\ c & a \end{pmatrix} + \beta \begin{pmatrix} 1 & 0 \\ 0 & 1 \end{pmatrix}, \quad \text{where } \alpha = \frac{y}{b}, \ \beta = x - \frac{ay}{b}.$$

The theorem is proved. $\qquad \square$

Remark 1.10 Theorem 1.1 also holds for matrices in $\mathcal{M}_2(\mathbb{Z}_p)$, where $p \geq 2$ is a prime number.

Corollary 1.1 *If $A \in \mathcal{M}_2(\mathbb{C})$ and $P \in \mathbb{C}[x]$, then there exist $a, b \in \mathbb{C}$ such that $P(A) = aI_2 + bA$.*

Proof If $A = kI_2$, for some $k \in \mathbb{C}$, then $P(A) = P(k)I_2$, so $a = P(k)$ and $b = 0$.

If $A \neq kI_2$, for any $k \in \mathbb{C}$, then since the matrices A and $P(A)$ commute the result follows based on part (b) of Theorem 1.1. $\square$

Corollary 1.2 Where do two centralizers intersect?

Let $X, Y \in \mathcal{M}_2(\mathbb{C})$ such that $XY \neq YX$. Then

$$\mathscr{C}(X) \cap \mathscr{C}(Y) = \{\alpha I_2 : \alpha \in \mathbb{C}\}.$$

Proof Since $XY \neq YX$ we get that $X \neq xI_2$, $\forall x \in \mathbb{C}$ and $Y \neq yI_2$, $\forall y \in \mathbb{C}$. If $Z \in \mathscr{C}(X) \cap \mathscr{C}(Y)$ we get, based on Theorem 1.1, that there exist $a, b, c, d \in \mathbb{C}$ such that $Z = aX + bI_2$ and $Z = cY + dI_2$. This implies $aX + bI_2 = cY + dI_2$. If $a \neq 0$ or $c \neq 0$ the previous equality would imply that $XY = YX$ which is impossible. Thus, $a = c = 0$ and $b = d = \alpha$, which in turns implies that $Z = \alpha I_2$. $\square$

Remark 1.11 Corollary 1.2 states that if a matrix $A \in \mathcal{M}_2(\mathbb{C})$ commutes with two noncommuting matrices, then A is of the form αI_2, $\alpha \in \mathbb{C}$.

Corollary 1.3 $A \in \mathcal{M}_2(\mathbb{C})$ *commutes with all*

(a) *nilpotent*
(b) *idempotent*
(c) *involutory*

matrices if and only if A is of the form αI_2, $\alpha \in \mathbb{C}$.

Proof Use Remark 1.11. $\square$

We mention that the basic algebraic formulae involving complex numbers also hold for commuting matrices.

Property 1.5 If $A, B \in \mathcal{M}_2(\mathbb{C})$ such that $AB = BA$, then:

(a) $A^m B^n = B^n A^m$, $\forall m, n \in \mathbb{N}$;
(b) $A^n - B^n = (A - B)(A^{n-1} + A^{n-2}B + \cdots + AB^{n-2} + B^{n-1})$;
(c) $A^{2n+1} + B^{2n+1} = (A + B)(A^{2n} - A^{2n-1}B + \cdots - AB^{2n-1} + B^{2n})$.

Theorem 1.2   The binomial theorem for matrices.

Let $n \geq 1$ be an integer. If $A, B \in \mathcal{M}_2(\mathbb{C})$ such that $AB = BA$, then

$$(A + B)^n = \sum_{k=0}^{n} \binom{n}{k} A^k B^{n-k}.$$

Proof The theorem can be proved by mathematical induction and by using the fact that $A^k B^p = B^p A^k$, for all $k, p \in \mathbb{N}$.   □

1.3   The set of complex numbers and matrices of order 2

In this section we establish an isomorphism between the field of complex numbers $\mathbb{C}$ and a special field of matrices of order 2. Let

$$\mathbb{C} = \{x + iy : x, y \in \mathbb{R}, \ i^2 = -1\}$$

and let

$$\mathcal{M}_{\mathbb{C}} = \left\{ A \in \mathcal{M}_2(\mathbb{R}) : A = \begin{pmatrix} x & -y \\ y & x \end{pmatrix} \right\}.$$

Theorem 1.3   An isomorphism between two special fields.

The following properties hold:

(a)   *if $A, B \in \mathcal{M}_{\mathbb{C}}$, then $A + B \in \mathcal{M}_{\mathbb{C}}$;*

(b)   *if $A, B \in \mathcal{M}_{\mathbb{C}}$, then $AB \in \mathcal{M}_{\mathbb{C}}$;*

(c)   *the matrix*

$$J = \begin{pmatrix} 0 & -1 \\ 1 & 0 \end{pmatrix},$$

verifies the equalities $J^2 = -I_2$ and $J^4 = I_2$;

Let

$$f : \mathbb{C} \to \mathcal{M}_{\mathbb{C}}, \quad f(x + iy) = \begin{pmatrix} x & -y \\ y & x \end{pmatrix}.$$

(continued)

Theorem 1.3 (continued)

Then,

(d) *f is a bijection and*

$$f^{-1} \begin{pmatrix} x & -y \\ y & x \end{pmatrix} = x + iy.$$

(e) $f(0) = O_2$;

(f) $f(1) = I_2$;

(g) $f(i) = J$;

(h) $f(z_1 + z_2) = f(z_1) + f(z_2), \ \forall z_1, z_2 \in \mathbb{C}$;

(i) $f(z_1 z_2) = f(z_1) f(z_2), \ \forall z_1, z_2 \in \mathbb{C}$;

(j) $f(z^n) = (f(z))^n, \ \forall z \in \mathbb{C}$;

(k) $f(\bar{z}) = (f(z))^T, \ \forall z \in \mathbb{C}$;

(l) $\det f(z) = |z|^2, \ \forall z \in \mathbb{C}$;

(m) $f\left(\dfrac{1}{z}\right) = (f(z))^{-1} = \dfrac{1}{|z|^2} (f(z))^T, \ \forall z \in \mathbb{C}$;

(n) $(\mathscr{M}_\mathbb{C}, +, \cdot)$ *is a field isomorphic to* $(\mathbb{C}, +, \cdot)$, *i.e., one has*

$$(\mathscr{M}_\mathbb{C}, +, \cdot) \cong (\mathbb{C}, +, \cdot) ;$$

(o) *Let U be the unit circle*

$$U = \{z \in \mathbb{C} : |z| = 1\} = \{\cos\alpha + i\sin\alpha : \alpha \in \mathbb{R}\}$$

and let $\mathscr{R}$ *be the set of rotation matrices in the plane*

$$\mathscr{R} = \left\{ R_\alpha = \begin{pmatrix} \cos\alpha & -\sin\alpha \\ \sin\alpha & \cos\alpha \end{pmatrix}, \ \alpha \in \mathbb{R} \right\} .$$

Then $f(\cos\alpha + i\sin\alpha) = R_\alpha$ *and* $f(U) = \mathscr{R}$;

(p) **the group of the rotations of a regular *n*-gon.**
   If $n \geq 2$ *is an integer and*

$$\mathscr{U}_n = \{z \in \mathbb{C} : z^n = 1\} \quad \text{and} \quad \mathscr{R}_n = \left\{ R_{\frac{2k\pi}{n}} : k = 0, 1, 2, \dots, n-1 \right\},$$

   then $(\mathscr{U}_n, \cdot) \cong (\mathscr{R}_n, \cdot)$.
   $(\mathscr{R}_n, \cdot)$, *the group of the rotations of a regular n-gon, is isomorphic to the multiplicative group of the nth roots of unity, which are also isomorphic to the cyclic group* $(\mathbb{Z}_n, +)$, *thus, any finite cyclic group is the group of the rotations of a regular n-gon.*

Proof The theorem which can be proved by straightforward calculations is left as an exercise to the interested reader.   □

The following formula is worth being mentioned

$$\begin{pmatrix} a & -b \\ b & a \end{pmatrix} = \sqrt{a^2 + b^2} \begin{pmatrix} \cos\alpha & -\sin\alpha \\ \sin\alpha & \cos\alpha \end{pmatrix},$$

where $\cos\alpha = \dfrac{a}{\sqrt{a^2 + b^2}}$ and $\sin\alpha = \dfrac{b}{\sqrt{a^2 + b^2}}$.

Example 1.2 **Skew involutory real matrices.**

We determine all skew involutory real matrices, i.e., matrices $A \in \mathcal{M}_2(\mathbb{R})$ such that $A^2 = -I_2$.

If $A = \begin{pmatrix} a & b \\ c & d \end{pmatrix} \in \mathcal{M}_2(\mathbb{R})$, we have, based on the matrix equation $A^2 = -I_2$, that

$$\begin{cases} a^2 + bc = -1 \\ b(a + d) = 0 \\ c(a + d) = 0 \\ d^2 + bc = -1 \end{cases} \quad \text{or} \quad \begin{cases} a^2 + bc = -1 \\ b(a + d) = 0 \\ c(a + d) = 0 \\ (a - d)(a + d) = 0. \end{cases}$$

We distinguish between the cases when $a + d \neq 0$ and $a + d = 0$.

When $a + d \neq 0$ we get that $b = c = 0$, $a = d$, and $a^2 = -1$, so there are no real matrices that verify the equation $A^2 = -I_2$.

When $a + d = 0$, we get, from our system of equations, that $a^2 + bc = -1$ and we obtain the matrices

$$\begin{pmatrix} a & b \\ -\dfrac{1 + a^2}{b} & -a \end{pmatrix}, \quad a \in \mathbb{R}, \ b \in \mathbb{R}^*.$$

In particular, for $a = 0$, $b = -1$ we get the skew involutory real matrix

$$J = \begin{pmatrix} 0 & -1 \\ 1 & 0 \end{pmatrix},$$

and for $a = b = 1$ we obtain another skew involutory real matrix

$$K = \begin{pmatrix} 1 & 1 \\ -2 & -1 \end{pmatrix}.$$

Remark 1.12 (1) For any fixed skew involutory real matrix $B \in \mathcal{M}_2(\mathbb{R})$, i.e., $B^2 = -I_2$, the centralizer of B

$$\mathscr{C}(B) = \{aI_2 + bB : a, b \in \mathbb{R}\},$$

together with the addition and the multiplication of matrices is a *commutative field* (check it!) and the following fields are isomorphic

$$(\mathscr{C}(B), +, \cdot) \cong (\mathbb{C}, +, \cdot) \cong (\mathcal{M}_\mathbb{C}, +, \cdot).$$

The function $f : \mathbb{C} \to \mathscr{C}(B), f(a + ib) = aI_2 + bB$ is a field isomorphism (prove it!).

(2) All skew involutory real matrices $A \in \mathcal{M}_2(\mathbb{R})$, i.e., $A^2 = -I_2$ are *similar to each other and also similar to the matrix*

$$J = \begin{pmatrix} 0 & -1 \\ 1 & 0 \end{pmatrix}.$$

If we consider

$$P = \begin{pmatrix} ab & -b \\ -(a^2 + 1) & 0 \end{pmatrix}, \quad a \in \mathbb{R}, \; b \in \mathbb{R}^*,$$

then

$$P^{-1} = -\frac{1}{b(a^2 + 1)} \begin{pmatrix} 0 & b \\ a^2 + 1 & ab \end{pmatrix}$$

and

$$P^{-1}AP = P^{-1} \begin{pmatrix} a & b \\ -\dfrac{1 + a^2}{b} & -a \end{pmatrix} P = J.$$

1.4   Problems

1.1 Let $A = \begin{pmatrix} 2 & i \\ i & 0 \end{pmatrix}$, where $i^2 = -1$. Prove that $A^n = \begin{pmatrix} n+1 & ni \\ ni & 1-n \end{pmatrix}$, $n \in \mathbb{N}$.

1.2 If $A \in \mathcal{M}_2(\mathbb{R})$ what is the possible number of negative entries of A^2?

1.3 Let $A = \begin{pmatrix} a & b \\ c & d \end{pmatrix} \in \mathcal{M}_2(\mathbb{R})$ be such that $bc \neq 0$ and there exists $n \geq 2$ an

integer such that $b_n c_n = 0$, where $A^n = \begin{pmatrix} a_n & b_n \\ c_n & d_n \end{pmatrix}$, $n \in \mathbb{N}$. Prove that $a_n = d_n$.

1.4 Determine all matrices in $\mathcal{M}_2(\mathbb{C})$ which commute with the matrix $A = \begin{pmatrix} 1 & 2 \\ 3 & 4 \end{pmatrix}$.

1.5 (a) Prove that $A \in \mathcal{M}_2(\mathbb{C})$ commutes with all symmetric matrices if and only if $A = \alpha I_2$, $\alpha \in \mathbb{C}$.

(b) Prove that $A \in \mathcal{M}_2(\mathbb{C})$ commutes with all circulant matrices if and only if A is a circulant matrix.

1.6 Involutory and nilpotent matrices do not commute.

Let $A \in \mathcal{M}_2(\mathbb{C})$, $A \neq \pm I_2$, be an involutory matrix and let $B \in \mathcal{M}_2(\mathbb{C})$, $B \neq O_2$, be a nilpotent matrix. Prove that $AB \neq BA$.

Moreover, if $C \in \mathcal{M}_2(\mathbb{C})$ commutes with both A and B, then $C = \alpha I_2$, $\alpha \in \mathbb{C}$.

1.7 Normal real matrices. Prove that $A \in \mathcal{M}_2(\mathbb{R})$ commutes with its transpose if and only if A is symmetric or A is a scalar multiple of a rotation matrix.

1.8 Find all matrices $A \in \mathcal{M}_2(\mathbb{C})$ such that $A^2 = O_2$.

1.9 Nilpotent real matrices. Let $A \in \mathcal{M}_2(\mathbb{R})$. Prove that $A^2 = O_2$ if and only if there exist $a \in \mathbb{R}$ and $\alpha \in [0, 2\pi)$ such that $A = a \begin{pmatrix} \cos \alpha & 1 + \sin \alpha \\ -1 + \sin \alpha & -\cos \alpha \end{pmatrix}$.

Observe $B = \begin{pmatrix} \cos \alpha & \sin \alpha \\ \sin \alpha & -\cos \alpha \end{pmatrix}$ is a reflection matrix and $C = \begin{pmatrix} 0 & 1 \\ -1 & 0 \end{pmatrix}$ is a rotation matrix of angle $\frac{3\pi}{2}$. So any nilpotent real matrix A can be written as $A = a(B + C)$. Moreover, this writing is unique (see problem **1.43**).

1.10 Determine the number of nilpotent matrices in $\mathcal{M}_2(\mathbb{Z}_p)$, where $p \geq 2$ is a prime number.

1.11 Find all matrices $A \in \mathcal{M}_2(\mathbb{R})$ such that $(I_2 + iA)^{-1} = I_2 - iA$.

Remark 1.13 The reader may prove that if $A, B \in \mathcal{M}_2 (\mathbb{R})$, then $(A+iB)^{-1} = A-iB$ if and only if $AB = BA$ and $A^2 + B^2 = I_2$. This is equivalent to finding matrices $A \in \mathcal{M}_2 (\mathbb{C})$ such that $A^{-1} = \overline{A}$.

1.12 Find all matrices $A \in \mathcal{M}_2 (\mathbb{C})$ such that $A^2 = I_2$.

1.13 Determine the number of involutory matrices in $\mathcal{M}_2 (\mathbb{Z}_p)$, where $p \geq 2$ is a prime number.

1.14 Find all matrices $A \in \mathcal{M}_2 (\mathbb{C})$ such that $A^2 = A$.

1.15 Determine the number of idempotent matrices in $\mathcal{M}_2 (\mathbb{Z}_p)$, where $p \geq 2$ is a prime number.

1.16 Prove that any matrix $X \in \mathcal{M}_2 (\mathbb{R})$ can be written as a linear combination of four orthogonal real matrices.

1.17 Complex orthogonal and skew orthogonal matrices.

(a) Find all matrices $A \in \mathcal{M}_2 (\mathbb{C})$ such that $AA^T = I_2$.
(b) Find all matrices $A \in \mathcal{M}_2 (\mathbb{C})$ such that $AA^T = -I_2$.

1.18 Let $A \in \mathcal{M}_2 (\mathbb{C})$. Prove that $A^2 = A$ if and only if $(2A - I_2)^2 = I_2$.

Remark 1.14 Problem **1.18** states that A is idempotent if and only if $2A - I_2$ is involutory.

1.19 Let $A, B \in \mathcal{M}_2 (\mathbb{C})$ be nonzero idempotent matrices. Prove that if $A + B$ is idempotent, then $A + B = I_2$ (see also problem **5.10**).

1.20 Let $A, B \in \mathcal{M}_2 (\mathbb{C})$ be nonzero idempotent matrices. Prove that if $A + B$ is involutory, then $A + B = I_2$ (see also problem **5.11**).

1.21 Let $A, B \in \mathcal{M}_2 (\mathbb{C})$ be such that A is idempotent and B is involutory. Prove that if $A + B$ is involutory, then $A = O_2$.

1.22 A trace equality on special classes of matrices.

(a) Prove that $\text{Tr}(AB) = \text{Tr}(A)\text{Tr}(B)$ for all involutory matrices $B \in \mathcal{M}_2 (\mathbb{C})$ if and only if $A = O_2$.
(b) Prove that $\text{Tr}(AB) = \text{Tr}(A)\text{Tr}(B)$ for all skew involutory matrices $B \in \mathcal{M}_2 (\mathbb{C})$ if and only if $A = O_2$.

(c) Prove that $\mathrm{Tr}(AB) = \mathrm{Tr}(A)\mathrm{Tr}(B)$ for all idempotent matrices $B \in \mathcal{M}_2(\mathbb{C})$ if and only if $A = O_2$.

1.23 Proving that two matrices are equal.

(a) Let $X \in \mathcal{M}_2(\mathbb{C})$ such that $X^2 = X$. Prove the matrix $I_2 + X$ is invertible and $(I_2 + X)^{-1} = I_2 - \frac{1}{2}X$.
(b) Let $A, B \in \mathcal{M}_2(\mathbb{C})$ such that $A - AB = B^2$ and $B - BA = A^2$. Prove that $A = B$.

1.24 *Inverses of various matrices.* Let $A \in \mathcal{M}_2(\mathbb{C})$.

(a) If $A^2 = A$ and $\alpha \in \mathbb{C}, \alpha \neq -1$, then $(I_2 + \alpha A)^{-1} = I_2 - \frac{\alpha}{\alpha+1}A$.
(b) If $A^2 = -A$ and $\alpha \in \mathbb{C}, \alpha \neq 1$, then $(I_2 + \alpha A)^{-1} = I_2 + \frac{\alpha}{\alpha-1}A$.
(c) If $A^2 = O_2$ and $\alpha \in \mathbb{C}$, then $(I_2 + \alpha A)^{-1} = I_2 - \alpha A$.

1.25 Let $A = \begin{pmatrix} 1 & 0 \\ 1 & 1 \end{pmatrix}$. Calculate A^n, $n \geq 1$.

1.26 Let $n \subset \mathbb{N}$ and let $A \in \mathcal{M}_2(\mathbb{C})$ such that $A + A^{-1} = -I_2$. Calculate $A^n + A^{-n}$.

1.27 Let $\lambda \in \mathbb{C}$ and let $J_2(\lambda)$ be the *Jordan cell of order* 2 corresponding to λ

$$J_2(\lambda) = \begin{pmatrix} \lambda & 1 \\ 0 & \lambda \end{pmatrix}.$$

Prove that $J_2^n(\lambda) = \begin{pmatrix} \lambda^n & n\lambda^{n-1} \\ 0 & \lambda^n \end{pmatrix}$, $n \in \mathbb{N}$.

1.28 *Two rotation matrices in disguise.*

(a) Calculate $\begin{pmatrix} 1 + \sqrt{3} & 1 - \sqrt{3} \\ \sqrt{3} - 1 & 1 + \sqrt{3} \end{pmatrix}^n$, $n \in \mathbb{N}$.

(b) Let $a, b \in \mathbb{R}$. Calculate $\begin{pmatrix} a & -b \\ b & a \end{pmatrix}^n$, $n \in \mathbb{N}$.

1.29 A Fibonacci matrix and Lucas numbers.

Let $(F_n)_{n \geq 0}$ be the Fibonacci sequence defined by the recurrence relation $F_0 = 0$, $F_1 = 1$ and $F_{n+1} = F_n + F_{n-1}$, $\forall n \geq 1$ and let $A = \begin{pmatrix} 1 & 1 \\ 1 & 0 \end{pmatrix}$.

(continued)

1.29 (continued)

Prove that:

(a) $A^{n+1} = A^n + A^{n-1}, \; \forall n \geq 1.$

(b) $A^n = \begin{pmatrix} F_{n+1} & F_n \\ F_n & F_{n-1} \end{pmatrix}, \; \forall n \geq 1.$

(c) $F_{n+m} = F_{n+1}F_m + F_n F_{m-1}, \; \forall m, n \in \mathbb{N}.$

(d) *Fibonacci quadratic and cubic identities*

   - $F_{n+1}F_{n-1} - F_n^2 = (-1)^n, \; n \geq 1$ *(Cassini's identity)*.
   - $F_{2n} = F_{n+1}^2 - F_{n-1}^2, \; n \geq 1.$
   - $F_{3n} = F_{n+1}^3 + F_n^3 - F_{n-1}^3, \; n \geq 1.$

(e) The *Lucas numbers* L_n are defined by the recurrence formula

$$L_0 = 2, \; L_1 = 1 \quad \text{and} \quad L_{n+2} = L_{n+1} + L_n, \; \forall n \geq 1.$$

   - Prove that $L_n = F_{n+1} + F_{n-1}, \; \forall n \in \mathbb{N}.$
   - Consider the recurrent system $X_{n+1} = AX_n, \; n \geq 0$, and show that if $X_0 = \begin{pmatrix} 3 \\ 1 \end{pmatrix}$ then, for all $n \geq 0$ one has

$$X_n = \begin{pmatrix} L_{n+2} \\ L_{n+1} \end{pmatrix}.$$

1.30 [23] Let $B(x) = \begin{pmatrix} x & 1 \\ 1 & x \end{pmatrix}$ and let $n \geq 2$ be an integer. Calculate the product

$$B(2)B(3) \cdots B(n).$$

Splendid determinant formulae

1.31 A special determinant formula.

(a) If $A, B \in \mathscr{M}_2 (\mathbb{C})$, then

$$\det(A + B) + \det(A - B) = 2 \det A + 2 \det B.$$

(continued)

1.31 (continued)

(b) Let $A_k \in \mathcal{M}_2(\mathbb{C})$, $k = \overline{1, n}$. Prove that

$$\sum \det(\pm A_1 \pm A_2 \pm \cdots \pm A_n) = 2^n \sum_{k=1}^{n} \det A_k,$$

where the sum is taken over all possible combinations of signs.

1.32 Let $A, B, C \in \mathcal{M}_2(\mathbb{C})$. Prove that

$$\det(A+B+C)+\det A+\det B+\det C = \det(A+B)+\det(B+C)+\det(A+C).$$

1.33 Let $A, B, C \in \mathcal{M}_2(\mathbb{C})$. Prove that

$$\det(A + B + C) + \det(-A + B + C) + \det(A - B + C) + \det(A + B - C)$$
$$= 4(\det A + \det B + \det C).$$

1.34 Let $n \geq 2$ be an integer. If $A_i \in \mathcal{M}_2(\mathbb{C})$, $i = \overline{1, n}$, then

$$\det\left(\sum_{i=1}^{n} A_i\right) = \sum_{1 \leq i < j \leq n} \det(A_i + A_j) - (n - 2)\sum_{i=1}^{n} \det A_i.$$

1.35 Let $n \geq 2$ be an integer, let $A_1, A_2, \ldots, A_n \in \mathcal{M}_2(\mathbb{C})$, and let $S = A_1 + A_2 + \cdots + A_n$. Prove that

$$\det(S - A_1) + \det(S - A_2) + \cdots + \det(S - A_n) = (n - 2)\det S + \sum_{i=1}^{n} \det A_i.$$

1.36 Prove that if $A, B, C \in \mathcal{M}_2(\mathbb{C})$ are such that $\det(A+B) = \det C$, $\det(B+C) = \det A$ and $\det(C + A) = \det B$, then $\det(A + B + C) = 0$.

1.37 Let $A \in \mathcal{M}_2(\mathbb{R})$ be a matrix such that $\det(A+A^T) = 8$ and $\det(A+2A^T) = 27$. Calculate $\det A$.

1.38 Let $A, B \in \mathcal{M}_2(\mathbb{C})$ such that $A^2 + B^2 + 2AB = O_2$ and $\det A = \det B$. Calculate $\det(A^2 - B^2)$.

1.39 Let $A \in \mathcal{M}_2(\mathbb{C}), A \neq O_2$, be a diagonal matrix with different diagonal entries. Prove that if $B \in \mathcal{M}_2(\mathbb{C})$ commutes with A, then B is also diagonal.

1.40 Let $\alpha \in \mathbb{R}$ and let $A_n = \begin{pmatrix} 1 & \frac{\alpha}{n} \\ -\frac{\alpha}{n} & 1 \end{pmatrix}^n$. Prove that:

(a) there exist two sequences $(a_n)_{n \geq 1}$ and $(b_n)_{n \geq 1}$ such that $A_n = \begin{pmatrix} a_n & b_n \\ -b_n & a_n \end{pmatrix}$;

(b) $\lim\limits_{n \to \infty} a_n = \cos \alpha$ and $\lim\limits_{n \to \infty} b_n = \sin \alpha$.

Remark 1.15 A simplified version of this problem can be found in [18, p. 76].

1.41 A unique decomposition of real matrices.

Let

$$\mathcal{S}_2(\mathbb{R}) = \left\{ A \in \mathcal{M}_2(\mathbb{R}) : A = A^T \right\}$$

and

$$\mathcal{A}_2(\mathbb{R}) = \left\{ A \in \mathcal{M}_2(\mathbb{R}) : A = -A^T \right\}.$$

Prove that:

(a) if $A, B \in \mathcal{S}_2(\mathbb{R})$, then $A + B \in \mathcal{S}_2(\mathbb{R})$ and if $A, B \in \mathcal{A}_2(\mathbb{R})$, then $A + B \in \mathcal{A}_2(\mathbb{R})$;
(b) $\mathcal{S}_2(\mathbb{R}) \cap \mathcal{A}_2(\mathbb{R}) = \{O_2\}$ (if a matrix is both symmetric and antisymmetric, then it is the zero matrix);
(c) $\forall M \in \mathcal{M}_2(\mathbb{R})$ there are $S \in \mathcal{S}_2(\mathbb{R})$ and $A \in \mathcal{A}_2(\mathbb{R})$, both unique, such that $M = S + A$ (any matrix $M \in \mathcal{M}_2(\mathbb{R})$ can be written uniquely as the sum of a symmetric and an antisymmetric, skew-symmetric, matrix).

1.42 Two matrix decompositions of complex matrices.

Prove that:

(a) Any matrix $A \in \mathcal{M}_2(\mathbb{C})$ can be written in exactly one way as $A = B + iC$, $B, C \in \mathcal{M}_2(\mathbb{R})$ (B is called the *real part* of A and C is the *imaginary part* of A);
(b) Any matrix $A \in \mathcal{M}_2(\mathbb{C})$ can be written in exactly one way as $A = H(A) + iK(A)$, in which both $H(A)$ and $K(A)$ are Hermitian; the representation $A = H(A) + iK(A)$ of a complex or a real matrix is called the *Toeplitz decomposition* [38, p. 227].

1.43 Rotation and reflection matrices and a direct sum.

Let $U(a, b) = \begin{pmatrix} a & -b \\ b & a \end{pmatrix}$, $a, b \in \mathbb{C}$ and let $V(\alpha, \beta) = \begin{pmatrix} \alpha & \beta \\ \beta & -\alpha \end{pmatrix}$, $\alpha, \beta \in \mathbb{C}$.

The following properties hold.

(1) $U(a, b)U(a', b') = U(aa' - bb', a'b + ab')$.

(2) $U^{-1}(a, b) = U\left(\dfrac{a}{a^2 + b^2}, -\dfrac{b}{a^2 + b^2}\right)$, $a^2 + b^2 \neq 0$.

(3) $U(a, b)V(\alpha, \beta) = V(a\alpha - b\beta, a\beta + b\alpha)$.

(4) $V(\alpha, \beta)U(a, b) = V(\alpha a + \beta b, \beta a - \alpha b)$.

(5) $U(\alpha, \beta)V(1, 0) = V(\alpha, \beta)$.

(6) $V(\alpha, \beta)V(\alpha', \beta') = U(\alpha\alpha' + \beta\beta', \alpha'\beta - \alpha\beta')$.

(7) $V^{-1}(\alpha, \beta) = V\left(\dfrac{\alpha}{\alpha^2 + \beta^2}, \dfrac{\beta}{\alpha^2 + \beta^2}\right)$, $\alpha^2 + \beta^2 \neq 0$.

(8) $V^{-1}\left(\dfrac{\alpha}{\sqrt{\alpha^2 + \beta^2}}, \dfrac{\beta}{\sqrt{\alpha^2 + \beta^2}}\right) = V\left(\dfrac{\alpha}{\sqrt{\alpha^2 + \beta^2}}, \dfrac{\beta}{\sqrt{\alpha^2 + \beta^2}}\right)$,

where $\alpha^2 + \beta^2 \neq 0$.

(9) $\mathscr{U} = \{U(a, b) : a, b \in \mathbb{C}\}$ and $\mathscr{V} = \{V(\alpha, \beta) : \alpha, \beta \in \mathbb{C}\}$ are vector spaces over $\mathbb{C}$.

(10) **A direct sum.** $\mathscr{M}_2(\mathbb{C}) = \mathscr{U} \oplus \mathscr{V}$. Any matrix $A = \begin{pmatrix} a & b \\ c & d \end{pmatrix} \in \mathscr{M}_2(\mathbb{C})$

has a unique writing as $A = U\left(\frac{a+d}{2}, \frac{c-b}{2}\right) + V\left(\frac{a-d}{2}, \frac{c+b}{2}\right)$.

(11) **Orthogonality.** The function $\langle \cdot, \cdot \rangle : \mathscr{M}_2(\mathbb{C}) \times \mathscr{M}_2(\mathbb{C}) \to \mathbb{C}$ defined by $\langle A, B \rangle = \mathrm{Tr}(AB^*)$ is an inner product on $\mathscr{M}_2(\mathbb{C})$ and $(\mathscr{M}_2(\mathbb{C}), \langle \cdot, \cdot \rangle)$ is an *Euclidean space*. If $U(a, b) \in \mathscr{U}$ and $V(\alpha, \beta) \in \mathscr{V}$, then $V^*(\alpha, \beta) = V\left(\overline{\alpha}, \overline{\beta}\right)$ and $\langle U(a, b), V(\alpha, \beta) \rangle = 0$. Thus, the subspaces $\mathscr{U}$ and $\mathscr{V}$ are *orthogonal* and $\mathscr{U}$ is the *orthogonal complement* of $\mathscr{V}$ in $\mathscr{M}_2(\mathbb{C})$, i.e., $\mathscr{U} = \mathscr{V}^\perp$ and $\mathscr{V} = \mathscr{U}^\perp$.

(12) **The geometric interpretation.** If $a, b \in \mathbb{R}$, then

$$U(a, b) = \sqrt{a^2 + b^2}\begin{pmatrix} \cos\theta & -\sin\theta \\ \sin\theta & \cos\theta \end{pmatrix},$$

where $\cos\theta = \frac{a}{\sqrt{a^2+b^2}}$, $\sin\theta = \frac{b}{\sqrt{a^2+b^2}}$, $\theta \in [0, 2\pi)$, so $U(a, b)$ is the matrix corresponding to the composition of the uniform scaling of factor $\sqrt{a^2 + b^2}$ with the rotation of angle θ.

If $\alpha, \beta \in \mathbb{R}$, then

$$V(\alpha, \beta) = \sqrt{\alpha^2 + \beta^2}\begin{pmatrix} \cos t & \sin t \\ \sin t & -\cos t \end{pmatrix},$$

(continued)

1.43 (continued)

where $\cos t = \frac{\alpha}{\sqrt{\alpha^2+\beta^2}}$, $\sin t = \frac{\beta}{\sqrt{\alpha^2+\beta^2}}$, $t \in [0, 2\pi)$, so $V(\alpha, \beta)$ is the matrix corresponding to the composition of the uniform scaling of factor $\sqrt{\alpha^2 + \beta^2}$ with the reflection across a line passing through the origin which makes an angle $\frac{t}{2}$ with the x-axis.

Nota bene. Any square 2×2 real matrix A can be written *uniquely* as the *linear combination* of a *rotation* matrix and a *reflection* matrix, i.e., $A = \lambda U(\cos \theta, \sin \theta) + \mu V(\cos t, \sin t)$, where $\lambda, \mu \in \mathbb{R}$.

1.44 Let $A, B, C \in \mathcal{M}_2 (\mathbb{C})$ such that $A^2 = BC$, $B^2 = CA$, $C^2 = AB$. Prove that:

(a) $A^3 = B^3 = C^3$;
(b) Give an example of three distinct matrices that satisfy the conditions of the problem.

1.45 Let $A = \begin{pmatrix} a & b \\ -b & a \end{pmatrix} \in \mathcal{M}_2 (\mathbb{R})$. The following statements are equivalent:

(a) there exists $n \in \mathbb{N}$ such that $A^n = I_2$;
(b) there exists $q \in \mathbb{Q}^*$ such that $a = \cos q\pi$ and $b = \sin q\pi$.

1.46 Let $A, B \in \mathcal{M}_2 (\mathbb{Z})$ such that $AB = BA$ and $\det A = \det B = 0$. Prove that, there exists $a \in \mathbb{Z}$ such that for any $n \in \mathbb{N}$ we have

$$\det(A^n - B^n) = -a^n \quad \text{and} \quad \det(A^n + B^n) = a^n.$$

1.47 Prove that, if $A, B, C \in \mathcal{M}_2 (\mathbb{R})$ verify the conditions $AB = BA$, $BC = CB$, $CA = AC$, and $\det(A^2 + B^2 + C^2 - AB - BC - CA) = 0$, then $\det(2A - B - C) = 3 \det(B - C)$.

1.48 [60] Determine all pairs (a, b) of real numbers for which there exists a unique 2×2 symmetric matrix M with real entries satisfying $\text{Tr}(M) = a$ and $\det M = b$.

1.5   A bouquet of group, ring, and field theory problems

The next problems establish a connection between various algebraic structures and square matrices. For a thorough exposition of the classical algebraic structures such as groups, rings, and fields as well as other related topics in abstract algebra the reader may wish to refer to the excellent book [17].

1.49 (a) Prove that the set $\mathcal{M} = \left\{ \begin{pmatrix} 2-a & a-1 \\ 2(1-a) & 2a-1 \end{pmatrix} : a \in \mathbb{R}^* \right\}$ together with the multiplication of matrices is an abelian group.

(b) Prove that $(\mathcal{M}, \cdot)$ is isomorphic to the multiplicative group $(\mathbb{R}^*, \cdot)$.

1.50 Prove that the set $\mathcal{M} = \left\{ \begin{pmatrix} x & 2y \\ y & x \end{pmatrix} : x, y \in \mathbb{Q}, \ x \neq 0 \ \text{or} \ y \neq 0 \right\}$ together with the multiplication of matrices is an abelian group.

1.51 Prove that the set $\mathcal{M} = \left\{ \begin{pmatrix} \cos \alpha & \sin \alpha \\ -\sin \alpha & \cos \alpha \end{pmatrix} : \alpha \in \mathbb{R} \right\}$ together with the multiplication of matrices is a group which is isomorphic to the multiplicative group of complex numbers of absolute value 1.

1.52 Prove that the set $\mathcal{M} = \left\{ \begin{pmatrix} \cos \alpha & 3\sin \alpha \\ -\frac{1}{3}\sin \alpha & \cos \alpha \end{pmatrix} : \alpha \in \mathbb{R} \right\}$ together with the multiplication of matrices is a group which is isomorphic to the multiplicative group of complex numbers of absolute value 1.

1.53 (a) Prove that the set $\mathcal{G} = \left\{ x + y\sqrt{5} : x \in \mathbb{Q}, \ y \in \mathbb{Q}, \ x^2 - 5y^2 = 1 \right\}$ together with the multiplication of numbers is an abelian group.

(b) Prove that the set $\mathcal{M} = \left\{ \begin{pmatrix} x & 2y \\ \frac{5}{2}y & x \end{pmatrix} : x, y \in \mathbb{Q} \ \text{and} \ x^2 - 5y^2 = 1 \right\}$ together with the multiplication of matrices is an abelian group.

(c) Prove that the function $f : \mathcal{G} \to \mathcal{M}, f(x + y\sqrt{5}) = \begin{pmatrix} x & 2y \\ \frac{5}{2}y & x \end{pmatrix}$ is a group isomorphism, i.e., $(\mathcal{G}, \cdot) \cong (\mathcal{M}, \cdot)$.

1.54 Let $d \in \mathbb{R}$ and let $\mathcal{M}_d = \left\{ \begin{pmatrix} a & db \\ b & a \end{pmatrix} : a, b \in \mathbb{R}, \ a^2 - db^2 \neq 0 \right\}$.

(a) Prove that the set $\mathcal{M}_d$ together with the multiplication of matrices is a group.

(b) Determine the values of d for which the group $(\mathcal{M}_d, \cdot)$ is isomorphic to $(\mathbb{C}^*, \cdot)$.

1.55 *Generators of the modular group* $\mathrm{SL}_2(\mathbb{Z})$.

Let $U = \begin{pmatrix} 1 & 1 \\ 0 & 1 \end{pmatrix}$, $V = \begin{pmatrix} 0 & -1 \\ 1 & 0 \end{pmatrix}$, $W = \begin{pmatrix} 1 & 0 \\ 1 & 1 \end{pmatrix}$ and $P = \begin{pmatrix} 0 & -1 \\ 1 & 1 \end{pmatrix}$, $Q = P^2 = \begin{pmatrix} -1 & -1 \\ 1 & 0 \end{pmatrix}$. Check that:

(a) $U^k = \begin{pmatrix} 1 & k \\ 0 & 1 \end{pmatrix}$, $k \in \mathbb{Z}$;

(b) $V^2 = -I_2$, $V^{-1} = -V$, $V^4 = I_2$;

(c) $W^k = \begin{pmatrix} 1 & 0 \\ k & 1 \end{pmatrix}$, $k \in \mathbb{Z}$, $W = UVU$;

(d) $P = VU = V^2 Q^2$, $P^3 = -I_2$;

(e) $Q = VUVU = VW$, $Q^2 = \begin{pmatrix} 0 & 1 \\ -1 & -1 \end{pmatrix}$, $Q^3 = I_2$;

(f) $U = V^{-1}P = -VP = VP^4 = VQ^2$.

These properties, which may be of independent interest for the reader, are used for proving that, a result far beyond the purpose of this book, the *modular group* $SL_2(\mathbb{Z})$ is generated by matrices U and V (see [43, Chapter 5]).

1.56 The Klein 4-group. Prove that the set $K_4 = \{I_2, S_x, S_y, S_0\}$ where

$$S_x = \begin{pmatrix} -1 & 0 \\ 0 & 1 \end{pmatrix}, \quad S_y = \begin{pmatrix} 1 & 0 \\ 0 & -1 \end{pmatrix} \quad \text{and} \quad S_0 = \begin{pmatrix} -1 & 0 \\ 0 & -1 \end{pmatrix},$$

together with the multiplication of matrices is an abelian group of four elements called the *Klein 4-group*. Also, $(K_4, \cdot)$ is not isomorphic to $(\mathcal{R}_4, \cdot)$, the group of the rotations of a square.

Nota bene. Geometrically, in 2D the Klein 4-group is the symmetry group of a rhombus and of a rectangle which are not squares, the four elements being the identity I_2, the horizontal reflection S_x, the vertical reflection S_y, and a 180 degree rotation (the reflection through the origin) S_0.

1.57 Let

$$\mathcal{M} = \left\{ \begin{pmatrix} a+b & b \\ c & a+c \end{pmatrix} : a, b, c \in \mathbb{R} \right\}.$$

Determine the set $\mathcal{G}$ of all orthogonal matrices from $\mathcal{M}$ and show that $(\mathcal{G}, \cdot)$ is isomorphic to the Klein 4-group (*Viergruppe*).

1.58 A group characterization of idempotent matrices.

Let $n \in \mathbb{N}$, $A \in \mathcal{M}_2(\mathbb{C})$, $A \neq O_2, I_2$ and let $\mathcal{M}_n = \{X \in \mathcal{M}_2(\mathbb{C}) : X^n = A\}$. Prove the following statements are equivalent:

(a) $(\mathcal{M}_n, \cdot)$ is a group;

(b) $A^2 = A$;

(c) $\mathcal{M}_n = \{sA, s \in \mathbb{C}, s^n = 1\} = \mathcal{U}_n A$, i.e., $(\mathcal{M}_n, \cdot) \cong (\mathcal{U}_n, \cdot)$.

1.59 *Two nonisomorphic groups.*

Let $G = \{A \in \mathcal{M}_2(\mathbb{C}) : \det A = \pm 1\}$ and let $S = \{A \in \mathcal{M}_2(\mathbb{C}) : \det A = 1\}$. Prove that G and S together with the multiplication of matrices are *nonisomorphic groups*. Note that S is the special linear group $SL_2(\mathbb{C})$.

1.60 Let $\mathscr{G}$ be a group of matrices from $(\mathscr{M}_2(\mathbb{C}), \cdot)$ whose identity element is different from I_2. Prove that $\mathscr{G}$ is isomorphic to a subgroup of $(\mathbb{C}^*, \cdot)$.

1.61 The rotation matrix R_α.

Let $\alpha \in \mathbb{R}$ and $M(x_M, y_M)$ be a point in the Cartesian plane. If we rotate counterclockwise the segment $[OM]$ by an angle α around the origin we get the segment $[ON]$. If $N(x_N, y_N)$, then one can check that

$$\begin{pmatrix} x_N \\ y_N \end{pmatrix} = \begin{pmatrix} \cos \alpha & -\sin \alpha \\ \sin \alpha & \cos \alpha \end{pmatrix} \begin{pmatrix} x_M \\ y_M \end{pmatrix}.$$

Because of this reason, the matrix $R_\alpha = \begin{pmatrix} \cos \alpha & -\sin \alpha \\ \sin \alpha & \cos \alpha \end{pmatrix}$ is called *the rotation matrix* of angle α.

Properties of the rotation matrix

(a) Prove that, for any $\alpha \in \mathbb{R}$ the matrix R_α is orthogonal.

(b) Let SO_2 be the set

$$SO_2 = \left\{ \begin{pmatrix} \cos \alpha & -\sin \alpha \\ \sin \alpha & \cos \alpha \end{pmatrix} : \alpha \in \mathbb{R} \right\}.$$

Prove that SO_2 together with the multiplication of matrices is an abelian group. Thus, SO_2, which is called *the special orthogonal group* consists of the orthogonal matrices whose determinant is 1.

(c) Prove that $R_\alpha R_\beta = R_{\alpha+\beta}$.

(d) Prove that $R_\alpha^{-1} = R_{-\alpha}$.

(e) Calculate R_α^n, $n \geq 1$.

    Nota bene. Recall that **Euler's totient function** φ counts the positive integers up to a given integer n that are relatively prime to n. If $n \geq 2$ is an integer, there are $\varphi(n)$ **real distinct pairwise commuting matrices having the same order** n. For example if $n = 8$, $R_{\frac{\pi}{8}}$, $R_{\frac{3\pi}{8}}$, $R_{\frac{5\pi}{8}}$, and $R_{\frac{7\pi}{8}}$ are distinct pairwise commuting matrices of order 8.

1.62 Two special subgroups of $\mathcal{M}_2(\mathbb{C})$.

Let $A \in \mathcal{M}_2(\mathbb{C})$, let $G(A) = \{X \in \mathcal{M}_2(\mathbb{C}) : \det(A + X) = \det A + \det X\}$, and let $H(A) = \{X \in \mathcal{M}_2(\mathbb{C}) : \mathrm{Tr}(AX) = \mathrm{Tr}(A)\mathrm{Tr}(X)\}$. Prove that $(G(A), +)$ and $(H(A), +)$ are subgroups of $(\mathcal{M}_2(\mathbb{C}), +)$.

Challenge problem. Prove that if $A, B \in \mathcal{M}_2(\mathbb{C})$ are two nonzero matrices, then the subgroups $(G(A), +)$ and $(H(B), +)$ are isomorphic.

1.63 Let $\theta \in \mathbb{R}$ and let $M(\theta) = \begin{pmatrix} \cosh\theta & \sinh\theta \\ \sinh\theta & \cosh\theta \end{pmatrix}$. Prove that:

(a) $\det M(\theta) = 1$;

(b) $M(\theta_1)M(\theta_2) = M(\theta_1 + \theta_2)$;

(c) $M^n(\theta) = M(n\theta), n \in \mathbb{N}$;

(d) **The hyperbolic group.** The set $\mathscr{H} = \{M(\theta) : \theta \in \mathbb{R}\}$ together with the multiplication of matrices is an abelian group.

1.64 The dihedral group D_{2n}. Let $n \geq 3$ be an integer. Prove that the set

$$D_{2n} = \left\{ \begin{pmatrix} \cos\theta & -\sin\theta \\ \sin\theta & \cos\theta \end{pmatrix}, \begin{pmatrix} -\cos\theta & \sin\theta \\ \sin\theta & \cos\theta \end{pmatrix} : \theta = 0, \frac{2\pi}{n}, \ldots, \frac{2(n-1)\pi}{n} \right\}$$

together with the multiplication of matrices is a group of $2n$ elements called the *dihedral group*, also known as the set of the symmetries of a regular n gon.

Nota bene. $(D_{2n}, \cdot)$ is not isomorphic to $(\mathscr{R}_{2n}, \cdot)$, the group of the rotations of a regular $2n$-gon.

A lovely presentation of dihedral groups can be found in [17, p. 23].

1.65 Prove that the ring $\mathcal{M}_2(\mathbb{R})$ contains a subring that is isomorphic to $\mathbb{C}$.

1.66 (a) Prove that the set $\mathcal{M} = \left\{ \begin{pmatrix} x & y \\ -y & x \end{pmatrix} : x, y \in \mathbb{Z} \right\}$ together with the addition and multiplication of matrices is a ring.

(b) Prove that $(\mathcal{M}, +, \cdot) \cong (\mathbb{Z}[i], +, \cdot)$, where $\mathbb{Z}[i]$ is the ring of *Gaussian integers*, i.e., the set of complex numbers $x + iy$ with $x, y \in \mathbb{Z}$.

1.67 Determine all functions $f, g : \mathbb{Z} \times \mathbb{Z} \to \mathbb{Z}$ such that $(\mathcal{M}, +, \cdot)$, where

$$\mathcal{M} = \left\{ \begin{pmatrix} x & y \\ f(x,y) & g(x,y) \end{pmatrix} : x, y \in \mathbb{Z} \right\},$$

is a subring of $(\mathcal{M}_2(\mathbb{Z}), +, \cdot)$ and $I_2 \in \mathcal{M}$.

1.68 [17, p. 251] Prove that the elements $\begin{pmatrix} 0 & 1 \\ 0 & 0 \end{pmatrix}$ and $\begin{pmatrix} 0 & 0 \\ 1 & 0 \end{pmatrix}$ are nilpotent elements of $\mathcal{M}_2(\mathbb{Z})$ whose sum is not nilpotent (note that these two matrices do not commute). Deduce that the set of nilpotent elements in the noncommutative ring $\mathcal{M}_2(\mathbb{Z})$ is not an ideal.

1.69 Determine all functions $f, g : \mathbb{Z} \times \mathbb{Z} \to \mathbb{Z}$ such that the set

$$\mathcal{M} = \left\{ \begin{pmatrix} x & f(x,y) \\ g(x,y) & y \end{pmatrix} : x, y \in \mathbb{Z} \right\}$$

together with the addition and the multiplication of matrices is a ring and $I_2 \in \mathcal{M}$.

1.70 Prove that the set $\mathcal{M} = \left\{ \begin{pmatrix} x & -3y \\ -y & x \end{pmatrix} : x, y \in \mathbb{Z} \right\}$ together with the addition and multiplication of matrices is a ring which is isomorphic to the ring

$$\mathcal{A} = \left\{ x + y\sqrt{3} : x, y \in \mathbb{Z} \right\}.$$

The ring isomorphism is of the form $x + y\sqrt{3} \longrightarrow \begin{pmatrix} x & -3y \\ -y & x \end{pmatrix}$.

1.71 Prove that the set $\mathcal{M} = \left\{ \begin{pmatrix} x & 4y \\ \frac{1}{2}y & x \end{pmatrix} : x, y \in \mathbb{Q} \right\}$ together with the addition and multiplication of matrices is a field which is isomorphic to the field

$$\mathcal{A} = \left\{ x + y\sqrt{2} : x, y \in \mathbb{Q} \right\}.$$

The field isomorphism is of the form $x + y\sqrt{2} \longrightarrow \begin{pmatrix} x & 4y \\ \frac{1}{2}y & x \end{pmatrix}$.

1.72 Determine all functions $f, g : \mathbb{Q} \times \mathbb{Q} \to \mathbb{Q}$ such that the set

$$\mathcal{M} = \left\{ \begin{pmatrix} x & f(x,y) \\ g(x,y) & y \end{pmatrix} : x, y \in \mathbb{Q} \right\}$$

together with the addition and the multiplication of matrices is a field.

1.73 Matrix Hamilton Quaternions.

Let $\mathcal{M}$ be the set of square matrices of the following form

$$\mathcal{M} = \{ m = aE + bI + cJ + dK : a, b, c, d \in \mathbb{R} \},$$

where

$$E = \begin{pmatrix} 1 & 0 \\ 0 & 1 \end{pmatrix}, \quad I = \begin{pmatrix} i & 0 \\ 0 & -i \end{pmatrix}, \quad J = \begin{pmatrix} 0 & 1 \\ -1 & 0 \end{pmatrix} \quad \text{and} \quad K = \begin{pmatrix} 0 & i \\ i & 0 \end{pmatrix}.$$

(a) Calculate $m\widetilde{m}$ and $\widetilde{m}m$ where $\widetilde{m} = aE - bI - cJ - dK$.

(b) Prove that $\mathcal{M}$ together with the addition and multiplication of matrices is a noncommutative field (*Matrix Hamilton Quaternions*). Historically, one of the first noncommutative rings was discovered in 1843 by Sir William Rowell Hamilton (1805–1865). A nice paper about quaternions describing what led Hamilton to his discovery is [7]. Quaternions, matrices of quaternions, and their properties as well as related problems are given in [62].

(c) Prove that the property "*a polynomial of degree n has at most n roots*" which holds when the coefficients of the polynomial function belong to a commutative field, fails to hold in $\mathcal{M}$. To show this, consider the polynomial function $x^2 + E$.

A *challenging problem* would be to solve in $\mathcal{M}$ the equation $x^2 + E = O_2$.

1.74 Hermitian and Pauli matrices.

(a) Prove the set of 2×2 Hermitian matrices is given by

$$\mathcal{H} = \left\{ \begin{pmatrix} u & \bar{v} \\ c & d \end{pmatrix} : a, d \in \mathbb{R}, \ c \in \mathbb{C} \right\}.$$

(continued)

1.74 (continued)

(b) The famous **Pauli matrices** σ_1, σ_2 and σ_3 are defined as follows

$$\sigma_1 = \begin{pmatrix} 0 & 1 \\ 1 & 0 \end{pmatrix}, \quad \sigma_2 = \begin{pmatrix} 0 & -i \\ i & 0 \end{pmatrix} \quad \text{and} \quad \sigma_3 = \begin{pmatrix} 1 & 0 \\ 0 & -1 \end{pmatrix}.$$

Prove the set of Pauli matrices together with the identity matrix I_2 form a basis for the real vector space $\mathcal{H}$ of 2×2 Hermitian matrices, so $\mathcal{H}$ is a vector space of dimension 4 over $\mathbb{R}$.

Observe that if $A \in \mathcal{H}$, then

$$A = \begin{pmatrix} a & x - iy \\ x + iy & d \end{pmatrix} = \frac{a+d}{2} I_2 + \frac{a-d}{2} \sigma_3 + x\sigma_1 + y\sigma_2, \quad a, d, x, y \in \mathbb{R}.$$

(c) Prove the real linear span of $\{I_2, i\sigma_1, i\sigma_2, i\sigma_3\}$ is isomorphic to the set $\mathcal{M}$, the Matrix Hamilton Quaternions, defined in problem **1.73**.

1.6 Solutions

1.1. Use mathematical induction.

1.2. A^2 can have one, two, or three negative entries. For example, A^2 can have one negative entry for $A = \begin{pmatrix} 2 & 1 \\ -1 & 3 \end{pmatrix}$, two negative entries for $A = \begin{pmatrix} 2 & -1 \\ -1 & 3 \end{pmatrix}$, and three negative entries for $A = \begin{pmatrix} -1 & 2 \\ -3 & -1 \end{pmatrix}$. It is not possible to have four negative entries!

1.3. Using the equality $AA^n = A^n A$ we get the relations

$$\begin{cases} aa_n + bc_n = a_n a + b_n c \\ ab_n + bd_n = a_n b + b_n d \\ ca_n + dc_n = c_n a + d_n c \\ cb_n + dd_n = c_n b + d_n d \end{cases}$$

which can be written as $bc_n = b_n c$, $(a-d)b_n = (a_n - d_n)b$, $(a-d)c_n = (a_n - d_n)c$, $cb_n = c_n b$. If $b_n = 0$ or $c_n = 0$ we get, since $b \neq 0$, $c \neq 0$, that $a_n - d_n = 0$ and this implies that $a_n = d_n$.

1.4. Let $X = \begin{pmatrix} a & b \\ c & d \end{pmatrix}$ be such that $AX = XA$. We have

$$\begin{pmatrix} 1 & 2 \\ 3 & 4 \end{pmatrix}\begin{pmatrix} a & b \\ c & d \end{pmatrix} = \begin{pmatrix} a & b \\ c & d \end{pmatrix}\begin{pmatrix} 1 & 2 \\ 3 & 4 \end{pmatrix},$$

and this implies that

$$\begin{cases} a + 2c = a + 3b \\ b + 2d = 2a + 4b \\ 3a + 4c = c + 3d \\ 3b + 4d = 2c + 4d. \end{cases}$$

We obtain the system

$$\begin{cases} 2c = 3b \\ 2(d - a) = 3b, \end{cases}$$

which has the solutions $a = x$, $b = 2y$, $c = 3y$, $d = x + 3y$, where $x, y \in \mathbb{C}$. Thus,

$$X = \begin{pmatrix} x & 2y \\ 3y & x + 3y \end{pmatrix} = yA + (x - y)I_2.$$

1.8. Let $A = \begin{pmatrix} a & b \\ c & d \end{pmatrix}$. The equation $A^2 = O_2$ implies that

$$\begin{cases} a^2 + bc = 0 \\ b(a + d) = 0 \\ c(a + d) = 0 \\ bc + d^2 = 0. \end{cases}$$

If $a + d \neq 0$, then $b = c = 0$, $a^2 = d^2 = 0$ and this implies $a = d = 0$, which contradicts $a + d \neq 0$. Thus, $a + d = 0$ and we get that $a = -d$. We look at the equation $a^2 + bc = 0$ and we consider the following two cases:

- if $b = 0$ we have that $a = 0$ and $A = \begin{pmatrix} 0 & 0 \\ c & 0 \end{pmatrix}$, where $c \in \mathbb{C}$;

- if $b \neq 0$ we have that $c = -\dfrac{a^2}{b}$ and $A = \begin{pmatrix} a & b \\ -\frac{a^2}{b} & -a \end{pmatrix}$, $a, b \in \mathbb{C}$, $b \neq 0$

1.10. Like in the solution of problem **1.8** from the equation $A^2 = O_2$, with $A = \begin{pmatrix} \widehat{a} & \widehat{b} \\ \widehat{c} & \widehat{d} \end{pmatrix}$, we get that $\widehat{a} + \widehat{d} = \widehat{0}$ and $\widehat{a}^2 + \widehat{bc} = \widehat{0}$. It follows that $\widehat{d} = -\widehat{a}$ and $\widehat{bc} = -\widehat{a}^2$.

If $\widehat{b} = \widehat{0}$, then $\widehat{a} = 0$ and $\widehat{c} \in \mathbb{Z}_p$ is arbitrary taken. Thus, we have p matrices of the form $\begin{pmatrix} \widehat{0} & \widehat{0} \\ \widehat{c} & \widehat{0} \end{pmatrix}$.

If $\widehat{b} \neq \widehat{0}$, then $\widehat{c} = -\widehat{b}^{-1}\widehat{a}^2$, where $\widehat{b}$ can be chosen in $p - 1$ ways and $\widehat{a}$ in p ways so that we have $p(p-1)$ matrices of the form $\begin{pmatrix} \widehat{a} & \widehat{b} \\ -\widehat{b}^{-1}\widehat{a}^2 & -\widehat{a} \end{pmatrix}, \widehat{a} \in \mathbb{Z}_p, \widehat{b} \in \mathbb{Z}_p \setminus \{\widehat{0}\}$.

Therefore, we have $p + p^2 - p = p^2$ nilpotent matrices in $\mathcal{M}_2(\mathbb{Z}_p)$.

1.11. From $(I_2 + iA)(I_2 - iA) = I_2$ we get that $A^2 = O_2$. It follows, based on the solution of problem **1.8**, that A is of the following form $A = \begin{pmatrix} 0 & 0 \\ c & 0 \end{pmatrix}$, where $c \in \mathbb{R}$,

or $A = \begin{pmatrix} a & b \\ -\frac{a^2}{b} & -a \end{pmatrix}$, $a, b \in \mathbb{R}, b \neq 0$.

1.12. Let $A = \begin{pmatrix} a & b \\ c & d \end{pmatrix}$. The equation $A^2 = I_2$ implies that

$$\begin{cases} a^2 + bc = 1 \\ b(a + d) = 0 \\ c(a + d) = 0 \\ bc + d^2 = 1. \end{cases}$$

If $a + d \neq 0$, then $b = c = 0$, $a^2 = d^2 = 1$ and it follows that $a = \pm 1$ and $d = \pm 1$. Since $a + d \neq 0$ we get that $a = 1, d = 1$ or $a = -1, d = -1$. It follows that

$$A = \begin{pmatrix} 1 & 0 \\ 0 & 1 \end{pmatrix} = I_2 \quad \text{or} \quad A = \begin{pmatrix} -1 & 0 \\ 0 & -1 \end{pmatrix} = -I_2.$$

If $a + d = 0$, then $d = -a$. If $b = 0$ we get that $a^2 = d^2 = 1$ and this implies $a = -1, d = 1$ or $a = 1, d = -1$. Thus,

$$A = \begin{pmatrix} -1 & 0 \\ c & 1 \end{pmatrix} \quad \text{or} \quad A = \begin{pmatrix} 1 & 0 \\ c & -1 \end{pmatrix}, \quad c \in \mathbb{C}.$$

If $b \neq 0$ we have that $c = \dfrac{1 - a^2}{b}$ and A has the form

$$A = \begin{pmatrix} a & b \\ \dfrac{1-a^2}{b} & -a \end{pmatrix}, \ a \in \mathbb{C}, \ b \in \mathbb{C}^*.$$

1.13. If $p \neq 2$, then a matrix $B \in \mathcal{M}_2(\mathbb{Z}_p)$ is idempotent ($B^2 = B$) if and only if the matrix $A = 2B - I_2$ is involutory ($A^2 = I_2$) which implies that, in this case, the number of involutory matrices is the same as the number of idempotent matrices, which is $p^2 + p + 2$ (see problem **1.15**).

When $p = 2$ we solve in $\mathcal{M}_2(\mathbb{Z}_2)$ the matrix equation $A^2 = I_2$. If $A = \begin{pmatrix} \widehat{a} & \widehat{b} \\ \widehat{c} & \widehat{d} \end{pmatrix}$, then we have that $\widehat{a}^2 + \widehat{bc} = \widehat{1}, \widehat{b}\left(\widehat{a}+\widehat{d}\right) = \widehat{0}, \widehat{c}\left(\widehat{a}+\widehat{d}\right) = \widehat{0}, \widehat{d}^2 + \widehat{bc} = \widehat{1}$ and these equations imply that $\widehat{d} = -\widehat{a}$.

If $\widehat{a} = \widehat{1}$ and $\widehat{bc} = \widehat{0}$, then we get the matrices

$$A_1 = \begin{pmatrix} \widehat{1} & \widehat{0} \\ \widehat{0} & \widehat{1} \end{pmatrix}, \quad A_2 = \begin{pmatrix} \widehat{1} & \widehat{0} \\ \widehat{1} & \widehat{1} \end{pmatrix}, \quad A_3 = \begin{pmatrix} \widehat{1} & \widehat{1} \\ \widehat{0} & \widehat{1} \end{pmatrix}.$$

If $\widehat{a} = \widehat{0}$, then $\widehat{d} = \widehat{0}$ and $\widehat{bc} = \widehat{1}$, so $\widehat{b} = \widehat{c} = \widehat{1}$. Thus, we obtain the matrix $A_4 = \begin{pmatrix} \widehat{0} & \widehat{1} \\ \widehat{1} & \widehat{0} \end{pmatrix}$. Therefore, when $p = 2$ there are 4 involutory matrices. Observe this is different than $p^2 + p + 2$ which is the number of involutory matrices when $p \neq 2$.

1.14. Let $A = \begin{pmatrix} a & b \\ c & d \end{pmatrix}$. The equation $A^2 = A$ implies that

$$\begin{cases} a^2 + bc = a \\ b(a+d) = b \\ c(a+d) = c \\ bc + d^2 = d \end{cases} \quad \text{or} \quad \begin{cases} a^2 + bc = a \\ b(a+d-1) = 0 \\ c(a+d-1) = 0 \\ (a-d)(a+d-1) = 0. \end{cases}$$

If $a + d - 1 \neq 0$, then $b = c = 0$, $a = d \in \{0, 1\}$. It follows that $A = O_2$ or $A = I_2$. If $a + d - 1 = 0$, the system reduces to the equation $a^2 + bc = a$. When $b \neq 0$, then

$$c = \frac{a - a^2}{b} \quad \text{and} \quad A = \begin{pmatrix} a & b \\ \dfrac{a - a^2}{b} & 1 - a \end{pmatrix}, \ a, b \in \mathbb{C}, \ b \neq 0.$$

When $b = 0$, then either $a = 0$ or $a = 1$ and this implies that

$$A = \begin{pmatrix} 0 & 0 \\ c & 1 \end{pmatrix} \quad \text{or} \quad A = \begin{pmatrix} 1 & 0 \\ c & 0 \end{pmatrix}, \ c \in \mathbb{C}.$$

1.15. Like in the solution of problem **1.14** we get matrices $A = O_2$, $A = I_2$ and then matrices $A = \begin{pmatrix} \widehat{a} & \widehat{b} \\ \widehat{c} & \widehat{d} \end{pmatrix}$ which satisfy the conditions $\widehat{a} + \widehat{d} = \widehat{1}$ and $\widehat{a}^2 + \widehat{bc} = \widehat{a}$.

If $\widehat{a} = \widehat{0}$, then $\widehat{d} = \widehat{1}$, $\widehat{bc} = \widehat{0}$, so $\widehat{b} = \widehat{0}$ or $\widehat{c} = \widehat{0}$. Thus, we have p matrices of the form $A_1 = \begin{pmatrix} \widehat{0} & \widehat{b} \\ \widehat{0} & \widehat{1} \end{pmatrix}$ and $p - 1$ matrices of the form $A_2 = \begin{pmatrix} \widehat{0} & \widehat{0} \\ \widehat{c} & \widehat{1} \end{pmatrix}$. Therefore, we have $2p - 1$ matrices since we have counted the matrix $\begin{pmatrix} \widehat{0} & \widehat{0} \\ \widehat{0} & \widehat{1} \end{pmatrix}$ once.

If $\widehat{a} = \widehat{1}$, then $\widehat{d} = \widehat{0}$ and $\widehat{bc} = \widehat{0}$. In the same way as above we get $2p - 1$ matrices of the form $A_3 = \begin{pmatrix} \widehat{1} & \widehat{b} \\ \widehat{0} & \widehat{0} \end{pmatrix}$ or $A_4 = \begin{pmatrix} \widehat{1} & \widehat{0} \\ \widehat{c} & \widehat{0} \end{pmatrix}$.

If $\widehat{a} \neq \widehat{0}, \widehat{1}$, then $\widehat{a}^2 - \widehat{a} \neq \widehat{0}$ and we get, from the two equations, that $\widehat{d} = \widehat{1} - \widehat{a}$, $\widehat{bc} = \widehat{a} - \widehat{a}^2$, so $\widehat{c} = \widehat{b}^{-1} (\widehat{a} - \widehat{a}^2)$. We obtain the matrices

$$A_5 = \begin{pmatrix} \widehat{a} & \widehat{b} \\ \widehat{b}^{-1} (\widehat{a} - \widehat{a}^2) & \widehat{1} - \widehat{a} \end{pmatrix}, \quad \widehat{a} \in \mathbb{Z}_p \setminus \{\widehat{0}, \widehat{1}\}.$$

Observe that $\widehat{a}$ is chosen in $p - 2$ ways and $\widehat{b} \neq \widehat{0}$ is taken in $p - 1$ ways, so we have $(p-1)(p-2)$ matrices of the form A_5.

In conclusion the number of idempotent matrices in $\mathcal{M}_2 (\mathbb{Z}_p)$ is $p^2 + p + 2$.

1.16. If $X = \begin{pmatrix} a & b \\ c & d \end{pmatrix}$, $A = \frac{1}{\sqrt{2}} \begin{pmatrix} 1 & 1 \\ -1 & 1 \end{pmatrix}$, $B = \frac{1}{\sqrt{2}} \begin{pmatrix} 1 & 1 \\ 1 & -1 \end{pmatrix}$ and $C = \begin{pmatrix} 0 & 1 \\ 1 & 0 \end{pmatrix}$, then $X = \alpha A + \beta B + \gamma C + \delta I_2$, where $\alpha = \frac{b-c}{\sqrt{2}}$, $\beta = \frac{a-d}{\sqrt{2}}$, $\gamma = \frac{c-a+b+d}{2}$ and $\delta = \frac{a+c+d-b}{2}$.

1.17. (a) $A = \begin{pmatrix} a & -b \\ b & a \end{pmatrix}$ or $A = \begin{pmatrix} -a & b \\ b & a \end{pmatrix}$, $a, b \in \mathbb{C}$, with $a^2 + b^2 = 1$.

(b) $A = \begin{pmatrix} a & -b \\ b & a \end{pmatrix}$ or $A = \begin{pmatrix} -a & b \\ b & a \end{pmatrix}$, $a, b \in \mathbb{C}$, with $a^2 + b^2 = -1$.

1.19. We have $A^2 = A$, $B^2 = B$. The equality $(A + B)^2 = A + B$ implies that $AB + BA = O_2$. We multiply this equality by A to the left respectively to the right and we get that $AB + ABA = O_2$ and $ABA + BA = O_2$. It follows that $AB = BA = O_2$.

If $A = \alpha I_2$, $\alpha \in \mathbb{C}^*$, we get, since $A^2 = A$, that $\alpha^2 = \alpha \Rightarrow \alpha = 1 \Rightarrow A = I_2 \Rightarrow B = BA = O_2$, which is impossible.

If $A \neq \alpha I_2$, $\alpha \in \mathbb{C}$, then we have, based on Theorem 1.1, that $B = aA + bI_2$, for some $a, b \in \mathbb{C}$. Since $B^2 = B$ we get that $(aA + bI_2)^2 = aA + bI_2 \Rightarrow (a^2 + 2ab - a)A = (b - b^2)I_2$. It follows that $a^2 + 2ab - a = 0$ and $b - b^2 = 0$. If $b = 0$, then $a = 0$ or $a = 1$. The case $a = b = 0$ implies that $B = O_2$, which is impossible. If $b = 0$ and $a = 1$, then $A = B \Rightarrow O_2 = AB = A^2 = A$, which is impossible. If $b = 1$, then $a = 0$ or $a = -1$. If $b = 1$ and $a = 0$, then $B = I_2 \Rightarrow A = AB = O_2$, which is impossible. If $b = 1$ and $a = -1$, then $A + B = I_2$.

1.23. (a) We have $(I_2 + X)(I_2 - \frac{1}{2}X) = I_2 + \frac{1}{2}X - \frac{1}{2}X^2 = I_2$.

(b) $A = AB + B^2$ and $B = BA + A^2$ and these imply $A + B = (A + B)^2$ and $(A - B)(I_2 + A + B) = O_2$. Since the matrix $I_2 + A + B$ is invertible, see part (a), the equality $(A - B)(I_2 + A + B) = O_2$ implies that $A = B$.

1.25. *Solution 1.* Observe that $A^n = \begin{pmatrix} 1 & 0 \\ n & 1 \end{pmatrix}$, $n \geq 1$ and prove it by induction.

Solution 2. We note that $A = I_2 + B$, where $B = \begin{pmatrix} 0 & 0 \\ 1 & 0 \end{pmatrix}$. Since $B^2 = O_2$ we have, based on the Binomial Theorem, that

$$A^n = (I_2 + B)^n = \binom{n}{0}I_2^n + \binom{n}{1}I_2^{n-1}B + \binom{n}{2}I_2^{n-2}B^2 + \cdots + \binom{n}{n}B^n$$

$$= I_2 + nB = \begin{pmatrix} 1 & 0 \\ n & 1 \end{pmatrix}.$$

1.27. The problem can be solved by mathematical induction.

1.28. (a) Observe that $1 + \sqrt{3} = 2\sqrt{2}\cos\frac{\pi}{12}$ and $\sqrt{3} - 1 = 2\sqrt{2}\sin\frac{\pi}{12}$. It follows that

$$\begin{pmatrix} 1 + \sqrt{3} & 1 - \sqrt{3} \\ \sqrt{3} - 1 & 1 + \sqrt{3} \end{pmatrix}^n = (2\sqrt{2})^n \begin{pmatrix} \cos\dfrac{n\pi}{12} & -\sin\dfrac{n\pi}{12} \\ \sin\dfrac{n\pi}{12} & \cos\dfrac{n\pi}{12} \end{pmatrix}.$$

(b) We have that

$$\begin{pmatrix} a & -b \\ b & a \end{pmatrix}^n = \left(\sqrt{a^2 + b^2}\right)^n \begin{pmatrix} \cos(n\alpha) & -\sin(n\alpha) \\ \sin(n\alpha) & \cos(n\alpha) \end{pmatrix},$$

where $\cos\alpha = \frac{a}{\sqrt{a^2+b^2}}$ and $\sin\alpha = \frac{b}{\sqrt{a^2+b^2}}$.

1.29. (a) We have $A = \begin{pmatrix} 1 & 1 \\ 1 & 0 \end{pmatrix}$, $A^2 = \begin{pmatrix} 2 & 1 \\ 1 & 1 \end{pmatrix}$ and it follows that $A^2 = A + I_2$. We multiply both sides of this equality by A^{n-1}, $n \geq 1$, and we get the Fibonacci matrix recurrence formula $A^{n+1} = A^n + A^{n-1}$.

(b) The formula $A^n = \begin{pmatrix} F_{n+1} & F_n \\ F_n & F_{n-1} \end{pmatrix}$, $n \geq 1$, can be proved by mathematical induction.

We give here a different approach. Let $A^n = \begin{pmatrix} a_n & b_n \\ c_n & d_n \end{pmatrix}$, $n \geq 1$. The recurrence formula $A^{n+1} = A^n + A^{n-1}$, from part (a) of the problem, implies that

$$\begin{cases} a_{n+1} = a_n + a_{n-1}, & a_1 = 1, \ a_2 = 2, \ n \geq 1 \\ b_{n+1} = b_n + b_{n-1}, & b_1 = 1, \ b_2 = 1, \ n \geq 1 \\ c_{n+1} = c_n + c_{n-1}, & c_1 = 1, \ c_2 = 1, \ n \geq 1 \\ d_{n+1} = d_n + d_{n-1}, & d_1 = 0, \ d_2 = 1, \ n \geq 1. \end{cases}$$

These recurrence relations imply $a_n = F_{n+1}$, $b_n = F_n$, $c_n = F_n$, $d_n = F_{n-1}$ and it follows that $A^n = \begin{pmatrix} F_{n+1} & F_n \\ F_n & F_{n-1} \end{pmatrix}, n \geq 1$.

(c) The matrix identity $A^{n+m} = A^n A^m$ implies that

$$\begin{pmatrix} F_{n+m+1} & F_{n+m} \\ F_{n+m} & F_{n+m-1} \end{pmatrix} = \begin{pmatrix} F_{n+1} & F_n \\ F_n & F_{n-1} \end{pmatrix} \begin{pmatrix} F_{m+1} & F_m \\ F_m & F_{m-1} \end{pmatrix}.$$

We look at entry $(1, 2)$ of this identity and we get that $F_{n+m} = F_{n+1}F_m + F_nF_{m-1}$, $n, m \in \mathbb{N}$.

(d) Since $\det(A^n) = (\det A)^n$ we get that

$$F_{n+1}F_{n-1} - F_n^2 = \det \begin{pmatrix} F_{n+1} & F_n \\ F_n & F_{n-1} \end{pmatrix} = \left(\det \begin{pmatrix} 1 & 1 \\ 1 & 0 \end{pmatrix} \right)^n = (-1)^n.$$

On the other hand, we have based on part (c) with $m = n$, that

$$\begin{aligned} F_{2n} &= F_{n+1}F_n + F_nF_{n-1} \\ &= F_n(F_{n+1} + F_{n-1}) \\ &= (F_{n+1} - F_{n-1})(F_{n+1} + F_{n-1}) \\ &= F_{n+1}^2 - F_{n-1}^2 \end{aligned}$$

and similarly one can also prove that $F_{2n+1} = F_{n+1}^2 + F_n^2$, $n \geq 0$.

We have, based on the previous quadratic formulae, that

$$\begin{aligned} F_{3n} = F_{2n+n} &= F_{2n+1}F_n + F_{2n}F_{n-1} \\ &= F_n(F_{n+1}^2 + F_n^2) + (F_{n+1}^2 - F_{n-1}^2)F_{n-1} \\ &= F_nF_{n+1}^2 + F_n^3 + F_{n+1}^2 F_{n-1} - F_{n-1}^3 \\ &= F_{n+1}^2(F_n + F_{n-1}) + F_n^3 - F_{n-1}^3 \\ &= F_{n+1}^3 + F_n^3 - F_{n-1}^3. \end{aligned}$$

(e) We have, based on part (b) that $\operatorname{Tr}(A^n) = F_{n+1} + F_{n-1}$. A calculation shows the eigenvalues of A are $\alpha = \frac{1+\sqrt{5}}{2}$ and $\beta = \frac{1-\sqrt{5}}{2}$. Since the eigenvalues of A^n are α^n and β^n we get that $\operatorname{Tr}(A^n) = \alpha^n + \beta^n = L_n = F_{n+1} + F_{n-1}$.

The equation $X_{n+1} = AX_n$ implies that $X_n = A^n X_0$. Thus

$$X_n = \begin{pmatrix} F_{n+1} & F_n \\ F_n & F_{n-1} \end{pmatrix} \begin{pmatrix} 3 \\ 1 \end{pmatrix} = \begin{pmatrix} 3F_{n+1} + F_n \\ 3F_n + F_{n-1} \end{pmatrix}.$$

It remains to prove that $L_{n+2} = 3F_{n+1} + F_n$, $n \geq 0$, and $L_{n+1} = 3F_n + F_{n-1}$, $n \geq 1$. These recurrence formulae can be proved easily by observing that the Lucas and Fibonacci sequences verify the formula $L_{n+1} = F_{n+2} + F_n$, for all $n \geq 0$.

1.30. *Solution 1.* Let $A(n) = B(2)B(3) \cdots B(n)$, $n \geq 2$, and we consider the sequences $(a_n)_{n\geq 2}$ and $(b_n)_{n\geq 2}$ such that

$$A(n) = \begin{pmatrix} a_n & b_n \\ b_n & a_n \end{pmatrix}.$$

Since $A(n + 1) = A(n)B(n + 1)$ we get that

$$A(n+1) = \begin{pmatrix} a_n & b_n \\ b_n & a_n \end{pmatrix} \begin{pmatrix} n+1 & 1 \\ 1 & n+1 \end{pmatrix} = \begin{pmatrix} a_n(n+1) + b_n & a_n + b_n(n+1) \\ a_n + b_n(n+1) & a_n(n+1) + b_n \end{pmatrix}.$$

This implies that

$$\begin{cases} a_{n+1} = a_n(n+1) + b_n \\ b_{n+1} = a_n + b_n(n+1), \end{cases}$$

for all $n \geq 2$. Adding and subtracting these two recurrence relations we get that

$$\begin{cases} a_{n+1} - b_{n+1} = n(a_n - b_n) \\ a_{n+1} + b_{n+1} = (n+2)(a_n + b_n), \end{cases}$$

for all $n \geq 2$. This implies that

$$\begin{cases} a_{n+1} = \dfrac{n!}{2} + \dfrac{(n+2)!}{4} \\ b_{n+1} = -\dfrac{n!}{2} + \dfrac{(n+2)!}{4}. \end{cases}$$

In conclusion

$$B(2)B(3) \cdots B(n) = \begin{pmatrix} \dfrac{(n-1)!}{2} + \dfrac{(n+1)!}{4} & -\dfrac{(n-1)!}{2} + \dfrac{(n+1)!}{4} \\ -\dfrac{(n-1)!}{2} + \dfrac{(n+1)!}{4} & \dfrac{(n-1)!}{2} + \dfrac{(n+1)!}{4} \end{pmatrix}, \quad n \geq 2.$$

Solution 2. We prove that

$$B(2)B(3)\cdots B(n) = \frac{(n-1)!}{4}\begin{pmatrix} n^2+n+2 & n^2+n-2 \\ n^2+n-2 & n^2+n+2 \end{pmatrix}.$$

A calculation shows that the eigenvalues of the matrix $B(x)$ are $x+1$ and $x-1$ with the corresponding eigenvectors $(\alpha\ \alpha)^T$ and $(-\beta\ \beta)^T$. Thus $B(x) = PJ_B(x)P^{-1}$, where $J_B(x)$ denotes the Jordan canonical form of the matrix $B(x)$ and P is the invertible matrix given below

$$J_B(x) = \begin{pmatrix} 1+x & 0 \\ 0 & x-1 \end{pmatrix}, \qquad P = \begin{pmatrix} 1 & -1 \\ 1 & 1 \end{pmatrix} \quad \text{and} \quad P^{-1} = \frac{1}{2}\begin{pmatrix} 1 & 1 \\ -1 & 1 \end{pmatrix}.$$

Thus,

$$B(2)B(3)\cdots B(n) = \begin{pmatrix} 1 & -1 \\ 1 & 1 \end{pmatrix}\begin{pmatrix} \dfrac{(n+1)!}{2} & 0 \\ 0 & (n-1)! \end{pmatrix}\frac{1}{2}\begin{pmatrix} 1 & 1 \\ -1 & 1 \end{pmatrix}$$

$$= \frac{(n-1)!}{4}\begin{pmatrix} n^2+n+2 & n^2+n-2 \\ n^2+n-2 & n^2+n+2 \end{pmatrix}.$$

Another solution of this problem can be found in [40].

1.31. (a) Let A_1, A_2 be the columns of A and B_1, B_2 be the columns of B. We have, based on Proposition 1.1, that

$$\det(A+B) = \det(A_1 + B_1 | A_2 + B_2)$$
$$= \det(A_1 | A_2 + B_2) + \det(B_1 | A_2 + B_2)$$
$$= \det(A_1 | A_2) + \det(A_1 | B_2) + \det(B_1 | A_2) + \det(B_1 | B_2)$$
$$= \det A + \det(A_1 | B_2) + \det(B_1 | A_2) + \det B$$

and

$$\det(A-B) = \det(A_1 - B_1 | A_2 - B_2)$$
$$= \det(A_1 | A_2 - B_2) - \det(B_1 | A_2 - B_2)$$
$$= \det(A_1 | A_2) - \det(A_1 | B_2) - \det(B_1 | A_2) + \det(B_1 | B_2)$$
$$= \det A - \det(A_1 | B_2) - \det(B_1 | A_2) + \det B.$$

Adding the previous equalities we get that part (a) of the problem is solved.

(b) We solve this part of the problem by mathematical induction. When $n = 1$ we need to prove that $\det A_1 + \det(-A_1) = \det A_1 + (-1)^2 \det A_1 = 2\det A_1$, which

holds trivially. Now we assume the formula holds for $n = \overline{1, p}, p \in \mathbb{N}$ and we prove
it for $n = p + 1$. We have

$$\sum \det(\pm A_1 \pm A_2 \pm \cdots \pm A_{p+1}) = \sum \det\left[(\pm A_1 \pm A_2 \pm \cdots \pm A_p) + A_{p+1}\right]$$

$$+ \sum \det\left[(\pm A_1 \pm A_2 \pm \cdots \pm A_p) - A_{p+1}\right]$$

$$\overset{\text{part (a)}}{=} 2 \sum \left[\det(\pm A_1 \pm A_2 \pm \cdots \pm A_p) + \det A_{p+1}\right]$$

$$= 2 \left(2^p \sum_{k=1}^{p} \det A_k + 2^p \det A_{p+1}\right)$$

$$= 2^{p+1} \sum_{k=1}^{p+1} \det A_k.$$

1.32. Let A_1, A_2 be the columns of A, B_1, B_2 be the columns of B and C_1, C_2 be the
columns of C respectively.
  We have

$$\det(A + B + C) = \det(A_1 + B_1 + C_1 | A_2 + B_2 + C_2)$$

$$= \det(A_1 | A_2) + \det(A_1 | B_2) + \det(A_1 | C_2)$$

$$+ \det(B_1 | A_2) + \det(B_1 | B_2) + \det(B_1 | C_2)$$

$$+ \det(C_1 | A_2) + \det(C_1 | B_2) + \det(C_1 | C_2).$$

On the other hand,

$$\det(A + B) = \det(A_1 + B_1 | A_2 + B_2)$$

$$= \det(A_1 | A_2) + \det(A_1 | B_2) + \det(B_1 | A_2) + \det(B_1 | B_2),$$

$$\det(B + C) = \det(B_1 + C_1 | B_2 + C_2)$$

$$= \det(B_1 | B_2) + \det(B_1 | C_2) + \det(C_1 | B_2) + \det(C_1 | C_2)$$

and

$$\det(A + C) = \det(A_1 + C_1 | A_2 + C_2)$$

$$= \det(A_1 | A_2) + \det(A_1 | C_2) + \det(C_1 | A_2) + \det(C_1 | C_2).$$

Putting all these together we get, since $\det A = \det(A_1 | A_2)$, $\det B = \det(B_1 | B_2)$ and
$\det C = \det(C_1 | C_2)$, that the problem is solved.
1.33. Let $S = \det(A + B + C) + \det(-A + B + C) + \det(A - B + C) + \det(A + B - C)$.
  We have, based on part (a) of problem **1.31**, that

$$\det(A + B + C) + \det(A + B - C) = 2\det(A + B) + 2\det C$$
$$\det(A + B + C) + \det(A - B + C) = 2\det(A + C) + 2\det B$$
$$\det(A + B + C) + \det(-A + B + C) = 2\det(B + C) + 2\det A.$$

Adding the previous equalities we get that

$$S + 2\det(A + B + C) = 2\det(A + B) + 2\det(B + C) + 2\det(C + A)$$
$$+ 2\det A + 2\det B + 2\det C,$$

and the result follows based on problem **1.32**.

1.34. We solve the problem by mathematical induction. Let $P(n)$ be the proposition

$$P(n): \quad \det\left(\sum_{i=1}^{n} A_i\right) = \sum_{1 \le i < j \le n} \det(A_i + A_j) - (n - 2)\sum_{i=1}^{n} \det A_i.$$

When $n = 2$ we have $\det(A_1 + A_2) = \det(A_1 + A_2)$, so there is nothing to prove. When $n = 3$ we need to prove that

$$P(3): \quad \det(A_1 + A_2 + A_3) = \det(A_1 + A_2) + \det(A_1 + A_3) + \det(A_2 + A_3)$$
$$- \det A_1 - \det A_2 - \det A_3,$$

which holds based on problem **1.32**.

Now we consider that $P(k)$ is true, for $k = 2, 3, \ldots, n$ and we prove that $P(n+1)$ is true. We have

$$\det(A_1 + A_2 + \cdots + A_{n-1} + A_n + A_{n+1}) = \det(B + A_n + A_{n+1})$$

$$\overset{P(3)\,\text{true}}{=} \det(B + A_n) + \det(B + A_{n+1}) + \det(A_n + A_{n+1}) - \det B - \det A_n - \det A_{n+1}$$

$$= \sum_{1 \le i < j \le n} \det(A_i + A_j) - (n - 2)\sum_{i=1}^{n} \det A_i + \sum_{\substack{1 \le i < j \le n+1 \\ i,j \ne n}} \det(A_i + A_j)$$

$$- (n - 2)\sum_{\substack{i=1 \\ i \ne n}}^{n+1} \det A_i + \det(A_n + A_{n+1}) - \sum_{1 \le i < j \le n-1} \det(A_i + A_j)$$

$$+ (n - 3)\sum_{i=1}^{n-1} \det A_i - \det A_n - \det A_{n+1}$$

$$= \sum_{1 \le i < j \le n+1} \det(A_i + A_j) - (2n - 4 - n + 3)\sum_{i=1}^{n} \det A_i - (n - 2)\det A_{n+1} - \det A_{n+1}$$

$$= \sum_{1 \le i < j \le n+1} \det(A_i + A_j) - (n - 1)\sum_{i=1}^{n+1} \det A_i.$$

1.35. We have, based on problem **1.34**, that

$$\det(S - A_1) = \sum_{2 \le i < j \le n} \det(A_i + A_j) - (n-3) \sum_{i=2}^{n} \det A_i.$$

The other $n - 1$ similar equalities also hold. Adding all these equalities we get that

$$\sum_{i=1}^{n} \det(S - A_i) = (n-2) \sum_{1 \le i < j \le n} \det(A_i + A_j) - (n-1)(n-3) \sum_{i=1}^{n} \det A_i.$$

On the other hand,

$$\det S = \sum_{1 \le i < j \le n} \det(A_i + A_j) - (n-2) \sum_{i=1}^{n} \det A_i.$$

Thus, we need to check that

$$(n-2) \sum_{1 \le i < j \le n} \det(A_i + A_j) - (n-1)(n-3) \sum_{i=1}^{n} \det A_i$$

$$= (n-2) \sum_{1 \le i < j \le n} \det(A_i + A_j) - (n-2)^2 \sum_{i=1}^{n} \det A_i + \sum_{i=1}^{n} \det A_i,$$

which holds since $(n-1)(n-3) = n^2 - 4n + 3 = (n-2)^2 - 1$.

1.36. We have, based on problem **1.32**, that

$$\det(A+B+C)=\det(A+B)+\det(B+C)+\det(C+A) - \det A - \det B - \det C=0.$$

1.37. We have, based on part (a) of problem **1.31**, that for $X, Y \in \mathcal{M}_2(\mathbb{C})$ one has that $\det(X+Y)+\det(X-Y) = 2\det X + 2\det Y$. If $X = A + A^T$ and $Y = A^T$ we get that $\det(A + 2A^T) + \det A = 2\det(A + A^T) + 2\det A^T$. This implies that $\det A = 11$.

1.38. We have, based on part (a) of problem **1.31**, that $\det(A^2 + B^2) + \det(A^2 - B^2) = 2\det(A^2) + 2\det(B^2)$. This implies, since $A^2 + B^2 = -2AB$, that $\det(-2AB) + \det(A^2 - B^2) = 4\det^2 A$. Thus $4\det^2 A + \det(A^2 - B^2) = 4\det^2 A$, which implies that $\det(A^2 - B^2) = 0$.

1.39. Use Theorem 1.1.

1.40. (a) Observe that

$$\begin{pmatrix} 1 & \dfrac{\alpha}{n} \\[2mm] -\dfrac{\alpha}{n} & 1 \end{pmatrix} = \frac{\sqrt{n^2+\alpha^2}}{n}\begin{pmatrix} \dfrac{n}{\sqrt{n^2+\alpha^2}} & \dfrac{\alpha}{\sqrt{n^2+\alpha^2}} \\[3mm] -\dfrac{\alpha}{\sqrt{n^2+\alpha^2}} & \dfrac{n}{\sqrt{n^2+\alpha^2}} \end{pmatrix}.$$

Let $\theta_n \in [0, \pi]$ be such that $\cos \theta_n = \frac{n}{\sqrt{n^2+\alpha^2}}$ and $\sin \theta_n = \frac{\alpha}{\sqrt{n^2+\alpha^2}}$. This implies that

$$A_n = \left(1 + \frac{\alpha^2}{n^2}\right)^{\frac{n}{2}}\begin{pmatrix} \cos(n\theta_n) & \sin(n\theta_n) \\ -\sin(n\theta_n) & \cos(n\theta_n) \end{pmatrix},$$

so $a_n = \left(1 + \frac{\alpha^2}{n^2}\right)^{\frac{n}{2}}\cos(n\theta_n)$ and $b_n = \left(1 + \frac{\alpha^2}{n^2}\right)^{\frac{n}{2}}\sin(n\theta_n)$.

(b) Since $\tan \theta_n = \frac{\alpha}{n}$ we get that $\theta_n = \arctan \frac{\alpha}{n}$. A calculation shows that

$$\lim_{n\to\infty}\left(1 + \frac{\alpha^2}{n^2}\right)^{\frac{n}{2}} = 1 \quad \text{and} \quad \lim_{n\to\infty} n \arctan \frac{\alpha}{n} = \alpha,$$

which imply that $\lim_{n\to\infty} a_n = \cos\alpha$ and $\lim_{n\to\infty} b_n = \sin\alpha$.

1.41. (a) If A and B are symmetric matrices, then $(A + B)^T = A^T + B^T = A + B$, which implies that $A + B$ is a symmetric matrix.

On the other hand, if A and B are antisymmetric matrices, then $(A + B)^T = A^T + B^T = -A - B = -(A + B)$ and this implies that $A + B$ is an antisymmetric matrix.

(b) Let $A \in \mathcal{S}_2(\mathbb{R}) \cap \mathcal{A}_2(\mathbb{R})$. This implies that $A^T = A$ and $A^T = -A$. Thus $2A = O_2 \Rightarrow A = O_2$.

(c) We have $M = \frac{M+M^T}{2} + \frac{M-M^T}{2}$. Now, one can prove that $\frac{M+M^T}{2} \in \mathcal{S}_2(\mathbb{R})$ and $\frac{M-M^T}{2} \in \mathcal{A}_2(\mathbb{R})$.

To prove the uniqueness assertion, observe that if $M = C + D$, with $C \in \mathcal{S}_2(\mathbb{R})$ and $D \in \mathcal{A}_2(\mathbb{R})$, then

$$S = \frac{M + M^T}{2} = \frac{C + D + (C + D)^T}{2} = \frac{C + C^T + D + D^T}{2} = C$$

and

$$A = \frac{M - M^T}{2} = \frac{(C + D) - (C + D)^T}{2} = \frac{C - C^T + D - D^T}{2} = D.$$

1.42. (a) Each $A \in \mathcal{M}_2(\mathbb{C})$ is written uniquely as $A = B + iC$, where $B = \frac{1}{2}(A + \bar{A})$ is the *real part* of A and $C = \frac{1}{2i}(A - \bar{A})$ is the *imaginary part* of A.

To prove the writing is unique, observe that if $A = E + iF$, with both $E, F \in \mathcal{M}_2(\mathbb{R})$, then

$$B = \frac{A + \overline{A}}{2} = \frac{E + iF + \overline{E + iF}}{2} = \frac{E + iF + E - iF}{2} = E$$

and

$$C = \frac{A - \overline{A}}{2i} = \frac{E + iF - \overline{E + iF}}{2i} = \frac{E + iF - E + iF}{2i} = F.$$

(b) Each $A \in \mathcal{M}_2(\mathbb{C})$ is written in exactly one way as $A = H(A) + iK(A)$, where $H(A) = \frac{1}{2}(A + A^*)$ is the *Hermitian part* of A and $iK(A) = \frac{1}{2}(A - A^*)$ is the *skew-Hermitian part* of A. One can check that both matrices $H(A)$ and $K(A)$ are Hermitian matrices.

To prove the uniqueness assertion, observe that if $A = E + iF$, with both E and F Hermitian, then

$$2H(A) = A + A^* = (E + iF) + (E + iF)^* = E + iF + E^* - iF^* = 2E$$

and

$$2iK(A) = A - A^* = (E + iF) - (E + iF)^* = E + iF - E^* + iF^* = 2iF.$$

1.44. (a) We have $A^3 = A \cdot A^2 = A(BC) = (AB)C = C^2 \cdot C = C^3$ and $B^3 = B \cdot B^2 = B(CA) = (BC)A = A^2 \cdot A = A^3$. It follows that $A^3 = B^3 = C^3$.

(b) Let $\epsilon \neq 1$ be a cubic root of unity, i.e., $\epsilon^2 + \epsilon + 1 = 0$. Matrices A, ϵA, and $\epsilon^2 A$ are distinct matrices which verify the conditions of the problem.

1.45. (a) $\Rightarrow$ (b) We assume that there exists $n \in \mathbb{N}$ such that $A^n = I_2$. Passing to determinants we get that $\det^n A = 1$ and since $\det A = a^2 + b^2$ we get that $\det A = a^2 + b^2 = 1$. This implies there exists $t \in \mathbb{R}$ such that $a = \cos t$ and $b = \sin t$. Thus

$$A = \begin{pmatrix} \cos t & \sin t \\ -\sin t & \cos t \end{pmatrix} \quad \Rightarrow \quad A^k = \begin{pmatrix} \cos kt & \sin kt \\ -\sin kt & \cos kt \end{pmatrix}, \quad k \in \mathbb{N}.$$

The equation $A^n = I_2$ implies that

$$\begin{pmatrix} \cos nt & \sin nt \\ -\sin nt & \cos nt \end{pmatrix} = \begin{pmatrix} 1 & 0 \\ 0 & 1 \end{pmatrix}.$$

It follows that $\cos nt = 1$ and $\sin nt = 0$ which implies that $nt = 2p\pi$, $p \in \mathbb{Z}$. Thus $t = \frac{2p}{n}\pi = q\pi$, where $q = \frac{2p}{n} \in \mathbb{Q}$ and consequently $a = \cos q\pi$ and $b = \sin q\pi$.

(b) $\Rightarrow$ (a) Let $a = \cos q\pi$ and $b = \sin q\pi$, where $q \in \mathbb{Q}^*$, i.e., $q = \frac{u}{v}$, $u \in \mathbb{Z}$ and $v \in \mathbb{N}$. Since $q = \frac{2u}{2v}$, we have that

$$A = \begin{pmatrix} \cos \dfrac{2u\pi}{2v} & \sin \dfrac{2u\pi}{2v} \\ -\sin \dfrac{2u\pi}{2v} & \cos \dfrac{2u\pi}{2v} \end{pmatrix},$$

and it follows that $A^{2v} = I_2$.

1.46. Let $f \in \mathbb{Z}[x]$ be the polynomial

$$f(x) = \det(A - xB) = \det A + \alpha x + (\det B)x^2 = \alpha x, \quad \alpha \in \mathbb{Z}.$$

Since $AB = BA$ we have

$$A^n - B^n = \prod_{k=0}^{n-1}(A - \epsilon_k B), \quad \epsilon_k^n = 1, \quad k = 0, 1, \ldots, n-1$$

and

$$A^n + B^n = \prod_{k=0}^{n-1}(A - \mu_k B), \quad \mu_k^n = -1, \quad k = 0, 1, \ldots, n-1.$$

Passing to determinants we get that

$$\det(A^n - B^n) = \prod_{k=0}^{n-1} f(\epsilon_k) = \prod_{k=0}^{n-1}(\alpha \epsilon_k) = \alpha^n(-1)^{n+1} = -(-\alpha)^n$$

and

$$\det(A^n + B^n) = \prod_{k=0}^{n-1} f(\mu_k) = \prod_{k=0}^{n-1}(\alpha \mu_k) = \alpha^n(-1)^n = (-\alpha)^n,$$

so $a = -\alpha$.

1.47. We have

$$M = A^2 + B^2 + C^2 - AB - BC - CA = A^2 - A(B+C) + \frac{1}{4}(B+C)^2 + \frac{3}{4}(B-C)^2$$

$$= \frac{1}{4}\left[(2A - B - C)^2 + (\sqrt{3}(B - C))^2\right].$$

Thus, $\det M = 0$ if and only if $\det\left[(2A - B - C)^2 + (\sqrt{3}(B - C))^2\right] = 0$.

Let $P = 2A - B - C$ and $Q = \sqrt{3}(B - C)$, $P, Q \in \mathcal{M}_2(\mathbb{R})$.

We prove that if $\det(P^2 + Q^2) = 0$ and $PQ = QP$, then $\det P = \det Q$. We have, since $P^2 + Q^2 = (P + iQ)(P - iQ)$, that $\det(P^2 + Q^2) = \det(P + iQ)\det(P - iQ)$.

On the other hand, $\det(P + iQ) = \det P + \alpha i + i^2 \det Q = \det P - \det Q + \alpha i$
and $\det(P - iQ) = \overline{\det(P + iQ)} = \det P - \det Q - \alpha i$. Thus, $\det(P^2 + Q^2) = (\det P - \det Q)^2 + \alpha^2 = 0$ implies that $\alpha = 0$ and $\det P = \det Q$. Therefore, $\det(2A - B - C) = \det((\sqrt{3}(B - C))$ which in turn implies that $\det(2A - B - C) = 3 \det(B - C)$.

1.48. We prove that $a^2 = 4b$. Let $M = \begin{pmatrix} x & z \\ z & y \end{pmatrix}$. The two conditions give us $x + y = a$
and $xy - z^2 = b$. Since these equations are symmetric in x and y, the matrix can be unique if $x = y$. This implies that $2x = a$ and $x^2 - z^2 = b$. Moreover, if (x, y, z) is a solution of this system of equations, then $(x, y, -z)$ is also a solution, so M can only be unique if $z = 0$. This means that $2x = a$ and $x^2 = b$, so $a^2 = 4b$.

If this is the case, then we prove that M is unique with the properties that $\text{Tr}(M) = a$ and $\det M = b$. If $x + y = a$ and $xy - z^2 = b$, then

$$(x - y)^2 + 4z^2 = (x + y)^2 + 4z^2 - 4xy = a^2 - 4b = 0,$$

so we must have $x = y$ and $z = 0$, which implies that $M = \begin{pmatrix} a/2 & 0 \\ 0 & a/2 \end{pmatrix}$.

The second solution is based on a technique involving the eigenvalues of M (see [60]).

1.49. (a) Let $J_1 = \begin{pmatrix} 2 & -1 \\ 2 & -1 \end{pmatrix}$ and $J_2 = \begin{pmatrix} -1 & 1 \\ -2 & 2 \end{pmatrix}$ and observe that $J_1^2 = J_1$, $J_2^2 = J_2$
and $J_1 J_2 = J_2 J_1 = O_2$.

First we prove that $\mathcal{M}$ is closed under the multiplication of matrices. If $A_a, A_b \in \mathcal{M}$, then $A_a = J_1 + aJ_2$ and $A_b = J_1 + bJ_2$. A calculation shows that $A_a A_b = (J_1 + aJ_2)(J_1 + bJ_2) = J_1^2 + bJ_1 J_2 + aJ_2 J_1 + abJ_2^2 = J_1 + abJ_2 = A_{ab} \in \mathcal{M}$.

Now we prove that $\mathcal{M}$ together with the multiplication of matrices is an abelian group.

- *associativity* $(A_a A_b)A_c = A_a(A_b A_c) = A_{abc}, \forall a, b, c \in \mathbb{R}^*$;
- *the identity* Since $I_2 = J_1 + J_2 = A_1 \Rightarrow A_a I_2 = I_2 A_a = A_a, \forall a \in \mathbb{R}^*$;
- *the inverse element* $A_a A_{1/a} = A_1 = I_2 = A_{1/a} A_a, \forall a \in \mathbb{R}^* \Rightarrow A_a^{-1} = A_{1/a}$;
- *commutativity* $A_a A_b = A_b A_a = A_{ab}, \forall a, b \in \mathbb{R}^*$.

(b) Let $f : \mathcal{M} \to \mathbb{R}^*$ be the function defined by $f(A_a) = a$. First, we observe that f is onto by definition. Now we prove that f is a one to one function. If $f(A_a) = f(A_b)$, then $a = b$ and this implies that $A_a = J_1 + aJ_2 = J_1 + bJ_2 = A_b$.

On the other hand $f(A_a A_b) = f(A_{ab}) = ab = f(A_a)f(A_b), \forall a, b \in \mathbb{R}^*$, which implies that f is an isomorphism.

1.50. First we prove that $\mathcal{M}$ is closed under the multiplication of matrices. Let
$A = \begin{pmatrix} \alpha & 2\beta \\ \beta & \alpha \end{pmatrix}$ and $B \begin{pmatrix} m & 2n \\ n & m \end{pmatrix}$, where $\alpha, \beta, m, n \in \mathbb{Q}, \alpha \neq 0$ or $\beta \neq 0, m \neq 0$ or $n \neq 0$. A calculation shows that

$$AB = \begin{pmatrix} \alpha m + 2\beta n & 2(\alpha n + \beta m) \\ \alpha n + \beta m & \alpha m + 2\beta n \end{pmatrix} = \begin{pmatrix} x & 2y \\ y & x \end{pmatrix},$$

where $x = \alpha m + 2\beta n$ and $y = \alpha n + \beta m$.

We prove that x and y cannot be both 0. If $x = y = 0$ we get that

$$\begin{cases} \alpha m + 2\beta n = 0 \\ \alpha n + \beta m = 0. \end{cases}$$

This is a homogeneous system of equations, in variables m and n, whose determinant is $\alpha^2 - 2\beta^2 \neq 0$. If $\alpha^2 - 2\beta^2 = 0 \Leftrightarrow (\alpha - \sqrt{2}\beta)(\alpha + \sqrt{2}\beta) = 0 \Leftrightarrow \alpha = \beta = 0$ which contradicts the fact that α and β cannot be both 0. Thus, $\alpha^2 - 2\beta^2 \neq 0$ and this implies that the system has the unique solution $m = n = 0$ which contradicts the hypothesis that m and n cannot be both 0. Therefore x and y cannot be both 0 and this implies that $\mathcal{M}$ is closed under the multiplication of matrices.

The reader should check that $(\mathcal{M}, \cdot)$ is an abelian group.

- *associativity* $(AB)C = A(BC)$, $\forall A, B, C \subset \mathcal{M}$;
- *the identity* $I_2 \in \mathcal{M}$ and $AI_2 = I_2A = A$, $\forall A \in \mathcal{M}$;
- *the inverse element* observe that

$$\begin{pmatrix} x & 2y \\ y & x \end{pmatrix}^{-1} - \begin{pmatrix} \dfrac{x}{x^2 - 2y^2} & -\dfrac{2y}{x^2 - 2y^2} \\ -\dfrac{y}{x^2 - 2y^2} & \dfrac{x}{x^2 - 2y^2} \end{pmatrix} \in \mathcal{M};$$

- *commutativity* $AB = BA$, $\forall A, B \in \mathcal{M}$.

1.51. To prove that $(\mathcal{M}, \cdot)$ is an abelian group can be done by direct computations. Let $U = \{z \in \mathbb{C} : |z| = 1\} = \{\cos\alpha + i\sin\alpha : \alpha \in \mathbb{R}\}$, be the set of complex numbers of absolute value 1. This set together with the multiplication of complex numbers is an abelian group (check it!).

The function $f : U \to \mathcal{M}$ defined by

$$f(\cos\alpha + i\sin\alpha) = \begin{pmatrix} \cos\alpha & \sin\alpha \\ -\sin\alpha & \cos\alpha \end{pmatrix}$$

is a group isomorphism.

1.52. The solution of this problem is similar to the solution of problem **1.51**.

1.53. (a) Let $z, z' \in \mathcal{G}$, $z = x + y\sqrt{5}$ and $z' = x' + y'\sqrt{5}$, where $x, y, x', y' \in \mathbb{Q}$ with $x^2 - 5y^2 = 1$, $x'^2 - 5y'^2 = 1$. A calculation shows that $zz' = (x + y\sqrt{5})(x' + y'\sqrt{5}) = xx' + 5yy' + (xy' + x'y)\sqrt{5} = X + Y\sqrt{5}$, where $X = xx' + 5yy' \in \mathbb{Q}$ and $Y = xy' + x'y \in \mathbb{Q}$. Also

$$X^2 - 5Y^2 = (xx' + 5yy')^2 - 5(xy' + x'y)^2$$
$$= x^2x'^2 + 25y^2y'^2 - 5x^2y'^2 - 5x'^2y^2$$
$$= x^2(x'^2 - 5y'^2) - 5y^2(x'^2 - 5y'^2)$$
$$= x^2 - 5y^2$$
$$= 1.$$

This implies that $\mathscr{G}$ is closed under the multiplication of numbers.

One can check that the multiplication of numbers is both associative and commutative. Since $1 = 1 + 0\sqrt{5}$ and $1^2 - 5 \cdot 0^2 = 1$ we get that the identity of $\mathscr{G}$ is 1. If $z = x + y\sqrt{5} \in \mathscr{G}$, with $x, y \in \mathbb{Q}$ and $x^2 - 5y^2 = 1$, then

$$\frac{1}{z} = \frac{1}{x + y\sqrt{5}} = \frac{x - y\sqrt{5}}{x^2 - 5y^2} = x - y\sqrt{5},$$

which implies, since $x, -y \in \mathbb{Q}$ and $x^2 - 5(-y)^2 = 1$, that $\frac{1}{z} \in \mathscr{G}$. Thus, the inverse of z is $\frac{1}{z}$. Putting all these together we get that $(\mathscr{G}, \cdot)$ is an abelian group.

(b) If $x, y, x', y' \in \mathbb{Q}$, $x^2 - 5y^2 = 1$, $x'^2 - 5y'^2 = 1$, then

$$\begin{pmatrix} x & 2y \\ \frac{5}{2}y & x \end{pmatrix} \begin{pmatrix} x' & 2y' \\ \frac{5}{2}y' & x' \end{pmatrix} = \begin{pmatrix} xx' + 5yy' & 2(x'y + y'x) \\ \frac{5}{2}(x'y + y'x) & xx' + 5yy' \end{pmatrix} = \begin{pmatrix} X & 2Y \\ \frac{5}{2}Y & X \end{pmatrix},$$

where $X = xx' + 5yy' \in \mathbb{Q}$ and $Y = xy' + x'y \in \mathbb{Q}$. Also, $X^2 - 5Y^2 = 1$ (see the calculations from part (a)). This implies that $\mathscr{M}$ is closed under the multiplication of matrices.

The multiplication of matrices is both associative and commutative. The unit matrix I_2 is the identity of $\mathscr{M}$ and the inverse of a matrix in $\mathscr{M}$ is given by

$$\begin{pmatrix} x & 2y \\ \frac{5}{2}y & x \end{pmatrix}^{-1} = \begin{pmatrix} x & -2y \\ -\frac{5}{2}y & x \end{pmatrix} \in \mathscr{M}.$$

Thus, $(\mathscr{M}, \cdot)$ is an abelian group.

(c) The function $f : \mathscr{G} \to \mathscr{M}$ defined by

$$f(x + y\sqrt{5}) = \begin{pmatrix} x & 2y \\ \frac{5}{2}y & x \end{pmatrix}$$

is a group isomorphism. We observe that f is onto by definition and that f is a one to one function is easy to check. To shows that f is a homomorphism we have

$$f((x+y\sqrt{5})(x'+y'\sqrt{5})) = \begin{pmatrix} xx'+5yy' & 2(xy'+x'y) \\ \dfrac{5}{2}(xy'+x'y) & xx'+5yy' \end{pmatrix}$$

and

$$f(x+y\sqrt{5})f(x'+y'\sqrt{5}) = \begin{pmatrix} x & 2y \\ \dfrac{5}{2}y & x \end{pmatrix}\begin{pmatrix} x' & 2y' \\ \dfrac{5}{2}y' & x' \end{pmatrix} = \begin{pmatrix} xx'+5yy' & 2(xy'+x'y) \\ \dfrac{5}{2}(xy'+x'y) & xx'+5yy' \end{pmatrix}$$

so $f((x+y\sqrt{5})(x'+y'\sqrt{5})) = f(x+y\sqrt{5})f(x'+y'\sqrt{5})$.

1.54. (b) First, we observe the identity of $(\mathcal{M}_d, \cdot)$ is I_2. If $f : \mathbb{C}^* \to \mathcal{M}_d$ is an isomorphism, then $f(1) = I_2$. If $f(i) = A = \begin{pmatrix} a & db \\ b & a \end{pmatrix}$, then $f(i^2) = f(-1) \neq f(1) = I_2$, so $A^2 \neq I_2$. On the other hand, $f(i^4) = f(1) = I_2$, so $A^4 = I_2$. A calculation shows that

$$A^2 = \begin{pmatrix} a^2+db^2 & d(2ab) \\ 2ab & a^2+db^2 \end{pmatrix}, \quad A^4 = \begin{pmatrix} (a^2+db^2)^2+4a^2b^2d & 4abd(a^2+db^2) \\ 4ab(a^2+db^2) & (a^2+db^2)^2+4a^2b^2d \end{pmatrix}.$$

The equation $A^4 = I_2$ implies that

$$\begin{cases} (a^2 + db^2)^2 + 4a^2b^2d = 1 \\ ab(a^2 + db^2) = 0. \end{cases}$$

- If $a^2 + db^2 = 0$ we get that $4a^2b^2d = 1 \Leftrightarrow 4a^2(-a^2) = 1 \Leftrightarrow 4a^4 = -1$, which does not have real solutions.
- If $b = 0$ we get that $a^4 = 1 \Rightarrow a = \pm 1$ and $A = \pm I_2$. This contradicts $A^2 = I_2$.
- If $a = 0$ we get that $(db^2)^2 = 1 \Rightarrow db^2 = \pm 1$. This equation also implies that $d \neq 0$. We have

$$A = \begin{pmatrix} 0 & db \\ b & 0 \end{pmatrix} \quad \text{and} \quad A^2 = \begin{pmatrix} db^2 & 0 \\ 0 & db^2 \end{pmatrix} \neq I_2$$

which implies that $db^2 = -1$, so $d < 0$ and $b = \pm \dfrac{1}{\sqrt{-d}}$.

We obtained the condition $d < 0$ is necessary.

To prove this condition is also sufficient we observe the function $f : \mathbb{C}^* \to \mathcal{M}_d$ defined by

$$f(x + iy) = \begin{pmatrix} x & d\alpha y \\ \alpha y & x \end{pmatrix},$$

where $x, y \in \mathbb{R}$ and $\alpha = \pm \dfrac{1}{\sqrt{-d}}$, is a group isomorphism.

1.55. The problem can be solved by direct computations.

1.56. Cayley's table for K_4 is given below

$\cdot$	I_2	S_x	S_y	S_0
I_2	I_2	S_x	S_y	S_0
S_x	S_x	I_2	S_0	S_y
S_y	S_y	S_0	I_2	S_x
S_0	S_0	S_y	S_x	I_2

Cayley's table for K_4

1.57. Let $a, b, c \in \mathbb{R}$ and let $A = \begin{pmatrix} a+b & b \\ c & a+c \end{pmatrix}$ be an orthogonal matrix.

We have $1 = \det I_2 = \det(AA^T) = \det^2 A$ which implies that $\det A = \pm 1$.

■ If $\det A = 1$, then $A^T = A^{-1} = A_*$, so

$$\begin{pmatrix} a+b & c \\ b & a+c \end{pmatrix} = \begin{pmatrix} a+c & -b \\ -c & a+b \end{pmatrix},$$

from which it follows that $b = c$ and $-b = c$. This implies that $b = c = 0$ and since $\det A = a^2$ we get that $a = \pm 1$. Therefore $A = \pm I_2$.

■ If $\det A = -1$, then $A^T = A^{-1} = -A_*$, so

$$\begin{pmatrix} a+b & c \\ b & a+c \end{pmatrix} = \begin{pmatrix} -a-c & b \\ c & -a-b \end{pmatrix},$$

from which it follows that $b = c$ and $a + b = -a - c$. This implies that $b = c$ and $a + b = 0$. Therefore $A = \begin{pmatrix} 0 & -a \\ -a & 0 \end{pmatrix}$ and since $\det A = -a^2 = -1$ we get that $a = \pm 1$. It follows that $A = \pm U$, where $U = \begin{pmatrix} 0 & 1 \\ 1 & 0 \end{pmatrix}$.

We obtained $\mathscr{G} = \{I_2, -I_2, U, -U\}$ and this group is isomorphic to the Klein 4-group K_4.

1.58. (a) $\Rightarrow$ (b) If E is the identity element of $\mathscr{M}_n$ and $X \in \mathscr{M}_n$, then $A = (EX)^n = E^n X^n = AA = A^2$.

(b) $\Rightarrow$ (c) Since $A^2 = A$ and $A \neq I_2$ we get that $\det A = 0$ and from $X^n = A$ we have that $\det X = 0$, for all $X \in \mathscr{M}_n$. The Cayley–Hamilton Theorem implies that for all $X \in \mathscr{M}_n$ there exists $t \in \mathbb{C}$ such that $X^2 = tX$. This implies that $A = X^n = t^{n-1} X$

and, since $A \neq O_2$, we get that $t \neq 0$, $X = sA$ with $s = t^{1-n}$. The equation $X^n = A$ implies that $s^n = 1$ and it follows that $\mathcal{M}_n \subseteq \{sA : s^n = 1\}$. To prove the other inclusion, if $X \in \{sA : s^n = 1\}$, then $X^n = s^n A = A$, so $\mathcal{M}_n \subseteq \{sA : s^n = 1\} = \mathcal{U}_n A$ and $f : \mathcal{U}_n \to \mathcal{M}_n, f(z) = zA$ is the isomorphism.

(c) $\Rightarrow$ (a) This implication holds by triviality.

1.59. It is easy to check that G and S together with the multiplication of matrices are groups. We assume that G is isomorphic to S. This implies, since any group isomorphism sends elements of G to elements of S of the same order and vice versa, that both groups have the same number of elements of order less than or equal to 2. Therefore, the equation $X^2 = I_2$ has the same number of solutions in both groups.

Let $X \in S$ such that $X^2 = I_2$. We have, based on the Cayley–Hamilton Theorem, that $X^2 - tX + I_2 = O_2$, where $t = \text{Tr}(X)$. It follows that $tX = 2I_2$, so $t \neq 0$ and $X = \frac{2}{t}I_2$. This implies that $t = \text{Tr}(X) = \text{Tr}\left(\frac{2}{t}I_2\right) = \frac{4}{t} \Rightarrow t = \pm 2$. Thus, $X \in \{-I_2, I_2\}$.

However, in G the equation $X^2 = I_2$ has also the solutions $X_a = \begin{pmatrix} 0 & a \\ \frac{1}{a} & 0 \end{pmatrix}$, $a \in \mathbb{C}^*$.

1.60. If $G = \{O_2\}$, then $G \cong \{1\}$, the unit subgroup of $(\mathbb{C}^*, \cdot)$. If $G \neq \{O_2\}$ we have that, if $O_2 \in G$, then $AO_2 = O_2$ and hence $G = \{O_2\}$. Thus, if $G \neq \{O_2\}$, then $O_2 \notin G$.

Let $A \in G$. Since $AE = A$ and $A^2 E = A^2$ we get based on the Cayley–Hamilton Theorem that $(tA - dI_2)E = tA - dI_2$, where $t = \text{Tr}(A)$ and $d = \det A$. It follows that $tA - dE = tA - dI_2$ and this implies that $d(E - I_2) = O_2$. Since $E \neq I_2$ one has that $\det A = 0$. Thus, $\det A = 0$, for all $A \in G$, and we have based on the Cayley–Hamilton Theorem that $A^2 = tA$.

Let A' be the symmetric element of A in group G. We have

$$A^2 A' = tAA' = tE \quad \Leftrightarrow \quad A(AA') = tE \quad \Leftrightarrow \quad AE = tE \quad \Leftrightarrow \quad A = tE = \text{Tr}(A)E.$$

This implies that $G \subset \{\alpha E, \alpha \in \mathbb{C}^*\}$. Let $f : G \to \mathbb{C}^*$ be the function defined by $f(A) = \alpha (= \text{Tr}(A))$. First, we note that f is well defined since if $A = \alpha E = \beta E$, from $E \neq O_2$, we get that $\alpha = \beta$.

If $A, B \in G$ such that $A = \alpha E$ and $B = \beta E$ we have

$$f(AB) = f(\alpha \beta E^2) = f(\alpha \beta E) = \alpha \beta = f(A)f(B),$$

which implies that f is a group homomorphism. Since f is injective (check it!) we get that $G \cong f(G) \leq (\mathbb{C}^*, \cdot)$.

1.61. (a) One can check, see part (d) of the problem, that $R_\alpha^{-1} = \begin{pmatrix} \cos\alpha & \sin\alpha \\ -\sin\alpha & \cos\alpha \end{pmatrix}$.

However, this implies that $R_\alpha^T = R_\alpha^{-1}$ which means that R_α is an orthogonal matrix.

(b) One should check that the following conditions hold:

- *associativity* $(R_\alpha R_\beta)R_\gamma = R_\alpha(R_\beta R_\gamma)$, $\forall \alpha, \beta, \gamma \in \mathbb{R}$;
- *the identity* $R_\alpha I_2 = I_2 R_\alpha = R_\alpha$, $\forall \alpha \in \mathbb{R}$ (note that $I_2 = R_0$);

- *the inverse element $R_\alpha R_{-\alpha} = R_{-\alpha} R_\alpha = I_2$, which implies that $R_\alpha^{-1} = R_{-\alpha}$;*
- *commutativity $R_\alpha R_\beta = R_\beta R_\alpha$, $\forall \alpha, \beta \in \mathbb{R}$.*

We leave these calculations to the interested reader.

(c) We have

$$R_\alpha R_\beta = \begin{pmatrix} \cos\alpha & -\sin\alpha \\ \sin\alpha & \cos\alpha \end{pmatrix} \begin{pmatrix} \cos\beta & -\sin\beta \\ \sin\beta & \cos\beta \end{pmatrix} = \begin{pmatrix} \cos(\alpha+\beta) & -\sin(\alpha+\beta) \\ \sin(\alpha+\beta) & \cos(\alpha+\beta) \end{pmatrix}$$

and

$$R_\beta R_\alpha = \begin{pmatrix} \cos\beta & -\sin\beta \\ \sin\beta & \cos\beta \end{pmatrix} \begin{pmatrix} \cos\alpha & -\sin\alpha \\ \sin\alpha & \cos\alpha \end{pmatrix} = \begin{pmatrix} \cos(\beta+\alpha) & -\sin(\beta+\alpha) \\ \sin(\beta+\alpha) & \cos(\beta+\alpha) \end{pmatrix}$$

and this implies that $R_\alpha R_\beta = R_\beta R_\alpha = R_{\alpha+\beta}$.

(d) We have, based on part (c), that $R_\alpha R_{-\alpha} = R_{-\alpha} R_\alpha = R_0 = I_2$ and this implies that $R_\alpha^{-1} = R_{-\alpha}$.

(e) Observe that

$$R_\alpha^n = \begin{pmatrix} \cos(n\alpha) & -\sin(n\alpha) \\ \sin(n\alpha) & \cos(n\alpha) \end{pmatrix}, \quad n \geq 1$$

and prove this formula by mathematical induction.

1.62. Clearly $O_2 \in G(A)$. We prove that if $X \in G(A)$, then $-X \in G(A)$. Since $\det(A+X) + \det(A-X) = 2\det A + 2\det X$ and $\det(A+X) = \det A + \det X$, we get that $\det(A-X) = \det A + \det X = \det A + \det(-X)$.

If $X, Y \in G(A)$ we prove that $X + Y \in G(A)$. Let $X, Y \in G(A)$. We have, based on problem **1.32**, that

$$\det(A+X+Y) = \det(A+X) + \det(A+Y) + \det(X+Y) - \det A - \det X - \det Y$$
$$= \det A + \det X + \det A + \det Y + \det(X+Y) - \det A - \det X - \det Y$$
$$= \det A + \det(X+Y),$$

and this implies that $X + Y \in G(A)$.

Now we prove that $(H(A), +)$ is a subgroup of $(\mathcal{M}_2(\mathbb{C}), +)$. First, we observe that $O_2 \in H(A)$. Second, we show that if $X, Y \in H(A)$, then $X - Y \in H(A)$. We have

$$\operatorname{Tr}(A(X-Y)) = \operatorname{Tr}(AX - AY)$$
$$= \operatorname{Tr}(AX) - \operatorname{Tr}(AY)$$
$$= \operatorname{Tr}(A)\operatorname{Tr}(X) - \operatorname{Tr}(A)\operatorname{Tr}(Y)$$
$$= \operatorname{Tr}(A)\left(\operatorname{Tr}(X) - \operatorname{Tr}(Y)\right)$$
$$= \operatorname{Tr}(A)\operatorname{Tr}(X-Y).$$

Remark 1.16 One can also prove that $(G'(A), +)$ is a subgroup of $(\mathcal{M}_2(\mathbb{C}), +)$, where $G'(A) = \{X \in \mathcal{M}_2(\mathbb{C}) : \det(A - X) = \det A + \det X\}$.

1.63. (a) $\det M(\theta) = \cosh^2 \theta - \sinh^2 \theta = 1$.

(b) We have

$$M(\theta_1)M(\theta_2) = \begin{pmatrix} \cosh \theta_1 & \sinh \theta_1 \\ \sinh \theta_1 & \cosh \theta_1 \end{pmatrix} \begin{pmatrix} \cosh \theta_2 & \sinh \theta_2 \\ \sinh \theta_2 & \cosh \theta_2 \end{pmatrix}$$

$$= \begin{pmatrix} \cosh \theta_1 \cosh \theta_2 + \sinh \theta_1 \sinh \theta_2 & \cosh \theta_1 \sinh \theta_2 + \sinh \theta_1 \cosh \theta_2 \\ \sinh \theta_1 \cosh \theta_2 + \cosh \theta_1 \sinh \theta_2 & \sinh \theta_1 \sinh \theta_2 + \cosh \theta_1 \cosh \theta_2 \end{pmatrix}$$

$$= \begin{pmatrix} \cosh(\theta_1 + \theta_2) & \sinh(\theta_1 + \theta_2) \\ \sinh(\theta_1 + \theta_2) & \cosh(\theta_1 + \theta_2) \end{pmatrix}$$

$$= M(\theta_1 + \theta_2).$$

(c) This follows based on part (b) combined to mathematical induction.

(d) The multiplication of matrices in $\mathcal{H}$ is both associative and commutative. The identity element is $M(0) = I_2$ and the inverse of $M(\theta)$ is the matrix $M(-\theta)$.

1.64. Observe that $D_{2n} = \left\{ R_\theta, SR_\theta : S = \begin{pmatrix} -1 & 0 \\ 0 & 1 \end{pmatrix}, R_\theta \in \mathcal{R}_n \right\}$, where $\mathcal{R}_n$ is the group of rotations introduced in Theorem 1.3.

We have the following relations:

- $R_{\theta_1}(SR_{\theta_2}) = SR_{\theta_2 - \theta_1}$;
- $(SR_{\theta_1})R_{\theta_2} = SR_{\theta_1 + \theta_2}$;
- $(SR_{\theta_1})(SR_{\theta_2}) = R_{\theta_2 - \theta_2}$.

1.65. The set

$$\mathcal{M}_{\mathbb{C}} = \left\{ A \in \mathcal{M}_2(\mathbb{R}) : A = \begin{pmatrix} x & -y \\ y & x \end{pmatrix} \right\}$$

is a subring of $(\mathcal{M}_2(\mathbb{R}), +, \cdot)$ which is isomorphic to $\mathbb{C}$ (see Theorem 1.3).

1.66. (a) This part of the problem can be solved by direct computations.

(b) Let $f : \mathbb{Z}[i] \to \mathcal{M}$ be the function defined by

$$f(x + iy) = \begin{pmatrix} x & y \\ -y & x \end{pmatrix}.$$

It is easy to see that f is a bijection. To prove that f is a ring homomorphism we check that

$$f((x+iy)+(x'+iy')) = f(x+x'+i(y+y'))$$

$$= \begin{pmatrix} x+x' & y+y' \\ -(y+y') & x+x' \end{pmatrix}$$

$$= \begin{pmatrix} x & y \\ -y & x \end{pmatrix} + \begin{pmatrix} x' & y' \\ -y' & x' \end{pmatrix}$$

$$= f(x+iy) + f(x'+iy')$$

and

$$f((x+iy)(x'+iy')) = f(xx'-yy'+i(xy'+yx'))$$

$$= \begin{pmatrix} xx'-yy' & xy'+yx' \\ -(xy'+yx') & xx'-yy' \end{pmatrix}$$

$$= \begin{pmatrix} x & y \\ -y & x \end{pmatrix} \begin{pmatrix} x' & y' \\ -y' & x' \end{pmatrix}$$

$$= f(x+iy)f(x'+iy').$$

1.67. Let $A(x,y) = \begin{pmatrix} x & y \\ f(x,y) & g(x,y) \end{pmatrix}$. We have, since $\mathcal{M}$ is closed under addition, that for $x_1, y_1, x_2, y_2 \in \mathbb{Z}$ there exist $x_3, y_3 \in \mathbb{Z}$ such that $A(x_1, y_1) + A(x_2, y_2) = A(x_3, y_3)$. This implies that

$$\begin{cases} f(x_1+x_2, y_1+y_2) = f(x_1, y_1) + f(x_2, y_2) \\ g(x_1+x_2, y_1+y_2) = g(x_1, y_1) + g(x_2, y_2). \end{cases} \tag{1.1}$$

If $x_1 = x_2 = y_1 = y_2 = 0$ we get that $f(0,0) = g(0,0) = 0$. If $x_1 = y_2 = 0$ and $x_2 = x, y_1 = y$ we get that

$$\begin{cases} f(x,y) = f(x,0) + f(0,y) = f_1(x) + f_2(y) \\ g(x,y) = g(x,0) + g(0,y) = g_1(x) + g_2(y), \end{cases}$$

where $f_1(x) = f(x,0), f_2(x) = f(0,x), g_1(x) = g(x,0), g_2(x) = g(0,x)$.

Letting $y_1 = y_2 = 0$ in (1.1), we get that

$$\begin{cases} f_1(x_1+x_2) = f_1(x_1) + f_1(x_2) \\ g_1(x_1+x_2) = g_1(x_1) + g_1(x_2) \end{cases} \quad \forall x_1, x_2 \in \mathbb{Z},$$

and by letting $x_1 = x_2 = 0$ in (1.1) we get the same relations hold for the functions f_2 and g_2, so f_1, f_2, g_1, g_2 are additive functions.

Let $h : \mathbb{Z} \to \mathbb{Z}$ be an additive function, i.e., $h(x + y) = h(x) + h(y)$, $\forall x, y \in \mathbb{Z}$. Then, $h(0) = 0$, $h(n) = nh(1)$, $h(-n) = -h(n)$, so $h(x) = xh(1)$, $\forall x \in \mathbb{Z}$. Therefore $f_1(x) = xf_1(1), f_2(x) = xf_2(1)$, $g_1(x) = xg_1(1)$ and $g_2(x) = xg_2(1)$. Since $I_2 \in \mathscr{M}$ we get that

$$I_2 \in \mathscr{M} \quad \Rightarrow \quad \begin{pmatrix} 1 & 0 \\ f(1,0) & g(1,0) \end{pmatrix} = \begin{pmatrix} 1 & 0 \\ 0 & 1 \end{pmatrix},$$

which implies that $f(1, 0) = 0$ and $g(1, 0) = 1$. This in turn implies that $f_1(1) = 0$ and $g_1(1) = 1$, so $f_1(x) = 0$, $\forall x \in \mathbb{Z}$ and $g_1(x) = x$, $\forall x \in \mathbb{Z}$.

Let $f_2(1) = a \in \mathbb{Z}$, $g_2(1) = b \in \mathbb{Z}$ and we have that

$$f(x, y) = ay \quad \text{and} \quad g(x, y) = x + by, \quad \forall (x, y) \in \mathbb{Z} \times \mathbb{Z}.$$

Now it is easy to check that these conditions are also sufficient.

1.68. Let $A = \begin{pmatrix} 0 & 1 \\ 0 & 0 \end{pmatrix}$ and $B = \begin{pmatrix} 0 & 0 \\ 1 & 0 \end{pmatrix}$. Then $A^2 = B^2 = O_2$, so both A and B are nilpotent elements.

On the other hand, $A + B = \begin{pmatrix} 0 & 1 \\ 1 & 0 \end{pmatrix}$ and we have $(A + B)^2 = I_2$. This implies that there is no $k \in \mathbb{N}$ such that $(A + B)^k = O_2$, so $A + B$ is not a nilpotent matrix. This also proves that the set of nilpotent matrices in the noncommutative ring $\mathscr{M}_2(\mathbb{Z})$ is not an ideal.

Nota bene. The problem states that the set of nilpotent elements in a noncommutative ring need not be an ideal. However, one can prove that if R is a commutative ring, then the set of nilpotent elements form an ideal called the *nilradical* of R and denoted by $\mathscr{N}(R)$.

1.69. Let $A(x, y) = \begin{pmatrix} x & f(x, y) \\ g(x, y) & y \end{pmatrix}$. We have, since $\mathscr{M}$ is closed under addition, that

$$\begin{cases} f(x_1 + x_2, y_1 + y_2) = f(x_1, y_1) + f(x_2, y_2) \\ g(x_1 + x_2, y_1 + y_2) = g(x_1, y_1) + g(x_2, y_2). \end{cases}$$

If $y_1 = y_2 = 0$ we get that $f(x_1 + x_2, 0) = f(x_1, 0) + f(x_2, 0)$. This implies that $f(n, 0) = nf(1, 0)$ and $f(-n, 0) = -nf(1, 0)$, $\forall n \in \mathbb{N}$. Since $f(0, 0) = 0$ we get that $f(k, 0) = ak$, $\forall k \in \mathbb{Z}$, where $a = f(1, 0) \in \mathbb{Z}$. Similarly we get that $f(0, k) = bk$, $\forall k \in \mathbb{Z}$, where $b = f(0, 1) \in \mathbb{Z}$. Thus, the functions f and g are of the following form

$$\begin{cases} f(x, y) = ax + by \\ g(x, y) = cx + dy \end{cases} \quad \forall (x, y) \in \mathbb{Z} \times \mathbb{Z}.$$

Since $I_2 \in \mathcal{M}$ we get that

$$\begin{pmatrix} 1 & 0 \\ 0 & 1 \end{pmatrix} = \begin{pmatrix} 1 & f(1,1) \\ g(1,1) & 1 \end{pmatrix},$$

and this implies that $f(1,1) = g(1,1) = 0 \Rightarrow a + b = c + d = 0$.

Now one can check that the set

$$\mathcal{M} = \left\{ \begin{pmatrix} x & a(x-y) \\ c(x-y) & y \end{pmatrix} : x, y \in \mathbb{Z} \right\}$$

together with the addition and the multiplication of matrices is a ring with unity.

In conclusion, the functions f and g are $f(x,y) = a(x-y)$, $\forall (x,y) \in \mathbb{Z} \times \mathbb{Z}$ and $g(x,y) = c(x-y)$, $\forall (x,y) \in \mathbb{Z} \times \mathbb{Z}$, where $a, c \in \mathbb{Z}$ are arbitrary constants.

1.70 and **1.71.** These two problems can be solved by direct computations.

1.72. Let $A(x,y) = \begin{pmatrix} x & f(x,y) \\ g(x,y) & y \end{pmatrix}$. Since $\mathcal{M}$ is closed under addition, we get that

$$\begin{cases} f(x_1 + x_2, y_1 + y_2) = f(x_1, y_1) + f(x_2, y_2) \\ g(x_1 + x_2, y_1 + y_2) = g(x_1, y_1) + g(x_2, y_2). \end{cases}$$

If $y_1 = y_2 = 0$ we get that $f(x_1 + x_2, 0) = f(x_1, 0) + f(x_2, 0)$. This implies the function $f_1(x) = f(x, 0)$, $\forall x \in \mathbb{Q}$ is an additive function, i.e., there exists $a \in \mathbb{Q}$ such that $f_1(x) = ax$, $\forall x \in \mathbb{Q}$. Similarly, we obtain that $f_2(y) = f(0, y)$, $g_1(x) = g(x, 0)$ and $g_2(y) = g_2(0, y)$ are additive functions on $\mathbb{Q}$, so $f_2(y) = by$, $g_2(x) = cx$ and $g_2(y) = dy$, $\forall x, y \in \mathbb{Q}$. These imply that $f(x, y) = ax + by$ and $g(x, y) = cx + dy$, $\forall x, y \in \mathbb{Q}$, where $a, b, c, d \in \mathbb{Q}$ are fixed constants.

The unit matrix I_2 should belong to $\mathcal{M}$ and this implies that there exist $x, y \in \mathbb{Q}$ such that $A(x, y) = I_2 \Leftrightarrow f(1, 1) = g(1, 1) = 0 \Rightarrow a + b = c + d = 0$.

Now one can check that the set

$$\mathcal{M} = \left\{ \begin{pmatrix} x & a(x-y) \\ c(x-y) & y \end{pmatrix} : x, y \in \mathbb{Q} \right\}$$

together with the addition and the multiplication of matrices is a ring. This ring is a field provided that every nonzero matrix in $\mathcal{M}$ has an inverse, i.e., $\det A(x, y) \neq 0$, when $(x, y) \neq (0, 0)$. This implies that $xy - ac(x - y)^2 \neq 0$, $\forall (x, y) \neq (0, 0)$.

Let $(x, y) \neq (0, 0)$ and consider the equation $acx^2 - (2ac + 1)xy + acy^2 = 0$. The condition $ac \neq 0$ is necessary. Otherwise, $(1, 0) \neq (0, 0)$ but $\det A(1, 0) = 0$. If $y = 0$ the previous equation implies that $x = 0$, which contradicts the fact that $(x, y) \neq (0, 0)$. If $y \neq 0$ we consider the equation

$$ac\left(\frac{x}{y}\right)^2 - (2ac+1)\frac{x}{y} + ac = 0,$$

which should not have rational solutions, so $\Delta = 4ac + 1 < 0$ or $\Delta \geq 0$ and $\sqrt{4ac+1} \notin \mathbb{Q}$.

In conclusion, $f(x,y) = a(x-y)$ and $g(x,y) = c(x-y)$, $\forall x, y \in \mathbb{Q}$, where $a, c \in \mathbb{Q}$ are constants such that either $4ac + 1 < 0$ or $4ac + 1 \geq 0$ and $\sqrt{4ac+1} \notin \mathbb{Q}$.

1.73. (a) If $m = aE + bI + cJ + dK$ and $m' = a'E + b'I + c'J + d'K$, then

$$m + m' = (a + a')E + (b + b')I + (c + c')J + (d + d')K \in \mathcal{M}$$

and

$$mm' = (aa' - bb' - cc' - dd')E + (ab' + ba' + cd' - dc')I$$
$$+ (ac' + ca' + db' - bd')J + (ad' + da' + bc' - cb')K \in \mathcal{M}.$$

We have $m\tilde{m} = (a^2 + b^2 + c^2 + d^2)E = \tilde{m}m$.

(b) One can check that $\mathcal{M}$ together with the addition and the multiplication of matrices is a ring with unity $E \in \mathcal{M}$, $E = 1E + 0I + 0J + 0K$. If $m \neq O_2$, i.e., at least one of the coefficients a, b, c or d is not zero, then $a^2 + b^2 + c^2 + d^2 \neq 0$ and

$$m\frac{1}{a^2 + b^2 + c^2 + d^2}\tilde{m} = \frac{1}{a^2 + b^2 + c^2 + d^2}\tilde{m}m = E.$$

Hence

$$m^{-1} = \frac{1}{a^2 + b^2 + c^2 + d^2}\tilde{m} \in \mathcal{M}.$$

Therefore $(\mathcal{M}, +, \cdot)$ is a field. This is a noncommutative field since $IJ \neq JI$.

(c) The polynomial $p(x) = x^2 + E$ has the roots $I, J, K, -I, -J$, and $-K$, so it has at least six roots. In fact, it can be shown that p has an infinite number of roots. To prove this observe that an element in $\mathcal{M}$ is a matrix which has the following form

$$\begin{pmatrix} a + bi & c + di \\ -c + di & a - bi \end{pmatrix}, \quad a, b, c, d \in \mathbb{R}.$$

Thus, to solve the equation $x^2 + E = O_2$ one has to determine the real numbers a, b, c, d such that

$$\begin{pmatrix} a + bi & c + di \\ -c + di & a - bi \end{pmatrix}^2 = \begin{pmatrix} -1 & 0 \\ 0 & -1 \end{pmatrix}.$$

A calculation shows that the solutions of this equation are of the following form

$$\begin{pmatrix} bi & c+di \\ -c+di & -bi \end{pmatrix} = bI + cJ + dK,$$

where $b, c, d \in \mathbb{R}$ with $b^2 + c^2 + d^2 = 1$.

Chapter 2
The Cayley–Hamilton Theorem

If $A \in \mathcal{M}_2(\mathbb{C})$, then $A^2 - \mathrm{Tr}(A)A + (\det A)I_2 = O_2$.
Cayley–Hamilton

2.1 The Cayley–Hamilton Theorem

If $A = \begin{pmatrix} a & b \\ c & d \end{pmatrix} \in \mathcal{M}_2(\mathbb{C})$, then:

- the *characteristic polynomial* of A is defined by

$$f_A(x) = \det(A - xI_2) = x^2 - (a+d)x + ad - bc = x^2 - \mathrm{Tr}(A)x + \det A \in \mathbb{C}[x];$$

- the equation

$$f_A(\lambda) = 0 \Leftrightarrow \lambda^2 - \mathrm{Tr}(A)\lambda + \det A = 0$$

is called the *characteristic equation* of A;

- the solutions λ_1, λ_2 of the characteristic equation are called the *eigenvalues* of A and the set $\{\lambda_1, \lambda_2\}$ is called the *spectrum* of A and is denoted by $\mathrm{Spec}(A)$.
  It follows, based on Viète's formulae, that

$$\lambda_1 + \lambda_2 = \mathrm{Tr}(A) \quad \text{and} \quad \lambda_1\lambda_2 = \det A.$$

Next we give some properties of the eigenvalues of a matrix which can be proved by direct computation.

© Springer International Publishing AG 2017
V. Pop, O. Furdui, *Square Matrices of Order 2*, DOI 10.1007/978-3-319-54939-2_2

Let $\lambda \in \mathbb{C}$. Then:

- if λ is not an eigenvalue of A, then the system

$$\begin{cases} (a - \lambda)x + by = 0 \\ cx + (d - \lambda)y = 0 \end{cases} \quad \Leftrightarrow \quad AX = \lambda X, \text{ where } X = \begin{pmatrix} x \\ y \end{pmatrix},$$

  has only the trivial solution $(x, y) = (0, 0)$;
- if λ is an eigenvalue of A, then the system $AX = \lambda X$ has at least a nontrivial solution $X \neq 0$; such a solution is called an *eigenvector* of A corresponding to the eigenvalue λ;
- $\lambda \in \mathbb{C}$ is an eigenvalue of A if and only if there exists

$$X = \begin{pmatrix} x \\ y \end{pmatrix} \neq \begin{pmatrix} 0 \\ 0 \end{pmatrix} \quad \text{such that} \quad AX = \lambda X;$$

If λ_1, λ_2 are the eigenvalues of A, then:

- λ_1^n, λ_2^n are the eigenvalues of A^n, $n \in \mathbb{N}$;
- $P(\lambda_1), P(\lambda_2)$ are the eigenvalues of the matrix $P(A)$, for any polynomial function $P \in \mathbb{C}[x]$;
- $\frac{1}{\lambda_1}, \frac{1}{\lambda_2}$ are the eigenvalues of A^{-1}, if A is invertible $\det A = \det(A - 0I_2) \neq 0$, so 0 is not an eigenvalue of A.

The next theorem gives the eigenvalues of the sum and the product of two commuting matrices.

Theorem 2.1 The eigenvalues of the sum and the product of two commuting matrices.

If $A, B \in \mathscr{M}_2(\mathbb{C})$ are commuting matrices, then the eigenvalues of matrices $A + B$ and AB are of the following form

$$\lambda_{A+B} = \lambda_A + \lambda_B \quad \text{and} \quad \lambda_{AB} = \lambda_A \lambda_B.$$

Proof If $B = \alpha I_2$, then $A + B = A + \alpha I_2$ which has eigenvalues $\lambda_1 + \alpha$ and $\lambda_2 + \alpha$, where λ_1, λ_2 are the eigenvalues of A and α is the eigenvalue of B. On the other hand, $AB = \alpha A$ which has eigenvalues $\alpha\lambda_1$ and $\alpha\lambda_2$.

If $B \neq \alpha I_2$, $\alpha \in \mathbb{C}$, then $B \in \mathscr{C}(A)$ and we have, based on part (b) of Theorem 1.1, that $B = \alpha I_2 + \beta A$, for some $\alpha, \beta \in \mathbb{C}$. We have that $\lambda_B = \alpha + \beta\lambda_A$, $A + B = \alpha I_2 + (\beta + 1)A$ and $AB = \alpha A + \beta A^2$. It follows that $\lambda_{A+B} = \alpha + (\beta + 1)\lambda_A = \lambda_A + \lambda_B$ and $\lambda_A\lambda_B = \alpha\lambda_A + \beta\lambda_A^2 = \lambda_A\lambda_B$. $\qquad \square$

Remark 2.1 If $\alpha, \beta \in \mathbb{C}$, $i, k \in \mathbb{N}$ and $A, B \in \mathcal{M}_2(\mathbb{C})$ are commuting matrices, then the eigenvalues of matrices $\alpha A + \beta B$ and $A^i B^k$ are of the following form

$$\lambda_{\alpha A + \beta B} = \alpha \lambda_A + \beta \lambda_B \quad \text{and} \quad \lambda_{A^i B^k} = \lambda_A^i \lambda_B^k.$$

Now we are ready to discuss the celebrated Cayley–Hamilton Theorem which states that *any square matrix cancels its characteristic polynomial.*

Theorem 2.2   The Cayley–Hamilton Theorem.
  If $A \in \mathcal{M}_2(\mathbb{C})$, then $A^2 - \text{Tr}(A)A + (\det A)I_2 = O_2$.

Proof We prove the theorem by direct computation. Let $A = \begin{pmatrix} a & b \\ c & d \end{pmatrix}$.

A calculation shows that

$$A^2 = \begin{pmatrix} a^2 + bc & b(a+d) \\ c(a+d) & d^2 + bc \end{pmatrix}.$$

If $x = \text{Tr}(A) = a + d$, then

$$A^2 - \text{Tr}(A)A + (\det A)I_2 = \begin{pmatrix} a^2 + bc & b(a+d) \\ c(a+d) & d^2 + bc \end{pmatrix} - \begin{pmatrix} ax & bx \\ cx & dx \end{pmatrix}$$

$$+ \begin{pmatrix} ad - bc & 0 \\ 0 & ad - bc \end{pmatrix}$$

$$= \begin{pmatrix} a^2 + ad - ax & 0 \\ 0 & d^2 + ad - dx \end{pmatrix}$$

$$= O_2,$$

and the theorem is proved.   □

Historical note. The Cayley–Hamilton Theorem was first proved in 1853 in terms of linear functions of quaternions by Hamilton [36]. This corresponds to the special case of certain 4×4 real or 2×2 complex matrices. In 1858 Cayley stated it for 3×3 matrices and published a proof only for the 2×2 case [14]. "Not generally an excitable person, at the point of discovery Cayley declared the Cayley–Hamilton Theorem as "very remarkable" and generations of mathematicians have shared his delight" [35, p. 772]. However, it was Frobenius who proved the general case in 1878 [19].

Next we give some applications of the Cayley–Hamilton Theorem.

Lemma 2.1 *If $A \in \mathcal{M}_2(\mathbb{C})$ is invertible, then*

$$A^{-1} = \frac{1}{\det A}(\text{Tr}(A)I_2 - A) \quad \text{and} \quad \text{Tr}(A^{-1}) = \frac{\text{Tr}(A)}{\det A}.$$

Proof We have, based on the Cayley–Hamilton Theorem, that $A^2 - \text{Tr}(A)A + (\det A)I_2 = O_2$. We multiply this equality by A^{-1} and we get that $A - \text{Tr}(A)I_2 + (\det A)A^{-1} = O_2 \Rightarrow A^{-1} = \frac{1}{\det A}(\text{Tr}(A)I_2 - A)$. The second part of the lemma follows by passing to trace in the first formula. $\qquad\square$

The next lemma is about calculating powers of a square matrix of order 2 for the special cases when the determinant or the trace of the matrix are 0.

Lemma 2.2 The nth power of two special matrices.

(a) *If $A \in \mathcal{M}_2(\mathbb{C})$ such that $\det A = 0$, then*

$$A^n = (\text{Tr}(A))^{n-1}A, \quad \forall n \in \mathbb{N}.$$

(b) *If $A \in \mathcal{M}_2(\mathbb{C})$ such that $\text{Tr}(A) = 0$, then*

$$A^n = \begin{cases} (-\det A)^k I_2, & n = 2k, \ k \in \mathbb{N} \\ (-\det A)^{k-1}A, & n = 2k-1, \ k \in \mathbb{N}. \end{cases}$$

Proof (a) We have, based on Theorem 2.2, that $A^2 = \text{Tr}(A)A$. This implies

$$A^3 = A^2 A = (\text{Tr}A)AA = \text{Tr}(A)\text{Tr}(A)A = \text{Tr}^2(A)A.$$

Using mathematical induction we have that $A^n = (\text{Tr}(A))^{n-1}A, \ \forall n \in \mathbb{N}$.

(b) Since $\text{Tr}(A) = 0$ we get based on Theorem 2.2 that $A^2 + (\det A)I_2 = O_2$. This implies that $A^2 = -(\det A)I_2$ and the proof is completed by mathematical induction according to the cases when n is an even or an odd integer. $\qquad\square$

Lemma 2.3 *Let $A \in \mathcal{M}_2(\mathbb{C})$. The following statements are equivalent:*

(a) $A^2 = O_2$;
(b)   *There is $n \in \mathbb{N}$, $n \geq 2$ such that $A^n = O_2$.*

Proof The implication (a) $\Rightarrow$ (b) is clear. To prove that (b) $\Rightarrow$ (a) we observe that the eigenvalues of A are all equal to 0 and hence the characteristic polynomial of A is $f_A(x) = x^2$. This implies, based on Theorem 2.2, that $A^2 = O_2$ and the lemma is proved. $\qquad\square$

As a consequence of Lemma 2.3 we have that if $A^2 \neq O_2$, then no power of A can be zero. We record it as a lemma.

Lemma 2.4 *If $A \in \mathcal{M}_2(\mathbb{C})$ such that $A^2 \neq O_2$, then $A^n \neq O_2$ for any $n \in \mathbb{N}$.*

Lemma 2.5 A fact on nilpotent matrices.

Let $A, B \in \mathcal{M}_2(\mathbb{C})$. If A and B are nilpotent matrices and $AB = BA$, then both $A + B$ and $A - B$ are nilpotent matrices.

Proof The proof of the lemma is equivalent to proving that if A and B are commuting matrices such that $A^2 = O_2$ and $B^2 = O_2$, then $(A \pm B)^2 = O_2$. We have $(A \pm B)^2 = A^2 \pm 2AB + B^2 = \pm 2AB$ and this implies that $(A \pm B)^4 = 4A^2 B^2 = O_2$. Now the result follows based on Lemma 2.3. $\qquad\square$

Now we turn our attention to the applications of Theorem 2.2 related to determinants of special matrices of order 2.

Lemma 2.6 The determinant in terms of traces.

If $A \in \mathcal{M}_2(\mathbb{C})$, then

$$\det A = \frac{1}{2}\left[(\mathrm{Tr}(A))^2 - \mathrm{Tr}(A^2)\right]. \tag{2.1}$$

Proof We have, based on Theorem 2.2, that $A^2 - \mathrm{Tr}(A)A + (\det A)I_2 = O_2$. Passing to trace on both sides of the previous equality we get that

$$\mathrm{Tr}(A^2) - \mathrm{Tr}(A)\mathrm{Tr}(A) + 2\det A = 0,$$

and the lemma is proved. □

Remark 2.2 Another version of the Cayley–Hamilton Theorem, based on identity (2.1), has the following formulation

$$A^2 - \mathrm{Tr}(A)A + \frac{1}{2}\left[(\mathrm{Tr}(A))^2 \quad \mathrm{Tr}(A^2)\right]I_2 = O_2, \quad \forall A \in \mathcal{M}_2(\mathbb{C}).$$

Lemma 2.7 [45] A master determinant formula

If $A, B \in \mathcal{M}_2(\mathbb{C})$ and $x \in \mathbb{C}$, then

$$\det(A + xB) = \det A + (\mathrm{Tr}(A)\mathrm{Tr}(B) - \mathrm{Tr}(AB))x + (\det B)x^2. \tag{2.2}$$

Proof We have, based on formula (2.1), that

$$\begin{aligned}
\det(A + xB) &= \frac{1}{2}\left[(\mathrm{Tr}(A + xB))^2 - \mathrm{Tr}((A + xB)^2)\right] \\
&= \frac{1}{2}\left[(\mathrm{Tr}(A) + x\mathrm{Tr}(B))^2 - \mathrm{Tr}(A^2 + xAB + xBA + B^2x^2)\right] \\
&= \frac{1}{2}\left[(\mathrm{Tr}(A))^2 + 2\mathrm{Tr}(A)\mathrm{Tr}(B)x + (\mathrm{Tr}(B))^2x^2 - \mathrm{Tr}(A^2) \right. \\
&\quad \left. -2\mathrm{Tr}(AB)x - \mathrm{Tr}(B^2)x^2\right] \\
&= \frac{1}{2}\left[(\mathrm{Tr}(A))^2 - \mathrm{Tr}(A^2)\right] + (\mathrm{Tr}(A)\mathrm{Tr}(B) - \mathrm{Tr}(AB))x \\
&\quad + \frac{1}{2}\left[(\mathrm{Tr}(B))^2 - \mathrm{Tr}(B^2)\right]x^2 \\
&= \det A + (\mathrm{Tr}(A)\mathrm{Tr}(B) - \mathrm{Tr}(AB))x + (\det B)x^2,
\end{aligned}$$

and the lemma is proved. □

Corollary 2.1 *If $A, B \in \mathcal{M}_2(\mathbb{C})$, then*

$$\det(A + B) + \det(A - B) = 2 \det A + 2 \det B.$$

Proof This follows based on formula (2.2) with $x = 1$ respectively $x = -1$ and then by adding the two equalities. See also the solution of part (a) of problem **1.31** for a different approach. $\square$

Corollary 2.2 Determinant and trace identities.

  If $A, B \in \mathcal{M}_2(\mathbb{C})$, then:

(a) $\det(A + B) - \det A - \det B = \mathrm{Tr}(A)\mathrm{Tr}(B) - \mathrm{Tr}(AB)$;
(b) $\det(A - B) - \det A - \det B = \mathrm{Tr}(AB) - \mathrm{Tr}(A)\mathrm{Tr}(B)$;
(c) $\det(A + B) - \det(A - B) = 2\left(\mathrm{Tr}(A)\mathrm{Tr}(B) - \mathrm{Tr}(AB)\right)$.

Proof Parts (a) and (b) follow in view of formula (2.2) by taking $x = 1$ and $x = -1$ and part (c) follows by subtracting the equalities from parts (a) and (b). $\square$

Theorem 2.3 The polarized Cayley–Hamilton Theorem.

  If $A, B \in \mathcal{M}_2(\mathbb{C})$, then

$$AB + BA - \mathrm{Tr}(A)B - \mathrm{Tr}(B)A + \left[\mathrm{Tr}(A)\mathrm{Tr}(B) - \mathrm{Tr}(AB)\right] I_2 = O_2.$$

Proof Let $x \in \mathbb{R}$. We apply the Cayley–Hamilton Theorem to the matrix $A + xB$ and we have

$$(A + xB)^2 - \mathrm{Tr}(A + xB)(A + xB) + \det(A + xB)I_2 = O_2.$$

Since $(A + xB)^2 = A^2 + B^2 x^2 + x(AB + BA)$ we have, based on Lemma 2.7, that

$$A^2 + B^2 x^2 + (AB + BA)x - \left[\mathrm{Tr}(A)A + x\left(\mathrm{Tr}(B)A + \mathrm{Tr}(A)B\right) + x^2 \mathrm{Tr}(B)B\right]$$
$$+ \left[\det A + (\mathrm{Tr}(A)\mathrm{Tr}(B) - \mathrm{Tr}(AB))x + x^2 \det B\right] I_2 = O_2.$$

Letting $x = 1$ in the previous equality the theorem is proved. $\square$

Corollary 2.3 *Let $A, B, C \in \mathcal{M}_2(\mathbb{C})$. Then*:

(a) *Another polarized version of the Cayley–Hamilton Theorem*

$$2ABC = \mathrm{Tr}(A)BC + \mathrm{Tr}(B)AC + \mathrm{Tr}(C)AB - \mathrm{Tr}(AC)B$$
$$+ [\mathrm{Tr}(AB) - \mathrm{Tr}(A)\mathrm{Tr}(B)]\, C + [\mathrm{Tr}(BC) - \mathrm{Tr}(B)\mathrm{Tr}(C)]\, A$$
$$- [\mathrm{Tr}(ACB) - \mathrm{Tr}(AC)\mathrm{Tr}(B)]\, I_2;$$

(b) *A trace identity*

$$\mathrm{Tr}(ABC) = \mathrm{Tr}(A)\mathrm{Tr}(BC) + \mathrm{Tr}(B)\mathrm{Tr}(AC) + \mathrm{Tr}(C)\mathrm{Tr}(AB)$$
$$- \mathrm{Tr}(ACB) - \mathrm{Tr}(A)\mathrm{Tr}(B)\mathrm{Tr}(C).$$

Proof (a) We have, based on Theorem 2.3, that

$$2ABC = A(BC + CB) + (AB + BA)C - [B(AC) + (AC)B]$$
$$- A\left[\mathrm{Tr}(B)C + \mathrm{Tr}(C)B - (\mathrm{Tr}(B)\mathrm{Tr}(C) - \mathrm{Tr}(BC))\, I_2\right]$$
$$+ \left[\mathrm{Tr}(A)B + \mathrm{Tr}(B)A - (\mathrm{Tr}(A)\mathrm{Tr}(B) - \mathrm{Tr}(AB))\, I_2\right] C$$
$$- \left[\mathrm{Tr}(AC)B + \mathrm{Tr}(B)AC - (\mathrm{Tr}(B)\mathrm{Tr}(AC) - \mathrm{Tr}(ACB))\, I_2\right]$$
$$= \mathrm{Tr}(A)BC + \mathrm{Tr}(B)AC + \mathrm{Tr}(C)AB - \mathrm{Tr}(AC)B$$
$$+ [\mathrm{Tr}(AB) - \mathrm{Tr}(A)\mathrm{Tr}(B)]\, C + [\mathrm{Tr}(BC) - \mathrm{Tr}(B)\mathrm{Tr}(C)]\, A$$
$$- [\mathrm{Tr}(ACB) - \mathrm{Tr}(AC)\mathrm{Tr}(B)]\, I_2.$$

(b) This part of the corollary follows by applying the trace function to the equality in part (a).   □

Corollary 2.4 A polynomial with special coefficients.

If $A, B \in \mathcal{M}_2(\mathbb{C})$ and $x, y \in \mathbb{C}$, then

$$\det(xA + yB) = x^2 \det A + y^2 \det B + xy\left[\det(A + B) - \det A - \det B\right].$$

Proof If $x = 0$ we have nothing to prove. If $x \neq 0$ we let $\alpha = \frac{y}{x}$. We have,

$$\det(xA + yB) = x^2 \det (A + \alpha B)$$

$$\overset{\text{Lemma 2.7}}{=} x^2 \left[\det A + (\text{Tr}(A)\text{Tr}(B) - \text{Tr}(AB))\alpha + \alpha^2 \det B \right]$$

$$= x^2 \det A + xy(\text{Tr}(A)\text{Tr}(B) - \text{Tr}(AB)) + y^2 \det B$$

$$= x^2 \det A + xy \left[\det(A + B) - \det A - \det B \right] + y^2 \det B,$$

where the last equality follows from part (a) of Corollary 2.2.   □

Corollary 2.5 *If $A \in \mathcal{M}_2(\mathbb{C})$ and $x \in \mathbb{C}$, then $\det(A + xI_2) = \det A + \text{Tr}(A)x + x^2$.*

Proof This follows from Lemma 2.7 by taking $B = I_2$.   □

Lemma 2.8 *If $A \in \mathcal{M}_2(\mathbb{C})$, then $A + A_* = \text{Tr}(A)I_2$.*

Proof The lemma can be proved by direct calculations.   □

Lemma 2.9 *If $A, B \in \mathcal{M}_2(\mathbb{C})$, then*

$$\text{Tr}(A_*B) = \text{Tr}(AB_*) = \text{Tr}(A)\text{Tr}(B) - \text{Tr}(AB).$$

Proof We have, based on Lemma 2.8, that

$$A_* = \text{Tr}(A)I_2 - A \implies \text{Tr}(A_*B) = \text{Tr}(\text{Tr}(A)B - AB) = \text{Tr}(A)\text{Tr}(B) - \text{Tr}(AB)$$

and similarly $\text{Tr}(AB_*) = \text{Tr}(A)\text{Tr}(B) - \text{Tr}(AB)$.   □

The next corollary is a consequence of Lemma 2.7 and Lemma 2.9.

Corollary 2.6 *If $A, B \in \mathcal{M}_2(\mathbb{C})$ and $x \in \mathbb{C}$, then*

$$\det(A + xB) = \det A + \text{Tr}(AB_*)x + (\det B)x^2.$$

Lemma 2.10 *If $A, B \in \mathcal{M}_2(\mathbb{C})$, then $\det(AB - BA) = \text{Tr}(A^2B^2) - \text{Tr}((AB)^2)$.*

Proof We have, based on formula (2.2), that

$$\det(AB - BA) = \det(AB) - \left(\text{Tr}(AB)\text{Tr}(BA) - \text{Tr}(AB^2A) \right) + \det(BA).$$

We note that $\det(AB) = \det(BA)$, $\text{Tr}(AB) = \text{Tr}(BA)$ and $\text{Tr}(AB^2A) = \text{Tr}(A^2B^2)$. It follows that

$$\det(AB - BA) = 2\det(AB) - (\text{Tr}(AB))^2 + \text{Tr}(A^2B^2)$$

which combined to formula (2.1) proves the lemma.   ⊔

Lemma 2.11 *If $A, B \in \mathcal{M}_2(\mathbb{C})$, then*

$$\det(A - B)\det(A + B) = \det(A^2 - B^2) + \det(AB - BA).$$

Proof We have, based on Corollary 2.1, that

$$\det[(A^2 - B^2) + (AB - BA)] + \det[(A^2 - B^2) - (AB - BA)]$$
$$= 2\left[\det(A^2 - B^2) + \det(AB - BA)\right].$$

However

$$\det[(A^2 - B^2) + (AB - BA)] = \det[(A - B)(A + B)] = \det(A - B)\det(A + B)$$

and

$$\det[(A^2 - B^2) - (AB - BA)] = \det[(A + B)(A - B)] = \det(A + B)\det(A - B)$$

and the lemma is proved. □

Lemma 2.12 *If $A, B \in \mathcal{M}_2(\mathbb{C})$, then*

$$\det(A^2 + B^2) = \det(AB - BA) + (\det A - \det B)^2 + (\det(A + B) - \det A - \det B)^2.$$

Proof We apply Lemma 2.11 with B replaced by iB and we get that

$$\det(A - iB)\det(A + iB) = \det(A^2 + B^2) + \det[i(AB - BA)].$$

This implies that

$$\det(A^2 + B^2) = \det(AB - BA) + \det(A - iB)\det(A + iB). \qquad (2.3)$$

On the other hand, we have based on formula (2.2) that

$$\det(A + iB) = \det A - \det B + (\mathrm{Tr}(A)\mathrm{Tr}(B) - \mathrm{Tr}(AB))\,i$$

and

$$\det(A - iB) = \det A - \det B - (\mathrm{Tr}(A)\mathrm{Tr}(B) - \mathrm{Tr}(AB))\,i.$$

It follows that

$$\det(A - iB)\det(A + iB) = (\det A - \det B)^2 + (\mathrm{Tr}(A)\mathrm{Tr}(B) - \mathrm{Tr}(AB))^2$$

which combined to part (a) of Corollary 2.2 shows that

$$\det(A - iB)\det(A + iB) = (\det A - \det B)^2 + (\det(A + B) - \det A - \det B)^2. \quad (2.4)$$

Combining (2.3) and (2.4) the lemma is proved.   □

Theorem 2.4 Power matrix identities.

   Let λ_1, λ_2 be the eigenvalues of $A \in \mathcal{M}_2(\mathbb{C})$. The following identities hold:

(a) $(A - \lambda_1 I_2)^{2n} + (A - \lambda_2 I_2)^{2n} = (\lambda_2 - \lambda_1)^{2n} I_2, \ n \geq 1$;

(b) $(A - \lambda_1 I_2)^{2n-1} - (A - \lambda_2 I_2)^{2n-1} = (\lambda_2 - \lambda_1)^{2n-1} I_2, \ n \geq 1$.

Proof (a) We prove part (a) of the theorem by induction on n. Let $P(n)$ be the proposition

$$P(n): \quad (A - \lambda_1 I_2)^{2n} + (A - \lambda_2 I_2)^{2n} = (\lambda_2 - \lambda_1)^{2n} I_2.$$

First, we prove that $P(1)$ is true. We need to check the equality

$$(A - \lambda_1 I_2)^2 + (A - \lambda_2 I_2)^2 = (\lambda_2 - \lambda_1)^2 I_2$$

holds true. We have

$$(A - \lambda_1 I_2)^2 + (A - \lambda_2 I_2)^2 = A^2 - 2\lambda_1 A + \lambda_1^2 I_2 + A^2 - 2\lambda_2 A + \lambda_2^2 I_2$$

$$= 2[A^2 - (\lambda_1 + \lambda_2)A + \lambda_1 \lambda_2 I_2] + (\lambda_1^2 - 2\lambda_1 \lambda_2 + \lambda_2^2) I_2$$

$$= (\lambda_2 - \lambda_1)^2 I_2,$$

where the last equality follows based on the Cayley–Hamilton Theorem.

   Now we assume that $P(k)$ is true for $k = 1, 2, \ldots, n$ and we prove that $P(n + 1)$ is true. We have, since $P(1)$ and $P(n)$ hold true, that

$$(A - \lambda_1 I_2)^{2n+2} + (A - \lambda_2 I_2)^{2n+2}$$

$$= \left[(A - \lambda_1 I_2)^{2n} + (A - \lambda_2 I_2)^{2n}\right]\left[(A - \lambda_1 I_2)^2 + (A - \lambda_2 I_2)^2\right]$$

$$\quad - (A - \lambda_1 I_2)^{2n}(A - \lambda_2 I_2)^2 - (A - \lambda_2 I_2)^{2n}(A - \lambda_1 I_2)^2$$

$$= (\lambda_2 - \lambda_1)^{2n}(\lambda_2 - \lambda_1)^2 I_2$$

$$= (\lambda_2 - \lambda_1)^{2n+2} I_2,$$

since $(A - \lambda_1 I_2)(A - \lambda_2 I_2) = (A - \lambda_2 I_2)(A - \lambda_1 I_2) = O_2$.

(b) When $n = 1$ there is nothing to prove. Let $X = A - \lambda_1 I_2$ and $Y = A - \lambda_2 I_2$. We have

$$X^{2n-1} - Y^{2n-1} = (X^{2n-2} + Y^{2n-2})(X - Y) + X^{2n-2}Y - Y^{2n-2}X$$

$$\overset{(a)}{=} (\lambda_2 - \lambda_1)^{2n-2}(\lambda_2 - \lambda_1)I_2$$

$$= (\lambda_2 - \lambda_1)^{2n-1}I_2,$$

since $XY = YX = O_2$. The theorem is proved. $\qquad\square$

2.2   The eigenvalues of symmetric matrices

In this section we show that a real 2×2 symmetric matrix is diagonalizable and the invertible matrix $P \in \mathcal{M}_2(\mathbb{R})$ which diagonalizes A can be chosen to be an orthogonal matrix, i.e., $P^T = P^{-1}$. In fact P is a rotation matrix. This idea is used frequently in Chapters 4 and 6 for calculating double integrals over various domains and it is also used in Chapter 6 for reducing a conic to its canonical form.

Theorem 2.5   Symmetric matrices and their eigenvalues.

Let $A = \begin{pmatrix} a & b \\ b & d \end{pmatrix} \in \mathcal{M}_2(\mathbb{R})$.

(a)   *A has real eigenvalues*

$$\lambda_1 = \frac{a + d + \sqrt{(a-d)^2 + 4b^2}}{2} \quad \textit{and} \quad \lambda_2 = \frac{a + d - \sqrt{(a-d)^2 + 4b^2}}{2},$$

and $\lambda_1 = \lambda_2 = \lambda$ if and only if $A = \lambda I_2$.

(b)   *A is diagonalizable and the invertible matrix $P \in \mathcal{M}_2(\mathbb{R})$ which diagonalizes A can be chosen to be a rotation matrix. We have*

$$P^{-1}AP = \begin{pmatrix} \lambda_1 & 0 \\ 0 & \lambda_2 \end{pmatrix}, \quad \textit{where } P = R_\theta, \ \tan\theta = \frac{d - a + \sqrt{(a-d)^2 + 4b^2}}{2b}, \ b \neq 0.$$

Proof (a) This part follows by direct computation.

(b) The invertible matrix P has as columns the eigenvectors corresponding to the eigenvalues λ_1 and λ_2. If v_i, $i = 1, 2$, are the eigenvectors corresponding to the eigenvalues λ_i, $i = 1, 2$, then the systems $(A - \lambda_i I_2)v_i = 0$, $i = 1, 2$, imply that

$$v_1 = \begin{pmatrix} 1 \\ \frac{d-a+\sqrt{(a-d)^2+4b^2}}{2b} \end{pmatrix} \quad \text{and} \quad v_2 = \begin{pmatrix} \frac{a-d-\sqrt{(a-d)^2+4b^2}}{2b} \\ 1 \end{pmatrix}.$$

We divide these two vectors by their length

$$\|v_1\| = \|v_2\| = \sqrt{1 + \left(\frac{d-a+\sqrt{(a-d)^2+4b^2}}{2b} \right)^2}$$

and we take

$$P = \begin{pmatrix} \frac{v_1}{\|v_1\|} & \frac{v_2}{\|v_2\|} \end{pmatrix} = R_\theta, \quad \text{where} \quad \tan\theta = \frac{d-a+\sqrt{(a-d)^2+4b^2}}{2b}, \quad b \neq 0.$$

The theorem is proved.   □

2.3   The reciprocal of the Cayley–Hamilton Theorem

In this section we discuss the *reciprocal* of the Cayley–Hamilton Theorem.

Theorem 2.6 The reciprocal of the Cayley–Hamilton Theorem.
  Let $A \in \mathcal{M}_2(\mathbb{C})$ and let $a, b \in \mathbb{C}$ be such that $A^2 - aA + bI_2 = O_2$. If $A \notin \{\alpha I_2 : \alpha \in \mathbb{C}\}$, then $\mathrm{Tr}(A) = a$ and $\det A = b$.

Proof We have, based on the Cayley–Hamilton Theorem, that

$$A^2 - aA + bI_2 = O_2$$

$$A^2 - \mathrm{Tr}(A)A + (\det A)I_2 = O_2,$$

and it follows that $[a - \mathrm{Tr}(A)] A = (b - \det A)I_2$.
  If $a - \mathrm{Tr}(A) \neq 0$ we get that

$$A = \frac{b - \det A}{a - \mathrm{Tr}(A)} I_2,$$

which is a contradiction to $A \notin \{\alpha I_2 : \alpha \in \mathbb{C}\}$.
  If $a - \mathrm{Tr}(A) = 0$ we get that $b - \det A = 0$ and the theorem is proved.   □

Remark 2.3 It is worth mentioning that there do exist matrices $A \in \mathcal{M}_2(\mathbb{C})$ such that $A^2 - aA + bI_2 = O_2$, with $a \neq \mathrm{Tr}(A)$ and $b \neq \det A$. To see this we let $A = \alpha I_2$, where $\alpha \in \mathbb{C}$ verifies the equation $\alpha^2 - a\alpha + b = 0$. Then, $\mathrm{Tr}(A) = 2\alpha$, $\det A = \alpha^2$ and if a and b are such that $a^2 - 4b \neq 0$ and $b \neq 0$ one has $a \neq \mathrm{Tr}(A)$ and $b \neq \det A$.

2.4   The characteristic polynomial of matrices XY and YX

In this section we prove two fundamental results in matrix theory concerning the characteristic polynomial of matrices XY and YX.

Theorem 2.7   The characteristic polynomial of matrices XY and YX.

If $X, Y \in \mathcal{M}_2(\mathbb{C})$, then matrices XY and YX have the same characteristic polynomials, i.e., $f_{XY} = f_{YX}$.

Proof We have, since $\mathrm{Tr}(XY) = \mathrm{Tr}(YX)$ and $\det(XY) = \det(YX)$, that

$$f_{XY}(x) = x^2 - \mathrm{Tr}(XY)x + \det(XY) = x^2 - \mathrm{Tr}(YX)x + \det(YX) = f_{YX}(x).$$

Nota bene. Theorem 2.7 implies the following equality holds

$$\det(XY - \lambda I_2) = \det(YX - \lambda I_2), \quad \forall X, Y \in \mathcal{M}_2(\mathbb{C}), \ \forall \lambda \in \mathbb{C}.$$

The theorem also implies that matrices XY and YX have the same eigenvalues. $\square$

The next theorem is the reciprocal of Theorem 2.7.

Theorem 2.8 *If $A \in \mathcal{M}_2(\mathbb{C})$ verifies*

$$\det(XY - A) = \det(YX - A), \quad \forall \, X, Y \in \mathcal{M}_2(\mathbb{C}),$$

then there exists $a \in \mathbb{C}$ such that $A = aI_2$.

Proof Let $E_{i,j}$ be the matrix having the (i, j) entry equal to 1 and all the other entries equal to 0 and let $A = \begin{pmatrix} a & b \\ c & d \end{pmatrix} \in \mathcal{M}_2(\mathbb{C})$.

If $X = E_{1,2}$ and $Y = E_{2,2}$, then $XY = E_{1,2}$, $YX = O_2$ and the equality from the hypothesis of the theorem becomes

$$\det \begin{pmatrix} -a & 1-b \\ -c & -d \end{pmatrix} = \det \begin{pmatrix} -a & -b \\ -c & -d \end{pmatrix},$$

which implies that $c = 0$.

If $X = E_{2,1}$, $Y = E_{1,1}$, then $XY = E_{2,1}$ and $YX = O_2$. In this case the equality from the hypothesis of the theorem becomes

$$\det \begin{pmatrix} -a & -b \\ 1-c & -d \end{pmatrix} = \det \begin{pmatrix} -a & -b \\ -c & -d \end{pmatrix},$$

which implies that $b = 0$. Thus, $A = \begin{pmatrix} a & 0 \\ 0 & d \end{pmatrix}$.

If $X = E_{1,2}$, $Y = E_{2,1}$ we get that $XY = E_{1,1}$ and $YX = E_{2,2}$ and the condition

$$\det \begin{pmatrix} 1-a & 0 \\ 0 & -d \end{pmatrix} = \det \begin{pmatrix} -a & 0 \\ 0 & 1-d \end{pmatrix}$$

implies that $a = d$. Thus, $A = aI_2$ and the theorem is proved. $\square$

Now we give an application of the previous theorem.

Corollary 2.7 *If $A, B \in \mathcal{M}_2(\mathbb{C})$ are two invertible matrices such that*

$$\det(XAY + B) = \det(YBX + A), \quad \forall\, X, Y \in \mathcal{M}_2(\mathbb{C}), \tag{2.5}$$

then, there exists $a \in \mathbb{C}^*$ such that $A^2 = B^2 = aI_2$.

Proof If $Y = O_2$ we get that $\det A = \det B$. If $Y = I_2$ we get that $\det(XA + B) = \det(BX + A)$, $\forall X \in \mathcal{M}_2(\mathbb{C})$. We multiply, since $\det A = \det B$, the left-hand side of the previous equality to the left by $\det B$ and the right-hand side of the same equality to the right by $\det A$ and we get that

$$\det(BXA + B^2) = \det(BXA + A^2), \quad \forall\, X \in \mathcal{M}_2(\mathbb{C}). \tag{2.6}$$

Since matrices A and B are invertible, the function $f : \mathcal{M}_2(\mathbb{C}) \to \mathcal{M}_2(\mathbb{C})$ defined by $f(X) = BXA$ is onto and equality (2.6) implies that

$$\det(Z + B^2) = \det(Z + A^2), \quad \forall\, Z \in \mathcal{M}_2(\mathbb{C}). \tag{2.7}$$

Taking Z be equal to $O_2, E_{1,1}, E_{1,2}, E_{2,1}, E_{2,2}$ in (2.7) we get that $B^2 = A^2$. Now we multiply the left-hand side of equality (2.5) to the right by $\det B$ and the right-hand side to the right by $\det A$ and we get that

$$\det(XAYB + B^2) = \det(YBXA + A^2), \quad \forall\, X, Y \in \mathcal{M}_2(\mathbb{C}),$$

or

$$\det(X_1 Y_1 + C) = \det(Y_1 X_1 + C), \quad \forall\, X_1, Y_1 \in \mathcal{M}_2(\mathbb{C}),$$

where $C = A^2 - B^2$. It follows, based on Theorem 2.8, that there exists $a \in \mathbb{C}^*$ such that $C = aI_2$. Thus, $A^2 = B^2 = aI_2$ and the corollary is proved. $\square$

2.5   The Jordan canonical form

> **Theorem 2.9   The complex Jordan canonical form.**
>
> Let $A \in \mathcal{M}_2(\mathbb{C})$ and let λ_1, λ_2 be the eigenvalues of A. Then:
>
> (a)   if $\lambda_1 \neq \lambda_2$ or $A = \alpha I_2$, for some $\alpha \in \mathbb{C}$, there exists an invertible matrix $P \in \mathcal{M}_2(\mathbb{C})$ such that
>
> $$A = P \begin{pmatrix} \lambda_1 & 0 \\ 0 & \lambda_2 \end{pmatrix} P^{-1};$$
>
> (b)   if $\lambda_1 = \lambda_2 = \lambda$ and $A \neq \lambda I_2$, there exists an invertible matrix $P \in \mathcal{M}_2(\mathbb{C})$ such that
>
> $$A = P \begin{pmatrix} \lambda & 1 \\ 0 & \lambda \end{pmatrix} P^{-1}.$$

Proof (a) Let $\lambda_1 \neq \lambda_2$ be the eigenvalues of $A = \begin{pmatrix} a & b \\ c & d \end{pmatrix}$. We have, since the eigenvalues are distinct, that $(a-d)^2 + 4bc \neq 0$. Also, there exists $X_1 = \begin{pmatrix} x_1 \\ y_1 \end{pmatrix} \neq \begin{pmatrix} 0 \\ 0 \end{pmatrix}$ such that

$$AX_1 = \lambda_1 X_1 \tag{2.8}$$

and there exists $X_2 = \begin{pmatrix} x_2 \\ y_2 \end{pmatrix} \neq \begin{pmatrix} 0 \\ 0 \end{pmatrix}$ such that

$$AX_2 = \lambda_2 X_2. \tag{2.9}$$

Now, we note that $X_2 \neq \alpha X_1$, for all $\alpha \in \mathbb{C}$, i.e., the eigenvectors associated with the eigenvalues λ_1 and λ_2 are not proportional. Otherwise, if $X_2 = \alpha X_1$ for some $\alpha \in \mathbb{C}$ this would imply that $AX_2 = \alpha AX_1$ which in turn implies that $\lambda_2 X_2 = \alpha \lambda_1 X_1$. Thus, $\alpha(\lambda_2 - \lambda_1)X_1 = 0$ and since $X_1 \neq 0$ we get that $\lambda_1 = \lambda_2$, which contradicts $\lambda_1 \neq \lambda_2$. Therefore the eigenvectors X_1 and X_2 are not proportional and this implies that the matrix $P = (X_1 \mid X_2)$ is invertible.

Equalities (2.8) and (2.9) can be written as follows $A(X_1 \mid X_2) = (\lambda_1 X_1 \mid \lambda_2 X_2)$ or

$$AP = P \begin{pmatrix} \lambda_1 & 0 \\ 0 & \lambda_2 \end{pmatrix} \quad \Leftrightarrow \quad A = PJ_A P^{-1} \quad \text{where} \quad J_A = \begin{pmatrix} \lambda_1 & 0 \\ 0 & \lambda_2 \end{pmatrix}.$$

(b) Now we consider the case when the eigenvalues of A are equal, i.e., $\lambda_1 = \lambda_2 = \lambda$ and $A \neq \lambda I_2$. We choose the vector $X_1 = \begin{pmatrix} x_1 \\ y_1 \end{pmatrix} \neq \begin{pmatrix} 0 \\ 0 \end{pmatrix}$ such that $AX_1 = \lambda X_1$ and $X_1' = \begin{pmatrix} x_1' \\ y_1' \end{pmatrix}$ such that $AX_1' = \lambda X_1' + X_1$. We mention that while the vector X_1 is the eigenvector associated with λ, the vector X_1' is called the *generalized eigenvector* associated with the eigenvalue λ. Let $P = (X_1 \,|\, X_1')$ and we have

$$AP = A(X_1 \,|\, X_1') = (\lambda X_1 \,|\, \lambda X_1' + X_1)$$

or

$$AP = P \begin{pmatrix} \lambda & 1 \\ 0 & \lambda \end{pmatrix} \quad \Leftrightarrow \quad A = P J_A P^{-1} \quad \text{where} \quad J_A = \begin{pmatrix} \lambda & 1 \\ 0 & \lambda \end{pmatrix}.$$

The theorem is proved.   □

Remark 2.4 The matrices

$$J_A = \begin{pmatrix} \lambda_1 & 0 \\ 0 & \lambda_2 \end{pmatrix} \quad \text{or} \quad J_A = \begin{pmatrix} \lambda & 1 \\ 0 & \lambda \end{pmatrix},$$

are called the *Jordan canonical forms* of A. The columns X_1, X_2 or X_1, X_1' of P form a basis in $\mathcal{M}_2(\mathbb{C})$ called the **Jordan basis** corresponding to the matrix A and the matrix P is, according to Lemma 1.4, the matrix of passing from the canonical basis $\mathcal{B} = \{E_1, E_2\}$ to the Jordan basis.

Corollary 2.8 The Jordan canonical form of special matrices.

■ *All nilpotent matrices $A \in \mathcal{M}_2(\mathbb{C})$ with $A \neq O_2$ are of the following form*

$$A = P \begin{pmatrix} 0 & 1 \\ 0 & 0 \end{pmatrix} P^{-1},$$

   where P is any invertible matrix.

■ *All idempotent matrices $A \in \mathcal{M}_2(\mathbb{C})$ are $A = O_2$, $A = I_2$ or*

$$A = P \begin{pmatrix} 1 & 0 \\ 0 & 0 \end{pmatrix} P^{-1},$$

(continued)

Corollary 2.8 (continued)

where P is any invertible matrix.

■ *All involutory matrices $A \in \mathcal{M}_2(\mathbb{C})$ are $A = \pm I_2$ or*

$$A = P \begin{pmatrix} 1 & 0 \\ 0 & -1 \end{pmatrix} P^{-1},$$

where P is any invertible matrix.

■ *All skew involutory matrices $A \in \mathcal{M}_2(\mathbb{C})$ are $A = \pm i \cdot I_2$ or*

$$A = P \begin{pmatrix} 0 & -1 \\ 1 & 0 \end{pmatrix} P^{-1},$$

where P is any invertible matrix.

Now we discuss the real canonical form of a matrix $A \in \mathcal{M}_2(\mathbb{R})$. We have the following theorem.

Theorem 2.10   The real canonical form of a real matrix.

(a) *If $A \in \mathcal{M}_2(\mathbb{R})$ and λ_1, λ_2 are the real eigenvalues of A, then there exists $P \in \mathcal{M}_2(\mathbb{R})$ such that*

$$A = P \begin{pmatrix} \lambda_1 & 0 \\ 0 & \lambda_2 \end{pmatrix} P^{-1} \quad or \quad A = P \begin{pmatrix} \lambda & 1 \\ 0 & \lambda \end{pmatrix} P^{-1},$$

according to whether the eigenvalues of A are distinct or not.

(b) *If $A \in \mathcal{M}_2(\mathbb{R})$ and the eigenvalues of A are $\lambda_1 = \alpha + i\beta$ and $\lambda_2 = \alpha - i\beta$, $\alpha \in \mathbb{R}$ and $\beta \in \mathbb{R}^*$, then there exists an invertible matrix $P \in \mathcal{M}_2(\mathbb{R})$ such that*

$$A = P \begin{pmatrix} \alpha & \beta \\ -\beta & \alpha \end{pmatrix} P^{-1}.$$

Proof (a) The proof of part (a) is similar to the proof of Theorem 2.9.

(b) We mention that, in this case, the Jordan canonical form of A is given by

$$J_A = \begin{pmatrix} \alpha + i\beta & 0 \\ 0 & \alpha - i\beta \end{pmatrix},$$

and if $AZ = \lambda_1 Z, Z \neq 0$, then $A\overline{Z} = \overline{\lambda_1 Z} = \lambda_2 \overline{Z}$ and the invertible matrix $P_{\mathbb{C}}$, which verifies $A = P_{\mathbb{C}} J_A P_{\mathbb{C}}^{-1}$, would be $P_{\mathbb{C}} = (Z \mid \overline{Z})$.

If $Z = X + iY$, with X and Y real vectors, we have

$$AZ = \lambda_1 Z \quad \Leftrightarrow \quad A(X + iY) = (\alpha + i\beta)(X + iY)$$

and we obtain the following equalities $AX = \alpha X - \beta Y$ and $AY = \beta X + \alpha Y$.

We define the matrix $P = (X \mid Y)$ and we have that

$$AP = A(X \mid Y) = (AX \mid AY) = (\alpha X - \beta Y \mid \beta X + \alpha Y) = P \begin{pmatrix} \alpha & \beta \\ -\beta & \alpha \end{pmatrix}$$

or $A = P J_A^{\mathbb{R}} P^{-1}$, where the matrix

$$J_A^{\mathbb{R}} = \begin{pmatrix} \alpha & \beta \\ -\beta & \alpha \end{pmatrix}$$

is called the *real canonical form* of A. The theorem is proved. $\qquad\qquad\square$

Now we give the rational canonical form of a matrix $A \in \mathcal{M}_2(\mathbb{Q})$. We have the following theorem.

Theorem 2.11 The rational canonical form of a rational matrix.

(a) *If $A \in \mathcal{M}_2(\mathbb{Q})$ and λ_1, λ_2 are the rational eigenvalues of A, then there exists $P \in \mathcal{M}_2(\mathbb{Q})$ such that*

$$A = P \begin{pmatrix} \lambda_1 & 0 \\ 0 & \lambda_2 \end{pmatrix} P^{-1} \quad or \quad A = P \begin{pmatrix} \lambda & 1 \\ 0 & \lambda \end{pmatrix} P^{-1},$$

according to whether the eigenvalues of A are distinct or not.

(b) *If $A \in \mathcal{M}_2(\mathbb{Q})$ and the distinct eigenvalues of A are $\lambda_1, \lambda_2 \in \mathbb{C} \setminus \mathbb{Q}$, then $\lambda_1 = \alpha + \sqrt{\beta}$ and $\lambda_2 = \alpha - \sqrt{\beta}$, $\alpha \in \mathbb{Q}$, $\beta \in \mathbb{Q}^*$ and there exists an invertible matrix $P \in \mathcal{M}_2(\mathbb{Q})$ such that*

$$A = P \begin{pmatrix} \alpha & 1 \\ \beta & \alpha \end{pmatrix} P^{-1}.$$

Proof (a) The proof of part (a) is similar to the proof of Theorem 2.9.

(b) In this case the Jordan canonical form of A is given by

$$J_A = \begin{pmatrix} \alpha + \sqrt{\beta} & 0 \\ 0 & \alpha - \sqrt{\beta} \end{pmatrix}.$$

If $Z \neq 0$ is the eigenvector associated with the eigenvalue $\lambda_1 = \alpha + \sqrt{\beta}$, then $AZ = \lambda_1 Z$. Let $Z = X + \sqrt{\beta}Y$, where X and Y are rational vectors. A calculation shows that $A(X + \sqrt{\beta}Y) = (\alpha + \sqrt{\beta})(X + \sqrt{\beta}Y)$ implies

$$AX = \alpha X + \beta Y \quad \text{and} \quad AY = X + \alpha Y.$$

This in turn implies that $A(X - \sqrt{\beta}Y) = (\alpha - \sqrt{\beta})(X - \sqrt{\beta}Y)$, or $AZ' = \lambda_2 Z'$, where $Z' = X - \sqrt{\beta}Y$. The invertible matrix $P_{\mathbb{C}}$ which verifies $A = P_{\mathbb{C}}J_A P_{\mathbb{C}}^{-1}$ is given by $P_{\mathbb{C}} = (Z \,|\, Z')$.

Let $P = (X \,|\, Y) \in \mathcal{M}_2(\mathbb{Q})$. We have

$$AP = (AX \,|\, AY) = (\alpha X + \beta Y \,|\, X + \alpha Y) = (X \,|\, Y)\begin{pmatrix} \alpha & 1 \\ \beta & \alpha \end{pmatrix} = P\begin{pmatrix} \alpha & 1 \\ \beta & \alpha \end{pmatrix}$$

or $A = P J_A^{\mathbb{Q}} P^{-1}$, where the matrix

$$J_A^{\mathbb{Q}} = \begin{pmatrix} \alpha & 1 \\ \beta & \alpha \end{pmatrix}$$

is called the *rational canonical form* of A. The theorem is proved. $\qquad\square$

2.6 Problems

2.1 Let $A \in \mathcal{M}_2(\mathbb{C})$ with $\det A = 1$. Prove that $\det(A^2 + A - I_2) + \det(A^2 + I_2) = 5$.

2.2 Let $A \in \mathcal{M}_2(\mathbb{Z})$ with $\det A = 1$. Find $\mathrm{Tr}(A)$ if

$$\det(A^2 - 3A + I_2) + \det(A^2 + A - I_2) = -4.$$

2.3 Let $A \in \mathcal{M}_2(\mathbb{C})$ with $\mathrm{Tr}(A) = -1$. Prove that

$$\det(A^2 + 3A + 3I_2) - \det(A^2 + A) = 3.$$

2.4 Let $a \in \mathbb{Z}$, $a \neq \pm 1$ and let $A \in \mathcal{M}_2(\mathbb{Z})$. Prove the matrices $aA + (a+1)I_2$ and $aA - (a+1)I_2$ are invertible.

2.5 Prove that any matrix $A \in \mathcal{M}_2(\mathbb{C})$ is the sum of two invertible matrices.

2.6 [1] Let $A \in \mathcal{M}_2(\mathbb{C})$ with $\det A = 0$. Prove that there is a sequence of matrices $(A_n)_{n \in \mathbb{N}}$ such that $\det A_n \neq 0$ and $\lim_{n \to \infty} A_n = A$.

---

[1]The problem states that any singular matrix is the limit of a sequence of nonsingular matrices.

Remark 2.5 **The density of invertible matrices.** Problem **2.6** is used for proving that the set of invertible matrices is dense in the set of all matrices.

2.7 Let $A \in \mathcal{M}_2(\mathbb{C})$. Prove that A and A_* have the same characteristic polynomials (the same eigenvalues).

2.8 A *right stochastic* matrix is a square matrix with nonnegative real numbers with each row summing to 1. Prove that the eigenvalues of $A \in \mathcal{M}_2(\mathbb{C})$ are 1, the largest, and $\text{Tr}(A) - 1$, the smallest, and determine the corresponding eigenvectors.

2.9 Prove that if $A \in \mathcal{M}_2(\mathbb{C})$ has all its eigenvalues equal to 1, then A is similar to A^k for every positive integer k.

2.10 Let $n \in \mathbb{N}$. Prove that if $A, B \in \mathcal{M}_2(\mathbb{C})$ are similar matrices, then A^n and B^n are similar matrices. Does the reverse implication hold?

2.11 Let $A = \begin{pmatrix} 0 & 0 \\ 1 & 0 \end{pmatrix}$.

(a) Determine all matrices $B \in \mathcal{M}_2(\mathbb{R})$ which are similar to A.
(b) Prove that A and O_2 have the same characteristic polynomial but the matrices are not similar.

2.12 Two classes of special similar matrices.

(a) Prove that any matrix $A \in \mathcal{M}_2(\mathbb{C})$ is *similar to its transpose*.
(b) Prove that any matrix $A \in \mathcal{M}_2(\mathbb{C})$ is *similar to a symmetric* matrix.

   Nota bene. Part (b) of the problem reduces to the case of proving that any matrix of the form $\begin{pmatrix} \lambda & 1 \\ 0 & \lambda \end{pmatrix}$, $\lambda \in \mathbb{C}$, is similar to a *complex symmetric* matrix.

2.13 The transpose and the adjugate matrices are similar.

   Prove that there exists $P \in \mathcal{M}_2(\mathbb{C})$ such that $A_* = PA^TP^{-1}$, for all $A \in \mathcal{M}_2(\mathbb{C})$. Determine all matrices P with this property.

2.14 Any matrix $A \in \mathcal{M}_2(\mathbb{C})$ is the product of two *symmetric* matrices.

2.15 Let $n \in \mathbb{N}$, $n \geq 2$ and let $A \in \mathcal{M}_2(\mathbb{R})$. Prove that if A^n is a symmetric matrix which is not of the form αI_2, $\alpha \in \mathbb{R}$, then A is a symmetric matrix.

2.16 Let $\mathcal{M}$ be the set of matrices in $\mathcal{M}_2(\mathbb{C})$ which have the property that the absolute values of their eigenvalues is less than or equal to 1. Prove that if $A, B \in \mathcal{M}$ and $AB = BA$, then $AB \in \mathcal{M}$.

2.17 Let $A = \begin{pmatrix} 2 & 5 \\ -3 & 10 \end{pmatrix}$ and let $B = \begin{pmatrix} 3 & -2 \\ 4 & 9 \end{pmatrix}$. Prove that

$$A^n - B^n = \frac{7^n - 5^n}{2}(A - B), \quad \forall\, n \in \mathbb{N}.$$

2.18 Let $A, B \in \mathcal{M}_2(\mathbb{R})$ such that $AB = \begin{pmatrix} 5 & 2 \\ 7 & 3 \end{pmatrix}$. Prove that $BA + A^{-1}B^{-1} = 8I_2$.

2.19 Let $A = \begin{pmatrix} 1 & 3 \\ 3 & 10 \end{pmatrix}$ and let $(a_n)_{n \geq 0}$ be the sequence defined by the recurrence relation $a_{n+1} = 3a_n + a_{n-1}, n \geq 1, a_0 = 0, a_1 = 1$.

(a) Prove that $A^n = \begin{pmatrix} a_{2n-1} & a_{2n} \\ a_{2n} & a_{2n+1} \end{pmatrix}, n \geq 1$.

(b) If the sequences $(x_n)_{n \geq 0}$ and $(y_n)_{n \geq 0}$ verify the recurrence relation $\begin{pmatrix} x_{n+1} \\ y_{n+1} \end{pmatrix} = A\begin{pmatrix} x_n \\ y_n \end{pmatrix}, n \geq 0$, and $\begin{pmatrix} x_0 \\ y_0 \end{pmatrix} = \begin{pmatrix} 1 \\ 0 \end{pmatrix}$, prove that $x_{n+1}^2 + 3x_{n+1}y_{n+1} - y_{n+1}^2 = x_n^2 + 3x_ny_n - y_n^2$, for all $n \geq 0$.

(c) Prove that if the natural numbers $x, y \in \mathbb{N}$ verify the equation $x^2 + 3xy - y^2 = 1$, then there exists $n \in \mathbb{N}$ such that $(x, y) = (a_{2n-1}, a_{2n})$.

2.20 Let $A, B \in \mathcal{M}_2(\mathbb{R})$ such that there exists $n \in \mathbb{N}$ with $(AB - BA)^n = I_2$. Prove that $(AB - BA)^4 = I_2$ and n is an even integer.

2.21 Let $n \geq 2$ be an integer and let $A, B \in \mathcal{M}_2(\mathbb{C})$ such that $AB \neq BA$ and $(AB)^n = (BA)^n$. Prove that $(AB)^n = bI_2$, for some $b \in \mathbb{C}$.

2.22 Let $A \in \mathcal{M}_2(\mathbb{R})$ such that $\det(A^2 - A + I_2) = 0$.

(a) Prove that $A^2 - A + I_2 = O_2$.
(b) Calculate $\det(A^2 + \alpha A + \beta I_2)$, where $\alpha, \beta \in \mathbb{R}$.

2.23 Prove that any matrix $A \in \mathcal{M}_2(\mathbb{R})$ can be written $A = B^2 + C^2$, with $B, C \in \mathcal{M}_2(\mathbb{R})$. Does the result hold if we add the supplementary condition $BC = CB$?

2.24 Let $A \in \mathcal{M}_2(\mathbb{C})$. Prove that:

(a) $\Re(\det A) = \det \Re(A) - \det \Im(A)$;
(b) $\Im(\det A) = \det(\Re(A) + \Im(A)) - \det \Re(A) - \det \Im(A)$.

2.25 An extremum problem.

Let $\mathcal{M} = \{A = (a_{i,j}) \in \mathcal{M}_2(\mathbb{R}) : -1 \leq a_{i,j} \leq 1, \ \forall i,j = 1,2\}$. Prove that $\max_{A,B \in \mathcal{M}} \det(AB - BA) = 16$.

2.26 Let $A, B \in \mathcal{M}_2(\mathbb{C})$. Prove that if $\det(A+X) = \det(B+X)$, for all $X \in \mathcal{M}_2(\mathbb{C})$, then $A = B$.

2.27 Let $A, B \in \mathcal{M}_2(\mathbb{R})$. Prove that

$$\det(A^2 + B^2 + AB - BA) = \det(A^2 + B^2) + \det(AB - BA).$$

2.28 Let $A, B \in \mathcal{M}_2(\mathbb{R})$. Prove that $\det(A^2 + B^2) \geq \det(AB - BA)$.

2.29 Let $A, B \in \mathcal{M}_2(\mathbb{R})$. Prove that if $\det(AB + BA) \leq 0$, then $\det(A^2 + B^2) \geq 0$.

2.30 Let $A, B \in \mathcal{M}_2(\mathbb{R})$ such that $A^2 + B^2 = O_2$ and $AB = BA$. Prove that $\det(A + B) = \det A + \det B$.

2.31 Prove that if $A, B \in \mathcal{M}_2(\mathbb{R})$ such that $\det(A^2+B^2) = 0$ and $AB = BA$, then:

(a) $\det A = \det B$;
(b) if $\det A \neq 0$, then $A^2 + B^2 = O_2$.

Nota bene. If $A, B \in \mathcal{M}_2(\mathbb{R})$ are such that $\det(A^2 + B^2) = 0$ and $AB = BA$, then it does not follow that $A^2 + B^2 = O_2$.

2.32 Let $A, B \in \mathcal{M}_2(\mathbb{R})$. Prove that if $AB = BA$ and $\det(2A^2 - 3AB + 2B^2) = 0$, then $\det A = \det B$ and $\det(A + B) = \frac{7}{2} \det A$.

2.33 Let $A, B \in \mathcal{M}_2(\mathbb{Q})$ be two commuting matrices such that $\det A = 10$ and $\det(A + \sqrt{5}B) = 0$. Calculate $\det(A^2 - AB + B^2)$.

2.34 Prove that $\forall A, B \in \mathcal{M}_2(\mathbb{C})$ and $\forall a, b, c \in \mathbb{C}$ one has

$$\det(aAB + bBA + cI_2) = \det(aBA + bAB + cI_2).$$

2.35 Let $A \in \mathcal{M}_2(\mathbb{R})$ and let

$$f_A : \mathcal{M}_2(\mathbb{R}) \to \mathbb{R}, \quad f_A(X) = \det(X + A) - \det(X - A).$$

Prove that:

(a) $f_{aA} = af_A$, $a \in \mathbb{R}$;
(b) $f_{A+B} = f_A + f_B$, $B \in \mathcal{M}_2(\mathbb{R})$;
(c) there exist sequences $(x_n)_{n \geq 1}$ and $(y_n)_{n \geq 1}$ such that $f_{A^n} = x_n f_A + y_n f_{I_2}$.

2.36 Prove that the unique function $f : \mathcal{M}_2(\mathbb{C}) \to \mathbb{C}$ which verifies the conditions

(a) $f(XY) = f(X)f(Y)$, $\forall X, Y \in \mathcal{M}_2(\mathbb{C})$
(b) $f(X + I_2) = f(X) + f(I_2) + \text{Tr}(X)$, $\forall X \in \mathcal{M}_2(\mathbb{C})$,

is the determinant function $f(X) = \det X$.

2.37 Let $A \in \mathcal{M}_2(\mathbb{R})$ and let $f : \mathcal{M}_{2,1}(\mathbb{R}) \to \mathcal{M}_{2,1}(\mathbb{R})$ be the function defined by

$$f_A \begin{pmatrix} x \\ y \end{pmatrix} = A \begin{pmatrix} x \\ y \end{pmatrix}, \quad \begin{pmatrix} x \\ y \end{pmatrix} \in \mathcal{M}_{2,1}(\mathbb{R}).$$

Prove the following statements are equivalent:

(a) f_A is injective;
(b) f_A is surjective;
(c) $\det A \neq 0$.

2.38 Let $A \in \mathcal{M}_2(\mathbb{Z})$ and let $f : \mathcal{M}_{2,1}(\mathbb{Z}) \to \mathcal{M}_{2,1}(\mathbb{Z})$ be the function defined by

$$f_A \begin{pmatrix} x \\ y \end{pmatrix} = A \begin{pmatrix} x \\ y \end{pmatrix}, \quad \begin{pmatrix} x \\ y \end{pmatrix} \in \mathcal{M}_{2,1}(\mathbb{Z}).$$

Prove that:

(a) f_A is injective if and only if $\det A \neq 0$;
(b) f_A is surjective if and only if $\det A \in \{-1, 1\}$.

2.39 *The power function.*

Prove the function $f : \mathcal{M}_2(\mathbb{C}) \to \mathcal{M}_2(\mathbb{C})$, $f(X) = X^n$ is neither injective nor surjective for any $n \in \mathbb{N}$, $n \geq 2$.

2.40 *Non-surjective functions.*

(a) Prove the function $f : \mathcal{M}_2(\mathbb{R}) \to \mathcal{M}_2(\mathbb{R})$, $f(X) = X^{2016} + X^{2015}$ is not surjective.
(b) Prove the function $f : \mathcal{M}_2(\mathbb{R}) \to \mathcal{M}_2(\mathbb{R})$, $f(X) = I_2 + X + X^2 + \cdots + X^{2016}$ is not surjective.

2.41 Let $a, b, c, d \in (0, \infty)$ such that $ad - bc > 0$ and let $A = \begin{pmatrix} a & -b \\ -c & d \end{pmatrix} \in \mathcal{M}_2(\mathbb{R})$. We say that a matrix $X \in \mathcal{M}_2(\mathbb{R})$ is *positive* if all of its entries are positive real numbers and we use the notation $X > 0$. Prove that:

(a) for any positive matrix X' there is a positive matrix X such that $AX = X'$;
(b) there is $X > 0$ such that $X' = AX > 0$.

2.42 [58] Let $A = \begin{pmatrix} a & b \\ c & d \end{pmatrix} \in \mathcal{M}_2(\mathbb{R})$ with $a > 0$, $b > 0$, $c > 0$, $d > 0$. Prove that

A has an eigenvector $X = \begin{pmatrix} x \\ y \end{pmatrix}$ with $x > 0$ and $y > 0$.

2.43 Let $A, B \in \mathcal{M}_2(\mathbb{R})$ be matrices with strictly positive entries. Prove that $(AB)^2 = (BA)^2$ if and only if $AB = BA$.

2.44 Let $A \in \mathcal{M}_2(\mathbb{C})$ such that $\text{Tr}(A) = -1$ and $\det A = 1$. How many elements does the set $\{A^n : n \in \mathbb{N}\}$ have?

2.45 A 2016 Seemous problem.

Let $n \geq 2$ be an integer and let $\mathscr{P}_n = \{X^n : X \in \mathcal{M}_2(\mathbb{C})\}$. Prove that $\mathscr{P}_2 = \mathscr{P}_n, \forall n \geq 2$.

The problem generalizes part (a) of Problem 2 of Seemous 2016, Protaras, Cyprus.

2.46 Let $A, B \in \mathcal{M}_2(\mathbb{R})$ such that $\det A = \det B = 1$. Prove that:

(a) $\text{Tr}(AB) + \text{Tr}(A^{-1}B) = \text{Tr}(A)\text{Tr}(B)$;
(b) $\text{Tr}(BAB) + \text{Tr}(A) = \text{Tr}(B)\text{Tr}(AB)$.

2.47 (a) Let $A \in \mathcal{M}_2(\mathbb{R})$ be such that $\text{Tr}(A) > 2$. Prove that for any $n \in \mathbb{N}, A^n \neq I_2$.

(b) Let $r > 0$ and let $A \in \mathcal{M}_2(\mathbb{R})$ be such that $\text{Tr}(A) > 2r$. Prove that for any $n \in \mathbb{N}$, $A^n \neq r^n I_2$.

2.48 Let $A \in \mathcal{M}_2(\mathbb{R})$ with $\det A = 1$ and $|\text{Tr}(A)| < 2$. Prove that for any $n \geq 2$ we have $|\text{Tr}(A^n)| \leq 2$ and there exists $B_n \in \mathcal{M}_2(\mathbb{R})$ such that $\det B_n = 1, |\text{Tr}(B_n)| < 2$ and $|\text{Tr}(B_n^n)| = 2$.

2.49 Let $A, B \in \mathcal{M}_2(\mathbb{C})$ with $A \neq B$ and let $C = AB - BA$. Prove that C commutes with both A and B if and only if $C = O_2$, that is, if and only if A commutes with B.

2.50 Let $A \in \mathcal{M}_2(\mathbb{C})$ such that $\det(A - I_2) \in \mathbb{R}$ and there exists $n \in \mathbb{N}$ such that $A^n = I_2$. Prove that $\det(A - xI_2) \in \mathbb{R}$, for any $x \in \mathbb{R}$.

2.51 Let $A, B \in \mathcal{M}_2(\mathbb{C})$ and let $n \geq 2$ be a fixed integer. Prove that:

(a) If $(AB)^n = O_2$, then $(BA)^n = O_2$;
(b) If $(AB)^n = I_2$, then $(BA)^n = I_2$;
(c) If $AB \neq BA$, find the matrix $C \in \mathcal{M}_2(\mathbb{C})$ for which the following implication holds $(AB)^n = C \Rightarrow (BA)^n = C$.

2.52 Let $A, B \in \text{GL}_2(\mathbb{C})$ and $\alpha, \beta \in \mathbb{C}$ with $|\alpha| \neq |\beta|$ such that $\alpha AB + \beta BA = I_2$. Prove that $\det(AB - BA) = 0$.

2.53 Let $A, B, C \in \mathcal{M}_2(\mathbb{R})$ be matrices which commute one another with $\det C = 0$. Prove that $\det(A^2 + B^2 + C^2) \geq 0$.

2.54 Let $A_0, A_1, \ldots, A_n \in \mathcal{M}_2(\mathbb{R})$, $n \geq 2$, be nonzero matrices which verify the conditions $A_0 \neq aI_2$, $\forall a \in \mathbb{R}$ and $A_0 A_k = A_k A_0$, $\forall k = 1, 2, \ldots, n$. Prove that:

(a) $\det\left(\sum\limits_{k=1}^{n} A_k^2\right) \geq 0$;

(b) If $\det\left(\sum\limits_{k=1}^{n} A_k^2\right) = 0$ and $A_2 \neq aA_1$, for all $a \in \mathbb{R}$, then $\sum\limits_{k=1}^{n} A_k^2 = O_2$.

2.55 Let $a \in (-1, 1)$ and let $A \in \mathcal{M}_2(\mathbb{R})$ be such that $\det(A^4 - aA^3 - aA + I_2) = 0$. Prove that $\det A = 1$.

2.56 Let $B \in \mathcal{M}_2(\mathbb{C})$ be a nilpotent matrix. Prove that if $A \in \mathcal{M}_2(\mathbb{C})$ commutes with B, then $\det(A + B) = \det A$.

2.57 Let $A, B \in \mathcal{M}_2(\mathbb{C})$ be such that $AB = BA$. Prove that if there exist integers $m, n \in \mathbb{N}$ such that $A^m = O_2$ and $B^n = O_2$, then $AB = O_2$. The problem states that *if two nilpotent matrices commute their product is zero.*

2.58 *When is the sum (difference) of two nilpotent matrices a nilpotent matrix?*

Let $A, B \in \mathcal{M}_2(\mathbb{C})$ be two nonzero nilpotent matrices. Prove that $A + B$ is a nilpotent matrix if and only if both AB and BA are nilpotent matrices.

2.59 Let $A, B \in \mathcal{M}_2(\mathbb{C})$. Prove that $\mathrm{Tr}((AB)^2) = \mathrm{Tr}(A^2B^2) \Leftrightarrow (AB - BA)^2 = O_2$.

2.60 *When is the matrix $AB - BA$ nilpotent?*

(a) [28] If $A, B \in \mathcal{M}_2(\mathbb{C})$ are such that $2015AB - 2016BA = 2017I_2$, then $(AB - BA)^2 = O_2$.

(b) More generally, let $m, n, p \in \mathbb{R}$, $m \neq n$ and let $A, B \in \mathcal{M}_2(\mathbb{C})$ such that $mAB - nBA = pI_2$. Prove that $(AB - BA)^2 = O_2$.

2.61 Let $A, B \in \mathcal{M}_2(\mathbb{C})$. Prove that $(AB)^2 = AB^2A \Leftrightarrow (BA)^2 = BA^2B$.

2.62 Let $A, B \in \mathcal{M}_2(\mathbb{C})$. Prove that

$$\det(A - B)\det(A + B) = \det(A^2 - B^2) \quad \Leftrightarrow \quad (AB - BA)^2 = O_2.$$

2.63 Let $A, B \in \mathcal{M}_2(\mathbb{R})$. Prove that any two of the following statements imply the third one:

(a) $\det(A^2 + B^2) = 0$;
(b) $\det(AB - BA) = 0$;
(c) $\det A = \det B = \frac{1}{2}\det(A + B)$.

2.64 If $A, B \in \mathcal{M}_2(\mathbb{R})$ such that $\det(AB + BA) = \det(AB - BA)$, then $\det(A^2 + B^2) \geq 0$.

2.65 If $A, B \in \mathcal{M}_2(\mathbb{C})$ and $n \in \mathbb{N}$, then $\det(A^n + B^n \pm AB) = \det(A^n + B^n \pm BA)$.

2.66 If $A, B \in \mathcal{M}_2(\mathbb{C})$ and $A^2 + B^2 = AB$, then $(AB - BA)^2 = O_2$.

2.67 If $A, B \in \mathcal{M}_2(\mathbb{C})$ and $A^2 = O_2$, then $\det(AB - BA) = 0 \Leftrightarrow \det(A + B) = \det B$.

2.68 Let $A \in \mathcal{M}_2(\mathbb{R})$ such that $A^2 = O_2$. Prove that $\forall B \in \mathcal{M}_2(\mathbb{R})$ the following inequalities hold $\det(AB - BA) \leq 0 \leq \det(AB + BA)$.

2.69 Let $A, B \in \mathcal{M}_2(\mathbb{C}) \setminus \{O_2\}$ be such that $AB + BA = O_2$. Prove that if $\det(A - B) = 0$, then $\mathrm{Tr}(A) = \mathrm{Tr}(B) = 0$.

2.70 If $A, B \in \mathcal{M}_2(\mathbb{C})$ such that $\mathrm{Tr}(A)\mathrm{Tr}(B) = \mathrm{Tr}(AB)$, then

$$\det(A^2 + B^2 + AB) = \det(A^2 + B^2) + \det(AB).$$

2.71 The centralizer of a nilpotent matrix.

Let $A \in \mathcal{M}_2(\mathbb{C})$ and let $\mathscr{C}(A) = \{X \in \mathcal{M}_2(\mathbb{C}) : AX = XA\}$. Prove that

$$A^2 = O_2 \quad \Leftrightarrow \quad |\det(A + X)| \geq |\det X|, \quad \forall X \in \mathscr{C}(A).$$

2.72 (a) If $A, B \in \mathcal{M}_2(\mathbb{R})$ are matrices such that $(A - B)^{-1} = A^{-1} - B^{-1}$, then $\det A = \det B = \det(A - B)$.
  (b) Does the result hold if $A, B \in \mathcal{M}_2(\mathbb{C})$?

2.73 Let P be a polynomial function with real coefficients which does not have real roots and let $A \in \mathcal{M}_2(\mathbb{R})$ be such that $\det P(A) = 0$. Prove that $P(A) = O_2$.

2.74 [58, p. 145] Let $A, B \in \mathcal{M}_2(\mathbb{R})$ be matrices such that $A^2 = B^2 = I_2$ and $AB + BA = O_2$. Prove that there exists an invertible matrix $Q \in \mathcal{M}_2(\mathbb{R})$ such that

$$Q^{-1}AQ = \begin{pmatrix} 1 & 0 \\ 0 & -1 \end{pmatrix} \quad \text{and} \quad Q^{-1}BQ = \begin{pmatrix} 0 & 1 \\ 1 & 0 \end{pmatrix}.$$

2.75 Let $A, B \in \mathcal{M}_2(\mathbb{C})$ be such that $AB = O_2$. Prove that

$$\det(A + B)^n = \det(A^n + B^n), \quad \forall n \geq 1.$$

2.76 (a) Prove that there exist matrices $A, B \in \mathcal{M}_2(\mathbb{R})$ such that

$$\det(xA + yB) = x^2 + y^2, \quad \forall\, x, y \in \mathbb{R}.$$

(b) Prove that there do not exist matrices $A, B, C \in \mathcal{M}_2(\mathbb{R})$ such that

$$\det(xA + yB + zC) = x^2 + y^2 + z^2, \quad \forall\, x, y, z \in \mathbb{R}.$$

2.7 Solutions

2.1. Let $f_A(x) = \det(A - xI_2) = x^2 - tx + 1$, where $t = \text{Tr}(A)$. We have, based on Theorem 2.2, that $A^2 = tA - I_2$ and this implies that $A^2 + A - I_2 = (t + 1)A - 2I_2$ and $A^2 + I_2 = tA$. A calculation shows that

$$\det(A^2 + A - I_2) + \det(A^2 + I_2) = (t + 1)^2 f_A\left(\frac{2}{t + 1}\right) + t^2$$

$$= (t + 1)^2\left[\left(\frac{2}{t + 1}\right)^2 - \frac{2t}{t + 1} + 1\right] + t^2$$

$$= 5.$$

2.2. $\text{Tr}(A) = 3$. See the solution of problem **2.1**.

2.3. Let $f_A(x) = \det(A - xI_2) = x^2 + x + d$, where $d = \det A$. Theorem 2.2 shows that $A^2 = -A - dI_2$ and this implies $A^2 + 3A + 3I_2 = 2A - (d - 3)I_2$ and $A^2 + A = -dI_2$. It follows that

$$\det(A^2 + 3A + 3I_2) - \det(A^2 + A) = 4f_A\left(\frac{d - 3}{2}\right) - d^2$$

$$= 4\left[\left(\frac{d - 3}{2}\right)^2 + \frac{d - 3}{2} + d\right] - d^2$$

$$= 3.$$

2.4. We have, based on Corollary 2.4, that $\det(aA + (a + 1)I_2) = a^2 \det A + a(a + 1)\alpha + (a + 1)^2$, for some $\alpha \in \mathbb{Z}$. If $\det(aA + (a + 1)I_2) = 0$, then $a^2 \det A + a(a + 1)\alpha + (a + 1)^2 = 0 \Rightarrow a \,|\, (a + 1)^2$. It follows that $a^2 + 2a + 1 = ab$, for some $b \in \mathbb{Z}$. However, the last equality implies that $a \,|\, 1$, which contradicts $a \neq \pm 1$.

2.5. Let $\lambda \notin \text{Spec}(A)$, $\lambda \neq 0$ and let $B = A - \lambda I_2$ and $C = \lambda I_2$.

2.6. Since $\det A = 0$ we get that $0 \in \text{Spec}(A)$. Let λ_n be a sequence of real or complex numbers such that $\lim_{n \to \infty} \lambda_n = 0$ and $\lambda_n \notin \text{Spec}(A)$. Let $A_n = A - \lambda_n I_2$. We have $\lim_{n \to \infty} A_n = A$ and $\det(A_n) = \det(A - \lambda_n I_2) \neq 0$, since $\lambda_n \notin \text{Spec}(A)$.

2.7. We have

$$f_{A_*}(x) = \det(A_* - xI_2) = \begin{vmatrix} d - x & -b \\ -c & a - x \end{vmatrix} = x^2 - (a + d)x + ad - bc = f_A(x).$$

2.9. Let J_A be the Jordan canonical form of A and let P be the invertible matrix such that $A = P J_A P^{-1}$. If $J_A = I_2$ there is nothing to prove. If $J_A = \begin{pmatrix} 1 & 1 \\ 0 & 1 \end{pmatrix}$, then

$$A^k = P J_A^k P^{-1} = P \begin{pmatrix} 1 & k \\ 0 & 1 \end{pmatrix} P^{-1}. \text{ A calculation shows that } \begin{pmatrix} 1 & k \\ 0 & 1 \end{pmatrix} = J_A^k = Q^{-1} J_A Q,$$

where $Q = \begin{pmatrix} 1 & 0 \\ 0 & k \end{pmatrix}$. This implies that

$$A^k = P Q^{-1} J_A Q P^{-1} = \left(P Q P^{-1} \right)^{-1} P J_A P^{-1} \left(P Q P^{-1} \right) = \left(P Q P^{-1} \right)^{-1} A \left(P Q P^{-1} \right),$$

which implies that $A^k \sim A$.

2.10. $A \sim B \Rightarrow \exists P \in \text{GL}_2(\mathbb{C})$ such that $B = P^{-1} A P$. This implies that $B^n = P^{-1} A^n P$, which shows that $A^n \sim B^n$.

The reverse implication does not hold. Let $n = 2$ and let $A = \begin{pmatrix} 0 & 0 \\ 1 & 0 \end{pmatrix}$ and $B = O_2$. Then, $A^2 = B^2 = O_2$ so $A^2 \sim B^2$. However, A and B are not similar matrices.

2.12. (a) It suffices to solve the problem for Jordan canonical forms. If $J_A = \begin{pmatrix} \lambda_1 & 0 \\ 0 & \lambda_2 \end{pmatrix}$ there is nothing to prove. If $J_A = \begin{pmatrix} \lambda & 1 \\ 0 & \lambda \end{pmatrix}$, then we let $P = \begin{pmatrix} 0 & 1 \\ 1 & 0 \end{pmatrix}$ and we have that $P^{-1} J_A P = J_A^T$.

(b) It suffices to solve the problem for Jordan canonical forms. If $J_A = \begin{pmatrix} \lambda_1 & 0 \\ 0 & \lambda_2 \end{pmatrix}$ there is nothing to prove. If $J_A = \begin{pmatrix} \lambda & 1 \\ 0 & \lambda \end{pmatrix}$, then we need to find an invertible matrix $Q \in \mathcal{M}_2(\mathbb{C})$ and a symmetric matrix $B \in \mathcal{M}_2(\mathbb{C})$ such that $Q^{-1} J_A Q = B$. Let $Q = \begin{pmatrix} a & b \\ c & d \end{pmatrix}$ and let $\Delta = ad - bc \neq 0$. A calculation shows that

$$B = \frac{1}{\Delta} \begin{pmatrix} cd + \lambda \Delta & d^2 \\ -c^2 & cd + \lambda \Delta \end{pmatrix},$$

and since B is a symmetric matrix we get that $c^2 + d^2 = 0$. Observe that B cannot be a real matrix. Let $d = i, c = 1, a = i, b = 0$ and we have that

$$Q = \begin{pmatrix} i & 0 \\ 1 & i \end{pmatrix} \quad \text{and} \quad B = \begin{pmatrix} \lambda - i & 1 \\ 1 & \lambda + i \end{pmatrix}.$$

2.13. Let $A = \begin{pmatrix} a & b \\ c & d \end{pmatrix}$ and $A_* = \begin{pmatrix} d & -b \\ -c & a \end{pmatrix}$. If $P = \begin{pmatrix} 0 & -1 \\ 1 & 0 \end{pmatrix}$, then $PA^T P^{-1} = A_*$.

If Q is another matrix such that $QA^T Q^{-1} = A_*, \forall A \in \mathcal{M}_2(\mathbb{C})$, then $PA^T P^{-1} = QA^T Q^{-1} \Rightarrow (Q^{-1}P)A^T = A^T(Q^{-1}P), \forall A \in \mathcal{M}_2(\mathbb{C})$. This implies that $Q^{-1}P$ commutes with all matrices in $\mathcal{M}_2(\mathbb{C})$. It follows, based on Theorem 1.1, that $Q^{-1}P = \alpha I_2$, for some $\alpha \in \mathbb{C}$. This implies that $Q = \begin{pmatrix} 0 & -\beta \\ \beta & 0 \end{pmatrix}$, where $\beta \in \mathbb{C}^*$.

2.14. Let J_A be the Jordan canonical form of A and let P be the invertible matrix such that $A = PJ_A P^{-1}$. If $J_A = \begin{pmatrix} \lambda_1 & 0 \\ 0 & \lambda_2 \end{pmatrix}$, then $J_A = BC$, where

$$B = \begin{pmatrix} \lambda_1 & 0 \\ 0 & 1 \end{pmatrix} \quad \text{and} \quad C = \begin{pmatrix} 1 & 0 \\ 0 & \lambda_2 \end{pmatrix}.$$

We have $A = PJ_A P^{-1} = [PBP^T][(P^{-1})^T CP^{-1}]$, where PBP^T and $(P^{-1})^T CP^{-1}$ are symmetric matrices.

If $J_A = \begin{pmatrix} \lambda & 1 \\ 0 & \lambda \end{pmatrix}$, then $J_A = BC$, where

$$B = \begin{pmatrix} 1 & 1 \\ 1 & 0 \end{pmatrix} \quad \text{and} \quad C = \begin{pmatrix} 0 & \lambda \\ \lambda & 1 - \lambda \end{pmatrix}.$$

We have $A = PJ_A P^{-1} = [PBP^T][(P^{-1})^T CP^{-1}]$, where PBP^T and $(P^{-1})^T CP^{-1}$ are symmetric matrices.

2.16. Let $A, B \in \mathcal{M}$. If λ_A is an eigenvalue of A and λ_B is an eigenvalue of B, then since $AB = BA$ we get, based on Theorem 2.1, that $\lambda_{AB} = \lambda_A \lambda_B$. It follows that $|\lambda_{AB}| = |\lambda_A||\lambda_B| \leq 1$.

2.17. The eigenvalues of A and B are 7 and 5. We have, based on Theorem 3.1, that $A^n = 7^n D + 5^n C$, where $D = \frac{A - 5I_2}{2}$ and $C = \frac{7I_2 - A}{2}$ and $B^n = 7^n U + 5^n V$, where $U = \frac{B - 5I_2}{2}$ and $V = \frac{7I_2 - B}{2}$. It follows that $A^n - B^n = \frac{1}{2}(7^n - 5^n)(A - B)$.

2.18. The characteristic polynomial of the matrix BA is $f_{BA}(x) = f_{AB}(x) = x^2 - 8x + 1$. It follows, based on Theorem 2.2, that $(BA)^2 - 8BA + I_2 = O_2$. We multiply this identity by $(BA)^{-1} = A^{-1}B^{-1}$ and we get that $BA + A^{-1}B^{-1} = 8I_2$.

2.20. We have that $\text{Tr}(AB - BA) = 0$ and $\det^n(AB - BA) = 1 \Rightarrow \det(AB - BA) = \pm 1$. We apply the Cayley–Hamilton Theorem for the matrix $AB - BA$ and we get that $(AB - BA)^2 = \pm I_2$ which implies that $(AB - BA)^4 = I_2$.

By contradiction, we assume that $n = 2k + 1$. We have, since $(AB - BA)^n = I_2$, that $\det(AB - BA) = -1$ and it follows based on the Cayley–Hamilton Theorem that $(AB - BA)^2 = -I_2$. Thus, $I_2 = (AB - BA)^{2k+1} = (-I_2)^k(AB - BA)$ and this implies that $AB - BA = (-1)^k I_2$. This contradicts the fact that $\text{Tr}(AB - BA) = 0$.

2.21. Let $f \in \mathbb{C}[x]$ be the characteristic polynomial of matrices AB and BA. Dividing the polynomial x^n by f we get that there exist $Q \in \mathbb{C}[x]$ and $a, b \in \mathbb{C}$ such that $x^n = f(x)Q(x) + ax + b$. Replacing x by AB and BA we get that $(AB)^n = aAB + bI_2$ and $(BA)^n = aBA + bI_2$. Since $(AB)^n = (BA)^n$ and $AB \neq BA$ we get that $a = 0$ and this in turn implies that $(AB)^n = (BA)^n = bI_2$.

2.22. (a) Observe that $A^2 - A + I_2 = (A - \epsilon I_2)(A - \bar{\epsilon} I_2)$, where $\epsilon^2 - \epsilon + 1 = 0$, $\epsilon \in \mathbb{C} \setminus \mathbb{R}$. Since $\det(A^2 - A + I_2) = 0$ we have that either $\det(A - \epsilon I_2) = 0$ or $\det(A - \bar{\epsilon} I_2) = 0$. Let $f(x) = \det(A - xI_2) \in \mathbb{R}[x]$ be the characteristic polynomial of A. If ϵ or $\bar{\epsilon}$ is a root of f, then since f has real coefficients we get that both ϵ and $\bar{\epsilon}$ are roots of f. This implies that $f(x) = (x - \epsilon)(x - \bar{\epsilon}) = x^2 - x + 1$ and the Cayley–Hamilton Theorem implies that $A^2 - A + I_2 = O_2$.

(b) We have $\det(A^2 + \alpha A + \beta I_2) = \det[(\alpha + 1)A + (\beta - 1)I_2]$. We distinguish between the cases when $\alpha = -1$ and $\alpha \neq -1$.

If $\alpha = -1$, we have, based on part (a), that $A^2 - A + \beta I_2 = (\beta - 1)I_2$ and this implies that $\det(A^2 - A + \beta I_2) = (\beta - 1)^2$.

If $\alpha \neq -1$, we have that

$$\det(A^2 + \alpha A + \beta I_2) = (\alpha + 1)^2 \det\left(A - \frac{1 - \beta}{\alpha + 1}I_2\right)$$

$$= (\alpha + 1)^2 f\left(\frac{1 - \beta}{\alpha + 1}\right)$$

$$= (\alpha + 1)^2 \left[\left(\frac{1 - \beta}{\alpha + 1}\right)^2 - \frac{1 - \beta}{\alpha + 1} + 1\right]$$

$$= \alpha^2 + \beta^2 + \alpha\beta + \alpha - \beta + 1.$$

2.23. (a) Let $A = \alpha I_2$, where $\alpha \in \mathbb{R}$. If $\alpha < 0$, then $B = C = \sqrt{-\frac{\alpha}{2}}\begin{pmatrix} 0 & 1 \\ -1 & 0 \end{pmatrix}$. If $\alpha = 0$, then $B = C = O_2$. If $\alpha > 0$, then $B = C = \sqrt{\frac{\alpha}{2}}I_2$.

Now we consider the case when $A \neq \alpha I_2$, for all $\alpha \in \mathbb{R}$. Let $t = \text{Tr}(A)$, $d = \det A$ and we have that $A^2 - tA + dI_2 = O_2$. We determine $\alpha, \beta, \gamma \in \mathbb{R}$ such that $A = (\alpha A + \beta I_2)^2 + \gamma I_2$. A calculation shows that $A = (\alpha^2 t + 2\alpha\beta)A + (\beta^2 + \gamma - \alpha^2 d)I_2$. This implies that $\alpha^2 t + 2\alpha\beta = 1$ and $\beta^2 + \gamma - \alpha^2 d = 0$. We choose $\alpha = 1$, $\beta = \frac{1-t}{2}$ and $\gamma = d - \frac{(1-t)^2}{4}$.

- If $\gamma > 0$, then $B = \alpha A + \beta I_2$ and $C = \sqrt{\gamma}I_2$.
- If $\gamma = 0$, then $B = C = \frac{1}{\sqrt{2}}(\alpha A + \beta I_2)$.

- If $\gamma < 0$, then $B = \alpha A + \beta I_2$ and $C = \sqrt{-\gamma}\begin{pmatrix} 0 & 1 \\ -1 & 0 \end{pmatrix}$.

(b) If $BC = CB$, then $B^2 + C^2 = (B + iC)(B - iC)$ and $\det(B^2 + C^2) = |\det(B + iC)|^2 \geq 0$, so there is no matrix A with $\det A < 0$ such that $A = B^2 + C^2$ with $BC = CB$.

2.24. Observe that $A = \Re(A) + i\Im(A)$ and use Corollary 2.4 with $x = 1$ and $y = i$.

2.25. Using Lemma 2.10 we have that $\det(AB - BA) = \text{Tr}(A^2B^2) - \text{Tr}((AB)^2)$. A calculation based on Theorem 2.2 shows that $\text{Tr}(A^2B^2) - \text{Tr}((AB)^2) = -t_{AB}^2 + t_A t_B t_{AB} - t_A^2 d_B - d_A t_B^2 + 4d_A d_B$, which is a quadratic function in t_{AB}. The discriminant of this function is $\Delta = (t_A^2 - 4d_A)(t_B^2 - 4d_B)$ and its maximum value is $\frac{\Delta}{4} = \frac{1}{4}(t_A^2 - 4d_A)(t_B^2 - 4d_B)$. If $A = \begin{pmatrix} a & b \\ c & d \end{pmatrix} \in \mathcal{M}_2(\mathbb{R})$, then $t_A^2 - 4d_A = (a - d)^2 + 4bc \leq 8$, since $a, b, c, d \in [-1, 1]$. It follows that $\max_{A,B \in \mathcal{M}} \det(AB - BA) = 16$, with equality when $A = \begin{pmatrix} 1 & -1 \\ -1 & 1 \end{pmatrix}$ and $B = \begin{pmatrix} -1 & -1 \\ -1 & 1 \end{pmatrix}$.

2.26. Let $E_{i,j}$, $i, j = 1, 2$ be the matrix having the (i, j) entry equal to 1 and all the other entries equal to 0 and let $X = O_2, E_{1,1}, E_{1,2}, E_{2,1}$ and $E_{2,2}$.

2.27. and 2.28. We have

$$|\det(A + iB)|^2 = \det(A + iB)\det(A - iB) = \det\left[A^2 + B^2 - i(AB - BA)\right].$$

It follows, based on Corollary 2.4 with $x = 1$ and $y = -i$, that

$$\det\left[A^2 + B^2 - i(AB - BA)\right] = \det(A^2 + B^2) - \det(AB - BA)$$
$$- i\left[\det(A^2 + B^2 + AB - BA) - \det(A^2 + B^2) - \det(AB - BA)\right].$$

Since $\det\left[A^2 + B^2 - i(AB - BA)\right] = |\det(A + iB)|^2 \geq 0$ we get that $\det(A^2 + B^2) \geq \det(AB - BA)$ and $\det(A^2 + B^2 + AB - BA) - \det(A^2 + B^2) - \det(AB - BA) = 0$.

2.29. Let $f(x) = \det\left[A^2 + B^2 + x(AB + BA)\right] \in \mathbb{R}[x]$. We have

$$f(1) = \det(A + B)^2 = \det^2(A + B) \geq 0,$$
$$f(-1) = \det(A - B)^2 = \det^2(A - B) \geq 0,$$
$$f(x) = \det(A^2 + B^2) + \alpha x + x^2 \det(AB + BA), \text{ where } \alpha \in \mathbb{R}.$$

If $\det(AB + BA) = 0$, then f is a linear monotonic function (or the constant function) and since $f(0)$ is between $f(-1)$ and $f(1)$ we get that $f(0) \geq 0$.

If $\det(AB + BA) < 0$, then f is a quadratic function which has a maximum. Since $f(-1) \geq 0$ and $f(1) \geq 0$ we get that -1 and 1 are between the roots of the equation $f(x) = 0$ and, since 0 is between -1 and 1, we get that $f(0) \geq 0$.

2.30. We have $0 = \det(A^2 + B^2) = \det(A + iB)(A - iB) = \det(A + iB)\det(A - iB)$ and it follows that either $\det(A + iB) = 0$ or $\det(A - iB) = 0$. Using Corollary 2.4 we get that

$$\det(A \pm iB) = \det A - \det B \pm i\left[\det(A + B) - \det A - \det B\right].$$

We have, since $\det(A \pm iB) = 0$, that $\det A = \det B$ and $\det(A + B) = \det A + \det B$.

2.31. (a) For this part of the problem see the solution of problem **2.30.**

(b) If $\det A \neq 0$, then $A^2 + B^2 = A^2 (I_2 + C^2)$, where $C = A^{-1}B \in \mathcal{M}_2 (\mathbb{R})$. We have $\det(A^2 + B^2) = 0 \Leftrightarrow \det(I_2 + C^2) = 0 \Leftrightarrow \det(C + iI_2)(C - iI_2) = 0$. Using a technique as in the solution of problem **2.30** we get that $\det C = 1$ and $\det(C + I_2) = \det C + 1$. The last equality implies, since $\det(C + I_2) = \det C + \text{Tr}(C) + 1$, that $\text{Tr}(C) = 0$. The Cayley–Hamilton Theorem applied to matrix C shows that $C^2 + I_2 = O_2$ which implies $A^2 + B^2 = A^2(I_2 + C^2) = O_2$.

2.32. Let $\alpha = \frac{3+i\sqrt{7}}{4}$ and observe that $2A^2 - 3AB + 2B^2 = 2(A - \alpha B)(A - \overline{\alpha}B)$. We have $0 = \det(2A^2 - 3AB + 2B^2) = 4|\det(A - \alpha B)|^2$ which implies $\det(A - \alpha B) = 0$. We have, based on Corollary 2.4, that

$$\det(A - \alpha B) = \det A + \alpha^2 \det B - \alpha \left[\det(A + B) - \det A - \det B\right]$$

and it follows, since $\alpha^2 = \frac{3}{2}\alpha - 1$, that

$$\det(A - \alpha B) = \det A - \det B + \alpha \left[\det A + \frac{5}{2}\det B - \det(A + B)\right] = 0.$$

Since $\alpha \notin \mathbb{R}$ we have that $\det A + \frac{5}{2}\det B - \det(A + B) = 0$ and $\det A = \det B$. However, this implies that $\det(A + B) = \frac{7}{2}\det A$.

Remark 2.6 If $A, B \in \mathcal{M}_2 (\mathbb{R})$ are commuting matrices with $\det A \neq 0$ or $\det B \neq 0$ and $\det(2A^2 - 3AB + 2B^2) = 0$, then $2A^2 - 3AB + 2B^2 = O_2$ (for a proof see the solution of problem **2.31**).

2.33. $\det(A^2 - AB + B^2) = 124$.

2.34. Let $f(x, y) = \det(xAB + yBA + cI_2)$ and $g(x, y) = \det(xBA + yAB + cI_2)$. We note that both f and g are polynomials of degree less than or equal to 2 in variables x and y of the following form

$$f(x, y) = a_{11}x^2 + a_{12}xy + a_{22}y^2 + a_1x + a_2y + a_3$$
$$g(x, y) = b_{11}x^2 + b_{12}xy + b_{22}y^2 + b_1x + b_2y + b_3.$$

Since $f(x, y) = g(y, x)$ we get that $a_{11} = b_{22}, a_{12} = b_{12}, a_{22} = b_{11}, a_1 = b_2$, $a_2 = b_1$, and $a_3 = b_3$. It follows that

$$f(x, y) = a_{11}x^2 + a_{12}xy + a_{22}y^2 + a_1x + a_2y + a_3$$
$$g(x, y) = a_{22}x^2 + a_{12}xy + a_{11}y^2 + a_2x + a_1y + a_3.$$

We have

$$f(x, 0) = \det(xAB + cI_2) = x^2 \det(AB) + cx\text{Tr}(AB) + c^2$$
$$g(x, 0) = \det(xBA + cI_2) = x^2 \det(BA) + cx\text{Tr}(BA) + c^2,$$

so $f(x, 0) = g(x, 0)$, $\forall x \in \mathbb{C}$ and it follows that $a_{11} = a_{22}$ and $a_1 = a_2$. Thus,

$$f(x, y) = g(x, y) = a_{11}(x^2 + y^2) + a_{12}xy + a_1(x + y) + a_3, \quad \forall x, y \in \mathbb{C}.$$

2.35. Let A_1, A_2 be the columns of A and let X_1, X_2 be the columns of X, i.e., $A = (A_1 \mid A_2)$ and $X = (X_1 \mid X_2)$. A calculation shows that

$$
\begin{aligned}
f_A(X) &= \det(A_1 + X_1 \mid A_2 + X_2) - \det(X_1 - A_1 \mid X_2 - A_2) \\
&= \det(A_1 \mid A_2) + \det(A_1 \mid X_2) + \det(X_1 \mid A_2) + \det(X_1 \mid X_2) \\
&\quad - \det(X_1 \mid X_2) + \det(X_1 \mid A_2) + \det(A_1 \mid X_2) - \det(A_1 \mid A_2) \\
&= 2\left[\det(A_1 \mid X_2) + \det(X_1 \mid A_2)\right].
\end{aligned}
$$

(a) We have

$$
\begin{aligned}
f_{aA}(X) &= 2\left[\det(aA_1 \mid X_2) + \det(X_1 \mid aA_2)\right] \\
&= 2a\left[\det(A_1 \mid X_2) + \det(X_1 \mid A_2)\right] \\
&= af_A(X).
\end{aligned}
$$

(b) We have

$$
\begin{aligned}
f_{A+B}(X) &= 2\left[\det(A_1 + B_1 \mid X_2) + \det(X_1 \mid A_2 + B_2)\right] \\
&= 2\left[\det(A_1 \mid X_2) + \det(X_1 \mid A_2)\right] + 2\left[\det(B_1 \mid X_2) + \det(X_1 \mid B_2)\right] \\
&= f_A(X) + f_B(X).
\end{aligned}
$$

(c) We have based on Theorem 3.2 that there exist sequences $(x_n)_{n \geq 1}$ and $(y_n)_{n \geq 1}$ such that $A^n = x_n A + y_n I_2$, $\forall n \geq 1$. It follows, based on parts (a) and (b), that

$$
f_{A^n} = f_{x_n A + y_n I_2} = f_{x_n A} + f_{y_n I_2} = x_n f_A + y_n f_{I_2}.
$$

We mention that $f_{I_2}(X) = 2\mathrm{Tr}(X)$.

2.37. We prove that (a) $\Leftrightarrow$ (c) and (b) $\Leftrightarrow$ (c). We have

$$
f_A\begin{pmatrix} x_1 \\ y_1 \end{pmatrix} = f_A\begin{pmatrix} x_2 \\ y_2 \end{pmatrix} \quad \Leftrightarrow \quad A\begin{pmatrix} x_1 - x_2 \\ y_1 - y_2 \end{pmatrix} = \begin{pmatrix} 0 \\ 0 \end{pmatrix},
$$

which is a homogeneous system of two equations and two variables. The system has only the zero solution $x_1 - x_2 = 0$, $y_1 - y_2 = 0$ if and only if $\det A \neq 0$. Thus,

$$
\det A \neq 0 \quad \Leftrightarrow \quad \begin{pmatrix} x_1 \\ y_1 \end{pmatrix} = \begin{pmatrix} x_2 \\ y_2 \end{pmatrix}.
$$

Now we consider the system $A\begin{pmatrix} x \\ y \end{pmatrix} = \begin{pmatrix} u \\ v \end{pmatrix}$, which has a solution for any $u, v \in \mathbb{R}$ if and only if $\det A \neq 0$ in which case $\begin{pmatrix} x \\ y \end{pmatrix} = A^{-1}\begin{pmatrix} u \\ v \end{pmatrix}$.

2.38. (a) See the solution of problem **2.37**.

(b) If $\det A = \pm 1$, then f_A is surjective (for a proof of this implication see the solution of problem **2.37**). Now we prove that if f_A is surjective, then $\det A = \pm 1$. We have, based on the surjectivity of f_A, that there exists $\begin{pmatrix} x_1 \\ y_1 \end{pmatrix} \in \mathcal{M}_{2,1}(\mathbb{Z})$ such that $f_A \begin{pmatrix} x_1 \\ y_1 \end{pmatrix} = \begin{pmatrix} 1 \\ 0 \end{pmatrix}$ and there exists $\begin{pmatrix} x_2 \\ y_2 \end{pmatrix} \in \mathcal{M}_{2,1}(\mathbb{Z})$ such that $f_A \begin{pmatrix} x_2 \\ y_2 \end{pmatrix} = \begin{pmatrix} 0 \\ 1 \end{pmatrix}$. It follows that

$$A \begin{pmatrix} x_1 & x_2 \\ y_1 & y_2 \end{pmatrix} = \begin{pmatrix} 1 & 0 \\ 0 & 1 \end{pmatrix},$$

so A is invertible and $A^{-1} = \begin{pmatrix} x_1 & x_2 \\ y_1 & y_2 \end{pmatrix} \in \mathcal{M}_2(\mathbb{Z})$. Since $A, A^{-1} \in \mathcal{M}_2(\mathbb{Z})$ and $AA^{-1} = I_2$ we have that $\det A \det (A^{-1}) = 1$ which implies $\det A = \pm 1$.

2.39. Let $A = \begin{pmatrix} 0 & 0 \\ 1 & 0 \end{pmatrix}$. Since $A^n = O_2 = O_2^n$ we get that f is not injective.

To prove that f is not surjective we let $B = \begin{pmatrix} 0 & 1 \\ 0 & 0 \end{pmatrix}$ and we prove the equation $X^n = B$ does not have solutions in $\mathcal{M}_2(\mathbb{C})$. If a solution $X \in \mathcal{M}_2(\mathbb{C})$ would exist, then $\det X = 0$ and we have based on the Cayley–Hamilton Theorem that $X^2 - tX = O_2$, where $t = \mathrm{Tr}(X)$. This implies that $X^n = t^{n-1}X \Rightarrow t^{n-1}X = B$. Passing to trace in this equality we get that $t^n = 0 \Rightarrow t = 0 \Rightarrow X^2 = O_2 \Rightarrow X^n = O_2$, which contradicts the fact that $X^n = B$.

2.40. (a) Let $g : \mathbb{R} \to \mathbb{R}$, $g(x) = x^{2016} + x^{2015}$ and let $Y = \begin{pmatrix} y & 0 \\ 0 & 0 \end{pmatrix}$, where $y < g\left(-\frac{2015}{2016}\right)$. The equation $f(X) = Y$ does not have solutions in $\mathcal{M}_2(\mathbb{R})$.

(b) Let $Y = \begin{pmatrix} -1 & 0 \\ 0 & 0 \end{pmatrix}$. The equation $f(X) = Y$ does not have solutions in $\mathcal{M}_2(\mathbb{R})$.

2.41. (a) $AX = X' \Leftrightarrow X = A^{-1}X'$, $A^{-1} = \dfrac{1}{ad - bc} \begin{pmatrix} d & b \\ c & a \end{pmatrix} > 0$. Thus, $A^{-1} > 0$, $X' > 0$ and these imply that $A^{-1}X' > 0$.

(b) Let X_1, X_2 be the columns of X and X_1', X_2' be the columns of X'. We have $X_1' = AX_1$ and $X_2' = AX_2$. If $X_1 = \begin{pmatrix} x \\ y \end{pmatrix}$, then $X_1' = \begin{pmatrix} ax - by \\ -cx + dy \end{pmatrix}$ and the conditions $X > 0$, $X' > 0$ give

$$\begin{cases} ax - by > 0 \\ -cx + dy > 0 \\ x > 0 \\ y > 0. \end{cases}$$

Each of these inequalities are, from a geometrical point of view, semiplans which we need to prove they intersect. The frontier of the first semiplan is a line of slope $m_1 = \frac{a}{b}$ and the frontier of the second semiplan has slope $m_2 = \frac{c}{d}$. These two semiplans intersect in the first quadrant if we have that $m_1 > m_2 \Leftrightarrow ad - bc > 0$, which holds. We can choose $X_1 = X_2$ with x and y a solution of the previous system of inequalities.

2.43. We have, based on the Cayley–Hamilton Theorem for matrices AB and BA, that

$$\begin{cases} (AB)^2 - \text{Tr}(AB)AB + \det(AB)I_2 = O_2 \\ (BA)^2 - \text{Tr}(BA)BA + \det(BA)I_2 = O_2. \end{cases}$$

Since $\text{Tr}(AB) = \text{Tr}(BA)$ and $\det(AB) = \det(BA)$ we get that $\text{Tr}(AB)(AB-BA) = O_2$. This implies, since $\text{Tr}(AB) > 0$, that $AB = BA$.

2.44. $\{I_2, A, A^2\}$.

2.45. Let $n \geq 2$. To prove that $\mathscr{P}_n \subseteq \mathscr{P}_2$ we need to prove that for any $X \in \mathcal{M}_2(\mathbb{C})$ there exists $Y \in \mathcal{M}_2(\mathbb{C})$ such that $X^n = Y^2$. Let J_X be the Jordan canonical form of X, let P be the invertible matrix such that $X = PJ_XP^{-1}$, and let $Y = PY_1P^{-1}$. The equation $X^n = Y^2$ becomes $J_X^n = Y_1^2$. We distinguish the following two cases.

If $J_X = \begin{pmatrix} \lambda_1 & 0 \\ 0 & \lambda_2 \end{pmatrix}$, then $J_X^n = \begin{pmatrix} \lambda_1^n & 0 \\ 0 & \lambda_2^n \end{pmatrix}$ and we choose $Y_1 = \begin{pmatrix} \mu_1 & 0 \\ 0 & \mu_2 \end{pmatrix}$, with $\mu_1, \mu_2 \subset \mathbb{C}$ such that $\mu_1^2 = \lambda_1^n$ and $\mu_2^2 = \lambda_2^n$.

If $J_X = \begin{pmatrix} \lambda & 1 \\ 0 & \lambda \end{pmatrix}$, then $J_X^n = \begin{pmatrix} \lambda^n & n\lambda^{n-1} \\ 0 & \lambda^n \end{pmatrix}$. If $Y_1 = \begin{pmatrix} a & b \\ 0 & a \end{pmatrix}$, then we have $Y_1^2 = \begin{pmatrix} a^2 & 2ab \\ 0 & a^2 \end{pmatrix}$ and we get the equations $a^2 = \lambda^n$ and $2ab = n\lambda^{n-1}$. If $\lambda = 0$ we take $a = b = 0$. If $\lambda \neq 0$, then we take $a \in \mathbb{C}^*$ such that $a^2 = \lambda^n$ and $b = \frac{n\lambda^{n-1}}{2a}$.

To prove the inclusion $\mathscr{P}_2 \subseteq \mathscr{P}_n$ we need to prove that for any $X \in \mathcal{M}_2(\mathbb{C})$ there exists $Y \in \mathcal{M}_2(\mathbb{C})$ such that $X^2 = Y^n$. Exactly as in the proof of the previous inclusion we pass to Jordan canonical form and we need to solve the equation $J_X^2 = Y_1^n$.

If $J_X = \begin{pmatrix} \lambda_1 & 0 \\ 0 & \lambda_2 \end{pmatrix}$ we take $Y_1 = \begin{pmatrix} \mu_1 & 0 \\ 0 & \mu_2 \end{pmatrix}$, with $\mu_1^n = \lambda_1^2$ and $\mu_2^n = \lambda_2^2$.

If $J_X = \begin{pmatrix} \lambda & 1 \\ 0 & \lambda \end{pmatrix}$, then $J_X^2 = \begin{pmatrix} \lambda^2 & 2\lambda \\ 0 & \lambda^2 \end{pmatrix}$ and we take $Y_1 = \begin{pmatrix} a & b \\ 0 & a \end{pmatrix}$, $Y_1^n = \begin{pmatrix} a^n & na^{n-1}b \\ 0 & a^n \end{pmatrix}$. We get the system of equations $a^n = \lambda^2$ and $na^{n-1}b = 2\lambda$.

If $\lambda = 0$ we let $a = b = 0$. If $\lambda \neq 0$ we let $a \in \mathbb{C}^*$ with $a^n = \lambda^2$ and $b = \frac{2\lambda}{na^{n-1}}$.

2.46. (a) We have, based on the Cayley–Hamilton Theorem, that $A^2 - \text{Tr}(A)A + I_2 = O_2$ and it follows that $A + A^{-1} = \text{Tr}(A)I_2$. This implies that $AB + A^{-1}B = \text{Tr}(A)B$ which in turn implies, by passing to trace, that $\text{Tr}(AB + A^{-1}B) = \text{Tr}(AB) + \text{Tr}(A^{-1}B) = \text{Tr}(\text{Tr}(A)B) = \text{Tr}(A)\text{Tr}(B)$.

(b) $B + B^{-1} = \text{Tr}(B)I_2 \Rightarrow BAB + BAB^{-1} = \text{Tr}(B)BA$. This implies, by passing to trace, that $\text{Tr}\left(BAB + BAB^{-1}\right) = \text{Tr}\left(\text{Tr}(B)BA\right) \Rightarrow \text{Tr}(BAB) + \text{Tr}(BAB^{-1}) = \text{Tr}(B)\text{Tr}(BA)$. Since $\text{Tr}(BAB^{-1}) = \text{Tr}(A)$ and $\text{Tr}(BA) = \text{Tr}(AB)$ we obtain that $\text{Tr}(BAB) + \text{Tr}(A) = \text{Tr}(B)\text{Tr}(AB)$.

2.47. (a) By way of contradiction we assume that there exists $n \in \mathbb{N}$ such that $A^n = I_2$. This implies that $\det(A^n) = \det^n A = 1$. If λ_1, λ_2 are the eigenvalues of A, then $\lambda_1 + \lambda_2 = \text{Tr}(A) > 2$, $\lambda_1 \lambda_2 = \det A = 1$ and it follows that $\lambda_1^n \lambda_2^n = 1$ and $\lambda_1^n + \lambda_2^n = \text{Tr}(A^n) = \text{Tr}(I_2) = 2$. This implies that $\lambda_1^n = \lambda_2^n = 1 \Rightarrow |\lambda_1| = |\lambda_2| = 1$. We have $2 = |\lambda_1| + |\lambda_2| \geq |\lambda_1 + \lambda_2| = \text{Tr}(A) > 2$, which is a contradiction.

(b) Let $A = rB$ and observe the problem reduces to part (a).

2.48. Let λ_1, λ_2 be the eigenvalues of A, i.e., the solutions of the equation $x^2 - \text{Tr}(A)x + \det A = 0$. Since $\Delta = \text{Tr}^2(A) - 4\det A < 0$ we get that $\lambda_1, \lambda_2 \in \mathbb{C} \setminus \mathbb{R}$ and $\lambda_2 = \overline{\lambda_1}$. On the other hand, $\lambda_1 \lambda_2 = 1$ which implies that $\lambda_{1,2} = \cos\alpha \pm i\sin\alpha$, with $\sin\alpha \neq 0$. We have $|\text{Tr}(A^n)| = |2\cos(n\alpha)| \leq 2$.

The matrix

$$B_n = \begin{pmatrix} \cos\frac{\pi}{n} & -\sin\frac{\pi}{n} \\ \sin\frac{\pi}{n} & \cos\frac{\pi}{n} \end{pmatrix}$$

verifies the conditions of the problem.

2.49. One implication is trivial. We prove that if C commutes with both A and B, then $C = O_2$. If A or B are of the form αI_2, $\alpha \in \mathbb{C}$, then there is nothing to prove. So we assume that both A and B are not of the form αI_2, $\alpha \in \mathbb{C}$. Since $AC = CA$ and $CB = BC$ we have, based on Theorem 1.1, that $C = \alpha_1 A + \beta_1 I_2$ and $C = \alpha_2 B + \beta_2 I_2$, for some $\alpha_1, \alpha_2, \beta_1, \beta_2 \in \mathbb{C}$. It follows that $\alpha_1 A + \beta_1 I_2 = \alpha_2 B + \beta_2 I_2$.

If $\alpha_1 = 0$ we get that $\beta_1 I_2 = \alpha_2 B + \beta_2 I_2$. If $\alpha_2 \neq 0$ we get that $B = \frac{\beta_1 - \beta_2}{\alpha_2}I_2$ which is impossible. Therefore $\alpha_2 = 0$ and $\beta_1 = \beta_2$. Since $\alpha_1 = 0$ we get that $C = \beta_1 I_2$. Also $\text{Tr}(C) = 0 \Rightarrow \beta_1 = 0 \Rightarrow C = O_2$.

If $\alpha_1 \neq 0$ we get that $A = \frac{\alpha_2}{\alpha_1}B + \frac{\beta_2 - \beta_1}{\alpha_1}I_2 = \delta B + \gamma I_2$, where $\delta = \frac{\alpha_2}{\alpha_1}$ and $\gamma = \frac{\beta_2 - \beta_1}{\alpha_1}$. It follows that $C = AB - BA = (\delta B + \gamma I_2)B - B(\delta B + \gamma I_2) = O_2$.

2.50. Let $f_A(x) = \det(A - xI_2) = x^2 - \text{Tr}(A)x + \det A$ be the characteristic polynomial of A and let λ_1, λ_2 be its roots. Since $A^n = I_2$ we get that $\lambda_1^n = \lambda_2^n = 1 \Rightarrow |\lambda_1| = |\lambda_2| = 1$ and the condition $\det(A - I_2) \in \mathbb{R}$ implies that $f_A(1) = 1 - (\lambda_1 + \lambda_2) + \lambda_1 \lambda_2 \in \mathbb{R}$.

We have

$$\lambda_1 \lambda_2 - (\lambda_1 + \lambda_2) \in \mathbb{R} \quad \Leftrightarrow \quad \overline{\lambda_1} \cdot \overline{\lambda_2} - (\overline{\lambda_1} + \overline{\lambda_2}) \in \mathbb{R}$$

$$\Leftrightarrow \quad \frac{1}{\lambda_1} \cdot \frac{1}{\lambda_2} - \frac{1}{\lambda_1} - \frac{1}{\lambda_2} \in \mathbb{R}$$

$$\Leftrightarrow \quad \frac{1 - \lambda_1 - \lambda_2}{\lambda_1 \lambda_2} \in \mathbb{R}.$$

Let $\lambda_1 + \lambda_2 = \lambda_1\lambda_2 + a$, $a \in \mathbb{R}$ and $\frac{1-\lambda_1\lambda_2-a}{\lambda_1\lambda_2} = b \in \mathbb{R}$. These imply that $\lambda_1\lambda_2 \in \mathbb{R}$ and $\lambda_1 + \lambda_2 \in \mathbb{R}$ and hence $f_A \in \mathbb{R}[x]$.

2.51. (a) We have $(AB)^n = O_2 \Rightarrow (AB)^2 = O_2 \Rightarrow B(AB)^2A = O_2 \Rightarrow (BA)^3 = O_2 \Rightarrow (BA)^2 = O_2 \Rightarrow (BA)^n = O_2$.

(b) Let $f = f_{AB} = f_{BA}$ be the characteristic polynomial of AB and BA. We have $x^n = Q(x)f(x) + ax + b$, where $Q \in \mathbb{R}[x]$ and $a, b \in \mathbb{R}$. Thus, $(AB)^n = I_2 \Leftrightarrow Q(AB)f(AB) + aAB + bI_2 = I_2 \Leftrightarrow aAB + (b-1)I_2 = O_2$.

If $a = 0$, then $b = 1$ and these imply that $(BA)^n = Q(BA)f(BA) + I_2 = I_2$.

If $a \neq 0$, then $AB = \frac{1-b}{a}I_2$, with $\frac{1-b}{a} \neq 0$, which implies, since A and B are invertible, that $AB = BA$.

(c) $(AB)^n = C \Leftrightarrow aAB + bI_2 = C$. On the other hand, $(BA)^n = C \Leftrightarrow aBA + bI_2 = C$ and, since $AB \neq BA$, we get $a = 0$ and $C = bI_2$.

2.52. $\alpha AB + \beta BA = I_2 \Rightarrow \alpha(AB - BA) = I_2 - (\alpha + \beta)BA$ and $\beta(BA - AB) = I_2 - (\alpha + \beta)AB$. Also,

$$\det(I_2 - xBA) = \det(A^{-1}I_2A - A^{-1}xABA)$$

$$= \det(A^{-1})\det(I_2 - xAB)\det A$$

$$= \det(I_2 - xAB).$$

It follows, by passing to determinants in the previous equalities, that

$$\alpha^2 \det(AB - BA) = \beta^2 \det(BA - AB) \quad \Leftrightarrow \quad (\alpha^2 - \beta^2)\det(AB - BA) = 0,$$

which implies, since $\alpha \neq \pm\beta$, that $\det(AB - BA) = 0$.

2.53. Let $f(x) = \det(A^2 + B^2 + C^2 + xBC) \in \mathbb{R}[x]$. We have $f(-2) = \det(A^2 + (B - C)^2) \geq 0$ and $f(2) = \det(A^2 + (B+C)^2) \geq 0$. A calculation, based on Corollary 2.4, shows that $f(x) = \det(A^2 + B^2 + C^2) + \alpha x + x^2 \det B \det C = \det(A^2 + B^2 + C^2) + \alpha x$, for some $\alpha \in \mathbb{R}$. Thus, f is a monotonic function being a polynomial of degree 1 and since 0 is between -2 and 2 we get that $f(0) = \det(A^2 + B^2 + C^2) \geq 0$.

2.55. We have $z^4 - az^3 - az + 1 = (z^2 - \alpha_1 z + 1)(z^2 - \alpha_2 z + 1)$, where $\alpha_{1,2} = \frac{a \pm \sqrt{a^2+8}}{2}$. Since $a \in (-1, 1)$ we get that $|\alpha_{1,2}| < 2$ which implies the equations $z^2 - \alpha_1 z + 1 = 0$ and $z^2 - \alpha_2 z + 1 = 0$ have only complex solutions. Their solutions are z_1, z_2 and z_3, z_4 such that $z_1 z_2 = z_3 z_4 = 1$. This implies, since z_1, z_2 and z_3, z_4 are complex conjugates, that $|z_1| = |z_2| = |z_3| = |z_4| = 1$. We have $\det\left[(A^2 - \alpha_1 A + I_2)(A^2 - \alpha_2 A + I_2)\right] = 0 \Rightarrow \det(A^2 - \alpha_1 A + I_2) = 0$ or $\det(A^2 - \alpha_2 A + I_2) = 0$. If $\det(A^2 - \alpha_1 A + I_2) = 0 \Leftrightarrow \det(A - z_1 I_2)(A - z_2 I_2) = |\det(A - z_1 I_2)|^2 = 0$. This implies $\det(A - z_1 I_2) = 0$. It follows, based on Corollary 2.5, that $0 = \det(A - z_1 I_2) = \det A - \text{Tr}(A)z_1 + z_1^2 = \det A - 1 + z_1(\alpha_1 - \text{Tr}(A))$ and this implies $\det A = 1$.

2.56. Since B is nilpotent we get that its eigenvalues are equal to 0. It follows, based on Theorem 2.1, that $\lambda_{A+B} = \lambda_A + \lambda_B = \lambda_A$ and $\mu_{A+B} = \mu_A + \mu_B = \mu_A$. This implies $\det(A + B) = \lambda_{A+B}\mu_{A+B} = \lambda_A\mu_A = \det A$.

2.57. If $m = 1$ or $n = 1$ the problem is trivial, so we assume that both m and n are greater than or equal to 2. Observe that $A^2 = B^2 = O_2$ and use problem **1.8**.

2.58. First we prove that if AB and BA are nilpotent matrices, then $A+B$ is a nilpotent matrix. We have $(A + B)^2 = A^2 + B^2 + AB + BA = AB + BA$ and $(A + B)^4 = (AB + BA)^2 = (AB)^2 + AB^2A + BA^2B + (BA)^2 = O_2$ and this implies, based on Lemma 2.3, that $(A + B)^2 = O_2$.

Now we prove that if $A + B$ is a nilpotent matrix, then both AB and BA are nilpotent matrices. We have $(A + B)^2 = O_2 \Rightarrow A^2 + AB + BA + B^2 = O_2 \Rightarrow AB = -BA$. This implies that $(AB)^2 = ABAB = A(-AB)B = -A^2B^2 = O_2$ and $(BA)^2 = BABA = B(-BA)A = -B^2A^2 = O_2$.

2.59. We apply the Cayley–Hamilton Theorem for the matrix $AB - BA$ and we get that

$$(AB - BA)^2 - \text{Tr}(AB - BA)(AB - BA) + \det(AB - BA)I_2 = O_2.$$

It follows, since $\text{Tr}(AB - BA) = 0$, that $(AB - BA)^2 + \det(AB - BA)I_2 = O_2$. This implies that $\det(AB - BA) = 0 \Leftrightarrow (AB - BA)^2 = O_2$. We have, based on Lemma 2.10, that $\det(AB - BA) = 0 \Leftrightarrow \text{Tr}(A^2B^2) = \text{Tr}((AB)^2)$.

2.61. We have $(AB)^2 = AB^2A \Rightarrow \text{Tr}((AB)^2) = \text{Tr}(AB^2A) = \text{Tr}(A^2B^2)$ and we get, based on problem **2.59**, that $(AB - BA)^2 = O_2$. This implies that $(AB)^2 - AB^2A + (BA)^2 - BA^2B = O_2$. Since $(AB)^2 = AB^2A$ we get that $(BA)^2 = BA^2B$. The other implication is solved similarly.

2.62. We have, based on Lemma 2.11, that

$$\det(A - B)\det(A + B) = \det(A^2 - B^2) \quad \Leftrightarrow \quad \det(AB - BA) = 0.$$

However, $(AB - BA)^2 = -\det(AB - BA)I_2$ and we have $(AB - BA)^2 = O_2 \Leftrightarrow \det(AB - BA) = 0$.

2.63. Use Lemma 2.12.

2.64. If $\det(AB - BA) \geq 0$ we have, based on Lemma 2.12, that $\det(A^2 + B^2) \geq 0$.

If $\det(AB - BA) \leq 0$, then $\det(AB + BA) \leq 0$. We have, based on Corollary 2.1, that

$$\det((A^2 + B^2) + (AB + BA)) + \det((A^2 + B^2) - (AB + BA))$$
$$= 2\det(A^2 + B^2) + 2\det(AB + BA).$$

However, $\det(A^2 + B^2 + AB + BA) = \det((A + B)^2) = \det^2(A + B)$ and $\det((A^2 + B^2) - (AB + BA)) = \det^2(A - B)$. It follows that $\det^2(A + B) + \det^2(A - B) = 2\det(A^2 + B^2) + 2\det(AB + BA)$. Since $\det(AB + BA) \leq 0$ we have $\det(A^2 + B^2) \geq 0$.

2.65. We have, based on problem **1.32**, that

$$\det(A^n + B^n + AB) = \det(A^n + B^n) + \det(A^n + AB) + \det(B^n + AB)$$
$$- \det(A^n) - \det(B^n) - \det(AB)$$
$$= \det(A^n + B^n) + \det A \det(A^{n-1} + B) + \det(B^{n-1} + A) \det B$$
$$- \det(A^n) - \det(B^n) - \det(BA)$$
$$= \det(A^n + B^n) + \det(A^{n-1} + B) \det A + \det B \det(B^{n-1} + A)$$
$$- \det(A^n) - \det(B^n) - \det(BA)$$
$$= \det(A^n + B^n) + \det(A^n + BA) + \det(B^n + BA)$$
$$- \det(A^n) - \det(B^n) - \det(BA)$$
$$= \det(A^n + B^n + BA).$$

Similarly one can prove that $\det(A^n + B^n - AB) = \det(A^n + B^n - BA)$.

2.66. We have, based on problem **2.65** with $n = 2$, that

$$\det(A^2 + B^2 - AB) = \det(A^2 + B^2 - BA).$$

Since $A^2 + B^2 - AB = O_2$ and $A^2 + B^2 - BA - AB$ BA we get that $\det(AB - BA) = 0 \Leftrightarrow (AB - BA)^2 = O_2$.

2.67. Since $A^2 = O_2$ we get that $\det A = 0$. We have, based on Lemma 2.12, that

$$\det(B^2) = \det(AB - BA) + (\det B)^2 + (\det(A + B) - \det B)^2,$$

which implies that $0 = \det(AB - BA) + (\det(A + B) - \det B)^2$. Now the equivalence to prove follows easily.

2.68. Since $A^2 = O_2$ we get that $\det A = 0$ and $\det(AB) = 0$, $\forall B \in \mathcal{M}_2(\mathbb{R})$. Using Lemma 2.7 we get that

$$\det(AB - BA) = \det(AB) - \text{Tr}(AB)\text{Tr}(BA) + \text{Tr}(AB^2A) + \det(BA)$$
$$= 2\det(AB) - (\text{Tr}(AB))^2 + \text{Tr}(A^2B^2)$$
$$= -(\text{Tr}(AB))^2$$
$$\leq 0.$$

Similarly one can prove that $\det(AB + BA) = (\text{Tr}(AB))^2 \geq 0$.

2.71. Let $A \in \mathcal{M}_2(\mathbb{C})$ with $A^2 = O_2$ and let $X \in \mathcal{M}_2(\mathbb{C})$ such that $AX = XA$. Then, $\det(AX - XA) = 0$ and we have, based on problem **2.67**, that $\det(A + X) = \det X$.

To prove the other implication we let x_1, x_2 be the solutions of the equation $\det(A + xI_2) = x^2 + \mathrm{Tr}(A)x + \det A = 0$. Since $x_1 I_2, x_2 I_2 \in \mathscr{C}(A)$ we get that $0 = |\det(A + x_i I_2)| \geq |x_i|^2$, $i = 1, 2$, which implies that $x_1 = x_2 = 0$. Thus, $\mathrm{Tr}(A) = \det A = 0 \Rightarrow A^2 = O_2$.

2.72. (a) $(A - B)(A^{-1} - B^{-1}) = I_2 \Rightarrow AB^{-1} + BA^{-1} = I_2 \Rightarrow A - B = AB^{-1}A$ and $B - A = BA^{-1}B$. It follows that

$$\frac{(\det B)^2}{\det A} = \det(BA^{-1}B) = \det(B - A) = \det(A - B) = \det(AB^{-1}A) = \frac{(\det A)^2}{\det B}$$

and this implies that $\det A = \det B$ and $\det(A - B) = \det A$.

We give below two matrices $A, B \in \mathscr{M}_2(\mathbb{R})$ which verify the condition of the problem. Let $\alpha, u, x \in \mathbb{R}$ with $\alpha, x \neq 0$ and let

$$A = \begin{pmatrix} x & \dfrac{\alpha x - \alpha^2 - x^2}{u} \\ u & \alpha - x \end{pmatrix} \quad \text{and} \quad B = \begin{pmatrix} x - \alpha & \dfrac{\alpha x - \alpha^2 - x^2}{u} \\ u & -x \end{pmatrix}.$$

Then

$$A^{-1} = \frac{1}{\alpha^2} \begin{pmatrix} \alpha - x & \dfrac{x^2 + \alpha^2 - \alpha x}{u} \\ -u & x \end{pmatrix} \quad \text{and} \quad B^{-1} = \frac{1}{\alpha^2} \begin{pmatrix} -x & \dfrac{x^2 + \alpha^2 - \alpha x}{u} \\ -u & x - \alpha \end{pmatrix}.$$

We have $\det A = \det B = \det(A - B) = \alpha^2$.

(b) The result does not hold. Let $\epsilon = \frac{1 + i\sqrt{3}}{2}$ and let

$$A = \begin{pmatrix} \epsilon & 0 \\ 0 & 1 \end{pmatrix} \quad \text{and} \quad B = \begin{pmatrix} 1 & 0 \\ 0 & \epsilon \end{pmatrix}.$$

Then $A^{-1} - B^{-1} = (A - B)^{-1}$ but $\det A \neq \det B$.

2.73. It suffices to consider that P is a polynomial of degree 2.

2.75. If A or B is invertible, then $B = O_2$ or $A = O_2$, so there is nothing to prove.

Now we assume that neither A nor B are invertible. Since $AB = O_2$ we get that

$$(A + B)^n = A^n + B^n + B\left(A^{n-2} + BA^{n-3} + \cdots + B^{n-2}\right)A = A^n + B^n + C.$$

All the factors which contain AB are equal to O_2. We have, based on problem **1.32**, that

$$\det(A + B)^n = \det(A^n + B^n + C)$$

$$= \det(A^n + B^n) + \det(A^n + C) + \det(B^n + C)$$

$$- \det(A^n) - \det(B^n) - \det C$$

$$= \det(A^n + B^n),$$

since $\det(A^n) = \det(B^n) = \det C = \det(A^n + C) = \det(B^n + C) = 0$.

2.76. (a) Let $A = I_2$ and $B = \begin{pmatrix} 0 & 1 \\ -1 & 0 \end{pmatrix}$. Then $\det(xA + yB) = \begin{vmatrix} x & y \\ -y & x \end{vmatrix} = x^2 + y^2$.

(b) For any matrices $A, B, C \in \mathcal{M}_2(\mathbb{R})$ we can choose the real numbers x_0, y_0, z_0 not all equal to 0 such that the first row of the matrix $x_0 A + y_0 B + z_0 C$ is zero. If the first row of the matrices A, B and C are $[a_1, a_2]$, $[b_1, b_2]$, and $[c_1, c_2]$ respectively, then the system

$$\begin{cases} a_1 x + b_1 y + c_1 z = 0 \\ a_2 x + b_2 y + c_2 z = 0, \end{cases}$$

has the nontrivial solution $(x_0, y_0, z_0) \neq (0, 0, 0)$. Then $\det(x_0 A + y_0 B + z_0 C) = 0$ and $x_0^2 + y_0^2 + z_0^2 \neq 0$.

2.8   Quickies

2.77 Let $A, B \in \mathcal{M}_2(\mathbb{C})$ such that $ABAB = O_2$. Does it follow that $BABA = O_2$?

2.78 Let $A \in \mathcal{M}_2(\mathbb{R})$ be a matrix such that $A^k \neq \lambda I_2, k \in \mathbb{N}$. Prove that if the matrix A^k has its $(1, 2)$ entry equal to 0, then the same property have all the matrices A^n, for all $n \subset \mathbb{N}$.

2.79 Do there exist matrices $A, B \in \mathcal{M}_2(\mathbb{Z})$ such that $\det(A + 2B) = 3$ and $\det(A + 5B) = 7$?

2.80 Let $A, B \in \mathcal{M}_2(\mathbb{Q})$ such that $\det A = 0$ and $\det(A + \sqrt{2}B) = 2$. Prove that $\det B = 1$ and $\det(A + \sqrt{p}B) = p$, for any prime number p.

2.81 Let $A, B \in \mathcal{M}_2(\mathbb{Q})$ such that $\det A = 1$ and $\det(A + \sqrt{3}B) = 4$. Prove that $\det B = 1$ and $\det(A + \sqrt{5}B) = 6$.

2.82 Let $A, B \in \mathcal{M}_2(\mathbb{Q})$ such that $\det(A + \sqrt[3]{3}B) = 3$. Prove that $\det(A + \sqrt{2}B) = 3$.

2.83 Let $A, B \in \mathcal{M}_2(\mathbb{C})$ such that $\mathrm{Tr}(AB) = 0$. Prove that $(AB)^2 = (BA)^2$.

2.84 Let $A, B \in \mathcal{M}_2(\mathbb{R})$ such that $A^2 + B^2 + 2AB = O_2$ and $\det(A^2 - B^2) = 0$. Prove that $\det(\mathrm{Tr}(A)A - \mathrm{Tr}(B)B) = 0$.

2.85 Let $n \in \mathbb{N}$. Prove that if $A \in \mathcal{M}_2(\mathbb{C})$ such that $\mathrm{Tr}(A) = 0$, then $\mathrm{Tr}(A^{2n+1}) = 0$.

2.86 Let $A \in \mathcal{M}_2(\mathbb{C})$ such that $A^k = A^{k+1}$, for some positive integer k. Prove that

$$\mathrm{Tr}(A) = \mathrm{Tr}(A^2) = \mathrm{Tr}(A^3) = \cdots = \mathrm{Tr}(A^n) = \ldots.$$

2.87 Let $A \in \mathcal{M}_2(\mathbb{C})$ such that there exists $n \in \mathbb{N}$ with $\mathrm{Tr}(A^n) = \mathrm{Tr}(A^{n+1}) = 0$. Prove that $A^2 = O_2$.

2.88 $A, B \in \mathcal{M}_2(\mathbb{C})$ have the same characteristic polynomials, and hence the same eigenvalues, if and only if $\text{Tr}(A^k) = \text{Tr}(B^k)$, for all $k = 1, 2$. Deduce that A is nilpotent if and only if $\text{Tr}(A^k) = 0$, for all $k = 1, 2$.

2.89 Jacobson's lemma.

Let $A, B \in \mathcal{M}_2(\mathbb{C})$ and let $C = AB - BA$. Prove that C is nilpotent if it commutes with either A or B.

2.90 *Are the matrices equal?*

Let $A, B \in \mathcal{M}_2(\mathbb{C})$ such that $\text{Tr}(A) = \text{Tr}(B)$ and $\det A = \det B$. Does it follow that $A = B$?

2.91 If $A, B \in \mathcal{M}_2(\mathbb{C})$ such that $\text{Tr}(A) = \text{Tr}(B)$, then $A(A - B)B = B(A - B)A$.

2.92 Let $A \in \mathcal{M}_2(\mathbb{C})$ such that $A^k = O_2$, $k \in \mathbb{N}$. Prove that $(I_2 - A)^{-1} = I_2 + A$.

2.93 For any $A, B \in \mathcal{M}_2(\mathbb{R})$ there exists $\alpha \in \mathbb{R}$ such that $(AB - BA)^2 = \alpha I_2$.

2.94 (a) Find two matrices $A, B \in \mathcal{M}_2(\mathbb{R})$ such that $A^2 + B^2 = \begin{pmatrix} 1 & 2 \\ 2 & 1 \end{pmatrix}$.

(b) Prove that any two matrices that verify the equality $A^2 + B^2 = \begin{pmatrix} 1 & 2 \\ 2 & 1 \end{pmatrix}$ do not commute.

2.9   Solutions

2.77. The answer is yes. Since $ABAB = O_2 \Rightarrow BABABA = O_2 \Rightarrow (BA)^3 = O_2 \Rightarrow (BA)^2 = O_2$.

2.78. We have, based on Theorem 3.2, that there exist sequences $(x_n)_{n \in \mathbb{N}}$ and $(y_n)_{n \in \mathbb{N}}$ such that $A^n = x_n A + y_n I_2$, for all $n \in \mathbb{N}$. When $n = k$ one has that $A^k = x_k A + y_k I_2$ which implies $A = \frac{1}{x_k}(A^k - y_k I_2) = \begin{pmatrix} * & 0 \\ * & * \end{pmatrix}$. Thus, if $n \in \mathbb{N}$ we get $A^n = x_n A + y_n I_2 = \begin{pmatrix} * & 0 \\ * & * \end{pmatrix}$.

2.79. Such matrices do not exist. We have, based on Lemma 2.7, that $\det(A \mid rR) - \det A + \alpha x + x^2 \det B$, where $\alpha \in \mathbb{Z}$. If $k \in \mathbb{Z}$ we have $\det(A + kB) = \det A + mk$, for some $m \in \mathbb{Z}$. Thus $7 = \det(A + 5B) = \det((A + 2B) + 3B) = \det(A + 2B) + 3m' = 3(1 + m')$, for some $m' \in \mathbb{Z}$, which is impossible.

2.80. Let $f(x) = \det(A + xB) = \alpha x + x^2 \det B$, where $\alpha \in \mathbb{Q}$ and $\det B \in \mathbb{Q}$. Since $f(\sqrt{2}) = 2$ we get that $\alpha + \sqrt{2} \det B = \sqrt{2}$ which implies that $\alpha = 0$ and $\det B = 1$. It follows that $\det(A + \sqrt{p}B) = f(\sqrt{p}) = p \det B = p$.

2.81. Let $f(x) = \det(A + xB) = 1 + \alpha x + x^2 \det B \in \mathbb{Q}[x]$, $\alpha \in \mathbb{Q}$. Since $f(\sqrt{3}) = 4$ we get that $1 + \alpha\sqrt{3} + 3 \det B = 4$ which implies, since $\det B \in \mathbb{Q}$, that $\alpha = 0$ and $\det B = 1$. Thus $f(x) = 1 + x^2$ and $\det(A + \sqrt{5}B) = f(\sqrt{5}) = 6$.

2.82. Let $f(x) = \det(A + xB) = \det A + \alpha x + x^2 \det B \in \mathbb{Q}[x]$, $\alpha \in \mathbb{Q}$. Since $f(\sqrt[3]{3}) = 3$ we get that $\det A + \alpha\sqrt[3]{3} + \sqrt[3]{9} \det B = 3$. It follows that $\det A = 3$ and $\alpha = \det B = 0$. Therefore $f(x) = 3$ and $\det(A + \sqrt{2}B) = f(\sqrt{2}) = 3$.

2.83. Since $\text{Tr}(AB) = \text{Tr}(BA) = 0$ we get, based on the Cayley–Hamilton Theorem, that $(AB)^2 = -\det(AB)I_2 = -\det(BA)I_2 = (BA)^2$.

2.85. Let λ_1, λ_2 be the eigenvalues of A. The eigenvalues of A^{2n+1} are $\lambda_1^{2n+1}, \lambda_2^{2n+1}$ and we have, since $\text{Tr}(A) = \lambda_1 + \lambda_2 = 0$, that $\text{Tr}(A^{2n+1}) = \lambda_1^{2n+1} + \lambda_2^{2n+1} = -\lambda_2^{2n+1} + \lambda_2^{2n+1} = 0$.

2.87. Let λ_1, λ_2 be the eigenvalues of A. We have $\text{Tr}(A^n) = \lambda_1^n + \lambda_2^n = 0$ and $\text{Tr}(A^{n+1}) = \lambda_1^{n+1} + \lambda_2^{n+1} = 0$. A calculation shows that $\lambda_1 = \lambda_2 = 0$ and Theorem 2.2 implies that $A^2 = O_2$.

2.88. Only one implication needs to be proved. Let λ_1, λ_2 be the eigenvalues of A and μ_1, μ_2 be the eigenvalues of B. An easy calculation shows that the system

$$\begin{cases} \lambda_1 + \lambda_2 = \mu_1 + \mu_2 \\ \lambda_1^2 + \lambda_2^2 = \mu_1^2 + \mu_2^2, \end{cases}$$

implies that $\lambda_1 = \mu_1$ and $\lambda_2 = \mu_2$ or $\lambda_1 = \mu_2$ and $\lambda_2 = \mu_1$. In both cases one has that A and B have the same characteristic polynomials.

To prove the second part of the problem observe that a matrix A is nilpotent if and only if $A^2 = O_2$ if and only if A and O_2 have the same characteristic polynomials.

2.89. We assume that $C = AB - BA$ commutes with A and we prove that C is nilpotent. We have $\text{Tr}(C) = \text{Tr}(AB - BA) = 0$ and $\text{Tr}(C^2) = \text{Tr}(C(AB - BA)) = \text{Tr}(CAB) - \text{Tr}(CBA) = \text{Tr}(ACB) - \text{Tr}(CBA) = 0$. It follows, based on problem **2.88**, that C is nilpotent.

2.90. The matrices are not equal. Let $A = \begin{pmatrix} 2 & 3 \\ 4 & 6 \end{pmatrix}$ and $B = \begin{pmatrix} 4 & 16 \\ 1 & 4 \end{pmatrix}$.

2.91. Theorem 2.2 implies that $A^2 = (\text{Tr}(A))A - (\det A)I_2$ and $B^2 = (\text{Tr}(B))B - (\det B)I_2$. We have, since $\text{Tr}(A) = \text{Tr}(B)$, that

$$A(A - B)B = A^2B - AB^2$$
$$= [(\text{Tr}(A))A - (\det A)I_2]B - A[(\text{Tr}(B))B - (\det B)I_2]$$
$$= A \det B - B \det A.$$

Similarly one can prove that $B(A - B)A = A \det B - B \det A$.

2.92. If $k = 1$ we get that $A = O_2$ and we have nothing to prove. Let $k \geq 2$. If λ is an eigenvalue of A we get that $\lambda^k = 0$ which implies that $\lambda = 0$. Thus, all the eigenvalues of A are 0. This in turn implies, based on Theorem 2.2, that $A^2 = O_2$. We have $(I_2 - A)(I_2 + A) = I_2 - A^2 = I_2 \Rightarrow (I_2 - A)^{-1} = I_2 + A$.

2.93. Let $X = AB - BA$. Since $\mathrm{Tr}(X) = 0$ we have, based on Theorem 2.2, that $X^2 = \alpha I_2$, where $\alpha = -\det(AB - BA)$.

2.94. (a) $A = \begin{pmatrix} 1 & 1 \\ 1 & 1 \end{pmatrix}$ and $B = \begin{pmatrix} 0 & 1 \\ -1 & 0 \end{pmatrix}$.

(b) If two such matrices commute, then $\det(A^2 + B^2) = |\det(A + iB)| \geq 0$ which contradicts $\det(A^2 + B^2) = \begin{vmatrix} 1 & 2 \\ 2 & 1 \end{vmatrix} = -3 < 0$.

Chapter 3
Applications of Cayley–Hamilton Theorem

The greatest mathematicians like Archimedes,
Newton, and Gauss have always been able
to combine theory and applications into one.

Felix Klein (1849–1925)

3.1 The nth power of a square matrix of order 2

In this section we prove a theorem which is about calculating the nth power of a matrix A in terms of both the entry values of A and the eigenvalues of A.

Theorem 3.1 *Let $A \in \mathcal{M}_2(\mathbb{C})$ and let λ_1, λ_2 be the eigenvalues of A.*

(a) *If $\lambda_1 \neq \lambda_2$, then for all $n \geq 1$ we have $A^n = \lambda_1^n B + \lambda_2^n C$, where*

$$B = \frac{A - \lambda_2 I_2}{\lambda_1 - \lambda_2} \quad \text{and} \quad C = \frac{A - \lambda_1 I_2}{\lambda_2 - \lambda_1}.$$

(b) *If $\lambda_1 = \lambda_2 = \lambda$, then for all $n \geq 1$ we have $A^n = \lambda^n B + n\lambda^{n-1} C$, where*

$$B = I_2 \quad \text{and} \quad C = A - \lambda I_2.$$

Proof Let $A = \begin{pmatrix} a & b \\ c & d \end{pmatrix}$ and we have $A^2 - \text{Tr}(A)A + (\det A)I_2 = O_2$, where $\text{Tr}(A) = a + d$ and $\det A = ad - bc$. We multiply the preceding identity by A^{n-1} and we get

$$A^{n+1} - \text{Tr}(A)A^n + (\det A)A^{n-1} = O_2$$

which implies that

$$A^{n+1} = \text{Tr}(A)A^n - (\det A)A^{n-1}. \tag{3.1}$$

© Springer International Publishing AG 2017

V. Pop, O. Furdui, *Square Matrices of Order 2*, DOI 10.1007/978-3-319-54939-2_3

Let $A^n = \begin{pmatrix} a_n & b_n \\ c_n & d_n \end{pmatrix}$. Using (3.1) we get that the following recurrence formulae hold:

$$a_{n+1} = \mathrm{Tr}(A)a_n - (\det A)a_{n-1}$$
$$b_{n+1} = \mathrm{Tr}(A)b_n - (\det A)b_{n-1}$$
$$c_{n+1} = \mathrm{Tr}(A)c_n - (\det A)c_{n-1}$$
$$d_{n+1} = \mathrm{Tr}(A)d_n - (\det A)d_{n-1}, \quad n \geq 2.$$

Thus, the sequences $(a_n)_{n\geq 1}$, $(b_n)_{n\geq 1}$, $(c_n)_{n\geq 1}$, and $(d_n)_{n\geq 1}$ verify the same recurrence relation

$$x_{n+1} = \mathrm{Tr}(A)x_n - (\det A)x_{n-1}, \quad n \geq 2$$

which has the characteristic equation $\lambda^2 - \mathrm{Tr}(A)\lambda + \det A = 0$.

We distinguish the following two cases.

- If $\lambda_1 \neq \lambda_2$ we get that $x_n = \alpha_x \lambda_1^n + \beta_x \lambda_2^n$, where $\alpha_x, \beta_x \in \mathbb{C}$. Thus

$$a_n = \alpha_a \lambda_1^n + \beta_a \lambda_2^n$$
$$b_n = \alpha_b \lambda_1^n + \beta_b \lambda_2^n$$
$$c_n = \alpha_c \lambda_1^n + \beta_c \lambda_2^n$$
$$d_n = \alpha_d \lambda_1^n + \beta_d \lambda_2^n.$$

These imply there exist matrices $B, C \in \mathscr{M}_2(\mathbb{C})$,

$$B = \begin{pmatrix} \alpha_a & \alpha_b \\ \alpha_c & \alpha_d \end{pmatrix} \quad \text{and} \quad C = \begin{pmatrix} \beta_a & \beta_b \\ \beta_c & \beta_d \end{pmatrix},$$

such that $A^n = \lambda_1^n B + \lambda_2^n C$.

- If $\lambda_1 = \lambda_2 = \lambda$ we get that $x_n = \alpha_x \lambda^n + \beta_x n\lambda^{n-1}$, where $\alpha_x, \beta_x \in \mathbb{C}$. Thus

$$a_n = \alpha_a \lambda^n + \beta_a n\lambda^{n-1}$$
$$b_n = \alpha_b \lambda^n + \beta_b n\lambda^{n-1}$$
$$c_n = \alpha_c \lambda^n + \beta_c n\lambda^{n-1}$$
$$d_n = \alpha_d \lambda^n + \beta_d n\lambda^{n-1}.$$

These imply there exist matrices $B, C \in \mathscr{M}_2(\mathbb{C})$,

$$B = \begin{pmatrix} \alpha_a & \alpha_b \\ \alpha_c & \alpha_d \end{pmatrix} \quad \text{and} \quad C = \begin{pmatrix} \beta_a & \beta_b \\ \beta_c & \beta_d \end{pmatrix},$$

such that $A^n = \lambda^n B + \lambda^{n-1} n C$.

Matrices B and C are determined by solving a system of linear matrix equations obtained by giving to n the values 0 and 1 and by considering that $A^0 = I_2$.

Remark 3.1 Theorem 3.1 has an equivalent statement.

If $n \in \mathbb{N}$, $A \in \mathcal{M}_2(\mathbb{C})$, and λ_1, λ_2 are the eigenvalues of A, then

$$A^n = \begin{cases} \frac{\lambda_1^n - \lambda_2^n}{\lambda_1 - \lambda_2}A - \det A \frac{\lambda_1^{n-1} - \lambda_2^{n-1}}{\lambda_1 - \lambda_2}I_2 & \text{if } \lambda_1 \neq \lambda_2 \\ n\lambda^{n-1}A - (n-1)\lambda^n I_2 & \text{if } \lambda_1 = \lambda_2 = \lambda. \end{cases}$$

Theorem 3.2 *If $A \in \mathcal{M}_2(\mathbb{C})$ there exist sequences $(x_n)_{n\geq 1}$ and $(y_n)_{n\geq 1}$ such that*

$$A^n = x_n A + y_n I_2, \quad \text{for all} \quad n \in \mathbb{N},$$

where the sequences $(x_n)_{n\geq 1}$ and $(y_n)_{n\geq 1}$ verify the recurrence relations:

$$x_{n+1} = \mathrm{Tr}(A)x_n - (\det A)x_{n-1}, \quad n \in \mathbb{N}$$
$$y_{n+1} = \mathrm{Tr}(A)y_n - (\det A)y_{n-1}, \quad n \in \mathbb{N}.$$

Proof If $A = \alpha I_2$, then $A^n = \alpha^n I_2$, $\mathrm{Tr}(A) = 2\alpha$, $\det A = \alpha^2$ and we take $x_n = \alpha^n$ and $y_n = 0$.

If $A \neq \alpha I_2$, we apply Theorem 1.1 and we have, since $A^n A = AA^n$, that $A^n = x_n A + y_n I_2$, $n \in \mathbb{N}$. From $A^{n+1} = x_{n+1}A + y_{n+1}I_2$ and

$$A^{n+1} = A^n A = x_n A^2 + y_n A = x_n [\mathrm{Tr}(A)A - (\det A)I_2] + y_n A$$

we obtain $x_{n+1} = x_n \mathrm{Tr}(A) + y_n$ and $y_{n+1} = -x_n \det A$. Since $y_n = -x_{n-1}\det A$ we have $x_{n+1} = x_n \mathrm{Tr}(A) - x_{n-1}\det A$ and similarly, we get that the same recurrence relation is verified by the sequence $(y_n)_{n\geq 1}$.

Remark 3.2 The characteristic equation of the sequences $(x_n)_{n\geq 1}$ and $(y_n)_{n\geq 1}$ is the characteristic equation of the matrix A

$$\lambda^2 - \mathrm{Tr}(A)\lambda + \det A = 0$$

having the solutions λ_1, λ_2, the eigenvalues of A.

■ If $\lambda_1 \neq \lambda_2$, a calculation shows that

$$x_n = \frac{\lambda_1^n - \lambda_2^n}{\lambda_1 - \lambda_2} \quad \text{and} \quad y_n = -\det A \frac{\lambda_1^{n-1} - \lambda_2^{n-1}}{\lambda_1 - \lambda_2}, \quad n \geq 1.$$

■ If $\lambda_1 = \lambda_2 = \lambda$, we have that $x_n = n\lambda^{n-1}$ and $y_n = -(n-1)\lambda^n$, $n \geq 1$.

3.1.1   Problems

3.1 Let $A = \begin{pmatrix} -2 & 4 \\ -5 & 7 \end{pmatrix}$. Calculate A^n, $n \in \mathbb{N}$.

3.2 Let $A = \begin{pmatrix} 1 & 3 \\ -3 & -5 \end{pmatrix}$. Calculate A^n, $n \in \mathbb{N}$.

3.3 Let $A = \begin{pmatrix} 1 & 3 \\ -1 & -2 \end{pmatrix}$. Calculate A^n, $n \in \mathbb{N}$.

3.4 Let $A = \begin{pmatrix} 3 & -2 \\ 2 & -1 \end{pmatrix}$. Calculate A^n, $n \in \mathbb{N}$.

3.5 Let $A = \begin{pmatrix} 3 & 1 \\ -1 & 1 \end{pmatrix}$. Calculate A^n, $n \in \mathbb{N}$.

3.6 Let $A = \begin{pmatrix} 1+i & 2-i \\ 2+i & 1-i \end{pmatrix}$. Calculate A^n, $n \in \mathbb{N}$.

3.7 Let $A = \begin{pmatrix} \widehat{4} & \widehat{2} \\ \widehat{2} & \widehat{2} \end{pmatrix} \in \mathcal{M}_2(\mathbb{Z}_5)$. Calculate A^n, $n \in \mathbb{N}$.

3.8 Let $A = \begin{pmatrix} 1 & -1 \\ -1 & 1 \end{pmatrix}$.

(a)   Prove that $A^n = 2^{n-1}A$, $\forall n \in \mathbb{N}$.
(b)   Calculate the sum $A + A^2 + \cdots + A^n$.

3.9 Let $A = \begin{pmatrix} a & 1 \\ 0 & a \end{pmatrix}$, $a \in \mathbb{R}$. Calculate $\det(A + A^2 + \cdots + A^n)$, $n \in \mathbb{N}$.

3.10 Prove that

$$\begin{pmatrix} \sqrt{3} & -1 \\ 1 & \sqrt{3} \end{pmatrix}^{12} = 2^{12} \begin{pmatrix} 1 & 0 \\ 0 & 1 \end{pmatrix}.$$

3.11 An invitation to circulant matrices.

■ A matrix of the form $C(a,b) = \begin{pmatrix} a & b \\ b & a \end{pmatrix}$, $a,b \in \mathbb{C}$, is called a *circulant* matrix. Let $\mathscr{C} = \{C(a,b) : a,b \in \mathscr{C}\}$ be the set of circulant matrices. Then:

(continued)

3.11 (continued)

(a) $C(1,0) = I_2, C(0,1) = \begin{pmatrix} 0 & 1 \\ 1 & 0 \end{pmatrix} = C, C^{2n} = I_2, C^{2n+1} = C, n \in \mathbb{N}$
and $C(a,b) = aI_2 + bC, a, b \in \mathbb{C}$.

(b) $\mathscr{C}$ is closed under addition and multiplication of matrices, i.e., if $A, B \in \mathscr{C}$, then $A + B \in \mathscr{C}, AB \in \mathscr{C}$ and the following formulae hold

$$C(a,b) + C(c,d) = C(a+c, b+d)$$
$$C(a,b)C(c,d) = C(ac + bd, ad + bc).$$

(c) **The nth power of a circulant matrix.** We have

$$C^n(a,b) = C\left(\frac{(a+b)^n + (a-b)^n}{2}, \frac{(a+b)^n - (a-b)^n}{2} \right), \quad n \in \mathbb{N}.$$

(d) The eigenvalues of C are $\lambda_1 = 1$ and $\lambda_2 = -1$ and the eigenvalues of the matrix $C(a,b)$ are $\mu_1 = a+b$ and $\mu_2 = a-b$. The Jordan canonical form of the matrix $C(a,b)$ is given by $J_{C(a,b)} = \begin{pmatrix} a+b & 0 \\ 0 & a-b \end{pmatrix}$ and the invertible matrix P which verifies the equality $J_{C(a,b)} = P^{-1}C(a,b)P$ is given by $P = \begin{pmatrix} 1 & -1 \\ 1 & 1 \end{pmatrix}$.

(e) If $n \in \mathbb{N}$ and A^n is a circulant matrix which is not of the form αI_2, $\alpha \in \mathbb{C}$, then A is also a circulant matrix.

(f) The matrix $C(a,b)$ is invertible if and only if $a^2 \neq b^2$ and

$$C^{-1}(a,b) = C\left(\frac{a}{a^2 - b^2}, -\frac{b}{a^2 - b^2} \right).$$

(g) $(\mathscr{C}, +, \cdot)$ is a commutative ring with unity, a subring of $(\mathscr{M}_2(\mathbb{C}), +, \cdot)$, in which the group of invertible elements $(U(\mathscr{C}), \cdot)$ consists of invertible circulant matrices.

(h) If $X, Y \in \mathscr{M}_2(\mathbb{C})$ are such that $XY = C(a,b), a^2 \neq b^2$, then X commutes with Y if and only if both X and Y are circulant matrices.

(i) The group $(\mathscr{C}, +)$ is a vector space over $\mathbb{C}$ of dimension 2 with canonical base $\mathscr{B}_{\mathscr{C}} = \{I_2, C\}$.

■ Let $a, b \in \mathbb{C}$ and let $D(a,b) = \begin{pmatrix} a & b \\ -b & -a \end{pmatrix} = aD_1 + bD_2$, where $D_1 = \begin{pmatrix} 1 & 0 \\ 0 & -1 \end{pmatrix}, D_2 = \begin{pmatrix} 0 & 1 \\ -1 & 0 \end{pmatrix}$ and let $\mathscr{D} = \{D(a,b) : a, b \in \mathbb{C}\}$.

(continued)

3.11 (continued)

Then:

(j) $D_1^2 = I_2, D_2^2 = -I_2, D_1 D_2 = C, D_2 D_1 = -C, CD_1 = -D_2, D_1 C = D_2, CD_2 = -D_1, D_2 C = D_1$.

(k) $D_1^{2n} = I_2, D_1^{2n+1} = D_1, D_2^{4n} = I_2, D_2^{4n+1} = D_2, D_2^{4n+2} = -I_2, D_2^{4n+3} = -D_2, n \in \mathbb{N}$.

(l) The matrix $D(a, b)$ is invertible if and only if $a^2 \neq b^2$ and

$$D^{-1}(a, b) = D\left(\frac{a}{a^2 - b^2}, \frac{b}{a^2 - b^2}\right).$$

(m) $(\mathscr{D}, +)$ is a vector space over $\mathbb{C}$ of dimension 2 with canonical basis $\mathscr{B}_{\mathscr{D}} = \{D_1, D_2\}$.

(n) If $A, B \in \mathscr{D}$, then $AB \in \mathscr{C}$.

(o) If $A \in \mathscr{C}$ and $B \in \mathscr{D}$, then $AB \in \mathscr{D}$ and $BA \in \mathscr{D}$.

(p) **A direct sum.** $\mathscr{M}_2(\mathbb{C}) = \mathscr{C} \oplus \mathscr{D}$. Any matrix $A = \begin{pmatrix} a & b \\ c & d \end{pmatrix} \in \mathscr{M}_2(\mathbb{C})$ has a unique writing as $A = C(x, y) + D(z, t)$, where $x = \frac{a+d}{2}, y = \frac{b+c}{2}, z = \frac{a-d}{2}$ and $t = \frac{b-c}{2}$.
Nota bene. Any matrix $A \in \mathscr{M}_2(\mathbb{C})$ can be written uniquely as *the sum of a circulant and a zero trace matrix.*

(q) **Orthogonality.** The function $\langle \cdot, \cdot \rangle : \mathscr{M}_2(\mathbb{C}) \times \mathscr{M}_2(\mathbb{C}) \to \mathbb{C}$ defined by $\langle A, B \rangle = \mathrm{Tr}(AB^*)$ is an inner product on $\mathscr{M}_2(\mathbb{C})$ and $(\mathscr{M}_2(\mathbb{C}), \langle \cdot, \cdot \rangle)$ is an *Euclidean space.*
If $C(a, b) \in \mathscr{C}$ and $D(c, d) \in \mathscr{D}$, then $D^*(c, d) = D(\bar{c}, -\bar{d})$ and $\langle C(a, b), D(c, d) \rangle = 0$. Thus, the subspaces $\mathscr{C}$ and $\mathscr{D}$ are *orthogonal.* Property (o) implies that $\mathscr{C}$ is the *orthogonal complement* of $\mathscr{D}$ in $\mathscr{M}_2(\mathbb{C})$, i.e., $\mathscr{C} = \mathscr{D}^\perp$ and $\mathscr{D} = \mathscr{C}^\perp$.
We also have

$$\mathscr{C} = \{C \in \mathscr{M}_2(\mathbb{C}) : \mathrm{Tr}(CD^*) = 0, \ \forall D \in \mathscr{D}\}$$

and

$$\mathscr{D} = \{D \in \mathscr{M}_2(\mathbb{C}) : \mathrm{Tr}(CD^*) = 0, \ \forall C \in \mathscr{C}\}.$$

Nota bene. $\mathscr{B}_{\mathscr{M}_2(\mathbb{C})} = \{I_2, C, D_1, D_2\}$ is an *orthogonal basis* of $\mathscr{M}_2(\mathbb{C})$ with $||I_2|| = ||C|| = ||D_1|| = ||D_2|| = \sqrt{2}$.

3.12 Double stochastic matrices.

(a) Prove that

$$\begin{pmatrix} a & 1-a \\ 1-a & a \end{pmatrix}^n = \frac{1}{2}\begin{pmatrix} 1+(2a-1)^n & 1-(2a-1)^n \\ 1-(2a-1)^n & 1+(2a+1)^n \end{pmatrix}, \quad a \in [0,1].$$

(b) Let $\theta \in \mathbb{R}$. Prove that

$$\begin{pmatrix} \cos^2\theta & \sin^2\theta \\ \sin^2\theta & \cos^2\theta \end{pmatrix}^n = \frac{1}{2}\begin{pmatrix} 1+\cos^n(2\theta) & 1-\cos^n(2\theta) \\ 1-\cos^n(2\theta) & 1+\cos^n(2\theta) \end{pmatrix}.$$

Remark 3.3 The matrices in parts (a) and (b) of problem **3.12** are called *double stochastic* matrices. A double stochastic matrix is a square matrix with nonnegative entries (representing a probability) with each row and column summing to 1.

3.13 Calculate A^n, where

$$A = \begin{pmatrix} 1+a & -a \\ -b & 1+b \end{pmatrix}, \quad a,b \in \mathbb{R}, \; n \in \mathbb{N}.$$

3.14 Let $\alpha \in \mathbb{R}^*$. Calculate

$$\begin{pmatrix} 1 & \alpha^2 \\ 1 & 1 \end{pmatrix}^n, \quad n \in \mathbb{N}.$$

3.15 Let $a \in \mathbb{C}$. Calculate

$$\begin{pmatrix} 2a & -a^2 \\ 1 & 0 \end{pmatrix}^n, \quad n \in \mathbb{N}.$$

3.16 The nth power of an L matrix.

Let $a, b \in \mathbb{R}$ with $a \neq b$ and $ab > 0$. Calculate $\begin{pmatrix} a & b \\ a & a \end{pmatrix}^n$.

3.17 Let $a, b \in \mathbb{R}$ such that $ab > 0$. Prove that

$$\begin{pmatrix} 1 & a \\ b & 1 \end{pmatrix}^n = \begin{pmatrix} \dfrac{(1+\sqrt{ab})^n + (1-\sqrt{ab})^n}{2} & \dfrac{a(1+\sqrt{ab})^n - a(1-\sqrt{ab})^n}{2\sqrt{ab}} \\ \dfrac{b(1+\sqrt{ab})^n - b(1-\sqrt{ab})^n}{2\sqrt{ab}} & \dfrac{(1+\sqrt{ab})^n + (1-\sqrt{ab})^n}{2} \end{pmatrix}.$$

3.18 Let $A = \begin{pmatrix} a & b \\ c & d \end{pmatrix} \in \mathcal{M}_2(\mathbb{R})$ be such that $a \neq d, b \neq c, b \neq 0, c \neq 0$.

If $A^n = \begin{pmatrix} a_n & b_n \\ c_n & d_n \end{pmatrix}, n \in \mathbb{N}$, prove that $\dfrac{b_n}{b} = \dfrac{c_n}{c} = \dfrac{a_n - d_n}{a - d}$.

3.19 Let $A = \begin{pmatrix} a & b \\ -b & a \end{pmatrix} \in \mathcal{M}_2(\mathbb{R})$ such that $a^2 + b^2 < 1$. Prove the matrix A^n, $n \in \mathbb{N}$, is of the following form

$$\begin{pmatrix} a_n & b_n \\ -b_n & a_n \end{pmatrix},$$

where $(a_n)_{n \geq 1}$ and $(b_n)_{n \geq 1}$ are sequences which converge to 0.

3.20 A *Pythagorean triple*[1] consists of three positive integers a, b and c such that $a^2 + b^2 = c^2$. Let (a, b, c) be a Pythagorean triple and let

$$A = \begin{pmatrix} a & -b \\ b & a \end{pmatrix} \in \mathcal{M}_2(\mathbb{Z}).$$

Let $(a_n)_{n \geq 1}$ and $(b_n)_{n \geq 1}$ be the sequences defined by $A^n = \begin{pmatrix} a_n & -b_n \\ b_n & a_n \end{pmatrix}, n \geq 1$.
Prove that $b_n \neq 0$, for all $n \geq 1$.

3.21 For $a \in \mathbb{R}$, we let $X_a = \begin{pmatrix} a & 1 \\ -1 & a \end{pmatrix}$ and let

$$X_a^n = \begin{pmatrix} a_n & b_n \\ -b_n & a_n \end{pmatrix}, \quad n \geq 1.$$

Prove there exists $a \in \mathbb{R}$ such that

$$b_1 < a_1, \quad b_2 < a_2, \quad b_3 < a_3, \quad \ldots, \quad b_{2016} < a_{2016} \text{ and } b_{2017} > a_{2017}.$$

3.22 Let a, b, c, d be real numbers in arithmetic progression. If

$$A = \begin{pmatrix} a & b \\ c & d \end{pmatrix} \quad \text{and} \quad A^n = \begin{pmatrix} a_n & b_n \\ c_n & d_n \end{pmatrix}, \quad n \in \mathbb{N},$$

prove the real numbers $b_n - a_n$, $c_n - d_n$, and $d_n - c_n$ are in arithmetic progression.

---

[1] A fundamental formula for generating Pythagorean triples given an arbitrary pair of positive integers m and n with $m > n$ is Euclid's formula. The formula states that the integers $a = m^2 - n^2$, $b = 2mn$, and $c = m^2 + n^2$ form a Pythagorean triple [15, p. 165].

3.23 Let $A \in \mathcal{M}_2(\mathbb{Z})$ be an invertible matrix such that $A^{-1} \in \mathcal{M}_2(\mathbb{Z})$. Let

$$A^n = \begin{pmatrix} a_n & b_n \\ c_n & d_n \end{pmatrix}, \quad n \geq 1.$$

Prove that $(a_n, b_n) = (a_n, c_n) = (b_n, d_n) = (c_n, d_n) = 1$, where (x, y) denotes the *greatest common divisor* of integers x and y.

3.24 Another Fibonacci matrix.

Let $(F_n)_{n \geq 0}$ be the Fibonacci sequence defined by the recurrence relation $F_0 = 0$, $F_1 = 1$ and $F_{n+1} = F_n + F_{n-1}$, $\forall n \geq 1$ and let $A = \begin{pmatrix} 0 & 1 \\ 1 & 1 \end{pmatrix}$.

Prove that:

(a) $A^n = \begin{pmatrix} F_{n-1} & F_n \\ F_n & F_{n+1} \end{pmatrix}$, $\forall n \geq 1$.

(b) *Two properties of the Fibonacci sequence.*

$$\begin{cases} F_{n+m-1} = F_n F_m + F_{n-1} F_{m-1}, & \forall\, m, n \geq 1 \\ F_{n-1} F_{n+1} - F_n^2 = (-1)^n, & n \geq 1. \end{cases}$$

(c) *The nth term of the Fibonacci sequence.*

$$F_n = \frac{1}{\sqrt{5}} \left[\left(\frac{1 + \sqrt{5}}{2} \right)^n - \left(\frac{1 - \sqrt{5}}{2} \right)^n \right], \quad \forall\, n \geq 0.$$

3.25 Let $A = \begin{pmatrix} 1 & 1 \\ 1 & 2 \end{pmatrix}$ and let $(F_n)_{n \geq 1}$ be the Fibonacci sequence defined by $F_1 = 1$, $F_2 = 1$ and $F_{n+1} = F_n + F_{n-1}$, $n \geq 2$.

(a) Prove that $A^n = \begin{pmatrix} F_{2n-1} & F_{2n} \\ F_{2n} & F_{2n+1} \end{pmatrix}$, $n \geq 1$.

(b) If the sequences $(x_n)_{n \geq 1}$ and $(y_n)_{n \geq 1}$ verify the recurrence relation $\begin{pmatrix} x_{n+1} \\ y_{n+1} \end{pmatrix} = A \begin{pmatrix} x_n \\ y_n \end{pmatrix}$, $n \geq 1$, and $\begin{pmatrix} x_1 \\ y_1 \end{pmatrix} = \begin{pmatrix} 1 \\ 1 \end{pmatrix}$, prove that $x_{n+1}^2 + x_{n+1} y_{n+1} - y_{n+1}^2 = x_n^2 + x_n y_n - y_n^2$, for all $n \geq 1$.

(c) Prove that if the natural numbers $x, y \in \mathbb{N}$ verify the equation $x^2 + xy - y^2 = 1$, then there exists $n \in \mathbb{N}$ such that $(x, y) = (F_{2n-1}, F_{2n})$.

3.26 Let $A = \begin{pmatrix} 1 & 1 \\ 0 & 1 \end{pmatrix}$. Determine the sequences $(x_n)_{n \geq 1}$ and $(y_n)_{n \geq 1}$ such that $A^n = x_n A + y_n I_2$, $n \in \mathbb{N}$ and calculate $\lim\limits_{n \to \infty} \dfrac{x_n}{y_n}$.

3.27 Let $A = \begin{pmatrix} a & b \\ c & d \end{pmatrix} \in \mathcal{M}_2(\mathbb{R})$ be a matrix such that $|\det A| \geq 1$ and let $A^n = \begin{pmatrix} a_n & b_n \\ c_n & d_n \end{pmatrix}$, $n \geq 1$. Prove the sequences $(a_n)_{n \geq 1}$, $(b_n)_{n \geq 1}$, $(c_n)_{n \geq 1}$, and $(d_n)_{n \geq 1}$ converge if and only if $A = I_2$.

3.1.2 Solutions

3.1. The characteristic equation of A is $\lambda^2 - 5\lambda + 6 = 0$, which implies that $\lambda_1 = 2$ and $\lambda_2 = 3$. It follows that $A^n = 2^n B + 3^n C$, where $B, C \in \mathcal{M}_2(\mathbb{R})$. We determine matrices B and C by letting $n = 0$ and $n = 1$ and it follows that

$$B = \begin{pmatrix} 5 & -4 \\ 5 & -4 \end{pmatrix} \quad \text{and} \quad C = \begin{pmatrix} -4 & 4 \\ -5 & 5 \end{pmatrix}.$$

Thus,

$$A^n = \begin{pmatrix} 5 \cdot 2^n - 4 \cdot 3^n & 4 \cdot 3^n - 4 \cdot 2^n \\ 5 \cdot 2^n - 5 \cdot 3^n & 5 \cdot 3^n - 4 \cdot 2^n \end{pmatrix}, \quad n \in \mathbb{N}.$$

3.2. $A^n = (-2)^{n-1} \begin{pmatrix} 3n-2 & 3n \\ -3n & -3n-2 \end{pmatrix}$, $n \in \mathbb{N}$.

3.3. The characteristic equation of A is $\lambda^2 + \lambda + 1 = 0$, so $A^2 + A + I_2 = O_2$. We multiply this equality by $A - I_2$ and we get that $A^3 = I_2$. Thus, $A^n = I_2$ if $n = 3k$, $A^n = A$ if $n = 3k+1$, and $A^n = A^2$ if $n = 3k+2$.

3.4. $A^n = \begin{pmatrix} 2n+1 & -2n \\ 2n & -2n+1 \end{pmatrix}$, $n \in \mathbb{N}$.

3.5. The characteristic equation of A is $(\lambda - 2)^2 = 0$, so we have $(A - 2I_2)^2 = O_2$.

Let $B = A - 2I_2 = \begin{pmatrix} 1 & 1 \\ -1 & -1 \end{pmatrix}$ and we observe that $B^2 = O_2$. We have, based on the Binomial Theorem, that

$$A^n = (2I_2 + B)^n = 2^n I_2 + n2^{n-1} B = \begin{pmatrix} 2^n + n2^{n-1} & n2^{n-1} \\ -n2^{n-1} & 2^n - n2^{n-1} \end{pmatrix}.$$

3.6. $A = I_2 + B$ where $B = \begin{pmatrix} i & 2-i \\ 2+i & -i \end{pmatrix}$ and $B^2 = 4I_2$. A calculation shows that if $k \geq 1$ is an integer, then $B^{2k-1} = 4^{k-1}B$ and $B^{2k} = 4^k I_2$. We have

$$
A^{2n} = \sum_{k=0}^{2n} \binom{2n}{k} B^k
$$

$$
= \sum_{i=0}^{n} \binom{2n}{2i} B^{2i} + \sum_{i=1}^{n} \binom{2n}{2i-1} B^{2i-1}
$$

$$
= \sum_{i=0}^{n} \binom{2n}{2i} 4^i I_2 + \sum_{i=1}^{n} \binom{2n}{2i-1} 4^{i-1} B
$$

$$
= \frac{3^{2n}+1}{2} I_2 + \frac{3^{2n}-1}{4} B
$$

and $A^{2n-1} = A^{2n-2}A$, $n \geq 1$.

3.7. $A^n = \begin{pmatrix} (2n+1)3^n & \widehat{4n3^n} \\ \widehat{2n3^{n-1}} & (4n+3)3^{n-1} \end{pmatrix}$, $n \in \mathbb{N}$.

3.8. (a) This part of the problem can be solved either by mathematical induction or by direct computation
   (b) We have

$$
A + A^2 + \cdots + A^n = A + 2A + 2^2 A + \cdots + 2^{n-1}A
$$

$$
= \left(1 + 2 + 2^2 + \cdots + 2^{n-1}\right) A
$$

$$
= (2^n - 1)A.
$$

3.9. We have $A^n = \begin{pmatrix} a^n & na^{n-1} \\ 0 & a^n \end{pmatrix}$. If $a = 1$, then

$$
\sum_{i=1}^{n} A^i = \sum_{i=1}^{n} \begin{pmatrix} 1 & i \\ 0 & 1 \end{pmatrix} = \begin{pmatrix} n & \frac{n(n+1)}{2} \\ 0 & n \end{pmatrix}
$$

which implies $\det(A + A^2 + \cdots + A^n) = n^2$.
  If $a \neq 1$, then

$$
\sum_{i=1}^{n} A^i = \sum_{i=1}^{n} \begin{pmatrix} a^i & ia^{i-1} \\ 0 & a^i \end{pmatrix} = \begin{pmatrix} \sum_{i=1}^{n} a^i & \sum_{i=1}^{n} ia^{i-1} \\ 0 & \sum_{i=1}^{n} a^i \end{pmatrix} = \begin{pmatrix} \frac{a(1-a^n)}{1-a} & \sum_{i=1}^{n} ia^{i-1} \\ 0 & \frac{a(1-a^n)}{1-a} \end{pmatrix}
$$

which implies $\det(A + A^2 + \cdots + A^n) = a^2 \left(\frac{1-a^n}{1-a}\right)^2$.

3.10. We have $\begin{pmatrix} \sqrt{3} & -1 \\ 1 & \sqrt{3} \end{pmatrix} = 2\begin{pmatrix} \cos\frac{\pi}{6} & -\sin\frac{\pi}{6} \\ \sin\frac{\pi}{6} & \cos\frac{\pi}{6} \end{pmatrix}$.

Thus, $\begin{pmatrix} \sqrt{3} & -1 \\ 1 & \sqrt{3} \end{pmatrix}^{12} = 2^{12}\begin{pmatrix} \cos 2\pi & -\sin 2\pi \\ \sin 2\pi & \cos 2\pi \end{pmatrix} = 2^{12}I_2$.

3.11. These properties of the circulant matrices can be checked by direct computation.

3.12. The matrix in part (a) is a circulant matrix and part (b) follows from part (a) by setting $a = \cos^2\theta$.

3.13. We note that $A = I_2 + B$, where $B = \begin{pmatrix} a & -a \\ -b & b \end{pmatrix}$. We have $B^2 = (a+b)B$ and $B^k = (a+b)^{k-1}B$, $\forall k \geq 1$. Thus,

$$A^n = (I_2 + B)^n$$

$$= I_2 + \sum_{k=1}^{n} \binom{n}{k} B^k$$

$$= I_2 + \left[\sum_{k=1}^{n} \binom{n}{k}(a+b)^{k-1}\right] B$$

$$= I_2 + \frac{(a+b+1)^n - 1}{a+b} B,$$

if $a+b \neq 0$. If $a+b = 0$ we have that $A^n = I_2 + nB$.

3.14. We have

$$\begin{pmatrix} 1 & \alpha^2 \\ 1 & 1 \end{pmatrix}^n = \frac{(1+\alpha)^n + (1-\alpha)^n}{2} I_2 + \frac{(1+\alpha)^n - (1-\alpha)^n}{2\alpha}\begin{pmatrix} 0 & \alpha^2 \\ 1 & 0 \end{pmatrix}.$$

3.15. $\begin{pmatrix} 2a & -a^2 \\ 1 & 0 \end{pmatrix}^n = a^{n-1}\begin{pmatrix} a(n+1) & -na^2 \\ n & a(-n+1) \end{pmatrix}$, $n \in \mathbb{N}$.

3.16. Observe that $\begin{pmatrix} a & b \\ a & a \end{pmatrix} = a\begin{pmatrix} 1 & \alpha^2 \\ 1 & 1 \end{pmatrix}$, where $\alpha^2 = \frac{b}{a}$ and use problem **3.14**.

3.18. Since $A^n A = AA^n$ we have that

$$\begin{pmatrix} a_n & b_n \\ c_n & d_n \end{pmatrix}\begin{pmatrix} a & b \\ c & d \end{pmatrix} = \begin{pmatrix} a & b \\ c & d \end{pmatrix}\begin{pmatrix} a_n & b_n \\ c_n & d_n \end{pmatrix},$$

and this implies

$$\begin{cases} aa_n + b_n c = aa_n + bc_n \\ a_n b + b_n d = ab_n + bd_n \\ c_n a + d_n c = ca_n + dc_n \\ c_n b + d_n d = cb_n + dd_n, \end{cases}$$

for all $n \geq 1$. From the first or the fourth equation we get that $b_n c = bc_n$, which implies that $\frac{b_n}{b} = \frac{c_n}{c}$, $\forall n \geq 1$. The second equation implies that $\frac{a_n - d_n}{a - d} = \frac{b_n}{b}$, $\forall n \geq 1$.

3.19. Observe that $A = \sqrt{a^2 + b^2} \begin{pmatrix} \cos\theta & \sin\theta \\ -\sin\theta & \cos\theta \end{pmatrix}$, where $\cos\theta = \dfrac{a}{\sqrt{a^2 + b^2}}$ and $\sin\theta = \dfrac{b}{\sqrt{a^2 + b^2}}$. It follows, proved it by mathematical induction, that

$$A^n = (a^2 + b^2)^{\frac{n}{2}} \begin{pmatrix} \cos n\theta & \sin n\theta \\ -\sin n\theta & \cos n\theta \end{pmatrix}, \quad n \geq 1.$$

Thus, $a_n = (a^2 + b^2)^{\frac{n}{2}} \cos n\theta$ and $b_n = (a^2 + b^2)^{\frac{n}{2}} \sin n\theta$ and, since $a^2 + b^2 < 1$, we have that $\lim\limits_{n\to\infty} a_n = \lim\limits_{n\to\infty} b_n = 0$.

3.20. Let $B = \frac{1}{c}A = \begin{pmatrix} x & -y \\ y & x \end{pmatrix}$, where $x = \frac{a}{c}$ and $y = \frac{b}{c}$. Since $a^2 + b^2 = c^2$ there exists $t \in [0, 2\pi)$ such that $x = \cos t \in \mathbb{Q}$ and $y = \sin t \in \mathbb{Q}$. This implies that

$$A^n = c^n \begin{pmatrix} \cos nt & -\sin nt \\ \sin nt & \cos nt \end{pmatrix},$$

so $a_n = c^n \cos nt$ and $b_n = c^n \sin nt$.

By contradiction, we assume that $b_n = 0$. This implies $\sin nt = 0$ and $\cos nt = \pm 1$, so $\cos 2nt = 2\cos^2 nt - 1 = 1$. We prove that if $\cos t \in \mathbb{Q}$ and $\cos 2nt = 1$, then $\cos t \in \{0, \pm 1, \pm\frac{1}{2}\}$. We need the following lemma.

Lemma 3.1 *There exists a monic polynomial of degree n, $P_n \in \mathbb{Z}[x]$ such that* $2\cos nt = P_n(2\cos t)$, $t \in \mathbb{R}$, $n \in \mathbb{N}$.

Proof We prove the lemma by induction on n. If $n = 1$ we let $P_1(x) = x$. If $n = 2$, then $P_2(x) = x^2 - 2$. Using the formula $2\cos(n + 1)t + 2\cos(n - 1)t = (2\cos t)(2\cos nt)$ we get that $P_{n+1}(x) + P_{n-1}(x) = xP_n(x)$ and this implies that if both P_n and P_{n-1} are monic polynomials, then P_{n+1} is a monic polynomial as well. This proves the lemma. $\square$

The equation $\cos(2nt) = 1$ implies, based on the previous lemma, that

$$2\cos(2nt) = (2\cos t)^{2n} + \cdots = 2 \quad \Leftrightarrow \quad x^{2n} + \cdots = 0,$$

where $x = 2\cos t \in \mathbb{Q}$. Since the rational roots of a monic polynomial with integer coefficients are integer roots we get that $2\cos t \in \mathbb{Z}$. This implies that $2\cos t \in \{0, \pm 1, \pm 2\} \Leftrightarrow \cos t \in \{0, \pm 1, \pm\frac{1}{2}\}$.

- if $\cos t = \frac{a}{c} = 0$ we get $a = 0$ which is a contradiction with $a \neq 0$.
- if $\cos t = \pm\frac{1}{2}$ we get that $\sin t = \pm\frac{\sqrt{3}}{2} \notin \mathbb{Q}$, which contradicts $\sin t \in \mathbb{Q}$.
- if $\cos t = \pm 1$ we have $\sin t = 0$ and, since $\sin t = \frac{b}{c}$, we get that $b = 0$ which contradicts $b \neq 0$.

Thus, our assumption that $b_n = 0$ is false and the problem is solved.

3.21. We write $X_a = \sqrt{1+a^2}\begin{pmatrix} \cos t & \sin t \\ -\sin t & \cos t \end{pmatrix}$, where $t \in [0, 2\pi)$ and $\cos t = \frac{a}{\sqrt{1+a^2}}$, $\sin t = \frac{1}{\sqrt{1+a^2}}$. This implies

$$X_a^n = \sqrt{1+a^2}^n\begin{pmatrix} \cos nt & \sin nt \\ -\sin nt & \cos nt \end{pmatrix}$$

and the conditions of the problem become $\cos nt > \sin nt$, for $n = 1, 2, \ldots, 2016$ and $\cos 2017t < \sin 2017t$.

We choose $t = \frac{\pi}{8066}$ and we let $b = \sin^2\frac{\pi}{8066}$. This implies, since $\sin t = \frac{1}{\sqrt{1+a^2}}$, that $a = \sqrt{\frac{1-b}{b}}$.

3.22. Let r be the ratio of the arithmetic progression a, b, c, d.

There exist sequences of real numbers $(\alpha_n)_{n \in \mathbb{N}}$ and $(\beta_n)_{n \in \mathbb{N}}$ such that $A^n = \alpha_n A + \beta_n I_2$. This implies

$$\begin{cases} a_n = \alpha_n a + \beta_n \\ b_n = \alpha_n b \\ c_n = \alpha_n c \\ d_n = \alpha_n d + \beta_n. \end{cases}$$

Thus, $b_n - a_n + d_n - c_n = (b_n + d_n) - (a_n + c_n) = \alpha_n(b - a + d - c) = 2r\alpha_n$ and $2(c_n - b_n) = 2\alpha_n(c - b) = 2r\alpha_n$, $\forall n \geq 1$.

3.23. The equality $AA^{-1} = I_2$ implies $\det A \det\left(A^{-1}\right) = 1$. Since $\det A, \det\left(A^{-1}\right) \in \mathbb{Z}$ we get that $\det A = \det\left(A^{-1}\right) \in \{-1, 1\}$. We have $\det(A^n) = \det^n A \in \{-1, 1\} \Leftrightarrow a_n d_n - b_n c_n \in \{-1, 1\}$. If $(a_n, b_n) = \alpha$, then α divides $a_n d_n - b_n c_n$, so α divides 1 or -1 and this implies $\alpha = 1$. Similarly, we have $(a_n, c_n) = (b_n, d_n) = (c_n, d_n) = 1$.

3.24. (a) This part of the problem can be solved either by mathematical induction or by direct computation (the eigenvalue technique).

(b) Since $A^{n+m} = A^n A^m$ we have that

$$\begin{pmatrix} F_{n+m-1} & F_{n+m} \\ F_{n+m} & F_{n+m+1} \end{pmatrix} = \begin{pmatrix} F_{n-1} & F_n \\ F_n & F_{n+1} \end{pmatrix} \begin{pmatrix} F_{m-1} & F_m \\ F_m & F_{m+1} \end{pmatrix}.$$

We look at $(1,1)$ entry of this identity and we have $F_{n+m-1} = F_n F_m + F_{n-1} F_{m-1}$, $\forall\, m, n \geq 1$. On the other hand, $\det(A^n) = \det^n A \Rightarrow F_{n-1} F_{n+1} - F_n^2 = (-1)^n$, $n \geq 1$.

(c) A calculation shows that the eigenvalues of A are $\alpha = \frac{1+\sqrt{5}}{2}$ and $\beta = \frac{1-\sqrt{5}}{2}$. It follows, based on Remark 3.1, that

$$\begin{pmatrix} F_{n-1} & F_n \\ F_n & F_{n+1} \end{pmatrix} = \begin{pmatrix} 0 & 1 \\ 1 & 1 \end{pmatrix}^n = \frac{\alpha^n - \beta^n}{\alpha - \beta} A + \frac{\alpha^{n-1} - \beta^{n-1}}{\alpha - \beta} I_2, \quad n \geq 1.$$

Looking at $(1,2)$ entry of this equality part (c) of the problem is solved.

3.25. The problem is about solving the diophantine equation $x^2 + xy - y^2 = 1$. This equation can be written equivalently as $(2x+y)^2 - 5y^2 = 4$ which is a Pell equation of the form $x^2 - dy^2 = k$.

3.26. Since $A = 1A + 0I_2$ we have $x_1 = 1$ and $y_1 = 0$. Also, $A^2 - 2A + I_2 = O_2$ which implies $A^2 = 2A - I_2$. Let $A^n = x_n A + y_n I_2$. We have

$$A^{n+1} = A^n A = (x_n A + y_n I_2)A = x_n A^2 + y_n A = x_n(2A - I_2) + y_n A$$

$$= (2x_n + y_n)A - x_n I_2 = x_{n+1} A + y_{n+1} I_2.$$

This implies $x_{n+1} = 2x_n + y_n$, $y_{n+1} = -x_n$, $\forall\, n \geq 1$. It follows that $x_{n+1} - 2x_n - x_{n-1} = 0$, $\forall\, n \geq 1$. The characteristic equation $r^2 - 2r + 1 = 0$ implies $r = 1$. Therefore, $x_n = \alpha + \beta n$, where $\alpha, \beta \in \mathbb{R}$. Since $x_1 = 1$ and $x_2 = 2$ we get that $\alpha = 0$, $\beta = 1$ which implies that $x_n = n$ and $y_n = -n + 1$. Thus, $\lim_{n \to \infty} \frac{x_n}{y_n} = -1$.

3.27. If the sequences $(a_n)_{n \geq 1}$, $(b_n)_{n \geq 1}$, $(c_n)_{n \geq 1}$, and $(d_n)_{n \geq 1}$ converge, then the sequence with the general term $a_n d_n - b_n c_n = \det^n A$ converges. This implies that $\det A \in (-1, 1]$ and since $|\det A| \geq 1$ we get that $\det A = 1$. Clearly, A verifies the equation $A^2 - (a + d)A + (\det A)I_2 = O_2$. We multiply this equation by A^{n-1} and we get that $A^{n+1} - (a + d)A^n + (\det A)A^{n-1} = O_2$. This implies that sequences $(a_n)_{n \geq 1}$, $(b_n)_{n \geq 1}$, $(c_n)_{n \geq 1}$ and $(d_n)_{n \geq 1}$ verify the recurrence relation $x_{n+1} - (a + d)x_n + (\det A)x_{n-1} = 0$. Passing to the limit in the preceding equality we get that $l_x(1 - a - d + \det A) = 0 \Leftrightarrow l_x(2 - a - d) = 0$, where $l_x = \lim_{n \to \infty} x_n$.

If $a + d \neq 2$ we get that $l_x = 0$ and this implies that sequences $(a_n)_{n \geq 1}$, $(b_n)_{n \geq 1}$, $(c_n)_{n \geq 1}$ and $(d_n)_{n \geq 1}$ converge to 0, which contradicts

$$\lim_{n \to \infty} (a_n d_n - b_n c_n) = \lim_{n \to \infty} (\det A)^n = 1.$$

Therefore, we must have $a + d = 2$ and we have $A^{n+1} - 2A^n + A^{n-1} = O_2$, $\forall n \geq 1$. Equivalently, $A^{n+1} - A^n = A^n - A^{n-1}$, $\forall n \geq 1$, which implies $A^n = I_2 + n(A - I_2)$, $\forall n \geq 1$. Hence, $a_n = 1 + n(a - 1)$, $b_n = nb$, $c_n = nc$ and $d_n = 1 + n(d - 1)$. These sequences converge if and only if $a = 1$, $b = 0$, $c = 0$ and $d = 1$, so $A = I_2$.

3.2 Sequences defined by systems of linear recurrence relations

In this section we bring into light a method for determining the general term of sequences defined by systems of linear recurrence relations.

Theorem 3.3 *Let*

$$A = \begin{pmatrix} a & b \\ c & d \end{pmatrix} \in \mathcal{M}_2(\mathbb{C})$$

and let $(x_n)_{n \geq 0}$ *and* $(y_n)_{n \geq 0}$ *be the sequences defined by the system of linear recurrence relations*

$$\begin{cases} x_{n+1} = ax_n + by_n \\ y_{n+1} = cx_n + dy_n, \quad n \geq 0. \end{cases} \tag{3.2}$$

Then

$$\begin{pmatrix} x_n \\ y_n \end{pmatrix} = A^n \begin{pmatrix} x_0 \\ y_0 \end{pmatrix}, \quad \forall n \geq 0.$$

Let λ_1, λ_2 *be the eigenvalues of A.*

- *If* $\lambda_1 \neq \lambda_2$, *then*

$$\begin{cases} x_n = \alpha \lambda_1^n + \beta \lambda_2^n \\ y_n = \gamma \lambda_1^n + \delta \lambda_2^n, \end{cases}$$

for some $\alpha, \beta, \gamma, \delta \in \mathbb{C}$.
- *If* $\lambda_1 = \lambda_2 = \lambda$, *then*

$$\begin{cases} x_n = \lambda^n(\alpha + \beta n) \\ y_n = \lambda^n(\gamma + \delta n), \end{cases}$$

for some $\alpha, \beta, \gamma, \delta \in \mathbb{C}$.

Proof The system (3.2) can be written in the following form

$$\begin{pmatrix} x_{n+1} \\ y_{n+1} \end{pmatrix} = \begin{pmatrix} a & b \\ c & d \end{pmatrix} \begin{pmatrix} x_n \\ y_n \end{pmatrix} \quad \text{or} \quad \begin{pmatrix} x_{n+1} \\ y_{n+1} \end{pmatrix} = A \begin{pmatrix} x_n \\ y_n \end{pmatrix}, \quad \forall n \geq 0.$$

It follows that

$$\begin{pmatrix} x_{n+1} \\ y_{n+1} \end{pmatrix} = A \begin{pmatrix} x_n \\ y_n \end{pmatrix} = A^2 \begin{pmatrix} x_n \\ y_n \end{pmatrix} = \cdots = A^{n+1} \begin{pmatrix} x_0 \\ y_0 \end{pmatrix}.$$

Thus

$$\begin{pmatrix} x_n \\ y_n \end{pmatrix} = A^n \begin{pmatrix} x_0 \\ y_0 \end{pmatrix}$$

and the problem reduces to the computation of A^n.

The second part of the theorem follows based on Theorem 3.1. $\qquad\square$

3.2.1 Problems

3.28 Find the general terms of the sequences $(x_n)_{n\in\mathbb{N}}$ and $(y_n)_{n\in\mathbb{N}}$ defined by the system of linear recurrence relations

$$\begin{cases} x_{n+1} = 3x_n + y_n \\ y_{n+1} = -x_n + y_n, \quad n \geq 1, \end{cases}$$

where $x_1 = 1$ and $y_1 = -2$.

3.29 Find the general terms of the sequences $(x_n)_{n\geq 0}$ and $(y_n)_{n\geq 0}$ defined by the system of linear recurrence relations

$$\begin{cases} x_{n+1} = x_n + 2y_n \\ y_{n+1} = -2x_n + 5y_n, \quad n \geq 0, \end{cases}$$

where $x_0 = 1$ and $y_0 = 2$.

3.30 Prove the sequences $(x_n)_{n\geq 0}$ and $(y_n)_{n\geq 0}$ defined by the system

$$\begin{cases} 2x_n = \sqrt{3}x_{n-1} + y_{n-1} \\ 2y_n = -x_{n-1} + \sqrt{3}y_{n-1}, \quad n \geq 1, \end{cases}$$

are *periodic* and have the same *period*.

3.31 (a) Find the general terms of sequences $(x_n)_{n\geq 0}$ and $(y_n)_{n\geq 0}$ defined by the system of linear recurrence relations

$$\begin{cases} x_{n+1} = \dfrac{x_n + 3y_n}{4} \\ y_{n+1} = \dfrac{3x_n + 2y_n}{5}, \end{cases}$$

where $x_0, y_0 \in \mathbb{R}$.

(b) Find $\lim_{n\to\infty} x_n$ and $\lim_{n\to\infty} y_n$.

3.32 Let $(a_n)_{n\geq 0}$ and $(b_n)_{n\geq 0}$ be the sequences defined by

$$a_0 = 1, \quad b_0 = 4, \quad a_{n+1} = \frac{a_n + 2b_n}{3}, \quad b_{n+1} = \frac{a_n + 3b_n}{4}, \quad \forall n \geq 0.$$

Prove that:

(a) the sequence $(c_n)_{n\in\mathbb{N}}$ defined by $c_n = b_n - a_n$ is a geometric progression;
(b) the sequence $(d_n)_{n\in\mathbb{N}}$ defined by $d_n = 3a_n + 8b_n$ is constant;
(c) calculate $\lim_{n\to\infty} a_n$ and $\lim_{n\to\infty} b_n$.

3.33 A geometric progression. Let $A = \begin{pmatrix} a & b \\ c & d \end{pmatrix} \in \mathcal{M}_2(\mathbb{C})$ and let $(x_n)_{n\in\mathbb{N}}$ and $(y_n)_{n\in\mathbb{N}}$ be the sequences defined by

$$\begin{cases} x_{n+1} = ax_n + by_n \\ y_{n+1} = cx_n + dy_n, \end{cases}$$

$n \in \mathbb{N}$. Prove that if $\lambda \in \mathbb{C}$ is an eigenvalue for A^T and $Z = \begin{pmatrix} \alpha \\ \beta \end{pmatrix}$ is the corresponding eigenvector, then the sequence $(u_n)_{n\in\mathbb{N}}$ defined by $u_n = \alpha x_n + \beta y_n$ is a geometric progression.

3.34 Let $(x_n)_{n\geq 1}$ and $(y_n)_{n\geq 1}$ be the sequences defined by

$$\begin{cases} x_n = -3x_{n-1} - y_{n-1} + n \\ y_n = x_{n-1} + y_{n-1} - 2, \end{cases}$$

for all $n \geq 2$ and $x_1 = y_1 = 1$. Find the general terms of the sequences $(x_n)_{n\geq 1}$ and $(y_n)_{n\geq 1}$.

3.35 Let $(x_n)_{n\geq 0}$ and $(y_n)_{n\geq 0}$ be the sequences defined by the system

$$\begin{cases} x_{n+1} = (1-a)x_n + ay_n \\ y_{n+1} = bx_n + (1-b)y_n, \quad n \geq 0, \end{cases}$$

where $a, b \in (0, 1)$ and $x_0, y_0 \in \mathbb{R}$. Calculate $\lim\limits_{n \to \infty} x_n$ and $\lim\limits_{n \to \infty} y_n$.

3.36 Study the convergence of the sequences $(x_n)_{n \geq 0}$ and $(y_n)_{n \geq 0}$ defined by the system of linear recurrence relations

$$\begin{cases} x_{n+1} = ax_n - by_n \\ y_{n+1} = bx_n + ay_n, \end{cases}$$

where $a, b, x_0, y_0 \in \mathbb{R}$ and $a^2 + b^2 \leq 1$.

3.37 Let $(t_n)_{n \geq 0}$ be a sequence of real numbers such that $t_n \in (0, 1)$, $\forall n \geq 0$ and there exists $\lim\limits_{n \to \infty} t_n \in (0, 1)$. Prove the sequences $(x_n)_{n \geq 0}$ and $(y_n)_{n \geq 0}$ defined by the recurrence relations

$$\begin{cases} x_{n+1} = t_n x_n + (1 - t_n)y_n \\ y_{n+1} = (1 - t_n)x_n + t_n y_n, \quad \forall n \geq 0, \end{cases}$$

are convergent and calculate their limits.

3.38 *An IMO 2013 shortlist problem.*

Let n be a positive integer and let $a_1, a_2, \ldots, a_{n-1}$ be arbitrary real numbers. Define the sequences $u_0, u_1, \ldots, u_n$ and $v_0, v_1, \ldots, v_n$ inductively by $u_0 = u_1 = v_0 = v_1 = 1$ and $u_{k+1} = u_k + a_k u_{k-1}$, $v_{k+1} = v_k + a_{n-k}v_{k-1}$, for $k = 1, \ldots, n-1$. Prove that $u_n = v_n$.

3.2.2 Solutions

3.28. $x_n = 2^{n-1} - (n-1)2^{n-2}$, $y_n = (n-1)2^{n-2} - 2^n$, $\forall n \geq 1$.

3.29. $x_n = 3^{n-1}(2n+3)$, $y_n = 3^{n-1}(2n+6)$, $\forall n \geq 0$.

3.30. $x_n = \left(\cos\dfrac{n\pi}{6}\right)x_0 + \left(\sin\dfrac{n\pi}{6}\right)y_0$, $y_n = -\left(\sin\dfrac{n\pi}{6}\right)x_0 + \left(\cos\dfrac{n\pi}{6}\right)y_0$, $n \geq 0$. Since $x_{n+12} = x_n$ and $y_{n+12} = y_n$, $\forall n \geq 0$, the sequences are periodic of period 12.

3.31. (a) A calculation shows that

$$x_n = \left[\frac{4}{9} + \left(-\frac{7}{20}\right)^n \frac{5}{9}\right]x_0 + \left[\frac{5}{9} - \left(-\frac{7}{20}\right)^n \frac{5}{9}\right]y_0$$

$$y_n = \left[\frac{4}{9} - \left(-\frac{7}{20}\right)^n \frac{4}{9}\right]x_0 + \left[\frac{5}{9} + \left(-\frac{7}{20}\right)^n \frac{4}{9}\right]y_0.$$

(b) $\lim\limits_{n\to\infty} x_n = \lim\limits_{n\to\infty} y_n = \dfrac{4x_0 + 5y_0}{9}$.

3.32. *Solution 1.* Let $A = \begin{pmatrix} \dfrac{1}{3} & \dfrac{2}{3} \\ \dfrac{1}{4} & \dfrac{3}{4} \end{pmatrix}$. We have

$$\begin{pmatrix} a_n \\ b_n \end{pmatrix} = A \begin{pmatrix} a_{n-1} \\ b_{n-1} \end{pmatrix} = \cdots = A^n \begin{pmatrix} a_0 \\ b_0 \end{pmatrix} = A^n \begin{pmatrix} 1 \\ 4 \end{pmatrix}.$$

The eigenvalues of A are $\lambda_1 = 1$, $\lambda_2 = \frac{1}{12}$ and we have, based on Theorem 3.1, that $A^n = B + \dfrac{1}{12^n} C$, $n \in \mathbb{N}$, where

$$B = \frac{1}{11} \begin{pmatrix} 3 & 8 \\ 3 & 8 \end{pmatrix} \quad \text{and} \quad C = \frac{1}{11} \begin{pmatrix} 8 & -8 \\ -3 & 3 \end{pmatrix}.$$

It follows that $a_n = \dfrac{1}{11}\left(35 - \dfrac{24}{12^n} \right)$ and $b_n = \dfrac{1}{11}\left(35 + \dfrac{9}{12^n} \right)$.

(a) $b_n - a_n = \dfrac{3}{12^n}$, $n \geq 0$, which is a geometric progression of ratio $\frac{1}{12}$.

(b) $3a_n + 8b_n = 35$, $n \geq 0$.

(c) $\lim\limits_{n\to\infty} a_n = \lim\limits_{n\to\infty} b_n = \dfrac{35}{11}$.

Solution 2. (a) We have

$$\frac{c_{n+1}}{c_n} = \frac{b_{n+1} - a_{n+1}}{b_n - a_n} = \frac{\frac{1}{4}a_n + \frac{3}{4}b_n - \frac{1}{3}a_n - \frac{2}{3}b_n}{b_n - a_n} = \frac{\frac{1}{12}b_n - \frac{1}{12}a_n}{b_n - a_n} = \frac{1}{12}.$$

(b) $d_{n+1} = 3a_{n+1} + 8b_{n+1} = a_n + 2b_n + 2a_n + 6b_n = 3a_n + 8b_n = d_n$, $\forall n \geq 0$. This implies $d_n = d_0 = 3a_0 + 8b_0 = 35$, $\forall n \geq 0$.

(c) We have, based on part (a), that $c_n = b_n - a_n = \dfrac{3}{12^n}$ and it follows that $\lim\limits_{n\to\infty} a_n = \lim\limits_{n\to\infty} b_n$. Using part (b) of the problem we get that $3 \lim\limits_{n\to\infty} a_n + 8 \lim\limits_{n\to\infty} b_n = 35$ and we have $\lim\limits_{n\to\infty} a_n = \lim\limits_{n\to\infty} b_n = \dfrac{35}{11}$.

3.33. We have $A^T Z = \lambda Z$ which implies that

$$\begin{cases} a\alpha + c\beta = \lambda\alpha \\ b\alpha + d\beta = \lambda\beta. \end{cases}$$

We calculate

$$\frac{u_{n+1}}{u_n} = \frac{\alpha x_{n+1} + \beta y_{n+1}}{\alpha x_n + \beta y_n}$$

$$= \frac{\alpha(a x_n + b y_n) + \beta(c x_n + d y_n)}{\alpha x_n + \beta y_n}$$

$$= \frac{(\alpha a + \beta c)x_n + (\alpha b + \beta d)y_n}{\alpha x_n + \beta y_n}$$

$$= \frac{\lambda \alpha x_n + \lambda \beta y_n}{\alpha x_n + \beta y_n}$$

$$= \lambda.$$

Thus, the sequence $(u_n)_{n\in\mathbb{N}}$ is a geometric progression of ratio λ which is also an eigenvalue of A.

3.34. Let $x_n = u_n + an + b$ and $y_n = v_n + cn + d$, $n \geq 1$. A calculation shows $a = 0$, $b = 3$, $c = 1$, and $d = -11$, so $x_n = u_n + 3$ and $y_n = v_n + n - 11$, $n \geq 1$. The system of recurrence relations becomes

$$\begin{cases} u_n = 3u_{n-1} - v_{n-1} \\ v_n = u_{n-1} + v_{n-1}, \end{cases}$$

for $n \geq 2$. Solving the system we obtain, after some calculations, that

- $x_n = \dfrac{-2\sqrt{3}-7}{2\sqrt{3}}(\sqrt{3}-1)^{n-1} + \dfrac{7-2\sqrt{3}}{2\sqrt{3}}(-1-\sqrt{3})^{n-1} + 3$

- $y_n = \dfrac{20+11\sqrt{3}}{2\sqrt{3}}(\sqrt{3}-1)^{n-1} + \dfrac{11\sqrt{3}-20}{2\sqrt{3}}(-1-\sqrt{3})^{n-1} + n - 11.$

3.35. We write the system in the matrix form

$$\begin{pmatrix} x_{n+1} \\ y_{n+1} \end{pmatrix} = \begin{pmatrix} 1-a & a \\ b & 1-b \end{pmatrix} \begin{pmatrix} x_n \\ y_n \end{pmatrix} \quad \text{or} \quad \begin{pmatrix} x_{n+1} \\ y_{n+1} \end{pmatrix} = A \begin{pmatrix} x_n \\ y_n \end{pmatrix},$$

where

$$A = \begin{pmatrix} 1-a & a \\ b & 1-b \end{pmatrix}.$$

It follows that $\begin{pmatrix} x_n \\ y_n \end{pmatrix} = A^n \begin{pmatrix} x_0 \\ y_0 \end{pmatrix}$. We calculate A^n. The eigenvalues of A are $\lambda_1 = 1$, $\lambda_2 = 1 - a - b$ and we note that $\lambda_1 \neq \lambda_2$ since $a + b \neq 0$. It follows that $A^n = B + (1 - a - b)^n C$, where

$$B = \frac{1}{a+b}\begin{pmatrix} b & a \\ b & a \end{pmatrix} \quad \text{and} \quad C = \frac{1}{a+b}\begin{pmatrix} a & -a \\ -b & b \end{pmatrix}.$$

Thus

$$A^n = \frac{1}{a+b}\begin{pmatrix} b+a(1-a-b)^n & a-a(1-a-b)^n \\ b-b(1-a-b)^n & a+b(1-a-b)^n \end{pmatrix},$$

which implies

$$x_n = \frac{1}{a+b}\{[b+a(1-a-b)^n]x_0 + [a-a(1-a-b)^n]y_0\}$$
$$y_n = \frac{1}{a+b}\{[b-b(1-a-b)^n]x_0 + [a+b(1-a-b)^n]y_0\}.$$

A calculation shows, since $|1-a-b| < 1$, that $\lim\limits_{n\to\infty} x_n = \lim\limits_{n\to\infty} y_n = \dfrac{bx_0 + ay_0}{a+b}$.

3.36. Let $U_n = \begin{pmatrix} x_n \\ y_n \end{pmatrix}$ and let $A = \begin{pmatrix} a & -b \\ b & a \end{pmatrix}$. Since $U_{n+1} = AU_n$ we get that $U_n = A^n U_0$. Let $r = \sqrt{a^2 + b^2}$ and let $t \in [0, 2\pi)$ such that $a = r\cos t$ and $b = r\sin t$. It follows that

$$A^n = r^n \begin{pmatrix} \cos nt & -\sin nt \\ \sin nt & \cos nt \end{pmatrix},$$

which implies that $x_n = r^n(x_0 \cos nt - y_0 \sin nt)$ and $y_n = r^n(x_0 \sin nt + y_0 \cos nt)$.

- If $r \in [0, 1)$, then $(x_n)_{n\geq 0}$ and $(y_n)_{n\geq 0}$ converge and $\lim\limits_{n\to\infty} x_n = \lim\limits_{n\to\infty} y_n = 0$.
- If $r = 1$ and $t \in \pi\mathbb{Q}$, then $(x_n)_{n\geq 0}$ and $(y_n)_{n\geq 0}$ are *periodic*. If $t = \frac{p}{q}\pi$, $(p,q) = 1$, the sequences $(x_n)_{n\geq 0}$ and $(y_n)_{n\geq 0}$ have the same period $2q$.
- If $r = 1$ and $t \in \pi(\mathbb{R} \setminus \mathbb{Q})$, then $(x_n)_{n\geq 0}$ and $(y_n)_{n\geq 0}$ are *dense* in the interval $\left[-\sqrt{x_0^2 + y_0^2}, \sqrt{x_0^2 + y_0^2}\right]$.

3.37. Let $U_n = \begin{pmatrix} x_n \\ y_n \end{pmatrix}$ and let $A_n = \begin{pmatrix} t_n & 1-t_n \\ 1-t_n & t_n \end{pmatrix}$. Since $U_{n+1} = A_n U_n, n \geq 0$, we have that $U_{n+1} = A_n A_{n-1} \cdots A_0 U_0$. We calculate the matrix product $A_n A_{n-1} \cdots A_0$. The eigenvalues of A_n are $\lambda_1 = 1$ and $\lambda_2 = 2t_n - 1$ and the corresponding eigenvectors are $X_1 = \begin{pmatrix} 1 \\ 1 \end{pmatrix}$ and $X_2 = \begin{pmatrix} -1 \\ 1 \end{pmatrix}$ (they are the same for all n). If

$$P = \begin{pmatrix} 1 & -1 \\ 1 & 1 \end{pmatrix} \quad \Rightarrow \quad A_n = P\begin{pmatrix} 1 & 0 \\ 0 & 2t_n - 1 \end{pmatrix}P^{-1}.$$

This implies that

$$A_n A_{n-1} \cdots A_0 = P \begin{pmatrix} 1 & 0 \\ 0 & s_n \end{pmatrix} P^{-1},$$

where $s_n = \displaystyle\prod_{k=0}^{n} (2t_k - 1)$.

If one of the terms of the sequence $(t_n)_{n\geq0}$ is $\frac{1}{2}$, i.e., $t_{n_0} = \frac{1}{2}$, then $s_n = 0$, $\forall n \geq n_0$. If all of the terms of $(t_n)_{n\geq0}$ are different from $\frac{1}{2}$, we obtain, since $\lim\limits_{n\to\infty} \frac{s_{n+1}}{s_n} = \lim\limits_{n\to\infty} (2t_{n+1} - 1) \in (-1, 1)$, that $\lim\limits_{n\to\infty} s_n = 0$ and this implies that

$$\lim_{n\to\infty} U_{n+1} = P \begin{pmatrix} 1 & 0 \\ 0 & 0 \end{pmatrix} P^{-1} \begin{pmatrix} x_0 \\ y_0 \end{pmatrix}.$$

Thus, $\lim\limits_{n\to\infty} x_n = \dfrac{x_0 + y_0}{2} = \lim\limits_{n\to\infty} y_n$.

3.38. For $k = 1, 2, \ldots, n-1$, let $x_{k+1} = u_{k+1} - u_k$, $y_{k+1} = v_{k+1} - v_k$ and let $A_k = \begin{pmatrix} 1 + a_k & -a_k \\ a_k & -a_k \end{pmatrix}$. The following relations hold

$$\begin{pmatrix} u_{k+1} \\ x_{k+1} \end{pmatrix} = A_k \begin{pmatrix} u_k \\ x_k \end{pmatrix} \quad \text{and} \quad \begin{pmatrix} v_{k+1} \\ y_{k+1} \end{pmatrix} = A_{n-k} \begin{pmatrix} v_k \\ y_k \end{pmatrix}$$

and it follows that

- $\begin{pmatrix} u_n \\ x_n \end{pmatrix} = A_{n-1} A_{n-2} \cdots A_1 \begin{pmatrix} u_1 \\ x_1 \end{pmatrix} = A_{n-1} A_{n-2} \cdots A_1 \begin{pmatrix} 1 \\ 0 \end{pmatrix}$

- $\begin{pmatrix} v_n \\ y_n \end{pmatrix} = A_1 A_2 \cdots A_{n-1} \begin{pmatrix} v_1 \\ y_1 \end{pmatrix} = A_1 A_2 \cdots A_{n-1} \begin{pmatrix} 1 \\ 0 \end{pmatrix}$.

This implies $u_n = v_n$.

Remark 3.4 If $a_1 = a_2 = \cdots = a_{n-1} = 1$, then we have $u_n = v_n = F_{n+1}$.

3.3   Sequences defined by homographic recurrence relations

In this section we discuss sequences defined by homographic recurrence relations.

Definition 3.1 The function $f : \mathbb{R} \setminus \left\{ -\dfrac{d}{c} \right\} \to \mathbb{R}$, $f(x) = \dfrac{ax + b}{cx + d}$, $a, b, c, d \in \mathbb{R}$ is called *a homographic function* and

$$M_f = \begin{pmatrix} a & b \\ c & d \end{pmatrix}$$

is the *matrix associated with* f.

- If $D \subset \mathbb{R}$ and $f, g : D \to \mathbb{R}$ are homographic functions, then $f \circ g$ and $f^n = \underbrace{f \circ f \circ \cdots \circ f}_{n \text{ functions}}, n \in \mathbb{N}$, are homographic functions and we have the following relations involving their associated matrices

$$M_{f \circ g} = M_f M_g \quad \text{and} \quad M_{f^n} = M_f^n, \ \ n \in \mathbb{N}.$$

Definition 3.2 A sequence defined by a recurrence relation $x_{n+1} = f(x_n)$, where f is a homographic function is called a *homographic sequence*. Thus, a homographic sequence is defined by the recurrence formula

$$x_{n+1} = \frac{a x_n + b}{c x_n + d}, \ \ n \geq 0, \ \ a, b, c, d \in \mathbb{R}.$$

- The sequence $(x_n)_{n \in \mathbb{N}}$ is well defined if $c x_n + d \neq 0$, for all $n \geq 0$.
- If $x_{n+1} = f(x_n), \ \forall \, n \geq 0$, then $x_n = f^n(x_0)$, where

$$f^n(x_0) = \underbrace{f \circ f \circ \cdots \circ f}_{n \text{ functions}}(x_0).$$

- If

$$f(x) = \frac{a x + b}{c x + d} \quad \text{and} \quad \begin{pmatrix} a & b \\ c & d \end{pmatrix}^n = \begin{pmatrix} a_n & b_n \\ c_n & d_n \end{pmatrix},$$

then

$$f^n(x) = \frac{a_n x + b_n}{c_n x + d_n}.$$

- If $x_{n+1} = f(x_n), n \geq 0$, then

$$x_n = \frac{a_n x_0 + b_n}{c_n x_0 + d_n}.$$

- The sequence $(x_n)_{n>0}$ is well defined, if some conditions are imposed upon its initial term x_0. More precisely, we determine the existence conditions of the sequence $(x_n)_{n\geq 0}$ from the expression of A^n, i.e., $c_n x_0 + d_n \neq 0$, $\forall n \geq 0$. This implies that $x_0 \neq -\dfrac{d_n}{c_n}$, for all $n \geq 0$. Thus, we need to determine the set

$$S = \left\{ -\frac{d_n}{c_n} : n \geq 0 \right\}$$

and the condition, the sequence $(x_n)_{n\geq 0}$ is well defined, is that $x_0 \in \mathbb{R} \setminus S$.

- To determine the general term of a sequence defined by a homographic recurrence relation one has to calculate the nth power of the matrix associated with the homographic function which defines the recurrence relation.

3.3.1 Problems

3.39 Let $f(x) = \dfrac{4x + 1}{2x + 3}$, $x \in \mathbb{R}$, be such that the function

$$f_n(x) = \underbrace{f \circ f \circ \cdots \circ f}_{n \text{ functions}}(x), \quad \forall n \in \mathbb{N},$$

is well defined. Determine f_n.

3.40 Let $f : (0, \infty) \to \mathbb{R}$, $f(x) = \dfrac{2x + 1}{x + 2}$. Calculate

$$f_n = \underbrace{f \circ f \circ \cdots \circ f}_{n \text{ functions}}, \quad \forall n \in \mathbb{N}.$$

3.41 Let $(x_n)_{n\geq 1}$ be the sequence defined by

$$x_1 = 1, \quad x_{n+1} = \frac{2 + x_n}{1 + x_n}, \quad \forall n \geq 1.$$

Prove the sequence $(x_n)_{n\geq 1}$ converges and find its limit.

3.42 Let $(x_n)_{n\geq 0}$ be the sequence defined by

$$x_0 = a > 0, \quad x_{n+1} = \frac{2x_n + 1}{2x_n + 3}, \quad \forall n \geq 0.$$

Determine the general term of the sequence $(x_n)_{n\geq 0}$ and calculate $\lim\limits_{n\to\infty} x_n$.

3.43 Let $a, x_0 \in \mathbb{R}$ and let $(x_n)_{n\geq 0}$ be the sequence defined by

$$x_{n+1} = \frac{2a x_n}{x_n + a}, \quad \forall n \geq 0.$$

Study the convergence of the sequence $(x_n)_{n\geq 0}$, when $a > 0$ and $x_0 > 0$.

3.44 Let $(x_n)_{n\geq 0}$ be the sequence defined by

$$x_0 > 0, \quad x_{n+1} = \frac{4}{x_n + 3}, \quad \forall n \geq 0.$$

Determine the general term of the sequence $(x_n)_{n\geq 0}$ and calculate $\lim_{n\to\infty} x_n$.

3.45 Calculate the limit of the sequence defined by

$$x_1 = \frac{1}{\pi^2}, \quad x_{n+1} = \frac{n^2 x_n}{x_n + n^2}, \quad \forall n \geq 1.$$

3.46 Study the convergence of the sequence

$$x_0 \in \mathbb{R} \setminus \mathbb{Q}, \quad x_{n+1} = 1 + \frac{1}{x_n}, \quad \forall n \geq 0.$$

3.47 Let $a \in \mathbb{R}$. Study the convergence of the sequence defined by

$$x_0 = 1, \quad x_{n+1} x_n + a(x_{n+1} - x_n) + 1 = 0, \quad \forall n \geq 0.$$

3.48 Let $(x_n)_{n\geq 0}$ be the sequence defined by

$$x_0 = 2 \quad \text{and} \quad x_{n+1} = \frac{2x_n + 1}{x_n + 2}, \quad \forall n \geq 0.$$

Prove the sequences $(x_n)_{n\geq 0}$ and $(x_0 + x_1 + \cdots + x_n - n)_{n\geq 0}$ converge.

3.49 Let $(a_n)_{n\geq 1}$ be the sequence of real numbers which verifies the recurrence relation

$$a_{n+1} a_n + 3a_{n+1} + a_n + 4 = 0, \quad \forall n \geq 1.$$

Determine all possible values of a_1 such that $a_{2016} \leq a_n$, for all $n \geq 1$.

3.50 Let $A = \begin{pmatrix} a & b \\ c & d \end{pmatrix} \in \mathcal{M}_2(\mathbb{Q})$ with the property that $bc \neq 0$ and there exists

$n \in \mathbb{N}$, $n \geq 2$ such that $b_n c_n = 0$, where $A^n = \begin{pmatrix} a_n & b_n \\ c_n & d_n \end{pmatrix}$, $n \in \mathbb{N}$.

(a) Prove that $a_n = d_n$.

(b) Study the convergence of the sequence $(x_n)_{n\geq 0}$ defined by the recurrence relation

$$x_0 \in \mathbb{R} \setminus \mathbb{Q}, \quad x_{n+1} = \frac{ax_n + b}{cx_n + d}, \quad n \geq 0.$$

3.51 A special sequence with arctangent sums.

Let $(x_n)_{n\geq 1}$ be the sequence defined by

$$x_1 = 1, \quad x_n = \frac{x_{n-1} + n}{1 - nx_{n-1}}, \quad n \geq 2.$$

Prove that:

(a) $x_n = \tan \sum_{k=1}^{n} \arctan k$;

(b) *Conjecture.* For $n \geq 5$, the value x_n is not an integer [3, Conjecture 1.2].

Remark 3.5 This sequence was studied in [3] where it was proved that $1 - nx_{n-1} \neq 0$ for $n > 1$, so $(x_n)_{n\geq 1}$ is well defined. Other special properties of this sequence, which are far beyond the goal of this book, are that x_n vanishes only when $n = 3$ and, for $n > 4$, the terms x_{n-1} and x_n cannot both be integers.

3.3.2 Solutions

3.39. $f_n(x) = \dfrac{(2^n + 2 \cdot 5^n)x + 5^n - 2^n}{(2 \cdot 5^n - 2^{n+1})x + 2^{n+1} + 5^n}$, $n \in \mathbb{N}$.

3.40. $f_n(x) = \dfrac{(3^n + 1)x + 3^n - 1}{(3^n - 1)x + 3^n + 1}$, $n \in \mathbb{N}$.

3.42. $x_n = \dfrac{(2 + 4^n)x_0 + 4^n - 1}{2(4^n - 1)x_0 + 2 \cdot 4^n + 1}$ and $\lim\limits_{n\to\infty} x_n = \frac{1}{2}$.

3.43. $x_n = \dfrac{2^n a x_0}{(2^n - 1)x_0 + a}$, $\forall n \in \mathbb{N}$. The sequence $(x_n)_{n\geq 0}$ converges and $\lim\limits_{n\to\infty} x_n = a$.

3.44. $x_n = \dfrac{[4^n + 4(-1)^n]x_0 + 4^{n+1} - 4(-1)^n}{[4^n - (-1)^n]x_0 + 4^{n+1} + (-1)^n}$ and $\lim\limits_{n\to\infty} x_n = 1$.

3.45. *Solution 1.* If $y_n = \dfrac{1}{x_n}$, then $y_{n+1} = \dfrac{1}{n^2} + y_n$, $\forall n \geq 1$, and it follows that $y_{n+1} = 1 + \dfrac{1}{2^2} + \cdots + \dfrac{1}{n^2} + \pi^2$. Thus, $x_n = \dfrac{1}{1 + \frac{1}{2^2} + \cdots + \frac{1}{(n-1)^2} + \pi^2}$ and $\lim\limits_{n\to\infty} x_n = \dfrac{6}{7\pi^2}$.

Solution 2. $x_{n+1} = f_n(x_n) = f_n \circ f_{n-1}(x_{n-1}) = \cdots = f_n \circ f_{n-1} \circ \cdots \circ f_1(x_1)$, where $f_n(x) = \dfrac{n^2 x}{x + n^2}$. Let $A_n = \begin{pmatrix} n^2 & 0 \\ 1 & n^2 \end{pmatrix}$. We have

$$f_n \circ f_{n-1} \circ \cdots \circ f_1(x_1) = \frac{ax_1 + b}{cx_1 + d},$$

where

$$\begin{pmatrix} a & b \\ c & d \end{pmatrix} = A_n A_{n-1} \cdots A_1 = (n!)^2 \begin{pmatrix} 1 & 0 \\ \sum\limits_{k=1}^{n} \frac{1}{k^2} & 1 \end{pmatrix}.$$

This implies that

$$x_{n+1} = \frac{x_1}{x_1 \sum\limits_{k=1}^{n} \frac{1}{k^2} + 1} \quad \text{and} \quad \lim_{n \to \infty} x_{n+1} = \frac{x_1}{\frac{\pi^2}{6} x_1 + 1} = \frac{6}{7\pi^2}.$$

3.46. $x_{n+1} = \dfrac{x_n + 1}{x_n}$, $\forall n \geq 0$. It follows that $x_n = \dfrac{F_{n+1} x_0 + F_n}{F_n x_0 + F_{n-1}}$, $\forall n \geq 1$, where $(F_n)_{n \geq 0}$ denotes the Fibonacci sequence. A calculation shows that $\lim\limits_{n \to \infty} x_n = \frac{1+\sqrt{5}}{2}$. We used that

$$F_n = \frac{1}{\sqrt{5}} \left[\left(\frac{1+\sqrt{5}}{2} \right)^n - \left(\frac{1-\sqrt{5}}{2} \right)^n \right], \quad n \geq 0.$$

3.47. $x_{n+1} = \dfrac{ax_n - 1}{x_n + a} \Rightarrow x_n = \dfrac{a_n x_0 + b_n}{c_n x_0 + d_n}$, where

$$\begin{pmatrix} a_n & b_n \\ c_n & d_n \end{pmatrix} = \begin{pmatrix} a & -1 \\ 1 & a \end{pmatrix}^n = \left(\sqrt{1 + a^2} \right)^n \begin{pmatrix} \cos nt & -\sin nt \\ \sin nt & \cos nt \end{pmatrix}$$

with $\tan t = \frac{1}{a}$, $a \neq 0$. If $a = 0$ the sequence is periodic of period 2, $x_{2n+1} = -1$ and $x_{2n} = 1$, for all $n \geq 0$. It follows that

$$x_n = \frac{\cos nt - \sin nt}{\cos nt + \sin nt} = \frac{1 - \tan nt}{1 + \tan nt}.$$

If $\dfrac{t}{\pi} \notin \mathbb{Q}$, the set $\{\tan nt : n \in \mathbb{N}\}$ is dense in $\mathbb{R}$ and the range of the function $f(x) = \dfrac{1-x}{1+x}$ is $\mathbb{R} \setminus \{-1\}$, so the sequence $(x_n)_{n \in \mathbb{N}}$ is dense in $\mathbb{R}$.

The expression $x_{n+1} = \dfrac{ax_n - 1}{x_n + a}$ is well defined since $x_n \neq -a$, $\forall n \geq 0$. Otherwise, if $x_n = -a$, for some n, then $ax_{n+1} + a(x_{n+1} + a) + 1 = 0 \Rightarrow a^2 + 1 = 0$, which is impossible since $a \in \mathbb{R}$.

3.50. (a) See the solution of problem **1.3.**

(b) Since $a, b, c, d \in \mathbb{Q}$ and $x_0 \in \mathbb{R} \setminus \mathbb{Q}$ we get that $x_k \in \mathbb{R} \setminus \mathbb{Q}$, $\forall k \geq 0$, and $cx_k + d \neq 0$. Let $f : \mathbb{R} \setminus \mathbb{Q} \to \mathbb{R} \setminus \mathbb{Q}$ be the function defined by $f(x) = \dfrac{ax + b}{cx + d}$. The recurrence relation $x_{k+1} = f(x_k)$ implies that $x_k = f^k(x_0)$, where $f^k = f \circ \cdots \circ f$. When $k = n$ we have, based on part (a), that $a_n = d_n$ and we also know that $b_n c_n = 0$ which implies (see the solution of problem 1.3) that $b_n = c_n = 0$. We obtain $x_n = \dfrac{a_n x_0}{a_n} = x_0$. It follows that

$$x_{n+1} = \frac{ax_n + b}{cx_n + d} = \frac{ax_0 + b}{cx_0 + d} = x_1, \quad x_{n+2} = \frac{ax_{n+1} + b}{cx_{n+1} + d} = \frac{ax_1 + b}{cx_1 + d} = x_2,$$

and $x_{n+k} = x_n$, $\forall k \in \mathbb{N}$. Such a sequence converges provided it is constant, so $x_0 = f(x_0) \Leftrightarrow x_0 = \dfrac{ax_0 + b}{cx_0 + d} \Leftrightarrow ax_0 + b = cx_0^2 + dx_0 \Leftrightarrow cx_0^2 + (d - a)x_0 - b = 0$. This equation has the solutions

$$x_0 = \frac{a - d \pm \sqrt{(d - a)^2 + 4bc}}{2c}, \quad (d - a)^2 + 4bc > 0.$$

Thus, the sequence converges provided that $(d - a)^2 + 4bc > 0$ and $(d - a)^2 + 4bc$ is not of the form q^2, where $q \subset \mathbb{Q}$.

3.51. (a) We prove this part of the problem by mathematical induction. Let $P(n)$ be the statement $x_n = \tan\left(\sum_{k=1}^{n} \arctan k\right)$. When $n = 1$ we get that $x_1 = \tan(\arctan 1) = 1$, so $P(1)$ is true. We prove that $P(n) \Rightarrow P(n + 1)$. We have

$$x_{n+1} = \frac{x_n + n + 1}{1 - (n + 1)x_n}$$

$$= \frac{\tan\left(\sum_{k=1}^{n} \arctan k\right) + n + 1}{1 - (n + 1)\tan\left(\sum_{k=1}^{n} \arctan k\right)}$$

$$= \frac{\tan\left(\sum_{k=1}^{n} \arctan k\right) + \tan(\arctan(n + 1))}{1 - \tan(\arctan(n + 1))\tan\left(\sum_{k=1}^{n} \arctan k\right)}$$

$$= \tan\left(\sum_{k=1}^{n} \arctan k + \arctan(n + 1)\right)$$

$$= \tan\left(\sum_{k=1}^{n+1} \arctan k\right).$$

Remarks and further comments. We mention that x_n can be expressed in terms of the *Stirling numbers of the first kind.* We have

$$x_n = f_n(x_{n-1}) = \cdots = f_n \circ f_{n-1} \circ \cdots \circ f_2(x_1) = f_n \circ f_{n-1} \circ \cdots \circ f_2(1),$$

where $f_n(x) = \dfrac{x+n}{-nx+1}$. Let $A_n = \begin{pmatrix} 1 & n \\ -n & 1 \end{pmatrix}$. It follows that

$$f_n \circ f_{n-1} \circ \cdots \circ f_2(x) = \frac{a_n x + b_n}{c_n x + d_n} \quad \text{and} \quad f_n \circ f_{n-1} \circ \cdots \circ f_2(1) = \frac{a_n + b_n}{c_n + d_n}$$

where

$$A_n A_{n-1} \cdots A_2 = \begin{pmatrix} a_n & b_n \\ c_n & d_n \end{pmatrix}.$$

A calculation shows the eigenvalues of A_n are $1 \pm ni$ and it follows that

$$A_n = \frac{1}{2} \begin{pmatrix} 1 & i \\ i & 1 \end{pmatrix} \begin{pmatrix} 1+ni & 0 \\ 0 & 1-ni \end{pmatrix} \begin{pmatrix} 1 & -i \\ -i & 1 \end{pmatrix},$$

which implies

$$A_n A_{n-1} \cdots A_2 = \frac{1}{2} \begin{pmatrix} 1 & i \\ i & 1 \end{pmatrix} \begin{pmatrix} \alpha & 0 \\ 0 & \beta \end{pmatrix} \begin{pmatrix} 1 & -i \\ -i & 1 \end{pmatrix}$$

$$= \frac{1}{2} \begin{pmatrix} \alpha + \beta & -\alpha i + \beta i \\ \alpha i - \beta i & \alpha + \beta \end{pmatrix},$$

where

$$\alpha = \prod_{k=2}^{n} (1 + ki) \quad \text{and} \quad \beta = \prod_{k=2}^{n} (1 - ki).$$

It follows that

$$x_n = \frac{\alpha(1-i) + \beta(1+i)}{\alpha(1+i) + \beta(1-i)} = \frac{\beta_1 - \alpha_1}{\beta_1 + \alpha_1} i,$$

where

$$\alpha_1 = \prod_{k=1}^{n} (1 + ki) \quad \text{and} \quad \beta_1 = \prod_{k=1}^{n} (1 - ki).$$

The Stirling numbers $s(n, k)$ of the first kind [59, p. 56] are defined by the formula

$$\prod_{k=1}^{n}(1 + kx) = \sum_{k=1}^{n+1}(-x)^{n+1-k}s(n + 1, k),$$

and this implies that

$$\alpha_1 = \sum_{k=1}^{n+1}(-i)^{n+1-k}s(n + 1, k) \quad \text{and} \quad \beta_1 = \sum_{k=1}^{n+1}i^{n+1-k}s(n + 1, k).$$

We consider the cases when n is an even, respectively an odd integer.

- *Case* $n = 2p$. We have $x_{2p} = \dfrac{\sum_{j=1}^{p}(-1)^{p-j+1}s(2p + 1, 2j)}{\sum_{j=1}^{p+1}(-1)^{p-j+1}s(2p + 1, 2j - 1)}$.

- *Case* $n = 2p - 1$. We have $x_{2p-1} = \dfrac{\sum_{j=1}^{p}(-1)^{p-j+1}s(2p, 2j - 1)}{\sum_{j=1}^{p}(-1)^{p-j}s(2p, 2j)}$.

3.4 Binomial matrix equations

In this section we solve the binomial equation $X^n = A$, where $A \in \mathcal{M}_2(\mathbb{C})$ and $n \geq 2$ is an integer.

Definition 3.3 Let $A \in \mathcal{M}_2(\mathbb{C})$ and let $n \geq 2$ be an integer. The equation $X^n = A$, where $X \in \mathcal{M}_2(\mathbb{C})$, is called the *binomial matrix equation*.

In general, for solving binomial matrix equations we need some simple properties which we record in the next lemma.

Lemma 3.2 *The following statements hold.*

(a) *If* $X \in \mathcal{M}_2(\mathbb{C})$ *and* $\det X = 0$, *then* $X^n = \operatorname{Tr}^{n-1}(X)X$, $n \geq 1$.

(b) *If* $X^n = A$, *then matrices* A *and* X *commute,* $AX = AA^n = A^{n+1} = A^nA = XA$.

(c) *If* $A \in \mathcal{M}_2(\mathbb{C})$, $A \neq aI_2$, $a \in \mathbb{C}$, *then matrix* X *which commutes with* A *has the following form* $X = \alpha A + \beta I_2$.

(d) *If* $X \in \mathcal{M}_2(\mathbb{C})$, $X = \begin{pmatrix} a & b \\ c & d \end{pmatrix}$, *then* $X^2 - (a + d)X + (ad - bc)I_2 = O_2$.

(e) *If* $X \in \mathcal{M}_2(\mathbb{C})$ *and if there exists* $n \geq 2$ *such that* $X^n = O_2$, *then* $X^2 = O_2$.

(f) *If the eigenvalues of* $A \in \mathcal{M}_2(\mathbb{C})$ *are distinct* $\lambda_1 \neq \lambda_2$, *then there exists a nonsingular matrix* P *such that*

$$P^{-1}AP = \begin{pmatrix} \lambda_1 & 0 \\ 0 & \lambda_2 \end{pmatrix}.$$

Proof These properties are elementary and left as an exercise to the interested reader. $\qquad\Box$

Theorem 3.4 **The equation** $X^n = A$, $n \geq 2$, $A \in \mathcal{M}_2(\mathbb{C})$ **with** $\det A = 0$.

Let $A \in \mathcal{M}_2(\mathbb{C})$ be such that $\det A = 0$ and let $n \geq 2$ be an integer.

(1) If $\mathrm{Tr}(A) \neq 0$ the equation $X^n = A$ has n solutions in $\mathcal{M}_2(\mathbb{C})$ given by

$$X_k = \frac{z_k}{\mathrm{Tr}(A)} A,$$

where z_k, $k = \overline{1, n}$, are the solutions of the equation $z^n = \mathrm{Tr}(A)$.

(2) If $\mathrm{Tr}(A) = 0$, then:

    (a) If $A \neq O_2$ and $A^2 = O_2$ the equation $X^n = A$ has no solutions in $\mathcal{M}_2(\mathbb{C})$, for $n \geq 2$;

    (b) If $A = O_2$, the solutions of the equation $X^n = A$ are

$$X_{a,b} = \begin{pmatrix} a & b \\ -\dfrac{a^2}{b} & -a \end{pmatrix}, \quad a \in \mathbb{C}, \; b \in \mathbb{C}^* \quad \text{and} \quad X_c = \begin{pmatrix} 0 & 0 \\ c & 0 \end{pmatrix}, \quad c \in \mathbb{C}.$$

Proof Since $X^n = A$, we get that $\det^n X = \det A = 0 \Rightarrow \det X = 0$. It follows, based on part (a) of Lemma 3.2, that $X^n = \mathrm{Tr}^{n-1}(X)X$. We obtain $\mathrm{Tr}^{n-1}(X)X = A$ which implies $\mathrm{Tr}^n(X) = \mathrm{Tr}(A)$.

We distinguish between the following cases.

(1) If $\mathrm{Tr}(A) \neq 0$, we get based on part (d) of Lemma 3.2, that $A^2 \neq O_2$, and the equation $\mathrm{Tr}^n(X) = \mathrm{Tr}(A)$ implies that $\mathrm{Tr}(X) \in \{t_1, t_2, \ldots, t_n\}$, where t_i, $i = \overline{1, n}$ are the solutions of the equation $z^n = \mathrm{Tr}(A)$.

    Thus, for $A \in \mathcal{M}_2(\mathbb{C})$, $A^2 \neq O_2$ and $\det A = 0$, the solutions of the matrix equation $X^n = A$ are

$$X_k = \frac{z_k}{\mathrm{Tr}(A)} A,$$

where z_k, $k = \overline{1, n}$, are the solutions of the equation $z^n = \mathrm{Tr}(A)$.

(2) If $\mathrm{Tr}(A) = 0$, then $A^2 = O_2$ and the equation $X^n = A$ implies that $X^{2n} = A^2 = O_2$ which combined to part (e) of Lemma 3.2 shows that $X^2 = O_2$.

    (a) Thus, if $A \neq O_2$ and $A^2 = O_2$ the equation $X^n = A$ has no solutions in $\mathcal{M}_2(\mathbb{C})$, for $n \geq 2$.

    (b) If $A = O_2$, then $X^n = O_2$ implies $X^2 = O_2$ which has the solutions (see problem **1.8**)

$$X_{a,b} = \begin{pmatrix} a & b \\ -\dfrac{a^2}{b} & -a \end{pmatrix}, \quad a \in \mathbb{C}, \ b \in \mathbb{C}^* \quad \text{and} \quad X_c = \begin{pmatrix} 0 & 0 \\ c & 0 \end{pmatrix}, \quad c \in \mathbb{C}.$$

The theorem is proved. □

Example 3.1 We solve the equation $X^4 = \begin{pmatrix} -1 & -2 \\ 1 & 2 \end{pmatrix}$.

Let $A = \begin{pmatrix} -1 & -2 \\ 1 & 2 \end{pmatrix}$. Then, $\det A = 0$, $\mathrm{Tr}(A) = 1$ and the equation $z^4 = 1$ has the following solutions, the *fourth roots of unity*, $\{1, -1, i, -i\}$. Thus, the solutions of $X^4 = A$ are matrices $\pm A, \pm iA$.

Example 3.2 Now we prove the equation $X^n = \begin{pmatrix} 0 & 1 \\ 0 & 0 \end{pmatrix}$ has no solutions for $n \geq 2$.

Squaring both sides of the equation we have $X^{2n} = O_2 \Rightarrow X^2 = O_2$ and, since $n \geq 2$, we obtain

$$X^n = O_2 \neq \begin{pmatrix} 0 & 1 \\ 0 & 0 \end{pmatrix}.$$

Example 3.3 We determine the matrices $X = \begin{pmatrix} a & b \\ -c & -d \end{pmatrix} \in \mathcal{M}_2(\mathbb{Z})$, where a, b, c, d are prime numbers, such that $X^2 = O_2$.

From the general solution of the equation $X^2 = O_2$ we get that

$$X_{a,b} = \begin{pmatrix} a & b \\ -\dfrac{a^2}{b} & -a \end{pmatrix}$$

and the condition $\dfrac{a^2}{b}$ is a prime number implies that $a = b$.
Thus, the solutions of our equation are

$$X = \begin{pmatrix} p & p \\ -p & -p \end{pmatrix} = p \begin{pmatrix} 1 & 1 \\ -1 & -1 \end{pmatrix},$$

where p is a prime number.

Theorem 3.5 **The equation** $X^n = aI_2$, $a \in \mathbb{C}^*$, $n \geq 2$.

Let $a \in \mathbb{C}^*$ and let $n \geq 2$ be an integer. The solutions of the equation $X^n = aI_2$ are given by

$$X = P \begin{pmatrix} a_i & 0 \\ 0 & a_j \end{pmatrix} P^{-1},$$

where P is any invertible matrix and a_i, $i = \overline{1, n}$, are the solutions of the equation $z^n = a$.

Proof We start by observing that if $X \in \mathcal{M}_2(\mathbb{C})$ is a solution of the equation $X^n = aI_2$, then the matrix $X_P = P^{-1}XP$ is also a solution, for any invertible matrix P. This can be proved as follows

$$X_P^n = \left(P^{-1}XP\right)\left(P^{-1}XP\right)\cdots\left(P^{-1}XP\right) = P^{-1}X^nP = P^{-1}\left(aI_2\right)P = aI_2.$$

We distinguish between the cases when the eigenvalues of X are distinct or not.

- If the eigenvalues of X are distinct, we have, based on part (f) of Lemma 3.2, that

$$X_P = \begin{pmatrix} \lambda_1 & 0 \\ 0 & \lambda_2 \end{pmatrix}$$

and the matrix equation becomes

$$\begin{pmatrix} \lambda_1^n & 0 \\ 0 & \lambda_2^n \end{pmatrix} = \begin{pmatrix} a & 0 \\ 0 & a \end{pmatrix}.$$

This implies $\lambda_1, \lambda_2 \in \{a_1, a_2, \ldots, a_n\}$, where a_i, $i = \overline{1, n}$, are the solutions of the equation $z^n = a$.

Thus, some of the solutions of the matrix equation $X^n = aI_2$ are given by

$$X = P \begin{pmatrix} a_i & 0 \\ 0 & a_j \end{pmatrix} P^{-1},$$

where $a_i \neq a_j$ are the arbitrary solutions of the equation $z^n = a$ and P is any invertible matrix.

- If the eigenvalues of X are equal $\lambda_1 = \lambda_2 = \lambda$, we have based on part (d) of Lemma 3.2 that $(X - \lambda I_2)^2 = O_2$. If $Y = X - \lambda I_2$, then $X = \lambda I_2 + Y$ with $Y^2 = O_2$. We have $X^n = \lambda^n I_2 + n\lambda^{n-1}Y$ and the equation $X^n = aI_2$ becomes $\lambda^n I_2 + n\lambda^{n-1}Y = aI_2$ which implies that $n\lambda^{n-1}Y = (a - \lambda^n)I_2$.

Since $a \neq 0$ we obtain that $\lambda \neq 0$ and $Y^2 = O_2$ combined to $Y = \dfrac{a - \lambda^n}{n\lambda^{n-1}}I_2$

implies that $a = \lambda^n$ and $Y = O_2$. Therefore $X = a_i I_2$, where a_i, $i = \overline{1, n}$, are the solutions of the equation $\lambda^n = a$.

In conclusion, the solutions of the equation $X^n = aI_2$ are

$$X = P \begin{pmatrix} a_i & 0 \\ 0 & a_j \end{pmatrix} P^{-1},$$

where a_i, a_j are the arbitrary solutions of the equation $z^n = a$ and P is any invertible matrix. □

Lemma 3.3 The nth roots of a special diagonal matrix.

Let $\alpha, \beta \in \mathbb{C}$ with $\alpha \neq \beta$ and let $n \geq 2$ be an integer. The solutions of the equation $X^n = \begin{pmatrix} \alpha & 0 \\ 0 & \beta \end{pmatrix}$ are given by $X = \begin{pmatrix} a & 0 \\ 0 & d \end{pmatrix}$, where $a, d \in \mathbb{C}$ with $a \neq d$ and $a^n = \alpha$ and $d^n = \beta$.

Proof Let $X = \begin{pmatrix} a & b \\ c & d \end{pmatrix} \in \mathcal{M}_2(\mathbb{C})$ such that $X^n = \begin{pmatrix} \alpha & 0 \\ 0 & \beta \end{pmatrix}$. Since X commutes with $\begin{pmatrix} \alpha & 0 \\ 0 & \beta \end{pmatrix}$ we get that $(\alpha - \beta)b = 0$ and $(\alpha - \beta)c = 0$. These imply, since $\alpha \neq \beta$, that $b = c = 0$. It follows that

$$X^n = \begin{pmatrix} a & 0 \\ 0 & d \end{pmatrix}^n = \begin{pmatrix} a^n & 0 \\ 0 & d^n \end{pmatrix} = \begin{pmatrix} \alpha & 0 \\ 0 & \beta \end{pmatrix},$$

and the lemma is proved. □

Nota bene. Lemma 3.3 states that, under certain conditions, the nth roots of a diagonal matrix are diagonal matrices.

Theorem 3.6 The equation $X^n = A$, when A has distinct eigenvalues.

Let $A \in \mathcal{M}_2(\mathbb{C})$ be a matrix which has distinct eigenvalues. The solutions of the equation $X^n = A$ are given by

$$X = P_A \begin{pmatrix} \alpha & 0 \\ 0 & \beta \end{pmatrix} P_A^{-1},$$

where P_A is the invertible matrix which verifies $P_A^{-1}AP_A = \begin{pmatrix} \lambda_1 & 0 \\ 0 & \lambda_2 \end{pmatrix}$ and $\alpha^n = \lambda_1$, $\beta^n = \lambda_2$, where $\lambda_1 \neq \lambda_2$ are the eigenvalues of A.

Proof Let $\lambda_1 \neq \lambda_2$ be the eigenvalues of A and let P_A be the invertible matrix which verifies

$$P_A^{-1}AP_A = \begin{pmatrix} \lambda_1 & 0 \\ 0 & \lambda_2 \end{pmatrix}.$$

We have

$$(P_A^{-1} X P_A)^n = P_A^{-1} X^n P_A = P_A^{-1} A P_A = \begin{pmatrix} \lambda_1 & 0 \\ 0 & \lambda_2 \end{pmatrix}.$$

This implies, based on Lemma 3.3, that $P_A^{-1} X P_A$ is a diagonal matrix, i.e.

$$P_A^{-1} X P_A = \begin{pmatrix} \alpha & 0 \\ 0 & \beta \end{pmatrix}.$$

It follows that

$$\begin{pmatrix} \alpha & 0 \\ 0 & \beta \end{pmatrix}^n = \begin{pmatrix} \lambda_1 & 0 \\ 0 & \lambda_2 \end{pmatrix}$$

and this implies that $\alpha^n = \lambda_1$ and $\beta^n = \lambda_2$.

Thus, the solutions of the equation are

$$X = P_A \begin{pmatrix} \alpha & 0 \\ 0 & \beta \end{pmatrix} P_A^{-1},$$

where P_A is the invertible matrix which verifies $P_A^{-1} A P_A = J_A$ and $\alpha^n = \lambda_1$, $\beta^n = \lambda_2$. The theorem is proved. $\square$

Another method for proving Theorem 3.6 is based on parts (b) and (c) of Lemma 3.2.

Corollary 3.1 The nth roots of an antidiagonal matrix.

Let $a, b \in \mathbb{R}$ such that $ab > 0$ and let $n \geq 2$ be an integer. The solutions, in $\mathcal{M}_2(\mathbb{C})$, of the equation

$$X^n = \begin{pmatrix} 0 & a \\ b & 0 \end{pmatrix}$$

are given by

$$X_{k,j} = \frac{\sqrt[2n]{ab}}{2} \begin{pmatrix} \epsilon_k + \mu_j & \frac{a}{\sqrt{ab}} (\epsilon_k - \mu_j) \\ \frac{\sqrt{ab}}{a} (\epsilon_k - \mu_j) & \epsilon_k + \mu_j \end{pmatrix},$$

where $\epsilon_k = \exp\left(\frac{2k\pi}{n} i\right)$, $k = \overline{0, n-1}$, are the nth roots of unity and $\mu_j = \exp\left(\frac{(2j+1)\pi}{n} i\right)$, $j = \overline{0, n-1}$, are the nth roots of -1.

Theorem 3.7 The equation $X^n = A$, where $A \neq aI_2$, $a \in \mathbb{C}$.

Let $A \in \mathcal{M}_2(\mathbb{C})$ be such that $A \neq aI_2$, $a \in \mathbb{C}$, and let $n \geq 2$ be an integer. Let $\lambda_1 \neq \lambda_2$ be the eigenvalues of A and let $\mu_1, \mu_2 \in \mathbb{C}$ be fixed such that $\mu_1^n = \lambda_1$ and $\mu_2^n = \lambda_2$. The solutions of the equation $X^n = A$ are given by

$$X = \alpha A + \beta I_2,$$

with

$$\alpha = \frac{\mu_1 \epsilon_k - \mu_2 \epsilon_p}{\lambda_1 - \lambda_2} \quad \text{and} \quad \beta = \frac{\mu_2 \epsilon_p \lambda_1 - \mu_1 \epsilon_k \lambda_2}{\lambda_1 - \lambda_2}, \quad \lambda_1 \neq \lambda_2,$$

where ϵ_k, ϵ_p are the nth roots of unity.

Proof We have, since $A \neq aI_2$, $a \in \mathbb{C}$, that the matrix X which verifies the equation $X^n = A$ commutes with A. This implies, based on Theorem 1.1, that $X = \alpha A + \beta I_2$, for some $\alpha, \beta \in \mathbb{C}$. If λ_1, λ_2 are the eigenvalues of A, then the eigenvalues of X are $\alpha \lambda_1 + \beta$ and $\alpha \lambda_2 + \beta$ and the eigenvalues of X^n are $(\alpha \lambda_1 + \beta)^n$ and $(\alpha \lambda_2 + \beta)^n$. The equation $X^n = A$ implies that $(\alpha \lambda_1 + \beta)^n = \lambda_1$ and $(\alpha \lambda_2 + \beta)^n = \lambda_2$.

Let $\mu_1 \in \mathbb{C}$ such that $\mu_1^n = \lambda_1$ and let $\mu_2 \in \mathbb{C}$ such that $\mu_2^n = \lambda_2$. The last two equations imply that

$$\begin{cases} \alpha \lambda_1 + \beta = \mu_1 \epsilon_k \\ \alpha \lambda_2 + \beta = \mu_2 \epsilon_p \end{cases}$$

where ϵ_k, ϵ_p are the nth roots of unity. Solving this system of equations we obtain the values of α and β as given above. The theorem is proved. $\square$

Lemma 3.4 The nth roots of a Jordan cell.

Let $\lambda \in \mathbb{C}^*$ and let $n \geq 2$ be an integer. The solutions of the equation $X^n = \begin{pmatrix} \lambda & 1 \\ 0 & \lambda \end{pmatrix} \in \mathcal{M}_2(\mathbb{C})$ are given by

$$X = \begin{pmatrix} a & \dfrac{1}{na^{n-1}} \\ 0 & a \end{pmatrix},$$

where $a \in \mathbb{C}$ with $a^n = \lambda$.

Proof Let $X = \begin{pmatrix} a & b \\ c & d \end{pmatrix}$. Since X commutes with $\begin{pmatrix} \lambda & 1 \\ 0 & \lambda \end{pmatrix}$ we obtain, after simple calculations, that $a = d$ and $c = 0$. It follows that $X = \begin{pmatrix} a & b \\ 0 & a \end{pmatrix}$ and the equation

$$X^n = \begin{pmatrix} a^n & na^{n-1}b \\ 0 & a^n \end{pmatrix} = \begin{pmatrix} \lambda & 1 \\ 0 & \lambda \end{pmatrix}$$

implies that $a^n = \lambda$ and $na^{n-1}b = 1$.   □

Nota bene. Lemma 3.4 states that, under certain conditions, the nth roots of a triangular matrix are triangular.

Theorem 3.8   The equation $X^n = A$, when A has equal nonzero eigenvalues.

Let $A \in \mathcal{M}_2(\mathbb{C})$ be a matrix which has equal nonzero eigenvalues such that $A \neq \alpha I_2$, $\alpha \in \mathbb{C}$, and let $n \geq 2$ be an integer. The solutions of the equation $X^n = A$ are given by

$$X = P_A \begin{pmatrix} a & \dfrac{1}{na^{n-1}} \\ 0 & a \end{pmatrix} P_A^{-1},$$

where P_A is the invertible matrix which verifies $P_A^{-1}AP_A = J_A$ and $a \in \mathbb{C}$ with $a^n = \lambda$.

Proof The theorem can be proved by using the same ideas as in the proof of Theorem 3.6 combined with Lemma 3.4.   □

Theorem 3.9   A special quadratic equation.

Let $a, b, c \in \mathbb{C}$, $a \neq 0$ and let $A \in \mathcal{M}_2(\mathbb{C})$. The quadratic equation

$$aX^2 + bX + cI_2 = A$$

reduces to an equation of the form $Y^2 = B$, for some $B \in \mathcal{M}_2(\mathbb{C})$.

Proof The equation $aX^2 + bX + cI_2 = A$ implies that

$$X^2 + \frac{b}{a}X + \frac{c}{a}I_2 = \frac{1}{a}A \Leftrightarrow \left(X + \frac{b}{2a}I_2\right)^2 = \frac{1}{a}A + \left(\frac{b^2}{4a^2} - \frac{c}{a}\right)I_2.$$

If Y and B are the matrices

$$Y = X + \frac{b}{2a}I_2 \quad \text{and} \quad B = \frac{1}{a}A + \frac{b^2 - 4ac}{4a^2}I_2,$$

the equation to solve becomes $Y^2 = B$.   □

The next theorem shows which *real matrices* admit *real square roots*.

Theorem 3.10 [4] *Let* $A \in \mathcal{M}_2(\mathbb{R})$ *be a given matrix. There are matrices* $S \in \mathcal{M}_2(\mathbb{R})$ *such that* $S^2 = A$ *if and only if* $\det A \geq 0$ *and, either* $A = -\sqrt{\det A}\,I_2$ *or* $\mathrm{Tr}(A) + 2\sqrt{\det A} > 0$. *Obviously, in the latter case,* $\mathrm{Tr}(A) + 2\sqrt{\det A} = 0$.

3.4.1 An artistry of binomial equations. The nth real roots of aI_2, $a \in \mathbb{R}^*$

In this section we solve in $\mathcal{M}_2(\mathbb{R})$ the equation $X^n = aI_2$, where $a \in \mathbb{R}^*$ and $n \in \mathbb{N}$. Let $X \in \mathcal{M}_2(\mathbb{R})$ be a solution of the equation $X^n = aI_2$.

■ First we consider the case when the eigenvalues of X are real.

(a) If n is odd, then $X = \sqrt[n]{a}I_2$, since the Jordan canonical form of a matrix X which verifies the equation $X^n = aI_2$ is a diagonal matrix and its eigenvalues verify the equation $\lambda^n = a$, which has the unique real solution $\lambda = \sqrt[n]{a}$.

(b) If n is even and $a > 0$, then

$$X_1 = \sqrt[n]{a}I_2, \quad X_2 = -\sqrt[n]{a}I_2 \quad \text{and} \quad X_3 = \sqrt[n]{a}P \begin{pmatrix} 1 & 0 \\ 0 & -1 \end{pmatrix} P^{-1},$$

where P is any arbitrary invertible matrix. We observe that $X_3 = \sqrt[n]{a}A$, with $A^2 = I_2$. The solution X_1 corresponds to the case when the eigenvalues of X are $\lambda_1 = \lambda_2 = \sqrt[n]{a}$, X_2 when $\lambda_1 = \lambda_2 = -\sqrt[n]{a}$ and X_3 when $\lambda_1 = \sqrt[n]{a}$ and $\lambda_2 = -\sqrt[n]{a}$ respectively.

■ Now we consider the case when $\lambda_1, \lambda_2 \in \mathbb{C} \setminus \mathbb{R}$. In this case $\lambda_1^n = \lambda_2^n = a$ and $\lambda_2 = \overline{\lambda_1}$. The complex canonical form of X is the matrix $\begin{pmatrix} \lambda_1 & 0 \\ 0 & \lambda_1 \end{pmatrix}$, where $\lambda_1 = \alpha + i\beta$, $\alpha, \beta \in \mathbb{R}$, $\beta \neq 0$, and the real canonical form of X is given by $\begin{pmatrix} \alpha & -\beta \\ \beta & \alpha \end{pmatrix}$. It follows that $X = P^{-1} \begin{pmatrix} \alpha & -\beta \\ \beta & \alpha \end{pmatrix} P$, where $P \in \mathcal{M}_2(\mathbb{R})$ is an invertible matrix. We obtain $X = \alpha I_2 + \beta B$, where $B \in \mathcal{M}_2(\mathbb{R})$ verifies the equation $B^2 = -I_2$.

Let $P = \begin{pmatrix} a & b \\ c & d \end{pmatrix}$ and let $\Delta = ad - bc \neq 0$. Then

$$P^{-1} \begin{pmatrix} \alpha & -\beta \\ \beta & \alpha \end{pmatrix} P = \alpha I_2 + \frac{\beta}{\Delta} \begin{pmatrix} -(ab + cd) & -(b^2 + d^2) \\ a^2 + c^2 & ab + cd \end{pmatrix} = \alpha I_2 + \beta B,$$

where $B = \dfrac{1}{\Delta} \begin{pmatrix} -(ab + cd) & -(b^2 + d^2) \\ a^2 + c^2 & ab + cd \end{pmatrix}$ and $B^2 = -I_2$.

It follows that the solutions of the equation $X^n = aI_2$, which have eigenvalues in $\mathbb{C} \setminus \mathbb{R}$, are of the following form:

(i) $X = \sqrt[n]{a} \left(\cos \dfrac{2k\pi}{n} I_2 + \sin \dfrac{2k\pi}{n} B \right)$, $k \in \{1, 2, \ldots, n-1\}$, $B \in \mathcal{M}_2(\mathbb{R})$ with $B^2 = -I_2$, for n odd or n even and $a > 0$.

(ii) $X = \sqrt[n]{-a}\left(\cos\dfrac{(2k+1)\pi}{n}I_2 + \sin\dfrac{(2k+1)\pi}{n}B\right)$, $k \in \{0, 1, \ldots, n-1\}$,

$B \in \mathcal{M}_2(\mathbb{R})$ with $B^2 = -I_2$, for n even and $a < 0$.

Conversely we prove that matrices in (i) and (ii) verify the equation $X^n = aI_2$.

(i)   We have

$$X^n = a\left(\cos\frac{2k\pi}{n}I_2 + \sin\frac{2k\pi}{n}B\right)^n$$

$$= a\sum_{j=0}^{n}\binom{n}{j}\sin^j\frac{2k\pi}{n}B^j\cos^{n-j}\frac{2k\pi}{n}I_2$$

$$= a\sum_{j=2l}\binom{n}{j}\sin^j\frac{2k\pi}{n}(-1)^l\cos^{n-j}\frac{2k\pi}{n}I_2$$

$$+ a\sum_{j=2l-1}\binom{n}{j}\sin^j\frac{2k\pi}{n}(-1)^{l-1}B\cos^{n-j}\frac{2k\pi}{n}I_2$$

$$= a\Re\left[\left(\cos\frac{2k\pi}{n} + i\sin\frac{2k\pi}{n}\right)^n\right]I_2 + a\Im\left[\left(\cos\frac{2k\pi}{n} + i\sin\frac{2k\pi}{n}\right)^n\right]B$$

$$= aI_2 + 0B$$

$$= aI_2.$$

(ii)   As in the previous case

$$X^n = -a\Re\left[\left(\cos\frac{(2k+1)\pi}{n} + i\sin\frac{(2k+1)\pi}{n}\right)^n\right]I_2$$

$$- a\Im\left[\left(\cos\frac{(2k+1)\pi}{n} + i\sin\frac{(2k+1)\pi}{n}\right)^n\right]B$$

$$= -a\left[(-1)I_2 + 0B\right]$$

$$= aI_2.$$

When $a = 1$ we have the following corollary.

Corollary 3.2 The nth real roots of I_2.

The solutions, in $\mathcal{M}_2(\mathbb{R})$, of the equation $X^n = I_2$ are of the following form

$$X = \cos \frac{2k\pi}{n} I_2 + \sin \frac{2k\pi}{n} B, \quad k \in \{0, 1, \ldots, n-1\},$$

where $B \in \mathcal{M}_2(\mathbb{R})$ with $B^2 = -I_2$ and if n is even we also have the matrices $-I_2$ and

$$X = P \begin{pmatrix} 1 & 0 \\ 0 & -1 \end{pmatrix} P^{-1},$$

where P is any arbitrary invertible matrix.

Remark 3.6 Observe that if $X = P \begin{pmatrix} 1 & 0 \\ 0 & -1 \end{pmatrix} P^{-1}$, then $X^2 = I_2$, so X is involutory. Conversely, if n is even, any involutory matrix X verifies the equation $X^n = I_2$. It follows that the solutions of the equation $X^n = I_2$ are

$$X = \cos \frac{2k\pi}{n} I_2 + \sin \frac{2k\pi}{n} B, \quad k = 0, 1, \ldots, n-1,$$

where $B \in \mathcal{M}_2(\mathbb{R})$ with $B^2 = -I_2$, for any n, and if n is even we also have the solutions $X = A$, $A \in \mathcal{M}_2(\mathbb{R})$, with $A^2 = I_2$. It follow from Example 1.2 that B is of the following form

$$B = \begin{pmatrix} a & b \\ -\dfrac{1+a^2}{b} & -a \end{pmatrix}, \quad a \in \mathbb{R}, \quad b \in \mathbb{R}^*.$$

We collect these calculations and state the following theorem.

Theorem 3.11 The nth real roots of aI_2.

The matrix $X \in \mathcal{M}_2(\mathbb{R})$ verifies the equation $X^n = aI_2$, $a \in \mathbb{R}^*$ if and only if X is of the following form:

■ *when n is odd or n is even and $a > 0$*

$$X = \sqrt[n]{a} \left(\cos \frac{2k\pi}{n} I_2 + \sin \frac{2k\pi}{n} B \right), \quad k \in \{0, 1, \ldots, n-1\},$$

(continued)

Theorem 3.11 (continued)

$B \in \mathcal{M}_2(\mathbb{R})$ with $B^2 = -I_2$. When n is even and $a > 0$ we also have the solutions

$$X = \sqrt[n]{a}A, \quad A \in \mathcal{M}_2(\mathbb{R}) \quad \text{with} \quad A^2 = I_2.$$

■ when n is even and $a < 0$

$$X = \sqrt[n]{-a}\left(\cos\frac{(2k+1)\pi}{n}I_2 + \sin\frac{(2k+1)\pi}{n}B\right), \quad k \in \{0, 1, \ldots, n-1\},$$

$B \in \mathcal{M}_2(\mathbb{R})$ with $B^2 = -I_2$.

The next corollary gives the nth real roots of $-I_2$.

Corollary 3.3 The nth real roots of $-I_2$.

The solutions, in $\mathcal{M}_2(\mathbb{R})$, of the equation $X^n = -I_2$ are of the following form:

■ for n odd

$$X = -\cos\frac{2k\pi}{n}I_2 - \sin\frac{2k\pi}{n}B, \quad k \in \{0, 1, \ldots, n-1\},$$

where $B \in \mathcal{M}_2(\mathbb{R})$ with $B^2 = -I_2$.
■ for n even

$$X = \cos\frac{(2k+1)\pi}{n}I_2 + \sin\frac{(2k+1)\pi}{n}B, \quad k \in \{0, 1, \ldots, n-1\},$$

where $B \in \mathcal{M}_2(\mathbb{R})$ with $B^2 = -I_2$.

Example 3.4 **The product is zero and the sum of their nth power is I_2.**

Let $n \in \mathbb{N}$. We determine matrices $A, B \in \mathcal{M}_2(\mathbb{R})$ such that

$$AB = O_2 \quad \text{and} \quad A^n + B^n = I_2.$$

We have $A^n = I_2 - B^n \Rightarrow O_2 = A^n B = B - B^{n+1} \Rightarrow B^{n+1} - B = O_2 \Rightarrow$ $\det B = 0$ or $\det^n B = 1$. Similarly, $A^{n+1} = A \Rightarrow \det A = 0$ or $\det^n A = 1$.
If $\det A \neq 0 \Rightarrow A$ is invertible $\Rightarrow B = O_2 \Rightarrow A^n = I_2$.
If $\det B \neq 0 \Rightarrow B$ is invertible $\Rightarrow A = O_2 \Rightarrow B^n = I_2$.

If $\det A = \det B = 0$, then $A^2 = t_A A$ and $B^2 = t_B B$, where $t_A = \mathrm{Tr}(A)$ and $t_B = \mathrm{Tr}(B)$. The equation $A^n + B^n = I_2$ implies that $t_A^{n-1} A + t_B^{n-1} B = I_2$.

Since $A^{n+1} = A$ we get that $(t_A^n - 1) A = O_2$. If $t_A^n \neq 1$, then $A = O_2 \Rightarrow B^n = I_2$, which contradicts $\det B = 0$. Thus, $t_A^n = 1 \Rightarrow t_A = \pm 1$. Similarly, we have $t_B = \pm 1$.

If $t_A = t_B = 1$ we obtain the matrices $A^2 = A$ and $B = I_2 - A$.

If $t_A = 1$ and $t_B = -1$ (this implies n is even), then $A^2 = A$ and $B^2 = -B$ and we obtain the matrices $A^2 = A$ and $B = A - I_2$.

If $t_A = -1$ and $t_B = 1$ (this implies n is even), then $A^2 = -A$ and $B^2 = B$ and we obtain the matrices $B^2 = B$ and $A = B - I_2$.

If $t_A = t_B = -1$ (this implies n is even) we obtain the matrices $A^2 = -A$ and $B = -A - I_2$.

3.4.2 Problems

3.52 *Quadratic binomial equations.*

(a) Let $\mathbb{N}_0 = \{0\} \cup \mathbb{N}$. Find all matrices $A \in \mathcal{M}_2(\mathbb{N}_0)$ such that $A^2 - 6A + 5I_2 = O_2$.
(b) Let $a, b \in \mathbb{N}_0$ with $a^2 - 4b < 0$. Prove the equation $A^2 - aA + bI_2 = O_2$ does not have solutions in $\mathcal{M}_2(\mathbb{N}_0)$.

3.53 Give an example of a matrix $A \in \mathcal{M}_2(\mathbb{C})$ that has exactly two square roots in $\mathcal{M}_2(\mathbb{C})$.

3.54 Find all $X \in \mathcal{M}_2(\mathbb{R})$ such that $X^2 = \begin{pmatrix} 6 & 5 \\ 10 & 11 \end{pmatrix}$.

3.55 (a) Find all matrices $A \in \mathcal{M}_2(\mathbb{R})$ such that $A^2 = \begin{pmatrix} 1 & 0 \\ d & 2 \end{pmatrix}$, where $d = \det A$.

(b) Find all matrices $A \in \mathcal{M}_2(\mathbb{R})$ such that $A^2 = \begin{pmatrix} 1 & 0 \\ t & 2 \end{pmatrix}$, where $t = \mathrm{Tr}(A)$.

3.56 Is there a real 2×2 matrix A such that

$$A^2 = \begin{pmatrix} -1 & 0 \\ 0 & -1-\epsilon \end{pmatrix}, \quad \epsilon > 0?$$

3.57 [58, p. 140] For which positive integer n is there a matrix $A \in \mathcal{M}_2(\mathbb{Z})$ such that $A^n = I_2$ and $A^k \neq I_2$ for $0 < k < n$?

3.58 Determine all matrices $A \in \mathcal{M}_2(\mathbb{R})$ such that $AA^T = \begin{pmatrix} 1 & 1 \\ 1 & 1 \end{pmatrix}$.

3.59 Find all matrices $A \in \mathcal{M}_2(\mathbb{R})$ such that $AA^T A = \begin{pmatrix} \alpha & \alpha \\ \alpha & \alpha \end{pmatrix}$, where $\alpha \in \mathbb{R}$.

3.60 [27] Let $A \in \mathcal{M}_2(\mathbb{R})$ such that $AA^T = \begin{pmatrix} a & b \\ b & a \end{pmatrix}$, where $a > b > 0$. Prove that

$AA^T = A^T A$ if and only if $A = \begin{pmatrix} \alpha & \beta \\ \beta & \alpha \end{pmatrix}$ or $A = \begin{pmatrix} \beta & \alpha \\ \alpha & \beta \end{pmatrix}$, where

$$\alpha = \frac{\pm\sqrt{a+b} \pm \sqrt{a-b}}{2} \quad \text{and} \quad \beta = \frac{\pm\sqrt{a+b} \mp \sqrt{a-b}}{2}.$$

3.61 If $A \in \mathcal{M}_2(\mathbb{Z})$ is such that $A^4 = I_2$, then either $A^2 = I_2$ or $A^2 = -I_2$.

3.62 If $A \in \mathcal{M}_2(\mathbb{Z})$ and there exists $n \in \mathbb{N}$ with $(n, 6) = 1$ such that $A^n = I_2$, then $A = I_2$.

3.63 Let $A \in \mathcal{M}_2(\mathbb{Q})$ be such that there exists $n \in \mathbb{N}$ with $A^n = -I_2$. Prove that either $A^2 = -I_2$ or $A^3 = -I_2$.

3.64 The order of an element in $\mathcal{M}_2(\mathbb{Q})$.

Let $A \in \mathcal{M}_2(\mathbb{Q})$ be such that there exists $n \in \mathbb{N}$ with $A^n = I_2$. Prove that $A^{12} = I_2$.

3.65 Determine all matrices $A \in \mathcal{M}_2(\mathbb{Z})$ such that $A^3 = \begin{pmatrix} 5 & 8 \\ 8 & 13 \end{pmatrix}$.

3.66 Solve in $\mathcal{M}_2(\mathbb{R})$ the equation $X^3 = \begin{pmatrix} 1 & -2 \\ 2 & -3 \end{pmatrix}$.

3.67 The real cubic roots of I_2.
(a) Determine all matrices $X \in \mathcal{M}_2(\mathbb{R})$ such that $X^3 = I_2$.
(b) Let $\epsilon \neq 1$ be a cubic root of unity. Determine all matrices $X \in \mathcal{M}_2(\mathbb{R})$ such that $X^2 + \epsilon X + \epsilon^2 I_2 = O_2$.

3.68 Find all $A \in \mathcal{M}_2(\mathbb{R})$ such that $AA^T A = I_2$.

3.69 Prove that there is no $A \in \mathcal{M}_2(\mathbb{Q})$ such that $A^4 + 15A^2 + 2I_2 = O_2$.

3.70 Let $\text{SL}_2(\mathbb{Z}) = \{X \in \mathcal{M}_2(\mathbb{Z}) : \det X = 1\}$.

(a) Prove that the equation $X^2 + X^{-2} = I_2$ has no solutions in $SL_2(\mathbb{Z})$.

(b) Prove that the equation $X^2 + X^{-2} = -I_2$ has solutions in $SL_2(\mathbb{Z})$ and determine the set $\{X^n + X^{-n} : X^2 + X^{-2} = -I_2,\ n \in \mathbb{N}\}$.

3.71 *A quintic equation with a unique solution.*

Prove that for any $a \in \mathbb{R}$ the equation

$$X^5 = \begin{pmatrix} a & 1-a \\ 1+a & -a \end{pmatrix}$$

has a unique solution in $\mathcal{M}_2(\mathbb{R})$.

3.72 Let $n \geq 1$ be an integer and let $A = \begin{pmatrix} \cos\alpha & \sin\alpha \\ -\sin\alpha & \cos\alpha \end{pmatrix}$. Find $\alpha \in \mathbb{R}$ such that $A^n = I_2$.

3.73 The nth real roots of the rotation matrix.

Let $t \in (0, \pi)$ be fixed. Find all solutions $X \in \mathcal{M}_2(\mathbb{R})$ of the equation

$$X^n = \begin{pmatrix} \cos t & -\sin t \\ \sin t & \cos t \end{pmatrix}.$$

3.74 Let $n \geq 2$ be an integer. Solve in $\mathcal{M}_2(\mathbb{C})$ the equation

$$X^n = \begin{pmatrix} 1 & 2 \\ 2 & 4 \end{pmatrix}.$$

3.75 Solve in $\mathcal{M}_2(\mathbb{C})$ the equation $X^n = \begin{pmatrix} 1 & a \\ 0 & 1 \end{pmatrix}$, $a \in \mathbb{C}^*$.

3.76 Let $A \in \mathcal{M}_2(\mathbb{C})$, $A \neq O_2$ and $\det A = 0$. Prove that the equation $X^n = A$, $n \geq 2$, has solutions if and only if $A^2 \neq O_2$.

3.77 Let $n \geq 2$ be an integer. Solve in $\mathcal{M}_2(\mathbb{C})$ the equation

$$X^n = \begin{pmatrix} a & b \\ b & a \end{pmatrix}, \quad a, b \in \mathbb{C},\ b \neq 0.$$

3.78 Let $n \in \mathbb{N}$, $n \geq 2$, $a \in \mathbb{R}$, and $b \in \mathbb{R}^*$. Solve in $\mathcal{M}_2(\mathbb{R})$ the equation

$$X^n = \begin{pmatrix} a & b \\ -b & a \end{pmatrix}.$$

3.79 Prove that the equation

$$X^n = \begin{pmatrix} 3 & -1 \\ 0 & 0 \end{pmatrix}, \quad n \in \mathbb{N}, \quad n \geq 2,$$

has no solutions in $\mathcal{M}_2(\mathbb{Q})$.

3.80 Solve in $\mathcal{M}_2(\mathbb{Z})$ the equation $X^3 - 3X = \begin{pmatrix} -7 & -9 \\ 3 & 2 \end{pmatrix}$.

3.81 Solve in $\mathcal{M}_2(\mathbb{R})$ the equation $X^3 + X^2 = \begin{pmatrix} 1 & 1 \\ 1 & 1 \end{pmatrix}$.

3.82 Two special equations with no solutions.
  (a) Prove that the equation $A^3 - A - I_2 = O_2$ has no solutions in $\mathcal{M}_2(\mathbb{Q})$.
  (b) Let $n \in \mathbb{N}, n \geq 2$. Prove that the equation $A^n - AC(0, 1) - I_2 = O_2$ has

no solutions in $\mathcal{M}_2(\mathbb{Q})$, where $C(0, 1) = \begin{pmatrix} 0 & 1 \\ 1 & 0 \end{pmatrix}$.

3.83 A jewel of binomial matrix theory.
  Let $n, k \geq 2$ be integers. Prove that the equation $A^n - A^k C(a, b) - I_2 = O_2$

has no solutions in $\mathcal{M}_2(\mathbb{Q})$, where $C(a, b) = \begin{pmatrix} a & b \\ b & a \end{pmatrix}$ with $a \geq 0$ and $b \geq 1$

integers. ($C(a, b)$ is the circulant matrix defined in problem **3.11**).

3.84 *A matrix equation with determinants and traces.*
  (a) Solve in $\mathcal{M}_2(\mathbb{Z})$ the equation

$$X^t + X = \begin{pmatrix} 2 & 0 \\ 3 & 2 \end{pmatrix}, \quad \text{where} \quad t = \text{Tr}(X).$$

  (b) Solve in $\mathcal{M}_2(\mathbb{Z})$ the equation

$$X^d + X = \begin{pmatrix} 2 & 0 \\ 3 & 2 \end{pmatrix}, \quad \text{where} \quad d = \det X.$$

3.85 Let $n \geq 3$ be an integer. Find $X \in \mathcal{M}_2(\mathbb{R})$ such that

$$X^n + X^{n-2} = \begin{pmatrix} 1 & -1 \\ -1 & 1 \end{pmatrix}.$$

3.86 Two matrix equations over $\mathcal{M}_2(\mathbb{Z})$.

(a) Let $n \in \mathbb{N}$. Solve in $\mathcal{M}_2(\mathbb{Z})$ the equation $X^{2n+1} + X = I_2$.
(b) Let $n \in \mathbb{N}$. Solve in $\mathcal{M}_2(\mathbb{Z})$ the equation $X^{2n+1} - X = I_2$.

3.87 [41] Find all prime numbers p such that there exists a 2×2 matrix A with integer entries, other than the identity matrix I_2, for which $A^p + A^{p-1} + \cdots + A = pI_2$.

3.88 Let $n \in \mathbb{N}$. Solve in $\mathcal{M}_2(\mathbb{R})$ the equation

$$A + A^3 + \cdots + A^{2n-1} = \begin{pmatrix} n & n^2 \\ 0 & n \end{pmatrix}.$$

3.89 Two cousin equations.

Let $m, n \geq 2$ be integers and let $A \in \mathcal{M}_2(\mathbb{C})$ be a given matrix. Prove that the equation $X^m = A$ has solutions in $\mathcal{M}_2(\mathbb{C})$ if and only if the equation $Y^n = A$ has solutions in $\mathcal{M}_2(\mathbb{C})$.

3.90 Viète's formulae for a quadratic matrix equation.

Let $A, B \in \mathcal{M}_2(\mathbb{C})$ be two given matrices and consider the quadratic equation in $\mathcal{M}_2(\mathbb{C})$

$$X^2 - AX + B = O_2.$$

Prove that if X_1, X_2 are two solutions of this equation and if the matrix $X_1 - X_2$ is invertible, then

$$\mathrm{Tr}(X_1 + X_2) = \mathrm{Tr}(A) \quad \text{and} \quad \det(X_1 X_2) = \det B.$$

3.91 Matrix delights in $\mathscr{M}_2\left(\mathbb{Z}_p\right)$.

Let p be a prime number. Prove that:

(a) $\begin{pmatrix} \widehat{a} & \widehat{b} \\ \widehat{0} & \widehat{a} \end{pmatrix}^p = \begin{pmatrix} \widehat{a} & \widehat{0} \\ \widehat{b} & \widehat{a} \end{pmatrix}^p = \widehat{a}I_2$.

(b) If $p \geq 3$, then $\begin{pmatrix} \widehat{a} & \widehat{b} \\ \widehat{b} & \widehat{a} \end{pmatrix}^p = \begin{pmatrix} \widehat{a} & \widehat{b} \\ \widehat{b} & \widehat{a} \end{pmatrix}$.

(c) If $\widehat{a} + \widehat{b} \neq \widehat{0}$, then $X^p = \begin{pmatrix} \widehat{a} & \widehat{b} \\ \widehat{a} & \widehat{b} \end{pmatrix}$ if and only if $X = \begin{pmatrix} \widehat{a} & \widehat{b} \\ \widehat{a} & \widehat{b} \end{pmatrix}$.

(d) If $\widehat{a} + \widehat{b} = \widehat{0}, \widehat{a} \neq \widehat{0}$, the equation $X^p = \begin{pmatrix} \widehat{a} & \widehat{b} \\ \widehat{a} & \widehat{b} \end{pmatrix}$ has no solutions in $\mathscr{M}_2\left(\mathbb{Z}_p\right)$.

(e) If $X \in \mathscr{M}_2\left(\mathbb{Z}_p\right)$ such that $\det X = \widehat{0}$ and $\operatorname{Tr}(X) \neq \widehat{0}$, then $X^p = X$.

(f) $\begin{pmatrix} \widehat{0} & \widehat{a} \\ \widehat{b} & \widehat{0} \end{pmatrix}^p = \begin{pmatrix} \widehat{0} & \widehat{a}^{\frac{p+1}{2}}\widehat{b}^{\frac{p-1}{2}} \\ \widehat{a}^{\frac{p-1}{2}}\widehat{b}^{\frac{p+1}{2}} & \widehat{0} \end{pmatrix}$, $p \geq 3$.

(g) $\begin{pmatrix} \widehat{a} & \widehat{b} \\ \widehat{a} & \widehat{a} \end{pmatrix}^p = \begin{pmatrix} \widehat{a} & \widehat{a}^{\frac{p-1}{2}}\widehat{b}^{\frac{p+1}{2}} \\ \widehat{a}^{\frac{p+1}{2}}\widehat{b}^{\frac{p-1}{2}} & \widehat{a} \end{pmatrix}$, $p \geq 3$.

(h) If $p \geq 5$ is a prime number, there are exactly p^2 matrices in $\mathscr{M}_2\left(\mathbb{Z}_p\right)$ which commute with $\begin{pmatrix} \widehat{1} & \widehat{2} \\ \widehat{3} & \widehat{4} \end{pmatrix}$.

(i) The number of invertible matrices in $\mathscr{M}_2\left(\mathbb{Z}_p\right)$ is $(p^2 - 1)(p^2 - p)$.

3.92 (a) Solve in $\mathscr{M}_2\left(\mathbb{Z}_5\right)$ the equation $X^5 = \begin{pmatrix} \widehat{0} & \widehat{1} \\ \widehat{2} & \widehat{0} \end{pmatrix}$.

(b) Let $p \geq 3$ be a prime number. Prove that the equation $X^p = \begin{pmatrix} \widehat{0} & \widehat{a} \\ \widehat{b} & \widehat{0} \end{pmatrix}$ has the unique solution

$$\begin{cases} X = \begin{pmatrix} \widehat{0} & \widehat{a} \\ \widehat{b} & \widehat{0} \end{pmatrix} & \text{if and only if } (\widehat{a}^{-1}\widehat{b})^{\frac{p-1}{2}} = \widehat{1} \\ X = \begin{pmatrix} \widehat{0} & \widehat{p-a} \\ \widehat{p-b} & \widehat{0} \end{pmatrix} & \text{if and only if } (\widehat{a}^{-1}\widehat{b})^{\frac{p-1}{2}} = -\widehat{1}. \end{cases}$$

3.93 Splendid binomial equations.

(a) [49] Solve in $\mathcal{M}_2(\mathbb{Z}_5)$ the equation

$$X^5 = \begin{pmatrix} \widehat{4} & \widehat{2} \\ \widehat{4} & \widehat{1} \end{pmatrix}.$$

(b) Let $p \geq 3$ be a prime number. Solve in $\mathcal{M}_2(\mathbb{Z}_p)$ the equation

$$X^p = \begin{pmatrix} \widehat{p-1} & \widehat{2} \\ \widehat{p-1} & \widehat{1} \end{pmatrix}.$$

3.94 Let $A \in \mathcal{M}_2(\mathbb{C})$. Prove that the equation $AX - XA = A$ has a solution in $\mathcal{M}_2(\mathbb{C})$ if and only if $A^2 = O_2$.

3.95 *Two binomial equations with symmetric terms.*

(a) Let $A \in \mathcal{M}_2(\mathbb{C})$. Prove that the equation $AX - XA = I_2$ does not have solutions in $\mathcal{M}_2(\mathbb{C})$.

(b) Prove that if the equation $AX + XA = I_2$ has solutions in $\mathcal{M}_2(\mathbb{C})$, then either A is invertible or $A^2 = O_2$.
Conversely, show that if A is invertible or $A^2 = O_2$ and $A \neq O_2$, then the equation $AX + XA = I_2$ has solutions in $\mathcal{M}_2(\mathbb{C})$.

3.96 Let $P \in \mathbb{C}[x]$ be a polynomial of degree n. Prove that the following statements are equivalent:

(a) the equation $P(X) = \begin{pmatrix} 1 & 1 \\ 0 & 1 \end{pmatrix}$ has n distinct solutions in $\mathcal{M}_2(\mathbb{C})$;

(b) the equation $P(x) = 1$ has n distinct solutions.

Is the problem true if the solutions of the equation $P(x) = 1$ are not distinct?

3.97 Let $A = \begin{pmatrix} a & b \\ c & d \end{pmatrix} \in \mathcal{M}_2(\mathbb{R})$ with $a + d \neq 0$. Prove that the matrix $B \in \mathcal{M}_2(\mathbb{R})$ commutes with A if and only if B commutes with A^2.

3.98 Let $m, n \in \mathbb{N}$ and let $A, B \in \mathcal{M}_2(\mathbb{R})$ be such that $A^m B^n = B^n A^m$. Prove that if A^m and B^n are not of the form λI_2, for some $\lambda \in \mathbb{R}$, then $AB = BA$.

3.99 Let $A, B \in \mathcal{M}_2(\mathbb{R})$ be such that $A^m B = A^m + B$, $m \in \mathbb{N}$. Prove that $AB = BA$.

3.4.3 Solutions

3.52. (a) If $A = \alpha I_2$ we get $\alpha = 1$ or $\alpha = 5 \Rightarrow A = I_2$ or $A = 5I_2$. If $A \neq \alpha I_2$ we have, based on Theorem 2.6, that $\text{Tr}(A) = 6$ and $\det A = 5$.

If $A = \begin{pmatrix} a & b \\ c & d \end{pmatrix} \in \mathcal{M}_2\,(\mathbb{N}_0)$, then $a + d = 6$ and $ad - bc = 5$. It follows that

$$A \in \left\{ \begin{pmatrix} 1 & 0 \\ c & 5 \end{pmatrix}, \begin{pmatrix} 1 & b \\ 0 & 5 \end{pmatrix}, \begin{pmatrix} 5 & 0 \\ c & 1 \end{pmatrix}, \begin{pmatrix} 5 & b \\ 0 & 1 \end{pmatrix}, \; b, c \in \mathbb{N}_0 \right\}$$

and

$$A \in \left\{ \begin{pmatrix} 2 & 3 \\ 1 & 4 \end{pmatrix}, \begin{pmatrix} 4 & 3 \\ 1 & 2 \end{pmatrix}, \begin{pmatrix} 2 & 1 \\ 3 & 4 \end{pmatrix}, \begin{pmatrix} 4 & 1 \\ 3 & 2 \end{pmatrix}, \begin{pmatrix} 3 & 4 \\ 1 & 3 \end{pmatrix}, \begin{pmatrix} 3 & 1 \\ 4 & 3 \end{pmatrix}, \begin{pmatrix} 3 & 2 \\ 2 & 3 \end{pmatrix} \right\}.$$

(b) Use Theorem 2.6.

3.53. $A = \begin{pmatrix} 1 & 0 \\ 0 & 0 \end{pmatrix}$.

3.54. *Solution 1.* The problem can be solved by direct computation.
Solution 2. Since $\det^2 X = \det\left(X^2\right) = 16$ we get that $\det X = \pm 4$.
Case 1. If $\det X = 4$ we have, based on Cayley–Hamilton Theorem, that $X^2 - \text{Tr}(X)X + 4I_2 = O_2$ and this implies, passing to trace, that $\text{Tr}(X^2) - (\text{Tr}(X))^2 + 4\text{Tr}(I_2) = 0$. We obtain $\text{Tr}(X) = \pm 5$ and $\pm 5X = X^2 + 4I_2$. Thus

$$X_{1,2} = \pm \begin{pmatrix} 2 & 1 \\ 2 & 3 \end{pmatrix}.$$

Case 2. If $\det X = -4$ we obtain, after some calculations similar to those in *Case 1*, that

$$X_{3,4} = \pm \frac{1}{3} \begin{pmatrix} 2 & 5 \\ 10 & 7 \end{pmatrix}.$$

3.55. (a) $\pm \begin{pmatrix} 1 & 0 \\ 2 - \sqrt{2} & \sqrt{2} \end{pmatrix}$ and $\pm \begin{pmatrix} 1 & 0 \\ 2 + \sqrt{2} & -\sqrt{2} \end{pmatrix}$.

(b) $\begin{pmatrix} 1 & 0 \\ 1 & \sqrt{2} \end{pmatrix}, \begin{pmatrix} -1 & 0 \\ 1 & -\sqrt{2} \end{pmatrix}, \begin{pmatrix} -1 & 0 \\ 1 & \sqrt{2} \end{pmatrix}$ and $\begin{pmatrix} 1 & 0 \\ 1 & -\sqrt{2} \end{pmatrix}$.

3.56. Let $B = \begin{pmatrix} -1 & 0 \\ 0 & -1 - \epsilon \end{pmatrix}$ and let $A \in \mathcal{M}_2\,(\mathbb{R})$ be such that $A^2 = B$. Since A and B commute we get that $A = \begin{pmatrix} u & 0 \\ 0 & d \end{pmatrix}$. This implies, since $A^2 = \begin{pmatrix} a^2 & 0 \\ 0 & d^2 \end{pmatrix}$, that

$a^2 = -1$ and $d^2 = -1 - \epsilon$. Since these equations do not have real solutions we have that there is no $A \in \mathcal{M}_2(\mathbb{R})$ such that $A^2 = B$.

3.57. The possible values of n are 2, 3, 4, and 6 (see [58, 528–529]).

3.58. $A = \begin{pmatrix} \cos\theta & \sin\theta \\ \cos\theta & \sin\theta \end{pmatrix}$, where $\theta \in \mathbb{R}$.

3.60. One implication is easy to prove. If $A = \begin{pmatrix} \alpha & \beta \\ \beta & \alpha \end{pmatrix}$ or $A = \begin{pmatrix} \beta & \alpha \\ \alpha & \beta \end{pmatrix}$, with

$$\alpha = \frac{\pm\sqrt{a+b} \pm \sqrt{a-b}}{2} \quad \text{and} \quad \beta = \frac{\pm\sqrt{a+b} \mp \sqrt{a-b}}{2}, \text{ then}$$

$$AA^T = A^T A = \begin{pmatrix} \alpha^2 + \beta^2 & 2\alpha\beta \\ 2\alpha\beta & \alpha^2 + \beta^2 \end{pmatrix} = \begin{pmatrix} a & b \\ b & a \end{pmatrix}.$$

Now we prove the other implication. First we note, since $\det(AA^T) = \det^2 A = a^2 - b^2 > 0$, that A is invertible. The equation $AA^T = \begin{pmatrix} a & b \\ b & a \end{pmatrix}$ implies that $A^T = A^{-1}\begin{pmatrix} a & b \\ b & a \end{pmatrix} = A^{-1}(aI_2 + bJ)$, where $J = \begin{pmatrix} 0 & 1 \\ 1 & 0 \end{pmatrix}$. The equation $AA^T = A^T A$ implies that $AA^T = aI_2 + bJ = (aA^{-1} + bA^{-1}J)A = A^T A$, and this in turn implies $bA^{-1}JA = bJ$ and, since $b \neq 0$, we get that $JA = AJ$ Let $A = \begin{pmatrix} x & y \\ u & v \end{pmatrix}$. Since $JA = AJ$ we get that $u = y$ and $v = x$, so $A = \begin{pmatrix} x & y \\ y & x \end{pmatrix}$. We have

$$AA^T = \begin{pmatrix} x^2 + y^2 & 2xy \\ 2xy & x^2 + y^2 \end{pmatrix} = \begin{pmatrix} a & b \\ b & a \end{pmatrix}$$

and this implies that $x^2 + y^2 = a$ and $2xy = b$. Since we have a symmetric system it is clear that the values of x and y could be interchanged. Adding and subtracting these equations we get that $(x + y)^2 = a + b$ and $(x - y)^2 = a - b$, and we have $x + y = \pm\sqrt{a+b}$ and $x - y = \pm\sqrt{a-b}$. Thus, $x = \dfrac{\pm\sqrt{a+b} \pm \sqrt{a-b}}{2}$ and $y = \dfrac{\pm\sqrt{a+b} \mp \sqrt{a-b}}{2}$.

3.61. If λ is an eigenvalue of A we have, since $A^4 = I_2$, that $\lambda^4 = 1$ which implies that $\lambda \in \{\pm 1, \pm i\}$. Let λ_1, λ_2 be the eigenvalues of A.

If $\lambda_1 = \pm 1$, then $\lambda_2 = \pm 1$ or $\lambda_2 = \mp 1$ and in all cases we have that $A^2 = I_2$.

If $\lambda_1 = i$, then $\lambda_2 = -i$ and we have that $A^2 = -I_2$.

3.62. First observe that $n = 6l + 1$ or $n = 6l + 5$, where $l \geq 0$ is an integer. If λ is an eigenvalue of A we have that $\lambda^n = 1 \Rightarrow \lambda \in \{\cos\frac{2k\pi}{n} + i\sin\frac{2k\pi}{n} : k = 0, 1, \ldots, n - 1\}$. Since the characteristic polynomial of A

has integer coefficients, then either the eigenvalues of A are equal to 1 or they are complex conjugate.

If $\lambda_1 = \lambda_2 = 1$ we have, based on the Cayley–Hamilton Theorem, that $(A - I_2)^2 = O_2$. It follows that $I_2 = A^n = (I_2 + A - I_2)^n = I_2 + n(A - I_2) \Rightarrow A = I_2$.

If the eigenvalues of A are complex conjugate, then we let $\lambda_1 = \cos \frac{2k\pi}{n} + i \sin \frac{2k\pi}{n}$ and $\lambda_2 = \cos \frac{2k\pi}{n} - i \sin \frac{2k\pi}{n}$, for some $k \in \{1, 2, \ldots, n-1\}$. Since $\mathrm{Tr}(A) = \lambda_1 + \lambda_2 = 2\cos \frac{2k\pi}{n} \in \mathbb{Z}$ we have that $2\cos \frac{2k\pi}{n} \in \{\pm 1, 0, \pm 2\}$.

If $2\cos \frac{2k\pi}{n} = -1 \Rightarrow \cos \frac{2k\pi}{n} = \cos \frac{2\pi}{3} \Rightarrow \frac{2k\pi}{n} = \frac{2\pi}{3} \Rightarrow n = 3k$, which contradicts $(n, 6) = 1$.

If $2\cos \frac{2k\pi}{n} = 1 \Rightarrow \cos \frac{2k\pi}{n} = \cos \frac{\pi}{3} \Rightarrow \frac{2k\pi}{n} = \frac{\pi}{3} \Rightarrow n = 6k$, which contradicts $(n, 6) = 1$.

If $2\cos \frac{2k\pi}{n} = 0 \Rightarrow \cos \frac{2k\pi}{n} = \cos \frac{\pi}{2} \Rightarrow \frac{2k\pi}{n} = \frac{\pi}{2} \Rightarrow n = 4k$, which contradicts $(n, 6) = 1$.

If $2\cos \frac{2k\pi}{n} = 2 \Rightarrow \cos \frac{2k\pi}{n} = 1$, which is impossible since $1 \le k \le n-1$.

If $2\cos \frac{2k\pi}{n} = -2 \Rightarrow \cos \frac{2k\pi}{n} = -1 \Rightarrow \frac{2k\pi}{n} = \pi \Rightarrow n = 2k$, which contradicts $(n, 6) = 1$.

3.63. Let $f_A(x) = \det(A - xI_2) \in \mathbb{Q}[x]$ be the characteristic polynomial of A and let λ_1, λ_2 be the eigenvalues of A. If λ is an eigenvalue of A we have, since $A^n = -I_2$, that $\lambda^n = -1$. It follows, since $f_A \in \mathbb{Q}[x]$, that either both eigenvalues of A are equal to -1 and n is odd or they are complex conjugate.

If $\lambda_1 = \lambda_2 = -1$ we have that $(A + I_2)^2 = O_2$. It follows that $-I_2 = A^n = (A + I_2 - I_2)^n = n(A + I_2) - I_2 \Rightarrow A = -I_2 \Rightarrow A^3 = -I_2$.

If the eigenvalues of A are complex conjugate, then we let $\lambda_1 = \cos t + i \sin t$ and $\lambda_2 = \cos t - i \sin t$, $\lambda_1, \lambda_2 \in \mathbb{C} \setminus \mathbb{R}$. Since $\lambda_1^n = -1$ we have that $\cos(nt) = -1$.

On the other hand, $\lambda_1 + \lambda_2 = 2\cos t = s \in \mathbb{Q}$. Using Lemma 3.1 we have that there exists a monic polynomial $P_n \in \mathbb{Z}[x]$ such that $2\cos(nt) = P_n(2\cos t)$. It follows that $2\cos t = s$ is a rational root of a monic polynomial with integer coefficients, hence s must be an integer. Since, $s \in [-2, 2]$, we get that $s \in \{\pm 1, 0, \pm 2\}$.

If $2\cos t = 2 \Rightarrow \lambda_1 = \lambda_2 = 1$, which is impossible since $1^n \ne -1$.

If $2\cos t = -2 \Rightarrow \lambda_1 = \lambda_2 = -1$ and this case was studied above.

If $2\cos t = 0 \Rightarrow A^2 + I_2 = O_2 \Rightarrow A^2 = -I_2$.

If $2\cos t = 1 \Rightarrow \mathrm{Tr}(A) = 2\cos t = 1$ and $\det A = \lambda_1 \lambda_2 = 1 \Rightarrow A^2 - A + I_2 = O_2 \Rightarrow (A + I_2)(A^2 - A + I_2) = O_2 \Rightarrow A^3 = -I_2$.

If $2\cos t = -1 \Rightarrow A^2 + A + I_2 = O_2 \Rightarrow (A - I_2)(A^2 + A + I_2) = O_2 \Rightarrow A^3 = I_2$, so $A^n \in \{I_2, A, A^2\}$. Since $A^n = -I_2$ we get that either $A = -I_2$ or $A^2 = -I_2$, which implies that either $A^3 = -I_2$ or $A^2 = -I_2$.

3.64. Let $f_A(x) = \det(A - xI_2) \in \mathbb{Q}[x]$ be the characteristic polynomial of A and let λ_1, λ_2 be the eigenvalues of A. We have, since $A^n = I_2$, that $\lambda_1^n = \lambda_2^n = 1$. It follows that either λ_1, λ_2 are real or they are complex conjugate.

If the eigenvalues of A are real, then $\lambda_1, \lambda_2 \in \{-1, 1\}$.

If $\lambda_1 = \lambda_2 = 1$, then we have that $(A - I_2)^2 = O_2 \Rightarrow I_2 = A^n = (I_2 + A - I_2)^n = I_2 + n(A - I_2) \Rightarrow A = I_2 \Rightarrow A^{12} = I_2$.

If $\lambda_1 = \lambda_2 = -1$, then n is even and $(A + I_2)^2 = O_2$. It follows that $I_2 = A^n = (-I_2 + (A + I_2))^n = I_2 - n(A + I_2) \Rightarrow A = -I_2 \Rightarrow A^{12} = I_2$.

If $\lambda_1 = 1$ and $\lambda_2 = -1$, then $A^2 - I_2 = O_2 \Rightarrow A^2 = I_2 \Rightarrow A^{12} = I_2$.

If the eigenvalues of A are complex conjugate we let $\lambda_{1,2} = \cos\alpha \pm i\sin\alpha$ and we have that $\mathrm{Tr}(A) = 2\cos\alpha = s \in \mathbb{Q}$, $\cos(n\alpha) = 1$ and $\det A = \lambda_1\lambda_2 = 1$.

We have based on Lemma 3.1 that there exists a monic polynomial $P_n \in \mathbb{Z}[x]$ such that $2\cos(n\alpha) = P_n(2\cos\alpha)$. Thus, $s = 2\cos\alpha$ is a root of a monic polynomial with integer coefficients, i.e., it is a solution of the equation $P_n(x) - 2 = 0$. Thus, s must be an integer and since $s \in [-2, 2]$ we have that $s \in \{\pm 2, \pm 1, 0\}$.

The cases when $s = -2$ or $s = 2$ imply that $\lambda_1 = \lambda_2 = -1$ or $\lambda_1 = \lambda_2 = 1$ which have been discussed above.

If $s = -1$, then $A^2 + A + I_2 = O_2 \Rightarrow (A - I_2)(A^2 + A + I_2) = O_2 \Rightarrow A^3 = I_2 \Rightarrow A^{12} = I_2$.

If $s = 1$, then $A^2 - A + I_2 = O_2 \Rightarrow (A + I_2)(A^2 - A + I_2) = O_2 \Rightarrow A^3 = -I_2 \Rightarrow A^{12} = I_2$.

If $s = 0$, then $A^2 + I_2 = O_2 \Rightarrow A^2 = -I_2 \Rightarrow A^{12} = I_2$.

The previous calculations show that the order of matrices in $\mathscr{M}_2(\mathbb{Q})$ could be 1, 2, 3, 4, and 6 respectively. Examples of matrices of such orders are given by

$A_1 = I_2$, with $A_1^1 = I_2$, $A_2 = -I_2$, with $A_2^2 = I_2$, $A_3 = \begin{pmatrix} -1 & 1 \\ -1 & 0 \end{pmatrix}$, with $A_3^3 = I_2$,

$A_4 = \begin{pmatrix} 0 & -1 \\ 1 & 0 \end{pmatrix}$, with $A_4^4 = I_2$ and $A_5 = \begin{pmatrix} 1 & 1 \\ 1 & 0 \end{pmatrix}$, with $A_5^6 = I_2$, which prove the order of matrices in $\mathrm{GL}_2(\mathbb{Q})$ are 1, 2, 3, 4, and 6 respectively.

For the case when A is a matrix with integer entries which verifies the conditions of the problem see [58, Problem 7.7.7, p. 145].

3.65. Since $\det^3 A = \det(A^3) = 1$ we get that $\det A = 1$. On the other hand, the Cayley–Hamilton Theorem implies that $A^2 = \mathrm{Tr}(A)A - I_2$ and it follows that $A^3 = \mathrm{Tr}(A)A^2 - A = (\mathrm{Tr}^2(A) - 1)A - \mathrm{Tr}(A)I_2$. Passing to trace in the previous equality we have $18 = \mathrm{Tr}(A^3) = \mathrm{Tr}^3(A) - 3\mathrm{Tr}(A)$. Thus, $\mathrm{Tr}(A) = 3$ and $A = \begin{pmatrix} 1 & 1 \\ 1 & 2 \end{pmatrix}$.

3.66. Since $\det^3 X = 1$ we get that $\det X = 1$. If $t = \mathrm{Tr}(A)$, then the Cayley–Hamilton Theorem implies $X^2 - tX + I_2 = O_2$ and $X^3 = tX^2 - X = (t^2 - 1)X - tI_2$. Passing to trace in this equality we have $t^3 - 3t + 2 = 0$, which implies that $t \in \{-2, 1\}$.

If $t = 1$ we get $X^3 = -I_2$ which is impossible since $X^3 = \begin{pmatrix} 1 & -2 \\ 2 & -3 \end{pmatrix}$.

If $t = -2$ we get, since $3X + 2I_2 = X^3 = \begin{pmatrix} 1 & -2 \\ 2 & -3 \end{pmatrix}$, that $X = \frac{1}{3}\begin{pmatrix} -1 & -2 \\ 2 & -5 \end{pmatrix}$.

3.67. (a) Let $X = \begin{pmatrix} a & b \\ c & d \end{pmatrix}$. Since $X^3 = I_2$ we get, by passing to determinants, that $\det X = 1$. We have, based on the Cayley–Hamilton Theorem, that $X^2 = (a+d)X - I_2$ and this implies $X^3 = (a + d)X^2 - X = (a + d)^2X - (a + d)I_2 - X = I_2$. Thus $[(a + d)^2 - 1]X = (1 + a + d)I_2$ which in turn implies that

$$\begin{cases} [(a+d)^2 - 1]a = 1 + a + d \\ [(a+d)^2 - 1]b = 0 \\ [(a+d)^2 - 1]c = 0 \\ [(a+d)^2 - 1]d = 1 + a + d. \end{cases}$$

Adding the first and the last equation we get $t^3 - 3t - 2 = 0$, where $t = a + d$, with solutions $t_1 = t_2 = -1$ and $t_3 = 2$.

If $a + d = -1$, then

$$X = \begin{pmatrix} a & b \\ c & -1-a \end{pmatrix}, \quad a \in \mathbb{R}, \quad bc = -1 - a - a^2.$$

If $a + d = 2$, then $a = d = 1$, $b = c = 0$ which implies that $X = I_2$.

(b) We get, since $X^2 + \epsilon X + \epsilon^2 I_2 = O_2$, that $X^3 = I_2$ and it follows, based on part (a), that $X = I_2$.

3.68. We have that $A^{-1} = AA^T$ and, since AA^T is a symmetric matrix, we get that A is also a symmetric matrix. The equation to solve becomes $A^3 = I_2$ and, since A is symmetric, we get that $A = I_2$.

3.71. If $A = \begin{pmatrix} a & 1-a \\ 1+a & -a \end{pmatrix}$, then $\det^5 X = \det A = -1$, so $\det X = -1$. Let $t = \mathrm{Tr}(X)$. We have, based on Cayley–Hamilton Theorem, that $X^2 - tX - I_2 = O_2$, $X^3 = tX^2 + X = (t^2 + 1)X + tI_2$, and $X^5 = X^2 X^3 = (t^4 + 3t^2 + 1)X + (t^3 + 2t)I_2$. Passing to trace in the last equality and using that $X^5 = A$ we get that $(t^4 + 3t^2 + 1)t + 2(t^3 + 2t) = 0 \Leftrightarrow t^5 + 5t^3 + 5t = 0$, which has the unique real solution $t = 0$. This implies that $X^2 = I_2$ and $X = X^5 = A$.

3.72. Since $A^n = \begin{pmatrix} \cos n\alpha & \sin n\alpha \\ -\sin n\alpha & \cos n\alpha \end{pmatrix}$, we get that $\cos n\alpha = 1$ and $\sin n\alpha = 0$. This implies that $n\alpha = 2k\pi$, $k \in \mathbb{Z}$ and $n\alpha = m\pi$, $m \in \mathbb{Z}$. Thus $2k = m$ and $\alpha = \dfrac{2k\pi}{n}$, $k \in \mathbb{Z}$.

3.73. Let

$$A = \begin{pmatrix} \cos t & -\sin t \\ \sin t & \cos t \end{pmatrix} \quad \text{and} \quad X = \begin{pmatrix} a & b \\ c & d \end{pmatrix}.$$

We have $X^{n+1} = AX = XA$ and this implies

$$\begin{cases} b \sin t = -c \sin t \\ -a \sin t = -d \sin t \end{cases} \overset{\sin t \neq 0}{\Longleftrightarrow} \begin{cases} a = d \\ b + c = 0, \end{cases}$$

so $X = \begin{pmatrix} a & -b \\ b & a \end{pmatrix}$. Since $X^n = A$ we get that $\det^n X = \det A = 1$, which implies that $\det X \in \{\pm 1\} \Leftrightarrow a^2 + b^2 \in \{\pm 1\}$, so $a^2 + b^2 = 1$. There exists $x \in \mathbb{R}$ such that $a = \cos x$ and $y = \sin x$ and this implies that

$$X = \begin{pmatrix} \cos x & -\sin x \\ \sin x & \cos x \end{pmatrix} \quad \text{and} \quad X^n = \begin{pmatrix} \cos nx & -\sin nx \\ \sin nx & \cos nx \end{pmatrix} = \begin{pmatrix} \cos t & -\sin t \\ \sin t & \cos t \end{pmatrix}.$$

Therefore $nx = t + 2k\pi$, $k \in \mathbb{Z}$. The equation has n solutions

$$X_k = \begin{pmatrix} \cos x_k & -\sin x_k \\ \sin x_k & \cos x_k \end{pmatrix}, \quad x_k = \frac{t + 2k\pi}{n}, \quad k = \overline{0, n-1}.$$

3.74. Let $A = \begin{pmatrix} 1 & 2 \\ 2 & 4 \end{pmatrix}$. The equation $X^n = A$ implies that $\det X = 0$ and this in turn implies $t^{n-1}X = A$, where $t = \text{Tr}(A)$. Passing to trace in this equation we get that $t^n = 5$. The solutions of this equation are $t_k = \sqrt[n]{5}\left(\cos\frac{2k\pi}{n} + i\sin\frac{2k\pi}{n}\right)$, $k = 0, 1, \ldots, n-1$. Thus, $X_k = \frac{1}{t_k^{n-1}}A$, $k = 0, 1, \ldots, n-1$, are the n solutions of the matrix equation.

3.75. If $A = \begin{pmatrix} 1 & a \\ 0 & 1 \end{pmatrix}$ we get, since $AX = XA$, that $X = \begin{pmatrix} \alpha & \beta \\ 0 & \alpha \end{pmatrix}$ and this implies that

$$X^n = \begin{pmatrix} \alpha^n & n\alpha^{n-1}\beta \\ 0 & \alpha^n \end{pmatrix} = \begin{pmatrix} 1 & a \\ 0 & 1 \end{pmatrix},$$

so $\alpha^n = 1$ and $n\alpha^{n-1}\beta = a$. It follows that

$$\alpha = \epsilon_k = \cos\frac{2k\pi}{n} + i\sin\frac{2k\pi}{n} \quad \text{and} \quad \beta_k = \frac{a\epsilon_k}{n}, \quad k = \overline{0, n-1}.$$

The equation has n solutions

$$X_k = \begin{pmatrix} \epsilon_k & \dfrac{a\epsilon_k}{n} \\ 0 & \epsilon_k \end{pmatrix}, \quad k = \overline{0, n-1}.$$

3.76. First we prove the implication "$\Rightarrow$." We have $X^n = A \Rightarrow \det X = 0 \Rightarrow X^n = t^{n-1}X$, where $t = \text{Tr}(X)$. Thus, $t^{n-1}X = A$. If, by way of contradiction, $A^2 = O_2$, then $t = 0$ and $t^{n-1}X = A = O_2$, which contradicts $A \neq O_2$.

Now we prove the implication "$\Leftarrow$." Solving the equation $X^n = A$ we get $\det X = 0 \Rightarrow X^n = t^{n-1}X$, where $t = \text{Tr}(X)$ and we have $t \neq 0$ since $X^2 \neq O_2$. It follows $t^{n-1}X = A$ and, by passing to trace, we get that $t^n = \text{Tr}(A)$ and observe $\text{Tr}(A) \neq 0$

since $A^2 \neq O_2$. We obtain the solutions $X_k = \frac{t_k}{\mathrm{Tr}(A)} A$, $k = \overline{0, n-1}$, where t_k are the solutions of the equation $t^n = \mathrm{Tr}(A)$.

3.77. Let $A = \begin{pmatrix} a & b \\ b & a \end{pmatrix}$. Since $X^{n+1} = X^n X = X X^n$ we get that $AX = XA$ and this implies, since $b \neq 0$, that $X = \begin{pmatrix} x & y \\ y & x \end{pmatrix}$. A calculation shows that

$$X^n = \begin{pmatrix} \dfrac{(x+y)^n + (x-y)^n}{2} & \dfrac{(x+y)^n - (x-y)^n}{2} \\ \dfrac{(x+y)^n - (x-y)^n}{2} & \dfrac{(x+y)^n + (x-y)^n}{2} \end{pmatrix}.$$

The equation $X^n = A$ implies that

$$\begin{cases} (x+y)^n + (x-y)^n = 2a \\ (x+y)^n - (x-y)^n = 2b \end{cases} \Rightarrow \begin{cases} (x+y)^n = a+b \\ (x-y)^n = a-b. \end{cases}$$

The solutions of this system are $x = \frac{\alpha+\beta}{2}$ and $y = \frac{\alpha-\beta}{2}$, where $\alpha, \beta \in \mathbb{C}$ with $\alpha^n = a+b$ and $\beta^n = a-b$. Thus,

$$X = \begin{pmatrix} \dfrac{\alpha+\beta}{2} & \dfrac{\alpha-\beta}{2} \\ \dfrac{\alpha-\beta}{2} & \dfrac{\alpha+\beta}{2} \end{pmatrix}.$$

The matrix equation has n^2 solutions in $\mathcal{M}_2(\mathbb{C})$.

3.78. Observe that $A = \begin{pmatrix} a & b \\ -b & a \end{pmatrix} = \sqrt{a^2+b^2} \begin{pmatrix} \cos t & \sin t \\ -\sin t & \sin t \end{pmatrix}$ and see the solution of problem **3.73**.

3.79. Let $A = \begin{pmatrix} 3 & -1 \\ 0 & 0 \end{pmatrix}$. We assume, by way of contradiction, that there is $X \in \mathcal{M}_2(\mathbb{Q})$ such that $X^n = A$, $n \geq 2$. This implies that $\det X = 0 \Rightarrow X^n = t^{n-1} X$, where $t = \mathrm{Tr}(X)$. The matrix equation becomes $t^{n-1} X = A$ and, by passing to trace in this equation, we get that $t^n = 3$. However, this equation does not have rational solutions.

3.80. Let $A = \begin{pmatrix} -7 & -9 \\ 3 & 2 \end{pmatrix}$, $t = \mathrm{Tr}(X) \in \mathbb{Z}$ and let $d = \det X \in \mathbb{Z}$. The Cayley–Hamilton Theorem implies that $X^2 - tX + dI_2 = O_2$ and $X^3 - 3X = (t^2 - d - 3)X - tdI_2$. Thus, $A = (t^2 - d - 3)X - tdI_2$. Passing to trace in this equation we get that $(t^2 - d - 3)t - 2td = -5 \Leftrightarrow t(t^2 - 3d - 3) = -5$ which implies that $t \in \{-5, -1, 1, 5\}$.

If $t = -5$ we get that $d = 7$ and these imply that $X \notin \mathcal{M}_2(\mathbb{Z})$.

If $t = 1$ we have that $d = 1$ and $X = \begin{pmatrix} 2 & 3 \\ -1 & -1 \end{pmatrix} \in \mathcal{M}_2(\mathbb{Z})$.

The cases when $t = -1$ or $t = 5$ lead to $d \notin \mathbb{Z}$.

Hence, the only solution of the cubic matrix equation is $X = \begin{pmatrix} 2 & 3 \\ -1 & -1 \end{pmatrix}$.

3.81. Let $A = \begin{pmatrix} 1 & 1 \\ 1 & 1 \end{pmatrix}$ and let $X \in \mathcal{M}_2(\mathbb{R})$ be such that $X^3 + X^2 = A$. We have

$AX = XA = X^4 + X^3$ and it follows that $X = \begin{pmatrix} x & y \\ y & x \end{pmatrix}$, $x, y \in \mathbb{R}$. Straightforward calculations imply that

$$\begin{cases} x^3 + 3xy^2 + x^2 + y^2 = 1 \\ 3x^2y + y^3 + 2xy = 1. \end{cases}$$

Subtracting these equations we obtain that $(x - y)^2(x - y + 1) = 0$ and it follows that $x = y$ or $x = y - 1$. We obtain the solutions

$$X_1 = \frac{1}{2}\begin{pmatrix} 1 & 1 \\ 1 & 1 \end{pmatrix} \quad \text{and} \quad X_2 = \begin{pmatrix} 0 & 1 \\ 1 & 0 \end{pmatrix}.$$

3.82. (a) Let λ_1, λ_2 be the eigenvalues of A. We apply Theorem 2.11. First we consider the case when $\lambda_1, \lambda_2 \in \mathbb{Q}$.

If $J_A = \begin{pmatrix} \lambda_1 & 0 \\ 0 & \lambda_2 \end{pmatrix}$, then $A^3 - A - I_2 = O_2$ implies that $J_A^3 - J_A - I_2 = O_2 \Rightarrow$
$\lambda_i^3 - \lambda_i - 1 = 0$, $i = 1, 2$. However, the equation $x^3 - x - 1 = 0$ does not have rational solutions.

If $J_A = \begin{pmatrix} \lambda & 1 \\ 0 & \lambda \end{pmatrix}$, $\lambda \in \mathbb{Q}$, then $J_A^3 - J_A - I_2 = O_2$ implies that $\lambda^3 - \lambda - 1 = 0$ and

$3\lambda^2 - 1 = 0$. These equations do not have rational solutions.

Now we consider the case when $\lambda_1, \lambda_2 \in \mathbb{C} \setminus \mathbb{Q}$, $\lambda_1 = \alpha + \sqrt{\beta}$ and $\lambda_2 = \alpha - \sqrt{\beta}$,

$\alpha \in \mathbb{Q}$, $\beta \in \mathbb{Q}^*$. Let $J_A = \begin{pmatrix} \alpha & 1 \\ \beta & \alpha \end{pmatrix}$ be the rational canonical form of A. The equation

$J_A^3 - J_A - I_2 = O_2$ implies that $\alpha^3 + 3\alpha\beta - \alpha - 1 = 0$ and $3\alpha^2 + \beta - 1 = 0$. It follows that $8\alpha^3 - 2\alpha + 1 = 0$, which does not have rational solutions.

(b) See the solution of problem **3.83**.

3.83. Without loosing the generality we consider that $n \geq k$. We have $A^k(A^{n-k} - C(a, b)) = I_2$, which implies that A^k and $A^{n-k} - C(a, b)$ are inverses one another, hence they commute. It follows that $(A^{n-k} - C(a, b))A^k = I_2$ and this implies that

$A^k C(a, b) = C(a, b)A^k$. A calculation shows that $A^k = \begin{pmatrix} x & y \\ y & x \end{pmatrix}$, $x, y \in \mathbb{Q}$. We have

$A^k = \alpha_k A + \beta_k I_2$, $\alpha_k, \beta_k \in \mathbb{Q}$. We distinguish between the cases when $\alpha_k = 0$ and $\alpha_k \neq 0$.

The case $\alpha_k = 0$. If $\alpha_k = 0$, then $A^k = \beta_k I_2 = C(x, y) \Rightarrow \beta_k = x$ and $y = 0$, so $A^k = xI_2$. Observe that $x \neq 0$, otherwise $A^k = O_2$ which contradicts the fact that A is invertible. The equation $A^n - A^k C(a, b) - I_2 = O_2$ implies, since $A^k = xI_2$, that $A^{n-k} = C\left(a + \frac{1}{x}, b\right)$. Since A commutes with $C\left(a + \frac{1}{x}, b\right)$ and $b \neq 0$ we get that A is also a circulant matrix. Let $A = C(u, v)$. The equation $A^k = xI_2$ implies that $(u + v)^k + (u - v)^k = 2x$ and $(u + v)^k - (u - v)^k = 0$. If $u + v = 0$ we get that $A = C(u, -u)$ which is not invertible. Thus, $u + v \neq 0$ and the equation $(u + v)^k - (u - v)^k = 0$ implies that $u - v = \pm(u + v)$.

If $u - v = u + v$, then $v = 0 \Rightarrow A = uI_2 \Rightarrow A^{n-k} = u^{n-k} I_2 = C\left(a + \frac{1}{x}, b\right)$, which contradicts $b \neq 0$.

If $u - v = -u - v$, then $u = 0 \Rightarrow A = C(0, v)$ and the equation $A^k = xI_2$ implies that k is even and $x = v^k > 0$. On the other hand,

$$C\left(a + \frac{1}{x}, b\right) = A^{n-k} = \begin{cases} v^{n-k} C(0, 1) & \text{if } n \text{ is odd} \\ v^{n-k} I_2 & \text{if } n \text{ is even.} \end{cases}$$

If n is an odd integer, then we have $a + \frac{1}{x} = 0 \Rightarrow x = -\frac{1}{a} < 0$, which contradicts $x > 0$.

If n is an even integer, then $b = 0$ which is impossible.

The case $\alpha_k \neq 0$. We have $A = \frac{1}{\alpha_k}\left(A^k - \beta_k I_2\right) = \frac{1}{\alpha_k}\left(C(x, y) - \beta_k I_2\right)$. Thus, A is a circulant matrix. Let $A = C(\theta, \delta)$, $\theta, \delta \in \mathbb{Q}$. Let

$$t_k = \frac{(\theta + \delta)^k + (\theta - \delta)^k}{2} \quad \text{and} \quad w_k = \frac{(\theta + \delta)^k - (\theta - \delta)^k}{2}.$$

The equation $A^n - A^k C(a, b) - I_2 = O_2$ implies that $t_n - at_k - bw_k - 1 = 0$ and $w_n - bt_k - aw_k = 0$. Adding these two equations we get that $(\theta + \delta)^n - (a + b)(\theta + \delta)^k - 1 = 0$. However, this equation does not have rational solutions.

3.84. (a) Let $A = \begin{pmatrix} 2 & 0 \\ 3 & 2 \end{pmatrix}$ and let $X \in \mathcal{M}_2(\mathbb{Z})$ be a solution of the matrix equation $X^t + X = A$, where $t = \text{Tr}(X)$. Since X commutes with A, a calculation shows that $X = \begin{pmatrix} x & 0 \\ u & x \end{pmatrix}$, $x, u \in \mathbb{Z}$.

On the other hand, $X^k = \begin{pmatrix} x^k & 0 \\ kx^{k-1}u & x^k \end{pmatrix}$, $k \in \mathbb{Z}$, and this implies that

$$X^t + X = \begin{pmatrix} x^t + x & 0 \\ tx^{t-1}u + u & x^t + x \end{pmatrix} = \begin{pmatrix} 2 & 0 \\ 3 & 2 \end{pmatrix}.$$

We obtain the equations $x^t + x = 2$ and $tx^{t-1}u + u = 3$ where $t = \text{Tr}(X) = 2x$.

A calculation shows that $x = u = 1$ and hence $X = \begin{pmatrix} 1 & 0 \\ 1 & 1 \end{pmatrix}$.

(b) Exactly as in part (a) we get the equations $x^d + x = 2$ and $dx^{d-1}u + u = 3$, where $d = x^2$, and these imply that the equation $X^d + X = A$ does not have solutions.

3.85. Let $X = \begin{pmatrix} a & b \\ c & d \end{pmatrix}$ be a solution of our equation. We have

$$X^n + X^{n-2} = X^{n-2}(X + iI_2)(X - iI_2)$$

and it follows that $\det X = 0$ or $\det(X + iI_2) = 0$ or $\det(X - iI_2) = 0$.

If $\det(X + iI_2) = 0$ we get that $(a + i)(d + i) - bc = 0 \Rightarrow ad - bc - 1 = 0$ and $a + d = 0$. We have $d = -a$, $bc = -1 - a^2$ and a calculation shows that

$$X^2 = \begin{pmatrix} a^2 + bc & b(a + d) \\ c(a + d) & d^2 + bc \end{pmatrix} = \begin{pmatrix} -1 & 0 \\ 0 & -1 \end{pmatrix} = -I_2,$$

and this implies $X^{n-2}(X^2 + I_2) = O_2$, which is a contradiction.

By a similar analysis we get that the case $\det(X - iI_2) = 0$ leads to a contradiction.

Now we study the case $\det X = 0$. The Cayley–Hamilton Theorem implies that $X^2 = (a + d)X \Rightarrow X^k = (a + d)^{k-1}X, \forall k \geq 1$. Thus,

$$X^n + X^{n-2} = \left[(a + d)^{n-1} + (a + d)^{n-3}\right]X = \begin{pmatrix} 1 & -1 \\ -1 & 1 \end{pmatrix}.$$

Let $a + d = t$ and it follows, from the previous equation, that

$$\begin{cases} a(t^{n-1} + t^{n-3}) = 1 \\ b(t^{n-1} + t^{n-3}) = -1 \\ c(t^{n-1} + t^{n-3}) = -1 \\ d(t^{n-1} + t^{n-3}) = 1. \end{cases}$$

Adding the first and the last equation we have $t^n + t^{n-2} - 2 = 0$.

Let $f : \mathbb{R} \to \mathbb{R}, f(x) = x^n + x^{n-2} - 2$ and we note that $f'(x) = x^{n-3}(nx^2 + n - 2)$. We study the cases when n is an even or an odd integer.

- n is an even integer. In this case we have that $f'(x) > 0$ on $(0, \infty)$ and $f'(x) < 0$ on $(-\infty, 0)$. Since $f(-1) = f(1) = 0$ we get that -1 and 1 are the unique real solutions of the equation $f(x) = 0$. A calculation shows that

$$t = 1 \quad \Rightarrow \quad X_1 = \frac{1}{2}\begin{pmatrix} 1 & -1 \\ -1 & 1 \end{pmatrix}$$

and

$$t = -1 \quad \Rightarrow \quad X_2 = -\frac{1}{2}\begin{pmatrix} 1 & -1 \\ -1 & 1 \end{pmatrix}.$$

■ *n is an odd integer.* We have $f'(x) > 0$ for $x \neq 0$ and 1 is the unique real solution of the equation $f(x) = 0$, which implies the unique solution of the matrix equation is

$$X = \frac{1}{2}\begin{pmatrix} 1 & -1 \\ -1 & 1 \end{pmatrix}.$$

3.86. (a) The equation has solutions if and only if $n = 3k + 2$, $k \geq 0$. In this case the equation is equivalent to $X^2 - X + I_2 = O_2$.

(b) The equation has no solutions in $\mathcal{M}_2(\mathbb{Z})$.

3.87. The only primes that qualify are 2 and 3 (see [46]).

3.88. $A = \begin{pmatrix} 1 & 1 \\ 0 & 1 \end{pmatrix}$.

3.89. We prove that the equation $X^m = A$ has solution in $\mathcal{M}_2(\mathbb{C})$ if and only if $A^2 \neq O_2$ or $A = O_2$ (the same conditions hold for the equation $Y^n = A$). Let J_A be the Jordan canonical form of A and let $P \in \mathcal{M}_2(\mathbb{C})$ be the invertible matrix such that $A = PJ_AP^{-1}$. Let $X \in \mathcal{M}_2(\mathbb{C})$ be a solution of the equation $X^m = A$ and let $X_1 = P^{-1}XP$. The matrix equation $X^m = A$ becomes $X_1^m = J_A$.

If the matrix J_A is diagonal, i.e., $J_A = \begin{pmatrix} \lambda_1 & 0 \\ 0 & \lambda_2 \end{pmatrix}$ clearly a solution is $X_1 = \begin{pmatrix} \mu_1 & 0 \\ 0 & \mu_2 \end{pmatrix}$, with $\mu_1^m = \lambda_1$ and $\mu_2^m = \lambda_2$.

If $J_A = \begin{pmatrix} \lambda & 1 \\ 0 & \lambda \end{pmatrix}$ then X_1 commutes with J_A, so $X_1 = \begin{pmatrix} a & b \\ 0 & a \end{pmatrix}$ and $X_1^m = \begin{pmatrix} a^m & ma^{m-1}b \\ 0 & a^m \end{pmatrix}$. We obtain the equations $a^m = \lambda$ and $ma^{m-1}b = 1$, which have solutions if and only if $\lambda \neq 0$ and when $\lambda = 0$ they do not have solutions. Thus, the only case when the equation $X^m = A$ does not have solutions is the case when the Jordan canonical form of A is $J_A = \begin{pmatrix} 0 & 1 \\ 0 & 0 \end{pmatrix}$ which corresponds to the case when $A^2 = O_2$ and $A \neq O_2$.

3.90. Since X_1 and X_2 are solutions we get that $X_1^2 - AX_1 + B = O_2$, $X_2^2 - AX_2 + B = O_2$ and by subtracting these equations we get that $X_1^2 - X_2^2 = A(X_1 - X_2)$. Let $Y = X_1 - X_2$ and we obtain that $A = (X_1^2 - X_2^2)Y^{-1}$. Let $Z = YX_2Y^{-1}$. Since $\text{Tr}(X_2) = \text{Tr}(Z)$ we have

$$\mathrm{Tr}(X_1 + X_2) = \mathrm{Tr}(X_1) + \mathrm{Tr}(X_2)$$
$$= \mathrm{Tr}(X_1) + \mathrm{Tr}(Z)$$
$$= \mathrm{Tr}(X_1 + Z)$$
$$= \mathrm{Tr}\left[(X_1 Y + Y X_2) Y^{-1}\right]$$
$$= \mathrm{Tr}\left[(X_1^2 - X_1 X_2 + X_1 X_2 - X_2^2) Y^{-1}\right]$$
$$= \mathrm{Tr}\left[(X_1^2 - X_2^2) Y^{-1}\right]$$
$$= \mathrm{Tr}(A).$$

On the other hand,

$$B = A X_1 - X_1^2$$
$$= (X_1^2 - X_2^2) Y^{-1} X_1 - X_1^2$$
$$= (X_1^2 - X_2^2 - X_1 Y) Y^{-1} X_1$$
$$= (X_1^2 - X_2^2 - X_1^2 + X_1 X_2) Y^{-1} X_1$$
$$= (X_1 - X_2) X_2 Y^{-1} X_1$$
$$- Y X_2 Y^{-1} X_1,$$

and by passing to determinants we get $\det B = \det(X_1 X_2)$.

3.91. (a) Let $B = \begin{pmatrix} \widehat{0} & \widehat{1} \\ \widehat{0} & \widehat{0} \end{pmatrix}$ and observe that $B^2 = O_2$. We have

$$\begin{pmatrix} \widehat{a} & \widehat{b} \\ \widehat{0} & \widehat{a} \end{pmatrix}^p = (\widehat{a} I_2 + \widehat{b} B)^p = \widehat{a}^p I_2 + \widehat{b}^p B^p = \widehat{a} I_2.$$

(b) Let $p = 2k + 1$, $k \geq 1$ and let $J = \begin{pmatrix} \widehat{0} & \widehat{1} \\ \widehat{1} & \widehat{0} \end{pmatrix}$. Observe that $J^2 = I_2$ and $J^p = J^{2k} J = J$. We have

$$\begin{pmatrix} \widehat{a} & \widehat{b} \\ \widehat{b} & \widehat{a} \end{pmatrix}^p = (\widehat{a} I_2 + \widehat{b} J)^p = \widehat{a}^p I_2 + \widehat{b}^p J^p = \widehat{a} I_2 + \widehat{b} J.$$

(c) The equation $X^p = \begin{pmatrix} \widehat{a} & \widehat{b} \\ \widehat{a} & \widehat{b} \end{pmatrix}$ implies that $\det^p X = \widehat{0} \Rightarrow \det X = \widehat{0}$. It follows,

based on Cayley–Hamilton Theorem that $X^2 = \widehat{t}X$, where $\widehat{t} = \mathrm{Tr}(X)$. This implies

that $X^p = \widehat{t}^{p-1}X = X \Rightarrow X = \begin{pmatrix} \widehat{a} & \widehat{b} \\ \widehat{a} & \widehat{b} \end{pmatrix}$.

Conversely, if $X = \begin{pmatrix} \widehat{a} & \widehat{b} \\ \widehat{a} & \widehat{b} \end{pmatrix}$, then $\det X = \widehat{0}$ and $\mathrm{Tr}(X) = \widehat{a} + \widehat{b} \neq \widehat{0}$. It follows

that $X^2 = \widehat{t}X \Rightarrow X^p = \widehat{t}^{p-1}X = X$.

(d) The equation $X^p = \begin{pmatrix} \widehat{a} & \widehat{b} \\ \widehat{a} & \widehat{b} \end{pmatrix}$ implies that $\det^p X = \widehat{0} \Rightarrow \det X = \widehat{0}$. Since

$\mathrm{Tr}(X) = \widehat{a} + \widehat{b} = \widehat{0}$ we get, based on the Cayley–Hamilton Theorem, that $X^2 = O_2$.
It follows that $X^p = O_2$, which contradicts $\widehat{a} \neq \widehat{0}$.

(e) Let $\widehat{t} = \mathrm{Tr}(X)$. If $X \in \mathcal{M}_2\left(\mathbb{Z}_p\right)$ such that $\det X = \widehat{0}$ and $\mathrm{Tr}(X) \neq \widehat{0}$, then
$X^2 = \widehat{t}X$. This implies that $X^p = \widehat{t}^{p-1}X = X$.

(f) Let $X = \begin{pmatrix} \widehat{0} & \widehat{a} \\ \widehat{b} & \widehat{0} \end{pmatrix}$ and observe that $X^2 = \widehat{ab}I_2$. Then

$$X^p = \left(X^2\right)^{\frac{p-1}{2}} X = \left(\widehat{ab}\right)^{\frac{p-1}{2}} X = \begin{pmatrix} \widehat{0} & \widehat{a}^{\frac{p+1}{2}}\widehat{b}^{\frac{p-1}{2}} \\ \widehat{a}^{\frac{p-1}{2}}\widehat{b}^{\frac{p+1}{2}} & \widehat{0} \end{pmatrix}.$$

(g) Let $Y = \begin{pmatrix} \widehat{0} & \widehat{b} \\ \widehat{a} & \widehat{0} \end{pmatrix}$. We have, based on part (f), that

$$\begin{pmatrix} \widehat{a} & \widehat{b} \\ \widehat{a} & \widehat{a} \end{pmatrix}^p = (\widehat{a}I_2 + Y)^p = \widehat{a}^pI_2 + Y^p = \widehat{a}I_2 + \begin{pmatrix} \widehat{0} & \widehat{a}^{\frac{p-1}{2}}\widehat{b}^{\frac{p+1}{2}} \\ \widehat{a}^{\frac{p+1}{2}}\widehat{b}^{\frac{p-1}{2}} & \widehat{0} \end{pmatrix}.$$

(i) There are exactly $p^2 - 1$ ways to choose the first line of the matrix such that it
is nonzero, then the second line can be chosen in any way except for the cases when
the second line is proportional to the first line, so there are $p^2 - p$ such possibilities.
It follows that there are $(p^2 - 1)(p^2 - p)$ invertible matrices in $\mathcal{M}_2\left(\mathbb{Z}_p\right)$.

3.92. (a) Let $A = \begin{pmatrix} \widehat{0} & \widehat{1} \\ \widehat{2} & \widehat{0} \end{pmatrix}$. Since X commutes with A we get that $X = \begin{pmatrix} \widehat{a} & \widehat{b} \\ \widehat{2b} & \widehat{a} \end{pmatrix} = $

$\widehat{a}I_2 + \widehat{b}A$. A calculation shows that $A^5 = \begin{pmatrix} \widehat{0} & \widehat{4} \\ \widehat{3} & \widehat{0} \end{pmatrix} = \widehat{4}A$. Using the Binomial Theorem

we get that

$$X^5 = \left(\widehat{a}I_2 + \widehat{b}A\right)^5 = \widehat{a}^5I_2 + \widehat{b}^5A^5 = \widehat{a}I_2 + \widehat{4b}A = \begin{pmatrix} \widehat{a} & \widehat{4b} \\ \widehat{3b} & \widehat{a} \end{pmatrix} = \begin{pmatrix} \widehat{0} & \widehat{1} \\ \widehat{2} & \widehat{0} \end{pmatrix}.$$

It follows that $\widehat{a} = \widehat{0}$ and $\widehat{b} = \widehat{4}$. Thus, $X = \widehat{4}A = \begin{pmatrix} \widehat{0} & \widehat{4} \\ \widehat{3} & \widehat{0} \end{pmatrix}$.

(b) Let $A = \begin{pmatrix} \widehat{0} & \widehat{a} \\ \widehat{b} & \widehat{0} \end{pmatrix}$. Since X commutes with A we have that

$$X = \begin{pmatrix} \widehat{x} & \widehat{y} \\ \widehat{a}^{-1}\widehat{b}\widehat{y} & \widehat{x} \end{pmatrix} = \widehat{x}I_2 + \widehat{y}\begin{pmatrix} \widehat{0} & \widehat{1} \\ \widehat{a}^{-1}\widehat{b} & \widehat{0} \end{pmatrix} = \widehat{x}I_2 + \widehat{y}B,$$

where $B = \begin{pmatrix} \widehat{0} & \widehat{1} \\ \widehat{a}^{-1}\widehat{b} & \widehat{0} \end{pmatrix}$. A calculation shows that $B^2 = \alpha I_2$, where $\alpha = \widehat{a}^{-1}\widehat{b}$.

Let $p = 2k + 1$, $k \geq 1$. Then $B^p = B^{2k}B = \alpha^k B = \left(\widehat{a}^{-1}\widehat{b}\right)^{\frac{p-1}{2}} B$.
We have

$$X^p = (\widehat{x}I_2 + \widehat{y}B)^p = \widehat{x}^p I_2 + \widehat{y}^p B^p = \widehat{x}I_2 + \widehat{y}\left(\widehat{a}^{-1}\widehat{b}\right)^{\frac{p-1}{2}} B$$

and it follows that

$$\begin{pmatrix} \widehat{x} & \widehat{y}\left(\widehat{a}^{-1}\widehat{b}\right)^{\frac{p-1}{2}} \\ \widehat{y}\left(\widehat{a}^{-1}\widehat{b}\right)^{\frac{p+1}{2}} & \widehat{x} \end{pmatrix} = \begin{pmatrix} \widehat{0} & \widehat{a} \\ \widehat{b} & \widehat{0} \end{pmatrix}.$$

Thus, $\widehat{x} = \widehat{0}, \widehat{y}\left(\widehat{a}^{-1}\widehat{b}\right)^{\frac{p-1}{2}} = \widehat{a}$ and $\widehat{y}\left(\widehat{a}^{-1}\widehat{b}\right)^{\frac{p+1}{2}} = \widehat{b}$. The last two equalities imply that $\widehat{y}^2 = \widehat{a}^2 \Rightarrow \widehat{y} = \widehat{a}$ or $\widehat{y} = -\widehat{a}$.

If $\widehat{y} = \widehat{a}$ we get that $\left(\widehat{a}^{-1}\widehat{b}\right)^{\frac{p-1}{2}} = \widehat{1}$ and $X = \widehat{a}B = \begin{pmatrix} \widehat{0} & \widehat{a} \\ \widehat{b} & \widehat{0} \end{pmatrix}$.

If $\widehat{y} = -\widehat{a}$ we get that $\left(\widehat{a}^{-1}\widehat{b}\right)^{\frac{p-1}{2}} = -\widehat{1}$ and

$$X = -\widehat{a}B = \begin{pmatrix} \widehat{0} & -\widehat{a} \\ -\widehat{b} & \widehat{0} \end{pmatrix} = \begin{pmatrix} \widehat{0} & \widehat{p-a} \\ \widehat{p-b} & \widehat{0} \end{pmatrix}.$$

The reverse implication is easy to check.

3.93. (a) *Solution 1.* Let $A = \begin{pmatrix} \widehat{4} & \widehat{2} \\ \widehat{4} & \widehat{1} \end{pmatrix}$. Since X commutes with A a calculation shows that

$$X = \begin{pmatrix} \widehat{a} & \widehat{b} \\ \widehat{2b} & \widehat{a+b} \end{pmatrix} = \widehat{a}I_2 + \widehat{b}B,$$

where $B = \begin{pmatrix} \widehat{0} & \widehat{1} \\ \widehat{2} & \widehat{1} \end{pmatrix}$. One can check that $B^5 = B$ and we have, based on the Binomial Theorem, that

$$X^5 = \left(\widehat{a}I_2 + \widehat{b}B\right)^5 = \widehat{a}^5 I_2 + \widehat{b}^5 B^5 = \widehat{a}^5 I_2 + \widehat{b}^5 B = \begin{pmatrix} \widehat{a}^5 & \widehat{b}^5 \\ \widehat{2b^5} & \widehat{a}^5 + \widehat{b}^5 \end{pmatrix} = \begin{pmatrix} \widehat{a} & \widehat{b} \\ \widehat{2b} & \widehat{a} + \widehat{b} \end{pmatrix}.$$

Thus, $\widehat{a} = \widehat{4}, \widehat{b} = \widehat{2}$, and $X = A$ is the unique solution of the equation.

Solution 2. Let $A = \begin{pmatrix} \widehat{4} & \widehat{2} \\ \widehat{4} & \widehat{1} \end{pmatrix}$. Since, $\mathrm{Tr}(A) = \widehat{0}$ and $\det A = \widehat{1}$, we have based on the Cayley–Hamilton Theorem that $A^2 + \widehat{1}I_2 = O_2 \Rightarrow A^2 = \widehat{4}I_2$. This implies that $A^5 = A$.

Let $Y \in \mathcal{M}_2(\mathbb{Z}_5)$ such that $X = Y + A$. First we observe that Y commutes with A. We have $X^6 = AX = A(Y + A) = AY + A^2$ and $X^6 = XA = (Y + A)A = YA + A^2$. It follows that $AY = YA$. Using the Binomial Theorem we have

$$X^5 = (Y + A)^5 = Y^5 + A^5 = Y^5 + A = A \quad \Rightarrow \quad Y^5 = O_2 \quad \Rightarrow \quad Y^2 = O_2.$$

Since matrices which commute with A are of the following form

$$\begin{pmatrix} \widehat{a} & \widehat{b} \\ \widehat{2b} & \widehat{a} + \widehat{b} \end{pmatrix}$$

and Y commutes with A we get that there are $\widehat{a}, \widehat{b} \in \mathbb{Z}_5$ such that

$$Y = \begin{pmatrix} \widehat{a} & \widehat{b} \\ \widehat{2b} & \widehat{a} + \widehat{b} \end{pmatrix}.$$

The equation $Y^2 = O_2$ implies that

$$\begin{pmatrix} \widehat{a}^2 + \widehat{2b}^2 & \widehat{b}\left(\widehat{b} + \widehat{2a}\right) \\ \widehat{2b}\left(\widehat{b} + \widehat{2a}\right) & \widehat{2b}^2 + \left(\widehat{a} + \widehat{b}\right)^2 \end{pmatrix} = O_2.$$

A calculation shows that $\widehat{a} = \widehat{b} = \widehat{0}$ which implies that $Y = O_2$. Thus, the only solution of the matrix equation $X^5 = A$ is $X = A$.

Solution 3. Let $\widehat{t} = \mathrm{Tr}(X)$ and let $\widehat{d} = \det X$. Since $X^5 = A$ we get $\widehat{d}^5 = \widehat{1} \Rightarrow \widehat{d} = 1$. We distinguish between the following two cases.

- $\mathrm{Tr}(X) = \widehat{0}$. We have, based on the Cayley–Hamilton Theorem, that $X^2 + \widehat{1}I_2 = O_2 \Rightarrow X^2 = \widehat{4}I_2 \Rightarrow X^4 = \widehat{1}I_2 \Rightarrow X^5 = X$. Thus, $X = A$ is the solution of our equation.

- $\text{Tr}(X) \neq \widehat{0}$. The Cayley–Hamilton Theorem implies that $X^2 = \widehat{t}X + \widehat{4}I_2 \Rightarrow$ $X^4 = \left(\widehat{t^3} + \widehat{3t}\right)X + \left(\widehat{4t^2} + \widehat{1}\right)I_2 \Rightarrow X^5 = \left(\widehat{t^4} + \widehat{2t^2} + \widehat{1}\right)X + \left(\widehat{4t^3} + \widehat{2t}\right)I_2.$ This implies that $\left(\widehat{t^4} + \widehat{2t^2} + \widehat{1}\right)X + \left(\widehat{4t^3} + \widehat{2t}\right)I_2 = A.$ Passing to trace in this equation we get that

$$\left(\widehat{t^4} + \widehat{2t^2} + \widehat{1}\right)\widehat{t} + \widehat{2}\left(\widehat{4t^3} + \widehat{2t}\right) = \widehat{0} \quad \Rightarrow \quad \widehat{t^5} = \widehat{0}.$$

This implies that $\widehat{t} = \widehat{0}$, which is impossible.

Thus, the only solution of the equation $X^5 = A$ is $X = A$.

(b) The equation has a unique solution given by

$$\begin{cases} X = \begin{pmatrix} \widehat{p-1} & \widehat{2} \\ \widehat{p-1} & \widehat{1} \end{pmatrix} & \text{if } p \equiv 1 \mod 4 \\[4mm] X = \begin{pmatrix} \widehat{1} & \widehat{p-2} \\ \widehat{1} & \widehat{p-1} \end{pmatrix} & \text{if } p \equiv 3 \mod 4. \end{cases}$$

Let $A = \begin{pmatrix} \widehat{p-1} & \widehat{2} \\ \widehat{p-1} & \widehat{1} \end{pmatrix}$. Since $\text{Tr}(A) = \widehat{0}$ and $\det A = \widehat{1}$ we have, based on the Cayley–Hamilton Theorem, that $A^2 = -\widehat{1}I_2 = \widehat{p-1}I_2$. Let $p = 2k+1, k \geq 1$. We have

$$A^p = A^{2k}A = \widehat{p-1}^k A = \widehat{p-1}^{\frac{p-1}{2}}A = \widehat{\alpha}A,$$

where $\widehat{\alpha} = \widehat{p-1}^{\frac{p-1}{2}}$. On the other hand,

$$(\widehat{\alpha}A)^p = \widehat{\alpha}^p A^p = \widehat{\alpha}A^p = \widehat{\alpha}^2 A = \widehat{p-1}^{p-1}A = A.$$

Since X commutes with A a calculation shows that

$$X = \begin{pmatrix} \widehat{a} & \widehat{b} \\ \widehat{\frac{p-1}{2}b} & \widehat{a+b} \end{pmatrix}.$$

Let $Y \in \mathcal{M}_2\left(\mathbb{Z}_p\right)$ be such that $X = Y + \widehat{\alpha}A$. First we observe that Y commutes with A. We have $X^{p+1} = X^pX = AX = A(Y + \widehat{\alpha}A) = AY + \widehat{\alpha}A^2$ and $X^{p+1} = XX^p = XA = (Y + \widehat{\alpha}A)A = YA + \widehat{\alpha}A^2$. These imply that $AY = YA$.

We apply the Binomial Theorem and we have that

$$X^p = (Y + \widehat{\alpha}A)^p = Y^p + (\widehat{\alpha}A)^p = Y^p + A = A \quad \Rightarrow \quad Y^p = O_2 \quad \Rightarrow \quad Y^2 = O_2.$$

Since Y commutes with A we let $Y = \begin{pmatrix} \widehat{a} & \widehat{b} \\ \widehat{\frac{p-1}{2}b} & \widehat{a+b} \end{pmatrix}$. A calculation shows that

$$Y^2 = \begin{pmatrix} \widehat{a}^2 + \widehat{\frac{p-1}{2}b^2} & \widehat{b}\left(\widehat{2a}+\widehat{b}\right) \\ \widehat{\frac{p-1}{2}b}\left(\widehat{2a}+\widehat{b}\right) & \widehat{\frac{p-1}{2}b^2} + (\widehat{a}+\widehat{b})^2 \end{pmatrix}.$$

Since $Y^2 = O_2$ we get that $\widehat{a}^2 + \widehat{\frac{p-1}{2}b^2} = \widehat{0}$ and $\widehat{b}\left(\widehat{2a}+\widehat{b}\right) = \widehat{0}$. The equation $\widehat{b}\left(\widehat{2a}+\widehat{b}\right) = \widehat{0}$ implies that $\widehat{b} = \widehat{0}$ or $\widehat{2a}+\widehat{b} = \widehat{0}$.

If $\widehat{b} = \widehat{0}$ we get from the first equation that $\widehat{a} = \widehat{0}$, so $Y = O_2$.

If $\widehat{b} = -\widehat{2a} = \widehat{p-2a}$, the first equation implies that $\widehat{a}^2\left(\widehat{1} + \widehat{\frac{p-1}{2}p} - 2^2\right) = \widehat{0} \Rightarrow \widehat{p-1}\widehat{a}^2 = 0 \Rightarrow \widehat{a} = \widehat{0} \Rightarrow Y = O_2$.

Thus, the solution of the matrix equation is $X = \widehat{p-1}^{\frac{p-1}{2}}A$. If $p = 4i + 1$ we have that $\widehat{p-1}^{\frac{p-1}{2}} = \widehat{1}$, so $X = A$ and if $p = 4i + 3$, then $\widehat{p-1}^{\frac{p-1}{2}} = \widehat{p-1}$, so $X = \widehat{p-1}A = \begin{pmatrix} \widehat{1} & \widehat{p-2} \\ \widehat{1} & \widehat{p-1} \end{pmatrix}$.

3.94. First we prove that if the equation $AX - XA = A$ has a solution, then $A^2 = O_2$. We have, $\text{Tr}(A) = \text{Tr}(AX - XA) = 0$ and $\text{Tr}(A^2) = \text{Tr}[A(AX - XA)] = \text{Tr}(A^2X) - \text{Tr}(AXA) = \text{Tr}(AXA) - \text{Tr}(AXA) = 0$. It follows, based on problem **2.88**, that A is nilpotent and the implication is proved.

Now we prove that if $A^2 = O_2$, then the equation $AX - XA = A$ has a solution in $\mathcal{M}_2(\mathbb{C})$. Let $A = \begin{pmatrix} a & b \\ c & d \end{pmatrix}$ with $A^2 = O_2$. A calculation shows (see problem **1.8**) that $A_1 = \begin{pmatrix} 0 & 0 \\ c & 0 \end{pmatrix}$, $c \in \mathbb{C}$, or $A_2 = \begin{pmatrix} a & b \\ -\frac{a^2}{b} & -a \end{pmatrix}$, $a, b \in \mathbb{C}, b \neq 0$.

If $A = A_1$ the equation $AX - XA = A$ has the solution $X_1 = \begin{pmatrix} 1 & 0 \\ 0 & 0 \end{pmatrix}$ and if $A = A_2$ the equation has the solution $X_2 = \begin{pmatrix} 0 & 0 \\ \frac{a}{b} & 1 \end{pmatrix}$.

3.96. (a) $\Rightarrow$ (b) Let $P(x) = a_nx^n + a_{n-1}x^{n-1} + \cdots + a_1x + a_0$, $a_n \neq 0$ and let X be a solution of the equation $P(X) = A$, where $A = \begin{pmatrix} 1 & 1 \\ 0 & 1 \end{pmatrix}$. Since X and A commute we get that $X = \begin{pmatrix} a & b \\ 0 & a \end{pmatrix}$, $a, b \in \mathbb{C}$. A calculation shows that, for $k \geq 1$, we have

$$X^k = \begin{pmatrix} a^k & ka^{k-1}b \\ 0 & a^k \end{pmatrix} \quad \text{and} \quad P(X) = \begin{pmatrix} P(a) & bP'(a) \\ 0 & P(a) \end{pmatrix} = \begin{pmatrix} 1 & 1 \\ 0 & 1 \end{pmatrix},$$

which implies that $P(a) = 1$ and $bP'(a) = 1$. Observe that since the equation $P(x) = 1$ has n distinct solutions we cannot have the situation that there exists $\alpha \in \mathbb{C}$ such that $P(\alpha) = 1$ and $P'(\alpha) = 0$. Therefore, the equation $P(x) = 1$ has n distinct solutions.

(b) $\Rightarrow$ (a) If the solutions of the equation $P(x) = 1$ are $x_1, x_2, \ldots, x_n$, then the solutions of the matrix equation $P(X) = A$ are $\begin{pmatrix} x_k & \dfrac{1}{P'(x_k)} \\ 0 & x_k \end{pmatrix}$, $k = 1, 2, \ldots, n$.

If the solutions of the equation $P(x) = 1$ are not distinct, then the statement of the problem is no longer valid. Let $P(x) = (x-1)^2 + 1 = x^2 - 2x + 2$ and we note that the equation $P(x) = 1$ has the double solution 1. However, there is no matrix $X = \begin{pmatrix} a & b \\ 0 & a \end{pmatrix} \in \mathcal{M}_2(\mathbb{C})$ such that $P(X) = A$, since this would imply that $a^2 - 2a + 2 = 1$ and $2b(a-1) = 1$.

3.97. The Cayley–Hamilton Theorem implies that $A^2 - tA + dI_2 = O_2$, where $t = \mathrm{Tr}(A)$ and $d = \det A$. Therefore

$$\begin{cases} BA^2 = tBA - dB \\ A^2 B = tAB - dB, \end{cases}$$

and this implies, since $t \neq 0$, that $BA^2 = A^2 B \Leftrightarrow tBA = tAB \Leftrightarrow BA = AB$.

3.98. There are real numbers $\alpha_m, \beta_m, u_n, v_n \in \mathbb{R}$, $\alpha_m \neq 0$, $u_n \neq 0$, such that $A^m = \alpha_m A + \beta_m I_2$ and $B^n = u_n B + v_n I_2$. We have

$$\begin{cases} A^m B^n = \alpha_m u_n AB + \alpha_m v_n A + \beta_m u_n B + \beta_m v_n I_2 \\ B^n A^m = u_n \alpha_m BA + u_n \beta_m B + v_n \alpha_m A + v_n \beta_m I_2, \end{cases}$$

and this implies that $A^m B^n = B^n A^m \Leftrightarrow \alpha_m u_n AB = u_n \alpha_m BA \overset{u_n \alpha_m \neq 0}{\Longleftrightarrow} AB = BA$.

3.99. $A^m B = A^m + B \Leftrightarrow (A^m - I_2)(B - I_2) = I_2$. This implies that matrices $A^m - I_2$ and $B - I_2$ are inverses one another, hence they commute. Therefore $(B - I_2)(A^m - I_2) = I_2 \Leftrightarrow BA^m = B + A^m$. This implies $A^m B = BA^m$. If $A^m = \alpha_m I_2$, for some $\alpha_m \in \mathbb{R}$, then the matrix equality $A^m B = A^m + B$ implies that $(\alpha_m - 1)B = \alpha_m I_2$. Observe that $\alpha_m \neq 1$, otherwise we get a contradiction. Thus, $B = \dfrac{\alpha_m}{\alpha_m - 1} I_2$ and this clearly implies $AB = BA$. If $A^m = \alpha_m A + \beta_m I_2$, with $\alpha_m, \beta_m \in \mathbb{R}$, $\alpha_m \neq 0$, then since $A^m B = BA^m$, we have $\alpha_m AB = \alpha_m BA$ and since $\alpha_m \neq 0$ we get that $AB = BA$.

3.5 Pell's diophantine equation

Let $d \geq 2$ be an integer which is not a perfect square.

Definition 3.4 The diophantine equation

$$x^2 - dy^2 = 1, \quad x, \ y \in \mathbb{Z}, \tag{3.3}$$

is called *Pell's equation*[2].

In what follows we are going to solve, in integers, Pell's equation. First, we observe that the pairs $(-1, 0)$ and $(1, 0)$ are solutions of equation (3.3) which are called the *trivial solutions*. On the other hand, if (x, y) is a solution of equation (3.3), then $(-x, y)$, $(x, -y)$, and $(-x, -y)$ are also solutions of the same equation. Thus, to solve Pell's equation it suffices to find its solutions in positive integers, i.e., the solutions of the following form $(x, y) \in \mathbb{N} \times \mathbb{N}$.

Let $(x, y) \in \mathbb{N} \times \mathbb{N}$ and let

$$A_{(x,y)} = \begin{pmatrix} x & dy \\ y & x \end{pmatrix},$$

where x and y are such that $\det A_{(x,y)} = x^2 - dy^2 = 1$.

Let S_P be the set of the solutions of the equation (3.3). We note that $(x, y) \in S_P$ if and only if $\det A_{(x,y)} = 1$ and $(x, y) \neq (1, 0)$ if and only if $A_{(x,y)} \neq I_2$.

If $(x_0, y_0) \in S_P$, $(x_0, y_0) \neq (1, 0)$, then $\det A_{(x_0, y_0)} = 1$ and it follows that

$$\det A_{(x_0, y_0)}^n = 1.$$

Let

$$A_{(x_0, y_0)}^n = \begin{pmatrix} x_n & dy_n \\ y_n & x_n \end{pmatrix} \quad \text{with} \quad x_n^2 - dy_n^2 = 1.$$

If

$$A_{(x_0, y_0)}^{n+1} = \begin{pmatrix} x_{n+1} & dy_{n+1} \\ y_{n+1} & x_{n+1} \end{pmatrix},$$

then

$$A_{(x_0, y_0)}^{n+1} = A_{(x_0, y_0)}^n A_{(x_0, y_0)} = \begin{pmatrix} x_n & dy_n \\ y_n & x_n \end{pmatrix} \begin{pmatrix} x_0 & dy_0 \\ y_0 & x_0 \end{pmatrix}$$

$$= \begin{pmatrix} x_0 x_n + dy_0 y_n & d(y_0 x_n + x_0 y_n) \\ y_0 x_n + x_0 y_n & x_0 x_n + dy_0 y_n \end{pmatrix}$$

---

[2]This equation which bears the name of Pell, due to a confusion originating with Euler, should have been designated as Fermat's equation [15, p. 341].

and

$$\det A^{n+1}_{(x_0,y_0)} = \det \left(A^n_{(x_0,y_0)} A_{(x_0,y_0)} \right) = \det A^n_{(x_0,y_0)} \det A_{(x_0,y_0)} = 1.$$

It follows that

$$\begin{cases} x_{n+1} = x_0 x_n + d y_0 y_n \\ y_{n+1} = y_0 x_n + x_0 y_n \end{cases} \quad \text{or} \quad \begin{cases} x_n = x_0 x_{n-1} + d y_0 y_{n-1} \\ y_n = y_0 x_{n-1} + x_0 y_{n-1} \end{cases}$$

for $n \geq 1$, where x_0, y_0 are given such that $(x_0, y_0) \neq (1, 0)$.

We note that if $(x_0, y_0) \in \mathbb{N} \times \mathbb{N}$, then we also have that $(x_n, y_n) \in \mathbb{N} \times \mathbb{N}$. In other words, if (x_0, y_0) is a solution of equation (3.3), then (x_n, y_n) is also a solution of equation (3.3).

The previous recurrence relations can be written as follows

$$\begin{pmatrix} x_n \\ y_n \end{pmatrix} = \begin{pmatrix} x_0 & d y_0 \\ y_0 & x_0 \end{pmatrix} \begin{pmatrix} x_{n-1} \\ y_{n-1} \end{pmatrix},$$

and this implies that

$$\begin{pmatrix} x_n \\ y_n \end{pmatrix} = \begin{pmatrix} x_0 & d y_0 \\ y_0 & x_0 \end{pmatrix}^n \begin{pmatrix} x_0 \\ y_0 \end{pmatrix}.$$

Thus,

$$\begin{cases} x_n = \dfrac{1}{2} \left[\left(x_0 + y_0 \sqrt{d} \right)^{n+1} + \left(x_0 - y_0 \sqrt{d} \right)^{n+1} \right] \\ y_n = \dfrac{1}{2\sqrt{d}} \left[\left(x_0 + y_0 \sqrt{d} \right)^{n+1} - \left(x_0 - y_0 \sqrt{d} \right)^{n+1} \right], \quad n \geq 0. \end{cases} \tag{3.4}$$

By a *fundamental solution* of Pell's equation we understand the pair (x_0, y_0), with $x_0, y_0 \in \mathbb{N}$, and $x_0^2 - d y_0^2 = 1$ such that x_0 is minimal if and only if y_0 is minimal, i.e., $x_0 + \sqrt{d} y_0$ is minimal among $x + \sqrt{d} y$, where (x, y) is a solution, in positive integers, of Pell's equation. We mention that the existence of the fundamental solution of Pell's equation can be proved.

Now, if we consider that (x_0, y_0) is the fundamental solution of Pell's equation we get that

$$S_P \subseteq \{(-1, 0), (1, 0), (x_n, y_n), (-x_n, y_n), (x_n, -y_n) : n \in \mathbb{N}\} = S.$$

Next we prove that $S \subseteq S_P$. If $(x, y) \in S \cap (\mathbb{N} \times \mathbb{N})$, we define $B = A_{(x,y)}$ and $B_1 = A^{-1} B$, where

$$A = A_{(x_0,y_0)} = \begin{pmatrix} x_0 & dy_0 \\ y_0 & x_0 \end{pmatrix}$$

and (x_0, y_0) is the fundamental solution. It follows that $\det B_1 = 1$ and

$$B_1 = \begin{pmatrix} x' & dy' \\ y' & x' \end{pmatrix} \quad \text{with} \quad \begin{cases} x' = x_0 x - dy_0 y \\ y' = x_0 y - y_0 x. \end{cases}$$

It follows that $x' < x$, $y' < y$, and $(x', y') \in \mathbb{N} \times \mathbb{N}$. We continue this algorithm and we get that $B_2 = A^{-1} B_1$, $B_3 = A^{-1} B_2, \ldots, B_k = A^{-1} B_{k-1} = I_2$. We have, by going backwards, that $A_{(x,y)} = A_{(x_0,y_0)}^k$ which implies, based on (3.4), that $(x, y) \in S_P$.

Thus, we have proved the following theorem.

Theorem 3.12 *The diophantine equation $x^2 - dy^2 = 1$, where $d \geq 2$ is an integer which is not a perfect square, has the following solutions in positive integers*

$$\begin{cases} x_n = \dfrac{1}{2}\left[\left(x_0 + y_0\sqrt{d}\right)^{n+1} + \left(x_0 - y_0\sqrt{d}\right)^{n+1}\right] \\ y_n = \dfrac{1}{2\sqrt{d}}\left[\left(x_0 + y_0\sqrt{d}\right)^{n+1} - \left(x_0 - y_0\sqrt{d}\right)^{n+1}\right], \quad n \geq 0, \end{cases}$$

where (x_0, y_0) is the fundamental solution.

Example 3.5 We solve in $\mathbb{Z} \times \mathbb{Z}$ the equation $x^2 - 2y^2 = 1$.

Since the fundamental solution of this equation is $(3, 2)$ we have, based on Theorem 3.12, that the equation has infinitely many solutions which are given by

$$\begin{cases} x_n = \dfrac{1}{2}\left[\left(3 + 2\sqrt{2}\right)^{n+1} + \left(3 - 2\sqrt{2}\right)^{n+1}\right] \\ y_n = \dfrac{1}{2\sqrt{2}}\left[\left(3 + 2\sqrt{2}\right)^{n+1} - \left(3 - 2\sqrt{2}\right)^{n+1}\right]; \quad n \geq 0, \end{cases}$$

and hence $S_P = \{(\pm x_n, \pm y_n) : n \in \mathbb{N}\} \cup \{(\pm 1, 0)\}$.

Remark 3.7 The solutions of Pell's equation can be used to approximate the square roots of natural numbers which are not perfect square. If (x_n, y_n), $n \geq 1$, are positive solutions of Pell's equation $x^2 - dy^2 = 1$, then

$$x_n - \sqrt{d}y_n = \frac{1}{x_n + \sqrt{d}y_n} \quad \Rightarrow \quad \frac{x_n}{y_n} - \sqrt{d} = \frac{1}{y_n(x_n + \sqrt{d}y_n)},$$

which implies that

$$\lim_{n \to \infty} \frac{x_n}{y_n} = \sqrt{d}.$$

Thus, the fractions $\dfrac{x_n}{y_n}$ approximate $\sqrt{d}$ by an error less than $\dfrac{1}{y_n^2}$.

Now we study the diophantine equation

$$ax^2 - by^2 = 1, \quad \text{where} \quad a, b \in \mathbb{N}. \tag{3.5}$$

Lemma 3.5 *If $ab = k^2$, $k \in \mathbb{N}$, $k \geq 2$, then the equation $ax^2 - by^2 = 1$ has no solutions in $\mathbb{N} \times \mathbb{N}$.*

Proof We prove the lemma by contradiction. We assume that the equation has a solution $(x_0, y_0) \in \mathbb{N} \times \mathbb{N}$. It follows that $ax_0^2 - by_0^2 = 1$ and this implies that a and b are relatively prime. The identity $ab = k^2$ implies that $a = k_1^2$ and $b = k_2^2$, with $k_1 k_2 = k$, $k_1, k_2 \in \mathbb{N}$. In this case, the equation becomes $k_1^2 x_0^2 - k_2^2 y_0^2 = 1$ or $(k_1 x_0 - k_2 y_0)(k_1 x_0 + k_2 y_0) = 1$ and this implies that $1 = k_1 x_0 + k_2 y_0 = k_1 x_0 - k_2 y_0 \Rightarrow y_0 = 0$, which contradicts $y_0 \in \mathbb{N}$. □

We define the *Pell resolvent* of $ax^2 - by^2 = 1$ the following diophantine equation

$$u^2 - abv^2 = 1. \tag{3.6}$$

Lemma 3.6 *If equation (3.5) has a nontrivial solution in $\mathbb{N} \times \mathbb{N}$, then it has infinitely many solutions.*

Proof Let (x_0, y_0) be a solution of equation (3.5). Since ab is not a perfect square, see Lemma 3.5, we get that equation (3.6) has infinitely many solutions in positive integers which are given by the formulae in Theorem 3.12.

We denote by (u_n, v_n), $n \in \mathbb{N}$, the general solution of equation (3.6). Let (x_n, y_n), $n \in \mathbb{N}$, where $x_n = x_0 u_n + b y_0 v_n$ and $y_n = y_0 u_n + a x_0 v_n$, and we observe that (x_n, y_n) are solutions of the equation $ax^2 - by^2 = 1$, since

$$ax_n^2 - by_n^2 = a(x_0 u_n + b y_0 v_n)^2 - b(y_0 u_n + a x_0 v_n)^2$$
$$= (ax_0^2 - by_0^2)(u_n^2 - abv_n^2)$$
$$= 1.$$

The lemma is proved. □

Theorem 3.13 *Let (A, B) be the minimal solution of equation (3.5). The general solution of equation (3.5) is given by (x_n, y_n), $n \in \mathbb{N}$, with*

$$\begin{cases} x_n = Au_n + bBv_n \\ y_n = Bu_n + aAv_n, \end{cases}$$

where (u_n, v_n), $n \in \mathbb{N}$, is the general solution of equation (3.6).

Proof We showed in the proof of Lemma 3.6 that if $(u_n, v_n), n \in \mathbb{N}$, are the solutions of equation (3.6), then $(x_n, y_n), n \in \mathbb{N}$, are the solutions of equation (3.5).

To prove the other implication, we show that if $(x_n, y_n), n \in \mathbb{N}$, are the solutions of equation (3.5), then $(u_n, v_n), n \in \mathbb{N}$, with

$$\begin{cases} u_n = aAx_n - bBy_n \\ v_n = Bx_n - Ay_n, \end{cases}$$

are solutions of equation (3.6).

We have,

$$\begin{aligned} u_n^2 - abv_n^2 &= (aAx_n - bBy_n)^2 - ab(Bx_n - Ay_n)^2 \\ &= (aA^2 - bB^2)(ax_n^2 - by_n^2) \\ &= 1, \end{aligned}$$

and the theorem is proved.   □

In the particular case when $b = 1$, the technique given in the previous results can be used to solve the diophantine equation

$$dx^2 - y^2 = 1, \tag{3.7}$$

which is called the *conjugate Pell equation*.

The general solution of equation (3.7) is given by

$$\begin{cases} x_n = Au_n + Bv_n \\ y_n = Bu_n + dAv_n, \end{cases} \tag{3.8}$$

where (A, B) is the fundamental solution of equation (3.7) and $(u_n, v_n), n \in \mathbb{N}$, are the solutions of Pell's equation $u^2 - dv^2 = 1$.

Remark 3.8 The sequences $(x_n)_{n \geq 1}$ and $(y_n)_{n \geq 1}$ defined recursively by (3.8) verify the interesting identity

$$y_n = \left\lfloor \sqrt{d}x_n \right\rfloor, \quad n \in \mathbb{N},$$

where $\lfloor x \rfloor$ denotes the floor of x.

To see this, we note that since (x_n, y_n) is the solution of Pell's conjugate equation $dx^2 - y^2 = 1$ we have that

$$(\sqrt{d}x_n + y_n)(\sqrt{d}x_n - y_n) - 1.$$

However, $x_n, y_n \in \mathbb{N}$ and it follows that $\sqrt{d}x_n + y_n > 1$. Therefore, $0 < \sqrt{d}x_n - y_n < 1 \implies y_n < \sqrt{d}x_n < y_n + 1$, which implies that $y_n = \left\lfloor \sqrt{d}x_n \right\rfloor$, $n \in \mathbb{N}$.

Example 3.6 We solve in $\mathbb{N} \times \mathbb{N}$ the equation $6x^2 - 5y^2 = 1$.

First, we observe that the fundamental solution of this equation is $(1, 1)$. Also, Pell's resolvent equation becomes $u^2 - 30v^2 = 1$, which has the fundamental solution $(11, 2)$. It follows that the general solution of Pell's resolvent equation is (u_n, v_n), where

$$\begin{cases} u_{n+1} = 11u_n + 60v_n \\ v_{n+1} = 2u_n + 11v_n, \quad n \in \mathbb{N}, \end{cases}$$

with $u_1 = 11$ and $v_1 = 2$.

Thus, the general solution of our equation is

$$\begin{cases} x_n = \dfrac{6 + \sqrt{30}}{12}(11 + 2\sqrt{30})^n + \dfrac{6 - \sqrt{30}}{12}(11 - 2\sqrt{30})^n \\ y_n = \dfrac{5 + \sqrt{30}}{12}(11 + 2\sqrt{30})^n + \dfrac{5 - \sqrt{30}}{12}(11 - 2\sqrt{30})^n. \end{cases}$$

3.5.1 Problems

3.100 Find all right angle triangles ABC with integer side lengths a, b, c, with $a > b$, $a > c$ such that the triangle with sides $a' = a + 4$, $b' = b + 3$, and $c' = c + 3$ is a right angle triangle.

3.101 Solve in $\mathbb{Z} \times \mathbb{Z}$ the equation $x^2 - 8y^2 = 1$.

3.102 Solve in $\mathbb{Z} \times \mathbb{Z}$ the equation $2x^2 - 6xy + 3y^2 + 1 = 0$.

3.103 Prove that for any nonzero integer k the equation $x^2 - 2kxy + y^2 = 1$ has an infinite number of solutions in $\mathbb{Z} \times \mathbb{Z}$.

3.104 [3] Find all positive integers n such that $\dbinom{n}{k-1} = 2\dbinom{n}{k} + \dbinom{n}{k+1}$ for some natural numbers $k < n$.

3.105 Prove that if $m = 2 + 2\sqrt{28n^2 + 1}$ is an integer for some $n \in \mathbb{N}$, then m is a perfect square.

---

[3] Problems **3.104**, **3.105** and **3.106** are taken from [16].

3.106 Prove that if n is an integer such that $3n + 1$ and $4n + 1$ are both perfect squares, then n is divisible by 56.

Chebyshev polynomials. For $-1 < t < 1$, let θ be such that $0 < \theta < \pi$ and $t = \cos\theta$ (i.e., $\theta = \arccos t$). Let T_n and U_n be the polynomials defined by

$$T_n(t) = \cos n\theta = \cos(n \arccos t)$$

and

$$U_n(t) = \frac{\sin n\theta}{\sin\theta} = \frac{\sin(n \arccos t)}{\sqrt{1 - t^2}}.$$

While these functions are initially defined on a restricted domain, they turn out to be polynomials in t and so they have meanings for all real values of t.

The polynomials T_n are called *Chebyshev polynomials of the first kind* and the U_n are called *Chebyshev polynomials of the second kind*. These polynomials are widely used in a variety of mathematical contexts and they have a number of remarkable properties (see [8, Section 3.4]).

3.107 [8, p. 39] **Chebyshev polynomials and Pell's equation.**

Prove the solution of the equation $x^2 - (t^2 - 1)y^2 = 1$, where t is a parameter, is of the form $(x_n, y_n) = (T_n(t), U_n(t))$.

3.5.2   Solutions

3.100. Since a, b, and c are Pythagorean numbers, let $a = m^2 + n^2$, $b = 2mn$, and $c = m^2 - n^2$, where $m, n \in \mathbb{N}$ and $m > n$. The equation $(a+4)^2 = (b+3)^2 + (c+3)^2$ implies that $4a = 3b + 3c + 1 \Rightarrow m^2 + 7n^2 - 6mn = 1$ or $(m - 3n)^2 - 2n^2 = 1$. Using the substitutions $m - 3n = x$ and $n = y$ we get the equation $x^2 - 2y^2 = 1$. This equation, which has the minimal solution $(3, 2)$, is solved in Example 3.5 and the general solution (x_k, y_k), $k \in \mathbb{N}$, implies that $m_k = x_k + 3y_k$ and $n_k = y_k, k \in \mathbb{N}$.

3.101. Since the minimal solution of the equation is $(3, 1)$ we get that

$$\begin{cases} x_n = \dfrac{1}{2}\left[(3 + 2\sqrt{2})^{n+1} + (3 - 2\sqrt{2})^{n+1}\right] \\ y_n = \dfrac{1}{4\sqrt{2}}\left[(3 + 2\sqrt{2})^{n+1} + (3 - 2\sqrt{2})^{n+1}\right], \quad n \geq 0. \end{cases}$$

The solution of Pell's equation is given by $\{(\pm x_n, \pm y_n) : n \in \mathbb{N}\} \cup \{(-1, 0), (1, 0)\}$.

3.102. The equation can be written in the following form $x^2 - 3(y - x)^2 = 1$. Using the substitutions $X = x$, $Y = y - x$ we get that $X^2 - 3Y^2 = 1$. The minimal solution of this equation is $(2, 1)$ and we get that

$$
\begin{cases}
X_n = \dfrac{1}{2}\left[(2 + \sqrt{3})^{n+1} + (2 - \sqrt{3})^{n+1}\right] \\
Y_n = \dfrac{1}{2\sqrt{3}}\left[(2 + \sqrt{3})^{n+1} - (2 - \sqrt{3})^{n+1}\right], \quad n \geq 0,
\end{cases}
$$

and $x_n = X_n$ and $y_n = x_n + Y_n = X_n + Y_n$, $n \geq 0$. The solution of the equation is $\{(\pm x_n, \pm y_n) : n \in \mathbb{N}\} \cup \{(1, 1), (-1, -1)\}$.

3.103. The equation can be written in the following form $(x - ky)^2 - (k^2 - 1)y^2 = 1$. Using the substitutions $x - ky = u$ and $y = v$ the equation becomes $u^2 - dv^2 = 1$, where $d = k^2 - 1$. If $k \neq 0$ or $k \neq \pm 1$, then d is a positive integer which is not a perfect square.

If $k = 1$ the equation becomes $(x - y)^2 = 1$ which has an infinite numbers of solutions given by $(p \pm 1, p), p \in \mathbb{Z}$.

If $k = -1$ we get that $(x + y)^2 = 1$ with solutions $(-p \pm 1, p), p \in \mathbb{Z}$.

If $|k| \geq 2$ we consider the matrix $A = \begin{pmatrix} k & k^2 - 1 \\ 1 & k \end{pmatrix}$ which has $\det A = 1$. We have $\det(A^n) = 1$, $A^n = \begin{pmatrix} u_n & dv_n \\ v_n & u_n \end{pmatrix}$ and $\det(A^n) = u_n^2 - dv_n^2 = 1$. Since $u_0 = k$, $v_0 = 1$ the equation $u^2 - dv^2 = 1$ has an infinite number of solutions $(u_n, v_n), n \in \mathbb{N}$, which generate an infinite number of solutions for the equation $x^2 - 2kxy + y^2 = 1$, which are given by $x_n = u_n + kv_n$ and $y_n = v_n, n \in \mathbb{N}$.

3.104.–3.106. See [16].

3.107. If $t = \pm 1$, then $x = \pm 1$, so the solutions of our equation are $(\pm 1, \alpha), \alpha \in \mathbb{R}$. Let $|t| < 1$. An obvious solution is $(t, 1)$. Other solutions can be obtained from

$$
x_n + \sqrt{t^2 - 1}\, y_n = (t + \sqrt{t^2 - 1})^n = (t + i\sqrt{1 - t^2})^n.
$$

Using the substitution $t = \cos\theta$ we get that

$$
x_n + i\sin\theta\, y_n = (\cos\theta + i\sin\theta)^n = \cos(n\theta) + i\sin(n\theta),
$$

and it follows that $x_n = \cos(n\theta) = T_n(t)$ and $y_n = \dfrac{\sin(n\theta)}{\sin\theta} = U_n(t)$.

If $|t| > 1$, then

$$
\begin{cases}
x_n + \sqrt{t^2 - 1}\, y_n = (t + \sqrt{t^2 - 1})^n \\
x_n - \sqrt{t^2 - 1}\, y_n = (t - \sqrt{t^2 - 1})^n
\end{cases}
$$

and it follows that

$$\begin{cases} x_n = \dfrac{1}{2}\left[(t + \sqrt{t^2 - 1})^n + (t - \sqrt{t^2 - 1})^n\right] \\ y_n = \dfrac{1}{2\sqrt{t^2 - 1}}\left[(t + \sqrt{t^2 - 1})^n - (t - \sqrt{t^2 - 1})^n\right], \quad n \geq 0. \end{cases}$$

The reader should check that $(x_n, y_n) = (T_n(t), U_n(t))$, $n \geq 0$.

Chapter 4
Functions of matrices. Matrix calculus

Sleepiness and fatigue are the enemies of learning.
Platon (427 B.C.–347 B.C.)

4.1   Sequences and series of matrices

Let $A \in \mathscr{M}_2(\mathbb{C})$ and let $f \in \mathbb{C}[x]$ be the polynomial function

$$f(x) = a_0 + a_1 x + a_2 x^2 + \cdots + a_n x^n.$$

The matrix $f(A) = a_0 I_2 + a_1 A + a_2 A^2 + \cdots + a_n A^n$ is called the *polynomial function f* evaluated at A. For any matrix A and any polynomial function f we can define the matrix $f(A)$.

We extend this definition to other functions, non-polynomial ones, extension which turns out to have applications to other branches of mathematics such as solving systems of differential equations and studying the stability of various phenomena modeled by systems of differential equations. The difficulty of this extension stands in the fact that if a numerical function f, defined on a set D, is given and A is a matrix, then to define the matrix $f(A)$ one needs some conditions that the matrix A should satisfy.

It turns out to be very useful to study limits of polynomial functions, so it is necessary to define the limit of a sequence of matrices.

Let $(A_n)_{n \in \mathbb{N}}$ be a sequence of matrices, $A_n = \left(a_{i,j}^{(n)} \right)_{i,j=1,2} \in \mathscr{M}_2(\mathbb{C})$.

Definition 4.1 We say that the sequence $(A_n)_{n \in \mathbb{N}}$ *is convergent* if the sequences $\left(a_{i,j}^{(n)} \right)_{n \in \mathbb{N}}$ are convergent for all $i,j = 1,2$. If $a_{i,j} = \lim\limits_{n \to \infty} a_{i,j}^{(n)}$, then the matrix $A = (a_{i,j})_{i,j=1,2}$ is called *the limit of the sequence* $(A_n)_{n \in \mathbb{N}}$ and we write $A = \lim\limits_{n \to \infty} A_n$. Sometimes the notation $A_\infty = \lim\limits_{n \to \infty} A_n$ is used.

The next proposition, whose proof is straightforward, gives the most elementary properties of limits of sequences of matrices.

© Springer International Publishing AG 2017
V. Pop, O. Furdui, *Square Matrices of Order 2*, DOI 10.1007/978-3-319-54939-2_4

Proposition 4.1 *If $A = \lim\limits_{n\to\infty} A_n$, $B = \lim\limits_{n\to\infty} B_n$ and $P \in \mathscr{M}_2(\mathbb{C})$ is an invertible matrix, then*:

(a) $\lim\limits_{n\to\infty} (\alpha A_n + \beta B_n) = \alpha A + \beta B$, $\alpha, \beta \in \mathbb{C}$;

(b) $\lim\limits_{n\to\infty} (A_n B_n) = AB$;

(c) $\lim\limits_{n\to\infty} (P^{-1} A_n P) = P^{-1} A P$.

Remark 4.1 We mention that the limit of a sequence of invertible matrices need not be invertible, i.e., if $(A_n)_{n\in\mathbb{N}}$ is a sequence of invertible matrices and $A = \lim\limits_{n\to\infty} A_n$ then, the matrix A need not be invertible (see the case of $\lim\limits_{n\to\infty} \frac{1}{n} I_2 = O_2$).

Let $(f_n)_{n\in\mathbb{N}}$ be a sequence of polynomials, $f_n \in \mathbb{C}[x]$ and let $A \in \mathscr{M}_2(\mathbb{C})$.

Theorem 4.1 *If J_A is the Jordan canonical form of A, then $\lim\limits_{n\to\infty} f_n(A)$ exists if and only if $\lim\limits_{n\to\infty} f_n(J_A)$ exists. In this case if P is the invertible matrix such that $J_A = P^{-1} A P$, then*

$$\lim\limits_{n\to\infty} f_n(A) = P\left(\lim\limits_{n\to\infty} f_n(J_A)\right) P^{-1}.$$

Proof For any polynomial function $f \in \mathbb{C}[x]$ we have $f(J_A) = P^{-1} f(A) P$. This follows based on the formula $J_A = P^{-1} A P$ which implies that $J_A^n = P^{-1} A^n P$. Now we apply part (c) of Proposition 4.1 to the equalities $f_n(J_A) = P^{-1} f_n(A) P$ and $f_n(A) = P f_n(J_A) P^{-1}$ and we obtain the simultaneous existence or nonexistence of the limits $\lim\limits_{n\to\infty} f_n(A)$ and $\lim\limits_{n\to\infty} f_n(J_A)$ and the relation between them.   □

Remark 4.2 Theorem 4.1 reduces the calculation of the limit of a sequence of polynomial functions of a given matrix A to the study of the limit of the polynomial function of the corresponding Jordan canonical form J_A.

We mention that a 2×2 matrix can have two Jordan cells of order 1, i.e., these are matrices of the form $J_\lambda = [\lambda]$ or a Jordan cell of order 2, $J_\lambda = \begin{pmatrix} \lambda & 1 \\ 0 & \lambda \end{pmatrix}$.

Theorem 4.2 *Let $\lambda \in \mathbb{C}$ and let*

$$J_\lambda = \begin{pmatrix} \lambda & 1 \\ 0 & \lambda \end{pmatrix}$$

be a Jordan cell of order 2. Then, for any polynomial function $f \in \mathbb{C}[x]$ we have

$$f(J_\lambda) = \begin{pmatrix} f(\lambda) & f'(\lambda) \\ 0 & f(\lambda) \end{pmatrix}.$$

Proof Let $g_n(\lambda) = \lambda^n$ and let $J_0 = \begin{pmatrix} 0 & 1 \\ 0 & 0 \end{pmatrix}$. First we note that $J_0^2 = O_2$ and we have

$$J_\lambda^n = (\lambda I_2 + J_0)^n = \lambda^n I_2 + \binom{n}{1}\lambda^{n-1}J_0 = g_n(\lambda)I_2 + g'_n(\lambda)J_0.$$

Thus, the theorem is valid for any polynomial of the form $f(x) = x^n$ and based on linearity it is also valid for any polynomial function $f \in \mathbb{C}[x]$. $\qquad\square$

Theorem 4.3 *The sequence of matrices* $(f_n(J_\lambda))_{n\in\mathbb{N}}$ *converges if and only if the numerical sequences* $(f_n(\lambda))_{n\in\mathbb{N}}$ *and* $(f'_n(\lambda))_{n\in\mathbb{N}}$ *converge. If* $\lim_{n\to\infty} f_n(\lambda) = f(\lambda)$ *and* $\lim_{n\to\infty} f'_n(\lambda) = f'(\lambda)$, *then*

$$f(J_\lambda) = \lim_{n\to\infty} f_n(J_\lambda) = \begin{pmatrix} f(\lambda) & f'(\lambda) \\ 0 & f(\lambda) \end{pmatrix}.$$

Proof We apply theorem 4.2. $\qquad\square$

Recall that for a matrix A its spectrum, denoted by $\mathrm{Spec}(A)$ is the set of all eigenvalues of A. In our case $\mathrm{Spec}(A) = \{\lambda_1, \lambda_2\} \subset \mathbb{C}$.

Definition 4.2 We say that the sequence $(f_n)_{n\in\mathbb{N}}$ is convergent on the spectrum of A if for any $\lambda_i \in \mathrm{Spec}(A)$, $i = 1, 2$, the limits $\lim_{n\to\infty} f_n(\lambda_i)$, $i = 1, 2$ and $\lim_{n\to\infty} f'_n(\lambda_i)$, $i = 1, 2$ exist and are finite. Moreover, if there exists a function f defined on a subset of $\mathbb{C}$ which contains $\mathrm{Spec}(A)$ and $\lim_{n\to\infty} f_n(\lambda_i) = f(\lambda_i)$ and $\lim_{n\to\infty} f'_n(\lambda_i) = f'(\lambda_i)$, for $i = 1, 2$, then the function f is called the limit of the sequence $(f_n)_{n\in\mathbb{N}}$ on the spectrum of A and we write $\lim_{\mathrm{Spec}(A)} f_n = f$.

Theorem 4.4 *The sequence of matrices* $(f_n(A))_{n\in\mathbb{N}}$ *converges if and only if the sequence of polynomials* $(f_n)_{n\in\mathbb{N}}$ *is convergent on the spectrum of* A. *If*

$$\lim_{\mathrm{Spec}(A)} f_k = f \quad then \quad \lim_{n\to\infty} f_n(A) = f(A).$$

Proof This follows based on theorems 4.1, 4.2, and 4.3 $\qquad\square$

Definition 4.3 If the sequence $(f_n)_{n\in\mathbb{N}}$ is convergent on the spectrum of A and $\lim_{\mathrm{Spec}(A)} f_n = f$, then the matrix $f(A) = \lim_{n\to\infty} f_n(A)$ is called the *function f of the matrix A*.

Remark 4.3 We have $\lim_{n\to\infty} f_n(A) = \left(\lim_{\mathrm{Spec}(A)} f_n\right)(A)$.

Remark 4.4 Let $D \subset \mathbb{C}$, let $f : D \to \mathbb{C}$ be the function which is the limit of the sequence of polynomials $(f_n)_{n\in\mathbb{N}}$, and let A be a matrix. Then, in order to define the matrix $f(A)$ it is necessary to verify the conditions $\mathrm{Spec}(A) \subset D$ and the convergence of the sequence $(f_n)_{n \in \mathbb{N}}$ on the spectrum of A. A possible algorithm for calculating the matrix $f(A)$ has, based on the previous theorems, the following steps:

(1) Determine the spectrum of A and check the convergence on the spectrum of A of the sequence $(f_n)_{n \in \mathbb{N}}$ to the function f;

(2) Determine the matrix P and the Jordan canonical form J_A of the matrix A;

(3) Determine $f(J_A)$;

(4) Write $f(A) = Pf(J_A)P^{-1}$.

Sometimes it is not necessary to use the Jordan canonical form in order to find $\lim\limits_{n \to \infty} f_n(A)$. Next, we consider the case when f is an analytic function, these are functions that can be written as power series.

Let $\sum\limits_{m=0}^{\infty} a_m z^m$ be a power series having the radius of convergence R, let $f_n(z) = \sum\limits_{m=0}^{n} a_m z^m$ and let $f(z)$ be the sum of the power series. We have

$$\lim_{n \to \infty} f_n(z) = f(z) \quad \text{for} \quad z \in D_R = \{z \in \mathbb{C} : |z| < R\}$$

$$\lim_{n \to \infty} f_n^{(i)}(z) = f^{(i)}(z), \ i \in \mathbb{N}, \ z \in D_R$$

(4.1)

and for $|z| > R$ the preceding limits do not exist, so the function f is defined only on D_R and eventually at some points on the circle $\mathscr{C}_R = \partial D_R = \{z \in \mathbb{C} : |z| = R\}$.

Definition 4.4 Let $A \in \mathscr{M}_2(\mathbb{C})$. The *spectral radius* of A is the real number defined by

$$\rho(A) = \max \{|\lambda_1|, |\lambda_2|\}.$$

We try to determine the conditions on which one could define the matrix $f(A)$. Clearly $f_n(z) = \sum\limits_{m=0}^{n} a_m z^m$ are polynomials and we have, based on Theorem 4.4, that the matrix $f(A) = \lim\limits_{n \to \infty} f_n(A)$ exists if and only if the sequence of polynomials $(f_n)_{n \in \mathbb{N}}$ converges, on the spectrum of A, to f.

Theorem 4.5 *Let R be the radius of convergence of the power series $f(z) = \sum\limits_{m=0}^{\infty} a_m z^m$ and let $A \in \mathscr{M}_2(\mathbb{C})$. Then:*

(a) If $\rho(A) < R$, i.e., all the eigenvalues of A belong to the disk D_R, then the series of matrices $\sum_{m=0}^{\infty} a_m A^m$ converges and the matrix $f(A)$ exists and is defined by

$$f(A) = \sum_{m=0}^{\infty} a_m A^m;$$

(b) If $\rho(A) = R$, i.e., there are eigenvalues of A on the circle $\mathscr{C}_R$, then the series of matrices $\sum_{m=0}^{\infty} a_m A^m$ converges if for any eigenvalue λ, with $|\lambda| = R$, the series

$$\sum_{m=0}^{\infty} a_m \lambda^m \quad and \quad \sum_{m=1}^{\infty} m a_m \lambda^{m-1} \tag{4.2}$$

converge.

Proof (a) Since any eigenvalue of A belongs to D_R we have, based on (4.1), that the sequence $(f_n)_{n \in \mathbb{N}}$, of the partial sums of the power series, $\sum_{m=0}^{\infty} a_m z^m$ converges on the spectrum of A to f and this implies in view of Theorem 4.4 that $f(A)$ exists and $f(A) = \sum_{m=0}^{\infty} a_m A^m$.

(b) The convergence of the sequence $(f_n)_{n \in \mathbb{N}}$, of the partial sums, on the spectrum of A reduces to the conditions (4.2). We have, based on (4.1), that these conditions hold for the eigenvalues of A in the convergence disk and they need to be studied for the eigenvalues of A on $\mathscr{C}_R$. □

Theorem 4.6 *Let f be a function which has the Taylor series expansion at z_0,*

$$f(z) = \sum_{n=0}^{\infty} \frac{f^{(n)}(z_0)}{n!}(z - z_0)^n, \quad |z - z_0| < R,$$

where $R \in (0, \infty]$. If $A \in \mathcal{M}_2(\mathbb{C})$ has eigenvalues $\lambda_1, \lambda_2 \in \mathbb{C}$ such that $|\lambda_i - z_0| < R, i = 1, 2$, then the matrix $f(A)$ has the eigenvalues $f(\lambda_1)$ and $f(\lambda_2)$.

Proof Since similar matrices have the same eigenvalues and $f(A)$ is similar to $f(J_A)$ the theorem follows based on Theorems 4.2 and 4.3.

Another "proof" is based on a formal computation. Let $X \neq 0$ be the eigenvector corresponding to the eigenvalue λ, i.e., $AX = \lambda X$. We have

$$f(A)X = \left(\sum_{n=0}^{\infty} \frac{f^{(n)}(z_0)}{n!}(A - z_0 I_2)^n\right) X = \sum_{n=0}^{\infty} \frac{f^{(n)}(z_0)}{n!}(\lambda - z_0)^n X = f(\lambda)X,$$

and the theorem is proved. □

4.2   Elementary functions of matrices

In this section we introduce the elementary functions of matrices that are used throughout this book.

■ **The polynomial function**
  If $f \in \mathbb{C}[x]$ is the polynomial function defined by $f(x) = a_0 + a_1 x + \cdots + a_n x^n$, $a_i \in \mathbb{C}$, $i = \overline{0, n}$, then

$$f(A) = a_0 I_2 + a_1 A + \cdots + a_n A^n, \quad A \in \mathcal{M}_2(\mathbb{C}).$$

■ **The exponential function**

$$e^A = \sum_{n=0}^{\infty} \frac{A^n}{n!}, \quad A \in \mathcal{M}_2(\mathbb{C}).$$

■ **The hyperbolic functions**
  cosine hyperbolic

$$\cosh A = \sum_{n=0}^{\infty} \frac{A^{2n}}{(2n)!}, \quad A \in \mathcal{M}_2(\mathbb{C}).$$

  sine hyperbolic

$$\sinh A = \sum_{n=0}^{\infty} \frac{A^{2n+1}}{(2n+1)!}, \quad A \in \mathcal{M}_2(\mathbb{C}).$$

■ **The trigonometric functions**
  cosine

$$\cos A = \sum_{n=0}^{\infty} \frac{(-1)^n}{(2n)!} A^{2n}, \quad A \in \mathcal{M}_2(\mathbb{C}).$$

  sine

$$\sin A = \sum_{n=0}^{\infty} \frac{(-1)^n}{(2n+1)!} A^{2n+1}, \quad A \in \mathcal{M}_2(\mathbb{C}).$$

(continued)

■ **The Neumann (geometric) series**

$$(I_2 - A)^{-1} = \sum_{n=0}^{\infty} A^n, \quad A \in \mathcal{M}_2(\mathbb{C}), \ \rho(A) < 1.$$

■ **The binomial series**

$$(I_2 - A)^{-\alpha} = \sum_{n=0}^{\infty} \frac{\Gamma(n+\alpha)}{\Gamma(\alpha)n!} A^n, \quad A \in \mathcal{M}_2(\mathbb{C}), \ \rho(A) < 1, \ \alpha > 0,$$

where Γ denotes the *Gamma* function.

■ **The logarithmic functions**

$$\ln(I_2 + A) = \sum_{n=1}^{\infty} \frac{(-1)^{n-1}}{n} A^n, \quad A \in \mathcal{M}_2(\mathbb{C}), \ \rho(A) < 1.$$

$$\ln(I_2 - A) = -\sum_{n=1}^{\infty} \frac{A^n}{n}, \quad A \in \mathcal{M}_2(\mathbb{C}), \ \rho(A) < 1.$$

■ **The power function**
   If $z \in \mathbb{C}^*$, then

$$z^A = e^{(\ln z)A} = \sum_{n=0}^{\infty} \frac{\ln^n z}{n!} A^n, \quad A \in \mathcal{M}_2(\mathbb{C}).$$

Nota bene. If a formula of the form $\Phi(f_1(z_1), f_2(z_2), \ldots, f_p(z_p)) = 0$ holds on $\mathbb{C}$, where f_i, $i = \overline{1,p}$, are some functions and $z_i \in \mathbb{C}$, $i = \overline{1,p}$, then if the matrices $A_i \in \mathcal{M}_2(\mathbb{C})$, $i = \overline{1,p}$, commute and $f_i(A_i)$, $i = \overline{1,p}$, exist we also have the matrix formula $\Phi(f_1(A_1), f_2(A_2), \ldots, f_p(A_p)) = O_2$.

Lemma 4.1 Properties of the exponential function.
   The following statements hold:

(a) *If $a \in \mathbb{C}$, then $e^{aI_2} = e^a I_2$;*

(b) **Euler's matrix formula.** *If $A \in \mathcal{M}_2(\mathbb{C})$, then $e^{iA} = \cos A + i \sin A$;*

(c) *If $A, B \in \mathcal{M}_2(\mathbb{C})$ commute, then $e^A e^B = e^B e^A = e^{A+B}$.*

Proof (a) We have

$$e^{aI_2} = \sum_{n=0}^{\infty} \frac{(aI_2)^n}{n!} = \left(\sum_{n=0}^{\infty} \frac{a^n}{n!} \right) I_2 = e^a I_2.$$

(b) We calculate

$$
\begin{aligned}
e^{iA} &= \sum_{n=0}^{\infty} \frac{(iA)^n}{n!} \\
&= \sum_{k=0}^{\infty} \frac{(iA)^{2k}}{(2k)!} + \sum_{k=1}^{\infty} \frac{(iA)^{2k-1}}{(2k-1)!} \\
&= \sum_{k=0}^{\infty} (-1)^k \frac{A^{2k}}{(2k)!} + i \sum_{k=1}^{\infty} (-1)^{k-1} \frac{A^{2k-1}}{(2k-1)!} \\
&= \cos A + i \sin A.
\end{aligned}
$$

(c) We have

$$
\begin{aligned}
e^A e^B &= \sum_{n=0}^{\infty} \frac{A^n}{n!} \sum_{m=0}^{\infty} \frac{B^m}{m!} \\
&= \sum_{n=0}^{\infty} \sum_{m=0}^{\infty} \frac{A^n B^m}{n!m!} \\
&= \sum_{n=0}^{\infty} \sum_{m=0}^{\infty} \frac{1}{(n+m)!} \binom{n+m}{n} A^n B^m \\
&= \sum_{k=0}^{\infty} \frac{1}{k!} \sum_{n+m=k} \binom{n+m}{n} A^n B^m \\
&= \sum_{k=0}^{\infty} \frac{(A+B)^k}{k!} \\
&= e^{A+B},
\end{aligned}
$$

and the lemma is proved.   □

Lemma 4.2 Properties of the trigonometric functions sine and cosine.
    Let $A \in \mathcal{M}_2(\mathbb{C})$. *The following statements hold*:

(a) $\sin A = \dfrac{e^{iA} - e^{-iA}}{2i}$;

(b) $\cos A = \dfrac{e^{iA} + e^{-iA}}{2}$;

(c) **The fundamental identity of matrix trigonometry**

$$
\sin^2 A + \cos^2 A = I_2;
$$

(d)   *Double trigonometric matrix formulae*

$$\sin(2A) = 2\sin A\cos A \quad and \quad \cos(2A) = 2\cos^2 A - I_2 = I_2 - 2\sin^2 A;$$

(e)   *If $A, B \in \mathcal{M}_2(\mathbb{C})$ commute, then $\sin(A + B) = \sin A\cos B + \cos A\sin B$;*

(f)   *If $A \in \mathcal{M}_2(\mathbb{C})$ is involutory and $k \in \mathbb{Z}$, then $\cos(k\pi A) = (-1)^k I_2$.*

Proof (a) and (b) First we prove that if $A \in \mathcal{M}_2(\mathbb{C})$, then $\sin(-A) = -\sin A$ and $\cos(-A) = \cos A$. We have

$$\sin(-A) = \sum_{n=1}^{\infty}(-1)^{n-1}\frac{(-A)^{2n-1}}{(2n-1)!} = -\sum_{n=1}^{\infty}(-1)^{n-1}\frac{A^{2n-1}}{(2n-1)!} = -\sin A$$

$$\cos(-A) = \sum_{n=0}^{\infty}(-1)^n\frac{(-A)^{2n}}{(2n)!} = \sum_{n=0}^{\infty}(-1)^n\frac{A^{2n}}{(2n)!} = \cos A.$$

It follows, based on part (b) of Lemma 4.1, that

$$e^{iA} = \cos A + i\sin A$$

$$e^{-iA} = \cos(-A) + i\sin(-A) = \cos A - i\sin A.$$

Solving the system for $\cos A$ and $\sin A$ we get that parts (a) and (b) of the lemma are proved.

(c) We have, based on parts (a) and (b), that

$$\sin^2 A + \cos^2 A = \left(\frac{e^{iA} - e^{-iA}}{2i}\right)^2 + \left(\frac{e^{iA} + e^{-iA}}{2}\right)^2$$

$$= -\frac{e^{2iA} - 2I_2 + e^{2iA}}{4} + \frac{e^{2iA} + 2I_2 + e^{2iA}}{4}$$

$$= I_2.$$

(d) We have, based on parts (a) and (b), that

$$2\sin A\cos A = 2\frac{e^{iA} - e^{-iA}}{2i}\cdot\frac{e^{iA} + e^{-iA}}{2} = \frac{e^{2iA} - e^{-2iA}}{2i} = \sin(2A)$$

$$2\cos^2 A - I_2 = 2\left(\frac{e^{iA} + e^{-iA}}{2}\right)^2 - I_2 = \frac{e^{2iA} + e^{-2iA}}{2} = \cos(2A).$$

The identity $\cos(2A) = I_2 - 2\sin^2 A$ is proved similarly.

(e) Since the matrices A and B commute, we have in view of part (c) of Lemma 4.1 that

$$\sin A \cos B + \cos A \sin B = \frac{e^{iA} - e^{-iA}}{2i} \cdot \frac{e^{iB} + e^{-iB}}{2} + \frac{e^{iA} + e^{-iA}}{2} \cdot \frac{e^{iB} - e^{-iB}}{2i}$$

$$= \frac{e^{i(A+B)} - e^{-i(A+B)}}{2i}$$

$$= \sin(A + B).$$

(f) We have, since $A^{2n} = I_2$ for any integer $n \geq 0$, that

$$\cos(k\pi A) = \sum_{n=0}^{\infty} (-1)^n \frac{(k\pi A)^{2n}}{(2n)!} = \left(\sum_{n=0}^{\infty} (-1)^n \frac{(k\pi)^{2n}}{(2n)!} \right) I_2 = \cos(k\pi) I_2 = (-1)^k I_2,$$

and the lemma is proved.   □

Lemma 4.3 Limits and derivatives. *Let* $A \in \mathcal{M}_2(\mathbb{C})$. *Then:*

(a) $\displaystyle \lim_{t \to 0} \frac{e^{At} - I_2}{t} = A$;

(b) $\displaystyle \lim_{t \to 0} \frac{\sin(At)}{t} = A$;

(c) $\displaystyle \lim_{t \to 0} \frac{I_2 - \cos(At)}{t^2} = \frac{A^2}{2}$;

(d) $(e^{At})' = A e^{At}$;

(e) $(\sin(At))' = A \cos(At)$;

(f) $(\cos(At))' = -A \sin(At)$.

Proof (a) Since $A = P J_A P^{-1}$ we have that

$$\frac{e^{At} - I_2}{t} = P \frac{e^{J_A t} - I_2}{t} P^{-1}. \tag{4.3}$$

Let λ_1 and λ_2 be the eigenvalues of A. Then

$$\frac{e^{J_A t} - I_2}{t} = \begin{cases} \begin{pmatrix} \dfrac{e^{\lambda_1 t} - 1}{t} & 0 \\ 0 & \dfrac{e^{\lambda_2 t} - 1}{t} \end{pmatrix} & \text{if } J_A = \begin{pmatrix} \lambda_1 & 0 \\ 0 & \lambda_2 \end{pmatrix}, \\[2em] \begin{pmatrix} \dfrac{e^{\lambda t} - 1}{t} & e^{\lambda t} \\ 0 & \dfrac{e^{\lambda t} - 1}{t} \end{pmatrix} & \text{if } J_A = \begin{pmatrix} \lambda & 1 \\ 0 & \lambda \end{pmatrix}. \end{cases} \tag{4.4}$$

Combining (4.3) and (4.4) and passing to the limit when $t \to 0$ we get that part (a) of the lemma is proved.

(b) Since

$$\sin(At) = \frac{e^{iAt} - e^{-iAt}}{2i},$$

we get that

$$\lim_{t \to 0} \frac{\sin(At)}{t} = \lim_{t \to 0} \frac{e^{iAt} - e^{-iAt}}{2it}$$

$$= \lim_{t \to 0} \frac{e^{iAt} - I_2}{2it} - \lim_{t \to 0} \frac{e^{-iAt} - I_2}{2it}$$

$$\stackrel{(a)}{=} \frac{iA - (-iA)}{2i}$$

$$= A.$$

(c) Since $2 \sin^2 \frac{At}{2} = I_2 - \cos(At)$ we get, based on part (b) of the lemma, that

$$\lim_{t \to 0} \frac{I_2 - \cos(At)}{t^2} = ? \lim_{t \to 0} \left(\frac{\sin \frac{At}{2}}{t} \right)^2 = \frac{A^2}{2}.$$

(d) Let $f(t) = e^{At}$. Then

$$f'(t) = \lim_{h \to 0} \frac{f(t+h) - f(t)}{h} = \lim_{h \to 0} \frac{e^{At} \left(e^{Ah} - I_2 \right)}{h} = e^{At} \lim_{h \to 0} \frac{e^{Ah} - I_2}{h} \stackrel{(a)}{=} Ae^{At}.$$

(e) We have, based on part (d) of the lemma, that

$$(\sin(At))' = \left(\frac{e^{iAt} - e^{-iAt}}{2i} \right)' = \frac{iAe^{iAt} - (-iA)e^{-iAt}}{2i}$$

$$= A \frac{e^{iAt} + e^{-iAt}}{2}$$

$$= A \cos(At).$$

(f) The proof of this part of the lemma is similar to the proof of part (e).   □

4.3   A novel computation of function matrices

In this section we give a technique, *presumably new?!*, for calculating the function matrix $f(A)$ where f is an analytic function and $A \in \mathcal{M}_2(\mathbb{C})$. We prove that $f(A)$ can be expressed as a linear combination of A and I_2. This method is different than the technique involving the Jordan canonical form of A.

Theorem 4.7   Expressing $f(A)$ as a linear combination of A and I_2. Let f *be a function which has the Taylor series expansion at* 0

$$f(z) = \sum_{n=0}^{\infty} \frac{f^{(n)}(0)}{n!} z^n, \quad |z| < R,$$

where $R \in (0, \infty]$ and let $A \in \mathcal{M}_2(\mathbb{C})$ be such that $\rho(A) < R$. Then:

$$f(A) = \begin{cases} \dfrac{f(\lambda_1) - f(\lambda_2)}{\lambda_1 - \lambda_2} A + \dfrac{\lambda_1 f(\lambda_2) - \lambda_2 f(\lambda_1)}{\lambda_1 - \lambda_2} I_2 & \text{if } \lambda_1 \neq \lambda_2 \\ f'(\lambda)A + (f(\lambda) - \lambda f'(\lambda))I_2 & \text{if } \lambda_1 = \lambda_2 = \lambda. \end{cases}$$

Proof First we consider the case when $\lambda_1 \neq \lambda_2$. We have, based on Theorem 3.1, that if $n \geq 1$ is an integer, then $A^n = \lambda_1^n B + \lambda_2^n C$, where

$$B = \frac{A - \lambda_2 I_2}{\lambda_1 - \lambda_2} \quad \text{and} \quad C = \frac{A - \lambda_1 I_2}{\lambda_2 - \lambda_1}.$$

It follows that

$$f(A) = \sum_{n=0}^{\infty} \frac{f^{(n)}(0)}{n!} A^n$$

$$= I_2 + \sum_{n=1}^{\infty} \frac{f^{(n)}(0)}{n!} \left(\lambda_1^n B + \lambda_2^n C \right)$$

$$= B + \sum_{n=1}^{\infty} \frac{f^{(n)}(0)}{n!} \lambda_1^n B + C + \sum_{n=1}^{\infty} \frac{f^{(n)}(0)}{n!} \lambda_2^n C$$

$$= \sum_{n=0}^{\infty} \frac{f^{(n)}(0)}{n!} \lambda_1^n B + \sum_{n=0}^{\infty} \frac{f^{(n)}(0)}{n!} \lambda_2^n C$$

$$- f(\lambda_1)B + f(\lambda_2)C$$

$$= \frac{f(\lambda_1) - f(\lambda_2)}{\lambda_1 - \lambda_2} A + \frac{\lambda_1 f(\lambda_2) - \lambda_2 f(\lambda_1)}{\lambda_1 - \lambda_2} I_2.$$

Now we consider the case when $\lambda_1 = \lambda_2 = \lambda$. Using Theorem 3.1 we have that if $n \geq 1$ is an integer, then $A^n = \lambda^n B + n\lambda^{n-1}C$, where $B = I_2$ and $C = A - \lambda I_2$. This implies that

$$f(A) = \sum_{n=0}^{\infty} \frac{f^{(n)}(0)}{n!} A^n$$

$$= I_2 + \sum_{n=1}^{\infty} \frac{f^{(n)}(0)}{n!} \left(\lambda^n B + n\lambda^{n-1}C\right)$$

$$= I_2 + \sum_{n=1}^{\infty} \frac{f^{(n)}(0)}{n!} \lambda^n B + \sum_{n=1}^{\infty} \frac{f^{(n)}(0)}{(n-1)!} \lambda^{n-1} C$$

$$= \sum_{n=0}^{\infty} \frac{f^{(n)}(0)}{n!} \lambda^n I_2 + \sum_{n=1}^{\infty} \frac{f^{(n)}(0)}{(n-1)!} \lambda^{n-1} (A - \lambda I_2)$$

$$= f(\lambda)I_2 + f'(\lambda)(A - \lambda I_2)$$

$$= f'(\lambda)A + (f(\lambda) - \lambda f'(\lambda))I_2.$$

The theorem is proved. $\qquad\qquad\qquad\qquad\qquad\qquad\qquad\qquad\qquad\qquad\qquad$ $\square$

Remark 4.5 More generally, one can prove that if f is a function which has the Taylor series expansion at z_0

$$f(z) = \sum_{n=0}^{\infty} \frac{f^{(n)}(z_0)}{n!} (z - z_0)^n, \quad |z - z_0| < R,$$

where $R \in (0, \infty]$ and $A \in \mathcal{M}_2(\mathbb{C})$ with $\lambda_1, \lambda_2 \in D(z_0, R)$, then

$$f(A) = \begin{cases} \dfrac{f(\lambda_1) - f(\lambda_2)}{\lambda_1 - \lambda_2} A + \dfrac{\lambda_1 f(\lambda_2) - \lambda_2 f(\lambda_1)}{\lambda_1 - \lambda_2} I_2 & \text{if } \lambda_1 \neq \lambda_2 \\ f'(\lambda)A + (f(\lambda) - \lambda f'(\lambda))I_2 & \text{if } \lambda_1 = \lambda_2 = \lambda. \end{cases}$$

The proof of the preceding formula, which is left as an exercise to the interested reader, follows the same steps as in the case when $z_0 = 0$.

Corollary 4.1 Let f be as in Theorem 4.7 and let $\alpha \in \mathbb{C}$.

(a) If $\lambda_1 \neq \lambda_2 \in \mathbb{C}$ are such that $|\lambda_i| < R$, $i = 1, 2$, then

(continued)

Corollary 4.1 (continued)

$$f\begin{pmatrix} \lambda_1 & \alpha \\ 0 & \lambda_2 \end{pmatrix} = \begin{pmatrix} f(\lambda_1) & \dfrac{f(\lambda_1)-f(\lambda_2)}{\lambda_1 - \lambda_2}\alpha \\ 0 & f(\lambda_2) \end{pmatrix}.$$

(b) *If $\lambda \in \mathbb{C}$ with $|\lambda| < R$, then*

$$f\begin{pmatrix} \lambda & \alpha \\ 0 & \lambda \end{pmatrix} = \begin{pmatrix} f(\lambda) & \alpha f'(\lambda) \\ 0 & f(\lambda) \end{pmatrix}.$$

Theorem 4.8 Expressing $f(A)$ **in terms of** $\mathrm{Tr}(A)$ **and** $\det A$.

Let f be a function which has the Taylor series expansion at 0

$$f(z) = \sum_{n=0}^{\infty} \frac{f^{(n)}(0)}{n!} z^n, \quad |z| < R,$$

where $R \in (0, \infty]$ and let $A \in \mathcal{M}_2(\mathbb{C})$ be such that $\rho(A) < R$.
If $t = \mathrm{Tr}(A)$, $d = \det A$ and $\Delta = t^2 - 4d$, then:

$$f(A) = \frac{f\left(\frac{t+\sqrt{\Delta}}{2}\right) - f\left(\frac{t-\sqrt{\Delta}}{2}\right)}{\sqrt{\Delta}} A + \frac{(t+\sqrt{\Delta})f\left(\frac{t-\sqrt{\Delta}}{2}\right) - (t-\sqrt{\Delta})f\left(\frac{t+\sqrt{\Delta}}{2}\right)}{2\sqrt{\Delta}} I_2$$

if $\Delta < 0$ or $\Delta > 0$ and

$$f(A) = f'\left(\frac{t}{2}\right) A + \left[f\left(\frac{t}{2}\right) - \frac{t}{2} f'\left(\frac{t}{2}\right)\right] I_2 \quad \text{if } \Delta = 0.$$

Proof This theorem follows from Theorem 4.7 since the eigenvalues of A are the solutions of the characteristic equation $\lambda^2 - t\lambda + d = 0$. □

Next we give a new proof, which is based on an application of Theorem 4.7, of a classical limit from matrix theory. More precisely we prove that if $A \in \mathcal{M}_2(\mathbb{C})$, then

$$\lim_{n\to\infty} \left(I_2 + \frac{A}{n} \right)^n = e^A.$$

Let $n \in \mathbb{N}$ and let f be the polynomial function defined by $f(x) = \left(1 + \frac{x}{n}\right)^n$. Let $A \in \mathcal{M}_2(\mathbb{C})$ and let λ_1, λ_2 be the eigenvalues of A. We have, based on Theorem 4.7, that

$$\left(I_2 + \frac{A}{n}\right)^n$$

$$= \begin{cases} \frac{\left(1+\frac{\lambda_1}{n}\right)^n - \left(1+\frac{\lambda_2}{n}\right)^n}{\lambda_1 - \lambda_2} A + \frac{\lambda_1\left(1+\frac{\lambda_2}{n}\right)^n - \lambda_2\left(1+\frac{\lambda_1}{n}\right)^n}{\lambda_1 - \lambda_2} I_2 & \text{if } \lambda_1 \neq \lambda_2 \\ \left(1 + \frac{\lambda}{n}\right)^{n-1} A + \left[\left(1 + \frac{\lambda}{n}\right)^n - \lambda\left(1 + \frac{\lambda}{n}\right)^{n-1}\right] I_2 & \text{if } \lambda_1 = \lambda_2 = \lambda. \end{cases}$$

Passing to the limit as $n \to \infty$ in the previous equality we get that

$$\lim_{n\to\infty} \left(I_2 + \frac{A}{n}\right)^n = \begin{cases} \dfrac{e^{\lambda_1} - e^{\lambda_2}}{\lambda_1 - \lambda_2} A + \dfrac{\lambda_1 e^{\lambda_2} - \lambda_2 e^{\lambda_1}}{\lambda_1 - \lambda_2} I_2 & \text{if } \lambda_1 \neq \lambda_2 \\ e^{\lambda} A + \left(e^{\lambda} - \lambda e^{\lambda}\right) I_2 & \text{if } \lambda_1 = \lambda_2 = \lambda \end{cases}$$

$$= e^A.$$

Another proof of this formula which uses the Jordan canonical form of A is given in the solution of problem **4.5**. The previous limit is a particular case of a more general result involving limits of functions of matrices.

Theorem 4.9 A general exponential limit.

If $f : \mathbb{C} \to \mathbb{C}$ is an entire function with $f(0) = 1$ and $A \in \mathcal{M}_2(\mathbb{C})$, then

$$\lim_{n\to\infty} f^n\left(\frac{A}{n}\right) = e^{f'(0)A}.$$

Proof Let λ_1, λ_2 be the eigenvalues of A, let J_A be the Jordan canonical form of A, and let P be the invertible matrix such that $A = P J_A P^{-1}$. We have $f\left(\frac{A}{n}\right) = Pf\left(\frac{J_A}{n}\right)P^{-1}$ which implies $f^n\left(\frac{A}{n}\right) = Pf^n\left(\frac{J_A}{n}\right)P^{-1}$.

We distinguish between the cases when $\lambda_1 \neq \lambda_2$ and $\lambda_1 = \lambda_2$.

Case $\lambda_1 \neq \lambda_2$. Let $J_A = \begin{pmatrix} \lambda_1 & 0 \\ 0 & \lambda_2 \end{pmatrix}$. We have, based on Corollary 4.1, that

$$f\left(\frac{J_A}{n}\right) = \begin{pmatrix} f\left(\frac{\lambda_1}{n}\right) & 0 \\ 0 & f\left(\frac{\lambda_2}{n}\right) \end{pmatrix}.$$

This implies

$$f^n\left(\frac{A}{n}\right) = P \begin{pmatrix} f^n\left(\frac{\lambda_1}{n}\right) & 0 \\ 0 & f^n\left(\frac{\lambda_2}{n}\right) \end{pmatrix} P^{-1}$$

and it follows that

$$\lim_{n\to\infty} f^n\left(\frac{A}{n}\right) = P\begin{pmatrix} e^{f'(0)\lambda_1} & 0 \\ 0 & e^{f'(0)\lambda_2} \end{pmatrix} P^{-1} = e^{f'(0)A}.$$

Case $\lambda_1 = \lambda_2 = \lambda$. In this case there are two possible subcases according to whether $J_A = \begin{pmatrix} \lambda & 0 \\ 0 & \lambda \end{pmatrix}$ or $J_A = \begin{pmatrix} \lambda & 1 \\ 0 & \lambda \end{pmatrix}$. If $J_A = \begin{pmatrix} \lambda & 0 \\ 0 & \lambda \end{pmatrix}$, then $A = \lambda I_2$. This implies that $f^n\left(\frac{A}{n}\right) = f^n\left(\frac{\lambda}{n}I_2\right) = f^n\left(\frac{\lambda}{n}\right)I_2$. Therefore,

$$\lim_{n\to\infty} f^n\left(\frac{A}{n}\right) = \lim_{n\to\infty} f^n\left(\frac{\lambda}{n}\right)I_2 = e^{f'(0)\lambda}I_2 = e^{f'(0)A}.$$

Now we consider the case when $J_A = \begin{pmatrix} \lambda & 1 \\ 0 & \lambda \end{pmatrix}$. An application of Corollary 4.1 shows that

$$f\left(\frac{J_A}{n}\right) = \begin{pmatrix} f\left(\frac{\lambda}{n}\right) & \frac{1}{n}f'\left(\frac{\lambda}{n}\right) \\ 0 & f\left(\frac{\lambda}{n}\right) \end{pmatrix},$$

which implies

$$f^n\left(\frac{A}{n}\right) = P\begin{pmatrix} f^n\left(\frac{\lambda}{n}\right) & f'\left(\frac{\lambda}{n}\right)f^{n-1}\left(\frac{\lambda}{n}\right) \\ 0 & f^n\left(\frac{\lambda}{n}\right) \end{pmatrix} P^{-1}.$$

Thus,

$$\lim_{n\to\infty} f^n\left(\frac{A}{n}\right) = P\begin{pmatrix} e^{f'(0)\lambda} & f'(0)e^{f'(0)\lambda} \\ 0 & e^{f'(0)\lambda} \end{pmatrix} P^{-1} = e^{f'(0)A},$$

and the theorem is proved.   □

Remark 4.6 We mention that Theorem 4.9 is a particular case of a more general result [37], which states that if $A, B \in \mathcal{M}_k(\mathbb{C})$ and f, g are entire functions with $f(0) = g(0) = 1$, then

$$\lim_{n\to\infty} \left(f\left(\frac{A}{n}\right)g\left(\frac{B}{n}\right)\right)^n = e^{f'(0)A+g'(0)B}. \tag{4.5}$$

When $f = g = \exp$, one obtains *Lie's famous product formula* for matrices. Herzog proves (4.5) by using a technique based on inequalities involving norms of matrices. However, when $g(x) = 1$ and $A \in \mathcal{M}_2(\mathbb{C})$, the proof of Theorem 4.9, which we believe is new in the literature, is based on an application of Corollary 4.1 combined to the Jordan canonical form theorem.

4.4 Explicit expressions of e^A, $\sin A$, and $\cos A$

In this section we give explicit expressions for the exponential function e^A and the trigonometric functions $\sin A$ and $\cos A$ in the case of a general matrix $A \in \mathcal{M}_2(\mathbb{C})$. These expressions are given in terms of both the entries and the eigenvalues of A.

Theorem 4.10 *Let $A \in \mathcal{M}_2(\mathbb{C})$ and let λ_1, λ_2 be the eigenvalues of A. Then*

$$e^A = \begin{cases} \dfrac{e^{\lambda_1} - e^{\lambda_2}}{\lambda_1 - \lambda_2} A + \dfrac{\lambda_1 e^{\lambda_2} - \lambda_2 e^{\lambda_1}}{\lambda_1 - \lambda_2} I_2 & \text{if } \lambda_1 \neq \lambda_2 \\ e^{\lambda}(A + (1 - \lambda)I_2) & \text{if } \lambda_1 = \lambda_2 = \lambda. \end{cases}$$

Proof This follows from Theorem 4.7. See also [10, Theorem 2.2, p. 1228]. □

Corollary 4.2 [9, p. 716] *If $A = \begin{pmatrix} a & b \\ 0 & d \end{pmatrix} \in \mathcal{M}_2(\mathbb{C})$, then*

$$e^A = \begin{cases} e^a \begin{pmatrix} 1 & b \\ 0 & 1 \end{pmatrix} & \text{if } a = d \\ \begin{pmatrix} e^a & \frac{e^a - e^d}{a - d} b \\ 0 & e^d \end{pmatrix} & \text{if } a \neq d. \end{cases}$$

Corollary 4.3 [9, p. 717] *Let $t \in \mathbb{C}$ and let $A = \begin{pmatrix} 0 & 1 \\ 0 & \alpha \end{pmatrix} \in \mathcal{M}_2(\mathbb{C})$. Then*

$$e^{tA} = \begin{cases} \begin{pmatrix} 1 & \frac{e^{\alpha t} - 1}{\alpha} \\ 0 & e^{\alpha t} \end{pmatrix} & \text{if } \alpha \neq 0 \\ \begin{pmatrix} 1 & t \\ 0 & 1 \end{pmatrix} & \text{if } \alpha = 0. \end{cases}$$

Corollary 4.4 [9, p. 717] *If $\theta \in \mathbb{R}$, $A = \begin{pmatrix} 0 & \theta \\ -\theta & 0 \end{pmatrix}$ and $B = \begin{pmatrix} 0 & \frac{\pi}{2} - \theta \\ -\frac{\pi}{2} + \theta & 0 \end{pmatrix}$, then*

$$e^A = \begin{pmatrix} \cos\theta & \sin\theta \\ -\sin\theta & \cos\theta \end{pmatrix} \quad \text{and} \quad e^B = \begin{pmatrix} \sin\theta & \cos\theta \\ -\cos\theta & \sin\theta \end{pmatrix}.$$

The next lemma gives the expressions for the exponential function of special matrices.

Lemma 4.4 *Let $A \in \mathscr{M}_2(\mathbb{C})$ and let $t \in \mathbb{C}$. The following statements hold*:

(a) *(nilpotent) If $A^2 = O_2$, then $e^{tA} = I_2 + tA$;*

(b) *(involutory) If $A^2 = I_2$, then $e^{tA} = (\cosh t)I_2 + (\sinh t)A$;*

(c) *(skew involutory) If $A^2 = -I_2$, then $e^{tA} = (\cos t)I_2 + (\sin t)A$;*

(d) *(idempotent) If $A^2 = A$, then $e^{tA} = I_2 + (e^t - 1)A$;*

(e) *If $A^2 = -A$, then $e^{tA} = I_2 + (1 - e^{-t})A$.*

Proof We prove only part (a) of the lemma and leave the proofs of the other parts as an exercise to the interested reader. First, we observe that since $A^2 = O_2$, then $A^n = O_2$, for all $n \geq 2$. Thus

$$e^{tA} = \sum_{n=0}^{\infty} \frac{(tA)^n}{n!} = I_2 + tA + \sum_{n=2}^{\infty} \frac{(tA)^n}{n!} = I_2 + tA,$$

and part (a) of the lemma is proved. $\square$

Let $J_2 \in \mathscr{M}_2(\mathbb{R})$ be the matrix

$$J_2 = \begin{pmatrix} 0 & 1 \\ -1 & 0 \end{pmatrix}.$$

Note that J_2 is skew symmetric and orthogonal, that is $J_2^T = -J_2 = J_2^{-1}$. Observe that, based on Theorem 1.3, this matrix corresponds to the complex number $-i$.

Lemma 4.5 *If $A = \begin{pmatrix} 0 & 1 \\ 1 & 0 \end{pmatrix}$, then*

$$e^{tA} = (\cosh t)I_2 + (\sinh t)A$$

and

$$e^{tJ_2} = (\cos t)I_2 + (\sin t)A.$$

Proof Use Theorem 4.7. $\square$

Theorem 4.11 *Let $A \in \mathscr{M}_2(\mathbb{C})$ and let λ_1, λ_2 be the eigenvalues of A. Then*

$$\sin A = \begin{cases} \dfrac{\sin \lambda_1 - \sin \lambda_2}{\lambda_1 - \lambda_2} A + \dfrac{\lambda_1 \sin \lambda_2 - \lambda_2 \sin \lambda_1}{\lambda_1 - \lambda_2} I_2 & \text{if } \lambda_1 \neq \lambda_2 \\ (\cos \lambda)A + (\sin \lambda - \lambda \cos \lambda)I_2, & \text{if } \lambda_1 = \lambda_2 = \lambda \end{cases}$$

and

$$\cos A = \begin{cases} \dfrac{\cos \lambda_1 - \cos \lambda_2}{\lambda_1 - \lambda_2} A + \dfrac{\lambda_1 \cos \lambda_2 - \lambda_2 \cos \lambda_1}{\lambda_1 - \lambda_2} I_2 & \text{if } \lambda_1 \neq \lambda_2 \\ -(\sin \lambda) A + (\cos \lambda + \lambda \sin \lambda) I_2 & \text{if } \lambda_1 = \lambda_2 = \lambda. \end{cases}$$

Proof This follows from Theorem 4.7.   □

4.5   Systems of first order differential equations with constant coefficients

In this section we solve systems of linear differential equations with constant coefficients by using a classical technique from matrix theory.

Let $A(t) = \begin{pmatrix} a(t) & b(t) \\ c(t) & d(t) \end{pmatrix}$, where $a, b, c, d : I \to \mathbb{R}$ are functions of t. Then,

$$A'(t) = \begin{pmatrix} a'(t) & b'(t) \\ c'(t) & d'(t) \end{pmatrix} \quad \text{and} \quad \int A(t)\mathrm{d}t = \begin{pmatrix} \int a(t)\mathrm{d}t & \int b(t)\mathrm{d}t \\ \int c(t)\mathrm{d}t & \int d(t)\mathrm{d}t \end{pmatrix}.$$

That is, by differentiating or integrating a matrix we mean to perform the operation on each of the matrix entries. It can be shown that the *product rule for derivatives in calculus holds for matrices whereas the power rule does not.*

Let $\mathscr{S}_0$ be the homogeneous system of linear differential equations with constant coefficients

$$\mathscr{S}_0 : \quad \begin{cases} x' = a_{11}x + a_{12}y \\ y' = a_{21}x + a_{22}y, \end{cases}$$

where $a_{i,j} \in \mathbb{R}$, $i, j = 1, 2$, and $x = x(t)$, $y = y(t)$ are the functions to be determined. Since the solutions of systems of differential equations with constant coefficients are defined on $\mathbb{R}$ in what follows we solve the system on $\mathbb{R}$.

Let $X(t) = \begin{pmatrix} x(t) \\ y(t) \end{pmatrix}$ and $A = \begin{pmatrix} a_{11} & a_{12} \\ a_{21} & a_{22} \end{pmatrix}$ and we observe that the system $\mathscr{S}_0$ becomes

$$\mathscr{S}_0 : \quad X'(t) = AX(t).$$

This implies that $X'(t) - AX(t) = O_2$ and by multiplying the system by the nonsingular matrix e^{-At} we get that

$$-\mathrm{e}^{-At}AX(t) + \mathrm{e}^{-At}X'(t) = O_2.$$

This implies, since $\left(e^{-At}\right)' = -Ae^{-At}$, that the system is equivalent to

$$\left(e^{-At}X(t)\right)' = O_2,$$

and it follows that the general solution of our system is

$$X(t) = e^{At}C,$$

where $C = \begin{pmatrix} c_1 \\ c_2 \end{pmatrix}$ is a constant vector.

If the initial condition $X(t_0) = X_0$, $t_0 \in \mathbb{R}$, is added to the system $\mathscr{S}_0$, then we have $X_0 = e^{At_0}C$ from which it follows that $C = e^{-At_0}X_0$. Thus, the solution of the *Cauchy problem* (or the system with initial condition) is

$$X(t) = e^{A(t-t_0)}X_0.$$

Now we turn our attention to the study of the nonhomogeneous systems of linear differential equations with constant coefficients.

We consider the system

$$\mathscr{S}: \quad \begin{cases} x' = a_{11}x + a_{12}y + f \\ y' = a_{21}x + a_{22}y + g, \end{cases}$$

where $f = f(t)$ and $g = g(t)$ are continuous functions, $t \in \mathbb{R}$.

Let $F(t) = \begin{pmatrix} f(t) \\ g(t) \end{pmatrix}$ and exactly as in the previous case we obtain that our system can be written into the matrix form

$$X'(t) = AX(t) + F(t).$$

This implies that $(X'(t) - AX(t) = F(t)$ and multiplying both sides of the system by the nonsingular matrix e^{-At} we get that

$$\left(e^{-At}X(t)\right)' = e^{-At}F(t)$$

and it follows that

$$e^{-At}X(t) = \int e^{-At}F(t)dt + C,$$

where C is a constant vector. Thus, the general solution of the nonhomogeneous system becomes

$$X(t) = e^{At} \left[\int e^{-At} F(t) dt \right] + e^{At} C.$$

If we add to the system $\mathscr{S}$ the initial condition $X(t_0) = X_0$, $t_0 \in \mathbb{R}$, then the solution of the Cauchy problem becomes, after simple computations which the reader is invited to check

$$X(t) = \int_{t_0}^{t} e^{A(t-u)} F(u) du + e^{A(t-t_0)} X_0.$$

Since for solving systems of linear differential equations with constant coefficients one needs to calculate the exponential matrix e^{At} we give below a simple algorithm for calculating e^{At}.

Algorithm for the computation of the exponential matrix e^{At}.

Step 1. Find the eigenvalues of A, determine the matrix J_A and the invertible matrix P which verifies $A = P J_A P^{-1}$.

Step 2. Observe that $e^{At} = P e^{J_A t} P^{-1}$.

Step 3. Determine $e^{J_A t}$ and e^{At}.

- If the eigenvalues of A are λ_1 and λ_2 and $J_A = \begin{pmatrix} \lambda_1 & 0 \\ 0 & \lambda_2 \end{pmatrix}$, then

$$e^{J_A t} = \begin{pmatrix} e^{\lambda_1 t} & 0 \\ 0 & e^{\lambda_2 t} \end{pmatrix}.$$

This implies

$$e^{At} = P \begin{pmatrix} e^{\lambda_1 t} & 0 \\ 0 & e^{\lambda_2 t} \end{pmatrix} P^{-1}.$$

- If the eigenvalues of A are $\lambda_1 = \lambda_2 = \lambda$ and $J_A = \begin{pmatrix} \lambda & 1 \\ 0 & \lambda \end{pmatrix}$, then

$$e^{J_A t} = e^{\lambda t} \begin{pmatrix} 1 & t \\ 0 & 1 \end{pmatrix}.$$

This implies

$$e^{At} = e^{\lambda t} P \begin{pmatrix} 1 & t \\ 0 & 1 \end{pmatrix} P^{-1}.$$

Remark 4.7 The solutions of the homogeneous system $\mathscr{S}_0 : X'(t) = AX(t), t \in \mathbb{R}$, are given by:

(a) If $\lambda_1, \lambda_2 \in \mathbb{R}$ and $\lambda_1 \neq \lambda_2$, then

$$\begin{cases} x(t) = \alpha_1 e^{\lambda_1 t} + \beta_1 e^{\lambda_2 t}, & \alpha_1, \beta_1 \in \mathbb{R}, \ t \in \mathbb{R} \\ y(t) = \alpha_2 e^{\lambda_1 t} + \beta_2 e^{\lambda_2 t}, & \alpha_2, \beta_2 \in \mathbb{R}, \ t \in \mathbb{R}; \end{cases}$$

(b) If $\lambda_1, \lambda_2 \in \mathbb{R}$ and $\lambda_1 = \lambda_2 = \lambda$ but $A \neq \lambda I_2$, then

$$\begin{cases} x(t) = e^{\lambda t} (\alpha_1 + \beta_1 t), & \alpha_1, \beta_1 \in \mathbb{R}, \ t \in \mathbb{R} \\ y(t) = e^{\lambda t} (\alpha_2 + \beta_2 t), & \alpha_2, \beta_2 \in \mathbb{R}, \ t \in \mathbb{R}; \end{cases}$$

(c) If $\lambda_1, \lambda_2 \in \mathbb{C}, \lambda_{1,2} = r \pm is, r, s \in \mathbb{R}, s \neq 0$, then

$$\begin{cases} x(t) = e^{rt} (\alpha_1 \cos st + \beta_1 \sin st), & \alpha_1, \beta_1 \in \mathbb{R}, \ t \in \mathbb{R} \\ y(t) = e^{rt} (\alpha_2 \cos st + \beta_2 \sin st), & \alpha_2, \beta_2 \in \mathbb{R}, \ t \in \mathbb{R}. \end{cases}$$

Definition 4.5 *The stability of homogeneous linear systems.*

Let $\mathscr{S}_0 : X'(t) = AX(t), t \in \mathbb{R}$, be a system of differential equations.

(a) The solution X_0 of $\mathscr{S}_0$ which verifies the initial condition $X_0(t_0) = C_0$ is called *stable (in the sense of Liapunov)* if for any $\epsilon > 0$ there exists $\delta(\epsilon) > 0$ such that for any $C \in \mathbb{R}^2$ with the property $||C - C_0|| < \delta$ we have $||X(t) - X_0(t)|| < \epsilon$ for any $t > t_0$, where $X(t)$ is the solution of the system $\mathscr{S}_0$ with the initial condition $X(t_0) = C$.
(b) The solution X_0 is called *unstable* if it is not stable.
(c) The solution X_0 is called *asymptotically stable* if it is stable and

$$\lim_{t \to \infty} ||X(t) - X_0(t)|| = 0,$$

for any solution X with the initial condition $X(t_0) = C$, where C is in a neighborhood from $\mathbb{R}^2$ of C_0.

Theorem 4.12 *All solutions of the system $\mathscr{S}_0$ have the same type of stability as the zero solution.*

Proof If X_{C_0} is a unique solution of the system with the initial condition $X_{C_0}(t_0) = C_0$ and X_C is the solution with the initial condition $X_C(t_0) = C$, then $X_C - X_{C_0}$ is the solution of the system with the initial condition $(X_C - X_{C_0})(t_0) = C - C_0$, so $X_C - X_{C_0} = X_{C-C_0}$. We have $||X_C(t) - X_{C_0}(t)|| = ||X_{C-C_0}(t) - 0|| = ||X_{C-C_0}(t)||$ and $\lim_{t \to \infty} ||X_C(t) - X_{C_0}(t)|| = 0 \Leftrightarrow \lim_{t \to \infty} ||X_{C-C_0}(t) - 0|| = 0.$ □

Remark 4.8 We mention that, based on the stability of the nonzero solution of system $\mathscr{S}_0$, stable, asymptotically stable or unstable, we say that the system $\mathscr{S}_0$ is stable, asymptotically stable, or unstable respectively.

For studying the stability of the system $\mathscr{S}_0$ we have based on remark 4.7 and Theorem 4.12 the following observations:

(a) If $\lambda_1, \lambda_2 \in \mathbb{R}$ and $\max\{\lambda_1, \lambda_2\} > 0$, then the function $f(t) = \alpha e^{\lambda_1 t} + \beta e^{\lambda_2 t}$, $t \geq t_0, \alpha^2 + \beta^2 \neq 0$, is unbounded;

(b) If $\lambda_1, \lambda_2 \in \mathbb{R}$ and $\max\{\lambda_1, \lambda_2\} < 0$, then $\lim_{t \to \infty} |\alpha e^{\lambda_1 t} + \beta e^{\lambda_2 t}| = 0$;

(c) If $\lambda_1 = \lambda_2 = 0$, then the function $f(t) = \alpha + \beta t, \beta \neq 0, t \geq t_0$ is unbounded;

(d) If $r > 0$ and $\alpha^2 + \beta^2 \neq 0$, then the function $f(t) = e^{rt}(\alpha \cos st + \beta \sin st)$, $t \geq t_0$ is unbounded;

(e) If $r < 0$, then $\lim_{t \to \infty} |e^{rt}(\alpha \cos st + \beta \sin st)| = 0$;

(f) The function $f(t) = \alpha \cos st + \beta \sin st, t \geq t_0$ is bounded.

4.6 The matrix Riemann zeta function

The celebrated zeta function of Riemann [61, p. 265] is a function of a complex variable defined by

$$\zeta(z) = \sum_{n=1}^{\infty} \frac{1}{n^z} = 1 + \frac{1}{2^z} + \cdots + \frac{1}{n^z} + \cdots, \quad \Re(z) > 1.$$

In this section we consider a matrix $A \in \mathscr{M}_2(\mathbb{C})$ and we introduce, hopefully for the first time in the literature?!, the matrix Riemann zeta function $\zeta(A)$ and discuss some of its properties. First, we define the *power matrix function* a^A, where $a \in \mathbb{C}^*$ and $A \in \mathscr{M}_2(\mathbb{C})$.

Definition 4.6 If $a \in \mathbb{C}^*$ and $A \in \mathscr{M}_2(\mathbb{C})$, then $a^A = e^{(\mathrm{Ln}\, a)A}$.

Remark 4.9 We mention that [42, p. 224] the function Ln, which is called the *logarithm*, is the multiple-valued function defined, for $z \in \mathbb{C}^*$, by $\mathrm{Ln}(z) = \ln|z| + i(\arg z + 2k\pi), k \in \mathbb{Z}$. The function $\ln z = \ln|z| + i \arg z$, where $\arg z \in (-\pi, \pi]$ is called the principal value. In what follows, throughout this book we consider, both in theory and problems, for the definition of the power matrix function the formula $a^A = e^{(\ln a)A}$ involving the principal value.

Theorem 4.13 *Let $a \in \mathbb{C}^*$ and let λ_1, λ_2 be the eigenvalues of $A \in \mathscr{M}_2(\mathbb{C})$. Then*

$$a^A = \begin{cases} \dfrac{a^{\lambda_1} - a^{\lambda_2}}{\lambda_1 - \lambda_2} A + \dfrac{\lambda_1 a^{\lambda_2} - \lambda_2 a^{\lambda_1}}{\lambda_1 - \lambda_2} I_2 & \text{if } \lambda_1 \neq \lambda_2 \\ \left(a^\lambda \ln a\right) A + a^\lambda (1 - \lambda \ln a) I_2 & \text{if } \lambda_1 = \lambda_2 = \lambda. \end{cases}$$

Proof The proof follows based on the formula $a^A = e^{(\ln a)A}$ combined to Theorem 4.7. □

Corollary 4.5 Power matrix function properties.

(a) *If $a \in \mathbb{C}^*$ and $\alpha \in \mathbb{C}$, then $a^{\alpha I_2} = a^{\alpha} I_2$.*
(b) *If $A \in \mathcal{M}_2(\mathbb{C})$, then $1^A = I_2$.*
(c) *If $a \in \mathbb{C}^*$, then $a^{O_2} = I_2$.*
(d) *If $A \in \mathcal{M}_2(\mathbb{C})$, then*

$$i^A = \cos\left(\frac{\pi}{2}A\right) + i\sin\left(\frac{\pi}{2}A\right).$$

(e) *If $a \in \mathbb{C}^*$ and $A \in \mathcal{M}_2(\mathbb{C})$ is a symmetric matrix, then a^A is also a symmetric matrix.*
(f) *If $AB = BA$, then $a^A a^B = a^{A+B}$, $a \in \mathbb{C}^*$.*
(g) *If $a, b \in \mathbb{C}^*$, then $a^A b^A = (ab)^A$.*

Corollary 4.6 Special power matrix functions.

(a) *If $a \in \mathbb{C}^*$ and $\alpha, \beta \in \mathbb{C}$, then*

$$a^{\begin{pmatrix} \alpha & 0 \\ 0 & \beta \end{pmatrix}} = \begin{pmatrix} a^{\alpha} & 0 \\ 0 & a^{\beta} \end{pmatrix}.$$

(b) *If $a \in \mathbb{C}^*$ and $\alpha, \beta \in \mathbb{C}$, then*

$$a^{\begin{pmatrix} \alpha & \beta \\ 0 & \alpha \end{pmatrix}} = a^{\alpha} \begin{pmatrix} 1 & \beta \ln a \\ 0 & 1 \end{pmatrix}.$$

(c) *If $a \in \mathbb{C}^*$ and $\alpha, \beta \in \mathbb{C}$, then*

$$a^{\begin{pmatrix} \alpha & 0 \\ \beta & \alpha \end{pmatrix}} = a^{\alpha} \begin{pmatrix} 1 & 0 \\ \beta \ln a & 1 \end{pmatrix}.$$

(d) *If $a \in \mathbb{C}^*$ and $\alpha \in \mathbb{C}$, then*

$$a^{\begin{pmatrix} \alpha & \alpha \\ \alpha & \alpha \end{pmatrix}} = \frac{1}{2}\begin{pmatrix} a^{2\alpha} + 1 & a^{2\alpha} - 1 \\ a^{2\alpha} - 1 & a^{2\alpha} + 1 \end{pmatrix}.$$

(e)   *If $a \in \mathbb{C}^*$ and $\alpha, \beta \in \mathbb{C}$, $\beta \neq 0$, then*

$$a^{\begin{pmatrix} \alpha & \beta \\ \beta & \alpha \end{pmatrix}} = \frac{1}{2} \begin{pmatrix} a^{\alpha+\beta} + a^{\alpha-\beta} & a^{\alpha+\beta} - a^{\alpha-\beta} \\ a^{\alpha+\beta} - a^{\alpha-\beta} & a^{\alpha+\beta} + a^{\alpha-\beta} \end{pmatrix}.$$

Definition 4.7   The Riemann zeta function of a 2×2 matrix.

Let $A \in \mathcal{M}_2(\mathbb{C})$ and let λ_1, λ_2 be its eigenvalues. *The Riemann zeta function* of the matrix A is defined by

$$\zeta(A) = \sum_{n=1}^{\infty} \left(\frac{1}{n}\right)^A, \quad \Re(\lambda_1) > 1, \quad \Re(\lambda_2) > 1.$$

By an abuse of notation we also use the writing $\zeta(A) = \sum_{n=1}^{\infty} \frac{1}{n^A}$. The next theorem gives the expression of $\zeta(A)$ in terms of both the entry values of A and the eigenvalues of A.

Theorem 4.14 *Let $A \in \mathcal{M}_2(\mathbb{C})$ and let λ_1, λ_2 be the eigenvalues of A with $\Re(\lambda_1) > 1$, $\Re(\lambda_2) > 1$. Then*

$$\zeta(A) = \begin{cases} \dfrac{\zeta(\lambda_1) - \zeta(\lambda_2)}{\lambda_1 - \lambda_2} A + \dfrac{\lambda_1 \zeta(\lambda_2) - \lambda_2 \zeta(\lambda_1)}{\lambda_1 - \lambda_2} I_2 & \text{if } \lambda_1 \neq \lambda_2 \\ \zeta'(\lambda) A + (\zeta(\lambda) - \lambda \zeta'(\lambda)) I_2 & \text{if } \lambda_1 = \lambda_2 = \lambda. \end{cases}$$

Proof We have, based on Theorem 4.13, that

$$\left(\frac{1}{n}\right)^A = \begin{cases} \dfrac{\frac{1}{n^{\lambda_1}} - \frac{1}{n^{\lambda_2}}}{\lambda_1 - \lambda_2} A + \dfrac{\frac{\lambda_1}{n^{\lambda_2}} - \frac{\lambda_2}{n^{\lambda_1}}}{\lambda_1 - \lambda_2} I_2 & \text{if } \lambda_1 \neq \lambda_2 \\ \left(\dfrac{1}{n^{\lambda}} \ln \dfrac{1}{n}\right) A + \dfrac{1}{n^{\lambda}}\left(1 - \lambda \ln \dfrac{1}{n}\right) I_2 & \text{if } \lambda_1 = \lambda_2 = \lambda. \end{cases}$$

If $\lambda_1 \neq \lambda_2$, then

$$\zeta(A) = \frac{1}{\lambda_1 - \lambda_2} \left(\sum_{n=1}^{\infty} \frac{1}{n^{\lambda_1}} - \sum_{n=1}^{\infty} \frac{1}{n^{\lambda_2}}\right) A + \frac{1}{\lambda_1 - \lambda_2} \left(\lambda_1 \sum_{n=1}^{\infty} \frac{1}{n^{\lambda_2}} - \lambda_2 \sum_{n=1}^{\infty} \frac{1}{n^{\lambda_1}}\right) I_2$$

$$= \frac{\zeta(\lambda_1) - \zeta(\lambda_2)}{\lambda_1 - \lambda_2} A + \frac{\lambda_1 \zeta(\lambda_2) - \lambda_2 \zeta(\lambda_1)}{\lambda_1 - \lambda_2} I_2.$$

If $\lambda_1 = \lambda_2 = \lambda$, then

$$\zeta(A) = \left(\sum_{n=1}^{\infty} \frac{1}{n^\lambda} \ln \frac{1}{n} \right) A + \sum_{n=1}^{\infty} \frac{1}{n^\lambda} \left(1 - \lambda \ln \frac{1}{n} \right) I_2$$

$$= \zeta'(\lambda)A + \left(\zeta(\lambda) - \lambda\zeta'(\lambda) \right) I_2,$$

and the theorem is proved.   □

Corollary 4.7 *If $A \in \mathcal{M}_2(\mathbb{C})$ is a symmetric matrix, then $\zeta(A)$ is also a symmetric matrix.*

Corollary 4.8 *Let $a \in \mathbb{C}$ be such that $\Re(a) > 1$. Then $\zeta(aI_2) = \zeta(a)I_2$.*

Corollary 4.9 Special matrix zeta functions.

(a) *If $a, b \in \mathbb{C}$ such that $\Re(a) > 1$, $\Re(b) > 1$ and $a \neq b$, then*

$$\zeta \begin{pmatrix} a & 0 \\ 0 & b \end{pmatrix} = \begin{pmatrix} \zeta(a) & 0 \\ 0 & \zeta(b) \end{pmatrix}.$$

(b) *If $a, b \in \mathbb{C}$ such that $\Re(a) > 1$, then*

$$\zeta \begin{pmatrix} a & b \\ 0 & a \end{pmatrix} = \begin{pmatrix} \zeta(a) & b\zeta'(a) \\ 0 & \zeta(a) \end{pmatrix}.$$

(c) *If $a, b \in \mathbb{C}$ such that $\Re(a) > 1$, then*

$$\zeta \begin{pmatrix} a & 0 \\ b & a \end{pmatrix} = \begin{pmatrix} \zeta(a) & 0 \\ b\zeta'(a) & \zeta(a) \end{pmatrix}.$$

(d) *If $a, b \in \mathbb{C}$ such that $\Re(a \pm b) > 1$, then*

$$\zeta \begin{pmatrix} a & b \\ b & a \end{pmatrix} = \frac{1}{2} \begin{pmatrix} \zeta(a+b) + \zeta(a-b) & \zeta(a+b) - \zeta(a-b) \\ \zeta(a+b) - \zeta(a-b) & \zeta(a+b) + \zeta(a-b) \end{pmatrix}.$$

4.7   The matrix Gamma function

Euler's Gamma function Γ [61, p. 235] is a function of a complex variable defined by

$$\Gamma(z) = \int_0^\infty t^{z-1} e^{-t} dt, \quad \Re(z) > 0.$$

We extend this definition to square matrices of order 2.

Definition 4.8 The matrix Gamma function of a 2×2 matrix.

If $A \in \mathcal{M}_2(\mathbb{C})$ and λ_1, λ_2 are the eigenvalues of A with $\Re(\lambda_i) > 0, i = 1, 2$, then we define

$$\Gamma(A) = \int_0^\infty t^A t^{-1} e^{-t} dt = \int_0^\infty t^{A-I_2} e^{-t} dt.$$

The next theorem gives the expression of $\Gamma(A)$ in terms of both the entry values of A and the eigenvalues of A.

Theorem 4.15 *Let $A \in \mathcal{M}_2(\mathbb{C})$ and let λ_1, λ_2 be the eigenvalues of A with $\Re(\lambda_i) > 0, i = 1, 2$. Then*

$$\Gamma(A) = \begin{cases} \dfrac{\Gamma(\lambda_1) - \Gamma(\lambda_2)}{\lambda_1 - \lambda_2} A + \dfrac{\lambda_1 \Gamma(\lambda_2) - \lambda_2 \Gamma(\lambda_1)}{\lambda_1 - \lambda_2} I_2 & \text{if} \quad \lambda_1 \neq \lambda_2 \\ \Gamma'(\lambda)A + \left(\Gamma(\lambda) - \lambda\Gamma'(\lambda)\right)I_2 & \text{if} \quad \lambda_1 = \lambda_2 - \lambda. \end{cases}$$

Proof We have, based on Theorem 4.13, that

$$t^A = \begin{cases} \dfrac{t^{\lambda_1} - t^{\lambda_2}}{\lambda_1 - \lambda_2} A + \dfrac{\lambda_1 t^{\lambda_2} - \lambda_2 t^{\lambda_1}}{\lambda_1 - \lambda_2} I_2 & \text{if} \quad \lambda_1 \neq \lambda_2 \\ \left(t^\lambda \ln t\right) A + t^\lambda (1 - \lambda \ln t) I_2 & \text{if} \quad \lambda_1 = \lambda_2 = \lambda. \end{cases}$$

If $\lambda_1 \neq \lambda_2$, then

$$\Gamma(A) = \frac{1}{\lambda_1 - \lambda_2} \left(\int_0^\infty t^{\lambda_1 - 1} e^{-t} dt - \int_0^\infty t^{\lambda_2 - 1} e^{-t} dt \right) A$$

$$+ \frac{1}{\lambda_1 - \lambda_2} \left(\lambda_1 \int_0^\infty t^{\lambda_2 - 1} e^{-t} dt - \lambda_2 \int_0^\infty t^{\lambda_1 - 1} e^{-t} dt \right) I_2$$

$$= \frac{\Gamma(\lambda_1) - \Gamma(\lambda_2)}{\lambda_1 - \lambda_2} A + \frac{\lambda_1 \Gamma(\lambda_2) - \lambda_2 \Gamma(\lambda_1)}{\lambda_1 - \lambda_2} I_2.$$

If $\lambda_1 = \lambda_2 = \lambda$, then

$$\Gamma(A) = \left(\int_0^\infty t^{\lambda - 1} e^{-t} \ln t\, dt \right) A + \left(\int_0^\infty t^{\lambda - 1} e^{-t} dt - \lambda \int_0^\infty t^{\lambda - 1} e^{-t} \ln t\, dt \right) I_2$$

$$= \Gamma'(\lambda)A + \left(\Gamma(\lambda) - \lambda\Gamma'(\lambda)\right) I_2,$$

and the theorem is proved. $\qquad\square$

Corollary 4.10 The Gamma function of special matrices.

(a) *If $a \in \mathbb{C}$ such that $\Re(a) > 0$, then $\Gamma(aI_2) = \Gamma(a)I_2$.*

(b) *If $\alpha, \beta \in \mathbb{C}$ such that $\Re(\alpha) > 0$, $\Re(\beta) > 0$, then*

$$\Gamma \begin{pmatrix} \alpha & 0 \\ 0 & \beta \end{pmatrix} = \begin{pmatrix} \Gamma(\alpha) & 0 \\ 0 & \Gamma(\beta) \end{pmatrix}.$$

(c) *If $\alpha, \beta \in \mathbb{C}$ such that $\Re(\alpha) > 0$, then*

$$\Gamma \begin{pmatrix} \alpha & \beta \\ 0 & \alpha \end{pmatrix} = \begin{pmatrix} \Gamma(\alpha) & \beta\Gamma'(\alpha) \\ 0 & \Gamma(\alpha) \end{pmatrix}.$$

(d) *If $\alpha, \beta \in \mathbb{C}$ such that $\Re(\alpha \pm \beta) > 0$, then*

$$\Gamma \begin{pmatrix} \alpha & \beta \\ \beta & \alpha \end{pmatrix} = \frac{1}{2} \begin{pmatrix} \Gamma(\alpha + \beta) + \Gamma(\alpha - \beta) & \Gamma(\alpha + \beta) - \Gamma(\alpha - \beta) \\ \Gamma(\alpha + \beta) - \Gamma(\alpha - \beta) & \Gamma(\alpha + \beta) + \Gamma(\alpha - \beta) \end{pmatrix}.$$

Lemma 4.6 A difference matrix equation.

If $A \in \mathcal{M}_2(\mathbb{C})$ with $\Re(\lambda_i) > 0$, $i = 1, 2$, then $\Gamma(I_2 + A) = A\Gamma(A)$.

Proof First we observe the eigenvalues of $I_2 + A$ are $1 + \lambda_1$ and $1 + \lambda_2$. Using Theorem 4.15 we have that

$$\Gamma(I_2 + A) = \begin{cases} \dfrac{\lambda_1\Gamma(\lambda_1) - \lambda_2\Gamma(\lambda_2)}{\lambda_1 - \lambda_2}A - \dfrac{\lambda_1\lambda_2\left(\Gamma(\lambda_1) - \Gamma(\lambda_2)\right)}{\lambda_1 - \lambda_2}I_2 & \text{if } \lambda_1 \neq \lambda_2 \\ \Gamma'(1 + \lambda)A - \lambda^2\Gamma'(\lambda)I_2 & \text{if } \lambda_1 = \lambda_2 = \lambda. \end{cases}$$

If $\lambda_1 \neq \lambda_2$ the Cayley–Hamilton Theorem implies that $A^2 - (\lambda_1 + \lambda_2)A + \lambda_1\lambda_2 I_2 = O_2$. We apply Theorem 4.15 and we have that

$$\begin{aligned} A\Gamma(A) &= \frac{\Gamma(\lambda_1) - \Gamma(\lambda_2)}{\lambda_1 - \lambda_2}A^2 + \frac{\lambda_1\Gamma(\lambda_2) - \lambda_2\Gamma(\lambda_1)}{\lambda_1 - \lambda_2}A \\ &= \frac{\lambda_1\Gamma(\lambda_1) - \lambda_2\Gamma(\lambda_2)}{\lambda_1 - \lambda_2}A - \frac{\lambda_1\lambda_2\left(\Gamma(\lambda_1) - \Gamma(\lambda_2)\right)}{\lambda_1 - \lambda_2}I_2. \end{aligned}$$

If $\lambda_1 \neq \lambda_2$ the Cayley–Hamilton Theorem implies that $A^2 - 2\lambda A + \lambda^2 I_2 = O_2$. We have, based on Theorem 4.15, that

$$A\Gamma(A) = \Gamma'(\lambda)A^2 + \left(\Gamma(\lambda) - \lambda\Gamma'(\lambda)\right)A = \Gamma'(1 + \lambda)A - \lambda^2\Gamma'(\lambda)I_2,$$

and the lemma is proved. $\square$

Corollary 4.11 A product of two Gamma functions.
Let α be a real number such that $0 < \alpha < 1$, let $\beta \in \mathbb{C}$ and let $A = \begin{pmatrix} \alpha & \beta \\ 0 & \alpha \end{pmatrix}$. Then

$$\Gamma(A)\Gamma(I_2 - A) = \frac{\pi}{\sin(\pi\alpha)} \begin{pmatrix} 1 & -\pi\beta\cot(\pi\alpha) \\ 0 & 1 \end{pmatrix}.$$

Proof Use part (c) of Corollary 4.10 and the formula $\Gamma(\alpha)\Gamma(1-\alpha) = \dfrac{\pi}{\sin(\pi\alpha)}$.   □

4.8 Problems

4.1 Prove that if $A = \begin{pmatrix} 0 & 1 \\ 1 & 0 \end{pmatrix}$, then

$$A^n = \begin{cases} \begin{pmatrix} 0 & 1 \\ 1 & 0 \end{pmatrix} & \text{if } n \text{ is odd} \\[2ex] \begin{pmatrix} 1 & 0 \\ 0 & 1 \end{pmatrix} & \text{if } n \text{ is even.} \end{cases}$$

Deduce that $A_\infty = \lim\limits_{n\to\infty} A^n$ does not exists. A clue to this behavior can be found by examining the eigenvalues, ± 1 of A.

Remark 4.10 The matrix A in the previous problem is called a *transition* or a *double stochastic* matrix. Such a matrix has the property that all its entries are greater than or equal to 0 and the sum of all its row and column entries is equal to 1 [34, p. 221].

4.2 Let $A \in \mathcal{M}_2(\mathbb{C})$. Prove that $\lim\limits_{n\to\infty} A^n = O_2$ if and only if $\rho(A) < 1$.

4.3 Let $A \in \mathcal{M}_2(\mathbb{C})$ such that $\rho(A) < 1$ and let $k \geq 1$ be an integer. Prove that $\lim\limits_{n\to\infty} n^k A^n = O_2$.

4.4 If $A, B \in \mathcal{M}_2(\mathbb{C})$ such that $AB = BA$ and $\lim\limits_{n\to\infty} A^n = O_2$ and $\lim\limits_{n\to\infty} B^n = O_2$, then $\lim\limits_{n\to\infty} (AB)^n = O_2$.

4.5 Let $A \in \mathcal{M}_2(\mathbb{C})$ and let $n \in \mathbb{N}$. Prove by using the Jordan canonical form of A that

$$\lim_{n\to\infty}\left(I_2+\frac{A}{n}\right)^n=e^A\quad\text{and}\quad\lim_{n\to\infty}\left(I_2-\frac{A}{n}\right)^n=e^{-A}.$$

4.6 [63, p. 339] Let $A=\begin{pmatrix}1-a & b\\a & 1-b\end{pmatrix}$ and $B=\begin{pmatrix}-a & b\\a & -b\end{pmatrix}$, where $0<a<1$,

$0<b<1$. Prove that $A^n=I_2+\dfrac{1-(1-a-b)^n}{a+b}B$, $n\in\mathbb{N}$, and calculate $\lim_{n\to\infty}A^n$.

Remark 4.11 Matrix A in the previous problem is called a *left stochastic* matrix or a *column stochastic* matrix. A left stochastic matrix is a square matrix with nonnegative entries with each column summing to 1.

4.7 [1] Let $B(x)=\begin{pmatrix}1 & x\\x & 1\end{pmatrix}$. Consider the infinite matrix product

$$M(t)=B(2^{-t})B(3^{-t})B(5^{-t})\cdots=\prod_p B(p^{-t}),\quad t>1,$$

where the product runs over all primes, taken in increasing order. Calculate $M(t)$.

4.8 Let

$$A(x)=\begin{pmatrix}x^{x^x} & 1\\(1-x)^3 & x^x\end{pmatrix}\in\mathcal{M}_2(\mathbb{R})\quad\text{and}\quad\text{let}\quad A^n(x)=\begin{pmatrix}a_n(x) & b_n(x)\\c_n(x) & d_n(x)\end{pmatrix},\quad n\in\mathbb{N}.$$

Calculate $\lim_{x\to 1}\dfrac{a_n(x)-d_n(x)}{c_n(x)}$.

4.9 Calculate

$$\begin{pmatrix}1+\frac{1}{n} & \frac{1}{n}\\\frac{1}{n} & 1\end{pmatrix}^n,\quad n\in\mathbb{N}.$$

4.10 Prove that

$$\lim_{n\to\infty}\begin{pmatrix}1 & \frac{1}{n}\\\frac{1}{n} & 1+\frac{1}{n^2}\end{pmatrix}^n=\begin{pmatrix}\cosh 1 & \sinh 1\\\sinh 1 & \cosh 1\end{pmatrix}.$$

4.11 Calculate

$$\lim_{n\to\infty}\begin{pmatrix}1-\frac{1}{n^2} & \frac{1}{n}\\\frac{1}{n} & 1+\frac{1}{n^2}\end{pmatrix}^n.$$

4.12 [29] Let $A = \begin{pmatrix} 3 & 1 \\ -4 & -1 \end{pmatrix}$. Calculate

$$\lim_{n \to \infty} \frac{1}{n} \left(I_2 + \frac{A^n}{n} \right)^n \quad \text{and} \quad \lim_{n \to \infty} \frac{1}{n} \left(I_2 - \frac{A^n}{n} \right)^n.$$

4.13 Let $A = \begin{pmatrix} 2 & -1 \\ -2 & 3 \end{pmatrix}$. Calculate

$$\lim_{n \to \infty} \left(\frac{\cos^2 A}{n + 1} + \frac{\cos^2 (2A)}{n + 2} + \cdots + \frac{\cos^2 (nA)}{2n} \right).$$

4.14 Gelfand's spectral radius formula, 1941.

Let $A = \begin{pmatrix} a & b \\ c & d \end{pmatrix} \in \mathcal{M}_2(\mathbb{C})$ and let $||A||$ be the *Frobenius* norm of A defined by

$$||A|| = \sqrt{\mathrm{Tr}(AA^*)} = \sqrt{|a|^2 + |b|^2 + |c|^2 + |d|^2}.$$

Prove that

$$\rho(A) = \lim_{n \to \infty} ||A^n||^{\frac{1}{n}}.$$

4.15 *Limits, nth roots and norms of matrices.* Let $A = \begin{pmatrix} 3 & -1 \\ 4 & -2 \end{pmatrix}$.

(a) Calculate

$$\lim_{n \to \infty} \frac{\sqrt[n]{||A(A + I_2)(A + 2I_2) \cdots (A + nI_2)||}}{n}.$$

(b) Calculate

$$\lim_{n \to \infty} \sqrt[n]{\frac{||A(A + 2I_2)(A + 4I_2) \cdots (A + 2nI_2)||}{n!}}.$$

(c) Calculate

$$\lim_{n \to \infty} \frac{||(A + (n + 1)I_2)(A + (n + 4)I_2) \cdots (A + (4n - 2)I_2)||}{||(A + nI_2)(A + (n + 3)I_2) \cdots (A + (4n - 3)I_2)||},$$

where $||A||$ denotes the norm of A defined in problem **4.14**.

4.16 *Pitagora in matrix form.* The *Frobenius* norm of a matrix $A \in \mathcal{M}_2(\mathbb{C})$, also known as the *Euclidean* norm or *Hilbert–Schmidt* norm, is defined by $||A|| = \mathrm{Tr}(AA^*)$. Prove that if $A \in \mathcal{M}_2(\mathbb{C})$, then $||A||^2 = ||\Re(A)||^2 + ||\Im(A)||^2$, where $\Re(A)$ and $\Im(A)$ denote the real part and the imaginary part of A respectively.

4.17 *A bella problema.* Let $A = \begin{pmatrix} 2 & 1 \\ -1 & 0 \end{pmatrix}$. Prove that:

(a) $A^n = \begin{pmatrix} n+1 & n \\ -n & 1-n \end{pmatrix}$, for $n \geq 1$;

(b) $e^A = eA$;

(c) $e^{Ax} = e^x \begin{pmatrix} x+1 & x \\ -x & 1-x \end{pmatrix}$, $x \in \mathbb{R}$.

4.18 Calculate e^A for the following matrices:

(a) $A = \begin{pmatrix} 3 & -1 \\ 1 & 1 \end{pmatrix}$;

(b) $A = \begin{pmatrix} 4 & -2 \\ 6 & -3 \end{pmatrix}$.

4.19 Prove that

$$e^{\begin{pmatrix} 0 & 1 \\ 1 & 0 \end{pmatrix}} = \begin{pmatrix} \cosh 1 & \sinh 1 \\ \sinh 1 & \cosh 1 \end{pmatrix}.$$

4.20 (a) Let $\alpha, \beta \in \mathbb{R}$. Prove that

$$e^{\begin{pmatrix} 0 & i\beta \\ i\beta & 0 \end{pmatrix}} = \begin{pmatrix} \cos \beta & i \sin \beta \\ i \sin \beta & \cos \beta \end{pmatrix}.$$

(b) Prove that

$$e^{\begin{pmatrix} \alpha & i\beta \\ i\beta & \alpha \end{pmatrix}} = e^\alpha \begin{pmatrix} \cos \beta & i \sin \beta \\ i \sin \beta & \cos \beta \end{pmatrix}.$$

4.21 A rotation matrix as an exponential.

If $\theta \in \mathbb{R}$ and $J_2 = \begin{pmatrix} 0 & 1 \\ -1 & 0 \end{pmatrix}$, then

(continued)

4.21 (continued)

$$e^{-\theta J_2} = \begin{pmatrix} \cos\theta & -\sin\theta \\ \sin\theta & \cos\theta \end{pmatrix}.$$

Remark 4.12 We note that, based on Theorem 1.3, this problem is the matrix version of the famous Euler's formula $e^{i\theta} = \cos\theta + i\sin\theta$, from complex analysis.

4.22 Let $A \in \mathcal{M}_2(\mathbb{C})$.

(a) Prove that if e^A is a triangular matrix which is not of the form αI_2, $\alpha \in \mathbb{C}$, then A is a triangular matrix.

(b) Show that if e^A is triangular, then A need not be triangular.

4.23 Prove that

$$e^{\begin{pmatrix} a & 1 \\ -1 & a \end{pmatrix}} = e^a \begin{pmatrix} \cos 1 & \sin 1 \\ -\sin 1 & \cos 1 \end{pmatrix}, \quad a \in \mathbb{R}.$$

4.24 Let $a, b \in \mathbb{R}$. Calculate e^A, where $A = \begin{pmatrix} a & b \\ -b & a \end{pmatrix}$.

4.25 [6, p. 205] Let $t \in \mathbb{R}$ and let $A(t) = \begin{pmatrix} t & t-1 \\ 0 & 1 \end{pmatrix}$. Prove that $e^{A(t)} = eA\left(e^{t-1}\right)$.

4.26 Let $A \in \mathcal{M}_2(\mathbb{C})$. Prove that if λ is an eigenvalue of A, then e^λ is an eigenvalue of e^A and $\det\left(e^A\right) = e^{\operatorname{Tr}(A)}$.

4.27 Commuting exponentials.

Let $A, B \in \mathcal{M}_2(\mathbb{C})$ be such that *both A and B have real eigenvalues*. Prove that if e^A *commutes* with e^B, then A *commutes* with B.

Nota bene. If A and B have complex eigenvalues the problem is no longer valid. If $A = \begin{pmatrix} 0 & \pi \\ -\pi & 0 \end{pmatrix}$ and $B = \begin{pmatrix} 0 & (7+4\sqrt{3})\pi \\ (-7+4\sqrt{3})\pi & 0 \end{pmatrix}$, then $e^A = e^B = -I_2$, $e^{A+B} = I_2$ and $AB \neq BA$ [9, p. 709].

More generally, if $(x_n, y_n) \neq (1, 0)$ are the solutions in positive integers of Pell's equation $x^2 - dy^2 = 1$, $d \in \mathbb{N}$, $d \geq 2$, and

$$A = \begin{pmatrix} 0 & \pi \\ -\pi & 0 \end{pmatrix} \quad \text{and} \quad B = \begin{pmatrix} 0 & (x_n + \sqrt{d}y_n)\pi \\ (-x_n + \sqrt{d}y_n)\pi & 0 \end{pmatrix},$$

(continued)

4.27 (continued)
then $e^A = e^B = -I_2$ and $AB \neq BA$. Moreover, if $1 + x_n = 2k^2$, $k \in \mathbb{N}$, then $e^{A+B} = I_2$, so $e^A e^B = e^B e^A \neq e^{A+B}$.

4.28 [25] **When is an exponential matrix an integer matrix?**

Let $A \in \mathcal{M}_2(\mathbb{Z})$. Prove that $e^A \in \mathcal{M}_2(\mathbb{Z})$ if and only if $A^2 = O_2$.

4.29 [26] **A gem of matrix analysis.**

Let $A \in \mathcal{M}_2(\mathbb{Z})$. Prove that:

- $\sin A \in \mathcal{M}_2(\mathbb{Z})$ if and only if $A^2 = O_2$;
- $\cos A \in \mathcal{M}_2(\mathbb{Z})$ if and only if $A^2 = O_2$.

4.30 [30] **Another gem of matrix analysis.**

Let $A \in \mathcal{M}_2(\mathbb{Q})$ be such that $\rho(A) < 1$. Prove that:

- $\ln(I_2 - A) \in \mathcal{M}_2(\mathbb{Q})$ if and only if $A^2 = O_2$;
- $\ln(I_2 + A) \in \mathcal{M}_2(\mathbb{Q})$ if and only if $A^2 = O_2$.

4.31 Let $A, B, C \in \mathcal{M}_2(\mathbb{R})$ be commuting matrices. Prove that if $\cos A + \cos B + \cos C = O_2$ and $\sin A + \sin B + \sin C = O_2$, then

(a) $\cos(2A) + \cos(2B) + \cos(2C) = O_2$;
(b) $\sin(2A) + \sin(2B) + \sin(2C) = O_2$;
(c) $\cos(3A) + \cos(3B) + \cos(3C) = 3\cos(A + B + C)$;
(d) $\sin(3A) + \sin(3B) + \sin(3C) = 3\sin(A + B + C)$.

4.32 Matrix delights. Prove that:

(a) $e^{\begin{pmatrix} a & b \\ b & a \end{pmatrix}} = e^a \begin{pmatrix} \cosh b & \sinh b \\ \sinh b & \cosh b \end{pmatrix}$, $a, b \in \mathbb{C}$;

(b) $e^{\begin{pmatrix} 0 & a \\ b & 0 \end{pmatrix}} = \begin{pmatrix} \cosh \sqrt{ab} & \frac{a}{\sqrt{ab}} \sinh \sqrt{ab} \\ \frac{b}{\sqrt{ab}} \sinh \sqrt{ab} & \cosh \sqrt{ab} \end{pmatrix}$, $a, b \in \mathbb{R}$, $ab > 0$;

(c) $e^{\begin{pmatrix} 1 & 1 \\ 1 & 0 \end{pmatrix}} = \frac{1}{\sqrt{5}} \begin{pmatrix} \alpha e^{\alpha} - \beta e^{\beta} & e^{\alpha} - e^{\beta} \\ e^{\alpha} - e^{\beta} & \alpha e^{\beta} - \beta e^{\alpha} \end{pmatrix}$, $\alpha = \frac{1+\sqrt{5}}{2}$, $\beta = \frac{1-\sqrt{5}}{2}$;

(d) $\sum_{n=1}^{\infty} \frac{1}{(2n-1)^A} = \left(I_2 - 2^{-A}\right) \zeta(A)$, where $A \in \mathcal{M}_2(\mathbb{C})$, $\Re(\lambda_1) > 1$, $\Re(\lambda_2) > 1$.

4.33 Functions of rotation matrices.

Let $\theta \in \mathbb{R}$ and let f be a function which has the Taylor series expansion at 0

$$f(z) = \sum_{n=0}^{\infty} \frac{f^{(n)}(0)}{n!} z^n, \quad |z| < R,$$

where $R > 1$. Prove that

$$f \begin{pmatrix} \cos\theta & -\sin\theta \\ \sin\theta & \cos\theta \end{pmatrix} = \frac{1}{2} \begin{pmatrix} f\left(e^{i\theta}\right) + f\left(e^{-i\theta}\right) & f\left(e^{-i\theta}\right) - f\left(e^{i\theta}\right) \\ f\left(e^{i\theta}\right) - f\left(e^{-i\theta}\right) & f\left(e^{i\theta}\right) + f\left(e^{-i\theta}\right) \end{pmatrix}.$$

4.34 Functions of circulant matrices.

(a) Let $\alpha, \beta \in \mathbb{R}$. Prove that

$$\cos \begin{pmatrix} \alpha & \beta \\ \beta & \alpha \end{pmatrix} = \begin{pmatrix} \cos\alpha\cos\beta & -\sin\alpha\sin\beta \\ -\sin\alpha\sin\beta & \cos\alpha\cos\beta \end{pmatrix}.$$

(b) Let f be a function which has the Taylor series expansion at 0

$$f(z) = \sum_{n=0}^{\infty} \frac{f^{(n)}(0)}{n!} z^n, \quad |z| < R,$$

where $R \in (0, \infty]$ and let $\alpha, \beta \in \mathbb{R}$ be such that $|\alpha \pm \beta| < R$. Prove that

$$f \begin{pmatrix} \alpha & \beta \\ \beta & \alpha \end{pmatrix} = \frac{1}{2} \begin{pmatrix} f(\alpha+\beta) + f(\alpha-\beta) & f(\alpha+\beta) - f(\alpha-\beta) \\ f(\alpha+\beta) - f(\alpha-\beta) & f(\alpha+\beta) + f(\alpha-\beta) \end{pmatrix}.$$

4.35   Complex numbers and matrix formulae. Courtesy of Theorem 1.3.

We identify, based on Theorem 1.3, the complex number $z = x + iy$, $x, y \in \mathbb{R}$, by the real matrix $A = \begin{pmatrix} x & -y \\ y & x \end{pmatrix}$.

(1)   The matrix corresponding to the complex conjugate $\bar{z} = x - iy$ is the adjugate matrix $A_* = \begin{pmatrix} x & y \\ -y & x \end{pmatrix}$.

(2) $(x + iy)(x - iy) = x^2 + y^2$ and

$$\begin{pmatrix} x & -y \\ y & x \end{pmatrix} \begin{pmatrix} x & y \\ -y & x \end{pmatrix} = (x^2 + y^2)I_2.$$

(3) $(x + iy)(a + ib) = xa - yb + i(xb + ya)$ and

$$\begin{pmatrix} x & -y \\ y & x \end{pmatrix} \begin{pmatrix} a & -b \\ b & a \end{pmatrix} = \begin{pmatrix} xa - yb & -(xb + ya) \\ xb + ya & xa - yb \end{pmatrix}.$$

(4) $\dfrac{1}{x + iy} = \dfrac{x - iy}{x^2 + y^2}$, $x^2 + y^2 \neq 0$ and

$$\begin{pmatrix} x & -y \\ y & x \end{pmatrix}^{-1} = \frac{1}{x^2 + y^2} \begin{pmatrix} x & y \\ -y & x \end{pmatrix}.$$

(5) $i^n = \cos\dfrac{n\pi}{2} + i \sin\dfrac{n\pi}{2}$ and

$$\begin{pmatrix} 0 & -1 \\ 1 & 0 \end{pmatrix}^n = \begin{pmatrix} \cos\frac{n\pi}{2} & -\sin\frac{n\pi}{2} \\ \sin\frac{n\pi}{2} & \cos\frac{n\pi}{2} \end{pmatrix}.$$

(6)   **de Moivre's formula.** $(\cos\theta + i \sin\theta)^n = \cos(n\theta) + i \sin(n\theta)$, $n \in \mathbb{N}$, $\theta \in \mathbb{R}$ and

$$\begin{pmatrix} \cos\theta & -\sin\theta \\ \sin\theta & \cos\theta \end{pmatrix}^n = \begin{pmatrix} \cos(n\theta) & -\sin(n\theta) \\ \sin(n\theta) & \cos(n\theta) \end{pmatrix}.$$

(7) $(x + iy)^n = \rho^n(\cos(n\theta) + i \sin(n\theta))$, $\rho = \sqrt{x^2 + y^2}$, $\theta \in [0, 2\pi)$ and

$$\begin{pmatrix} x & -y \\ y & x \end{pmatrix}^n = \rho^n \begin{pmatrix} \cos(n\theta) & -\sin(n\theta) \\ \sin(n\theta) & \cos(n\theta) \end{pmatrix}.$$

(continued)

4.35 (continued)

(8) **Euler's formula.** $e^{i\theta} = \cos\theta + i\sin\theta$, $\theta \in \mathbb{R}$ and

$$e^{\begin{pmatrix} 0 & -\theta \\ \theta & 0 \end{pmatrix}} = \begin{pmatrix} \cos\theta & -\sin\theta \\ \sin\theta & \cos\theta \end{pmatrix}.$$

(9) $e^{x+iy} = e^x(\cos y + i\sin y)$ and

$$e^{\begin{pmatrix} x & -y \\ y & x \end{pmatrix}} = e^x \begin{pmatrix} \cos y & -\sin y \\ \sin y & \cos y \end{pmatrix}.$$

(10) $\sinh(x + iy) = \sinh x \cos y + i\cosh x \sin y$ and

$$\sinh\begin{pmatrix} x & -y \\ y & x \end{pmatrix} = \begin{pmatrix} \sinh x \cos y & -\cosh x \sin y \\ \cosh x \sin y & \sinh x \cos y \end{pmatrix}.$$

(11) $\cosh(x + iy) = \cosh x \cos y + i\sinh x \sin y$ and

$$\cosh\begin{pmatrix} x & -y \\ y & x \end{pmatrix} = \begin{pmatrix} \cosh x \cos y & \sinh x \sin y \\ \sinh x \sin y & \cosh x \cos y \end{pmatrix}.$$

(12) $\sin(x + iy) = \sin x \cosh y + i\cos x \sinh y$ and

$$\sin\begin{pmatrix} x & -y \\ y & x \end{pmatrix} = \begin{pmatrix} \sin x \cosh y & -\cos x \sinh y \\ \cos x \sinh y & \sin x \cosh y \end{pmatrix}.$$

(13) $\cos(x + iy) = \cos x \cosh y - i\sin x \sinh y$ and

$$\cos\begin{pmatrix} x & -y \\ y & x \end{pmatrix} = \begin{pmatrix} \cos x \cosh y & \sin x \sinh y \\ -\sin x \sinh y & \cos x \cosh y \end{pmatrix}.$$

(14) If $f(x + iy) = u(x, y) + iv(x, y)$, then

$$f\begin{pmatrix} x & -y \\ y & x \end{pmatrix} = \begin{pmatrix} u(x, y) & -v(x, y) \\ v(x, y) & u(x, y) \end{pmatrix}.$$

4.36 Solve in $\mathcal{M}_2(\mathbb{C})$ the equation $e^A = \alpha I_2$, $\alpha \in \mathbb{C}^*$.

4.37 Solve in $\mathcal{M}_2(\mathbb{C})$ the equation $e^A = \begin{pmatrix} a & 0 \\ 0 & b \end{pmatrix}$, $a, b \in \mathbb{C}^*$, $a \neq b$.

4.38 A circulant exponential equation.

Solve in $\mathcal{M}_2\,(\mathbb{C})$ the equation $e^A = \begin{pmatrix} 0 & 1 \\ 1 & 0 \end{pmatrix}$.

4.39 A triangular exponential equation.

Solve in $\mathcal{M}_2\,(\mathbb{C})$ the equation $e^A = \begin{pmatrix} a & a \\ 0 & a \end{pmatrix}$, $a \in \mathbb{C}^*$.

4.40 Check the identity $\sin(2A) = 2\sin A \cos A$, where $A = \begin{pmatrix} \pi - 1 & 1 \\ -1 & \pi + 1 \end{pmatrix}$.

4.41 If $A \in \mathcal{M}_2\,(\mathbb{C})$ is an idempotent matrix and $k \in \mathbb{Z}$, then $\sin(k\pi A) = O_2$.

4.42 (a) Are there matrices $A \in \mathcal{M}_2\,(\mathbb{C})$ such that $\sin A = \begin{pmatrix} 1 & 2016 \\ 0 & 1 \end{pmatrix}$?

(b) Are there matrices $A \in \mathcal{M}_2\,(\mathbb{C})$ such that $\cosh A = \begin{pmatrix} 1 & \alpha \\ 0 & 1 \end{pmatrix}$, $\alpha \neq 0$?

4.43 Let $A \in \mathcal{M}_2\,(\mathbb{C})$ be a matrix such that $\det A = 0$. Prove that:

■ if $\mathrm{Tr}(A) = 0$, then $2^A = I_2 + (\ln 2)A$;

■ if $\mathrm{Tr}(A) \neq 0$, then $2^A = I_2 + \dfrac{2^{\mathrm{Tr}(A)} - 1}{\mathrm{Tr}(A)}A$.

4.44 *Divertimento.* Let $A = \begin{pmatrix} a & b \\ c & d \end{pmatrix} \in \mathcal{M}_2\,(\mathbb{R})$. Prove that

$$A^n = \begin{pmatrix} a^n & b^n \\ c^n & d^n \end{pmatrix}, \quad \forall n \in \mathbb{N}$$

if and only if A is of the following form

$$\begin{pmatrix} \alpha & 0 \\ \alpha & 0 \end{pmatrix}, \quad \begin{pmatrix} 0 & \alpha \\ 0 & \alpha \end{pmatrix}, \quad \begin{pmatrix} 0 & 0 \\ \alpha & \alpha \end{pmatrix}, \quad \begin{pmatrix} \alpha & \alpha \\ 0 & 0 \end{pmatrix}, \quad \begin{pmatrix} \alpha & 0 \\ 0 & \beta \end{pmatrix}, \quad \alpha, \beta \in \mathbb{R}.$$

Challenge problems.

(continued)

4.44 (continued)

- Find all matrices $A = \begin{pmatrix} a & b \\ c & d \end{pmatrix} \in \mathcal{M}_2(\mathbb{R})$ such that $\sin A = \begin{pmatrix} \sin a & \sin b \\ \sin c & \sin d \end{pmatrix}$.

  All matrices written above verify this equation. Are there any other?

- Find all matrices $A = \begin{pmatrix} a & b \\ c & d \end{pmatrix} \in \mathcal{M}_2(\mathbb{R})$ such that $\cos A = \begin{pmatrix} \cos a & \cos b \\ \cos c & \cos d \end{pmatrix}$.

4.45 [31] Are there matrices $A = \begin{pmatrix} a & b \\ c & d \end{pmatrix} \in \mathcal{M}_2(\mathbb{R})$ such that $e^A = \begin{pmatrix} e^a & e^b \\ e^c & e^d \end{pmatrix}$?

4.46 Derivatives of matrices.

(a) Let A and B be square matrices whose entries are differentiable functions. Prove that $(AB)' = A'B + AB'$.

(b) It can be shown that if A is a differentiable and invertible square matrix function, then A^{-1} is differentiable.

- Prove that $(A^{-1})' = -A^{-1}A'A^{-1}$.
- Calculate the derivative of $A^{-n} = (A^{-1})^n$, where $n \geq 2$ is an integer.

4.47 [6, p. 205] If $A(t)$ is a scalar function of t, the derivative of $e^{A(t)}$ is $e^{A(t)}A'(t)$. Calculate the derivative of $e^{A(t)}$ when $A(t) = \begin{pmatrix} 1 & t \\ 0 & 0 \end{pmatrix}$ and show that the result is not equal to either of the two products $e^{A(t)}A'(t)$ or $A'(t)e^{A(t)}$.

Systems of differential equations

4.48 Solve the system of differential equations

$$\begin{cases} x_1' = x_1 - 4x_2 \\ x_2' = -2x_1 + 3x_2. \end{cases}$$

4.49 Let $a > 0$ be a real number and let $A = \begin{pmatrix} 1 & 1 \\ -a & 1 \end{pmatrix}$.

(a) Calculate e^{At}, $t \in \mathbb{R}$.

(b) Solve the system of differential equations

$$\begin{cases} x' = x + y \\ y' = -ax + y. \end{cases}$$

Remark 4.13 This problem was inspired by [52, Example 5.2.5, p. 238].

4.50 Let $A = \begin{pmatrix} 2 & 1 \\ 3 & 4 \end{pmatrix}$.

(a) Calculate e^{At}, $t \in \mathbb{R}$.

(b) Solve the system of differential equations

$$\begin{cases} x' = 2x + y \\ y' = 3x + 4y, \end{cases}$$

with initial conditions $x(0) = 2$, $y(0) = 4$.

4.51 Let $A = \begin{pmatrix} -4 & 1 \\ -1 & -2 \end{pmatrix}$.

(a) Calculate e^{At}, $t \in \mathbb{R}$.

(b) Solve the system of differential equations

$$\begin{cases} x' = -4x + 2y \\ y' = -x - 2y + e^t, \end{cases}$$

with initial conditions $x(0) = 1$, $y(0) = 7$.

4.52 (a) Prove the solution of the linear differential system $\mathcal{S}_0$: $tX'(t) = AX(t)$, where $A \in \mathcal{M}_2(\mathbb{R})$ and $t > 0$ is given by $X(t) = t^A C$, where C is a constant vector.

(b) Determine the solution of the nonhomogeneous system of linear differential equations $\mathcal{S}$: $tX'(t) = AX(t) + F(t)$, F is a continuous function vector.

4.53 Solve the system of differential equations

$$\begin{cases} tx' = -x + 3y + 1 \\ ty' = x + y + 1, \end{cases}$$

with initial conditions $x(1) = \frac{1}{2}$ and $y(1) = \frac{3}{2}$.

Stability of homogeneous linear systems of differential equations

4.54 Discuss the stability of the system

$$\begin{cases} x' = ax + by \\ y' = cx + dy, \end{cases}$$

according to the values of the parameters $a, b, c, d \in \mathbb{R}$.

4.55 Discuss the stability of the system

$$\begin{cases} x' = -a^2x + ay \\ y' = x - y, \end{cases}$$

according to the values of the parameter $a \in \mathbb{R}$.

4.56 Discuss the stability of the system

$$\begin{cases} x' = -ax + (a-1)y \\ y' = x, \end{cases}$$

according to the values of the parameter $a \in \mathbb{R}$.

4.57 Discuss the stability of the system

$$\begin{cases} x' = by \\ y' = cx, \end{cases}$$

according to the values of the parameters $b, c \in \mathbb{R}$.

4.58 Discuss the stability of the system

$$\begin{cases} x' = -x + ay \\ y' = bx - y, \end{cases}$$

according to the values of the parameters $a, b \in \mathbb{R}$.

4.59 Discuss the stability of the system

$$\begin{cases} x' = ax + y \\ y' = bx + ay, \end{cases}$$

according to the values of the parameters $a, b \in \mathbb{R}$.

Series of matrices

4.60 Let $x \in \mathbb{R}$ and let $A = \begin{pmatrix} -1 & x \\ 0 & -1 \end{pmatrix}$. Prove that $\displaystyle\sum_{n=1}^{\infty} \frac{A^n}{n^2} = \begin{pmatrix} -\frac{\pi^2}{12} & x \ln 2 \\ 0 & -\frac{\pi^2}{12} \end{pmatrix}$.

Remark 4.14 This problem was inspired by [63, problem 12, p. 65].

4.61 Let $A \in \mathcal{M}_2(\mathbb{C})$. Prove that

$$\sum_{n=0}^{\infty} \frac{A^{3n}}{(3n)!} = \frac{1}{3}\left(e^A + 2e^{-\frac{A}{2}} \cos \frac{\sqrt{3}A}{2} \right).$$

4.62 Abel's summation by parts formula for matrices.

Let $(a_n)_{n \geq 1}$ be a sequence of complex numbers, $(B_n)_{n \geq 1} \in \mathcal{M}_2(\mathbb{C})$ be a sequence of matrices and let $A_n = \sum_{k=1}^{n} a_k$. Then:

(a) *the finite version*

$$\sum_{k=1}^{n} a_k B_k = A_n B_{n+1} + \sum_{k=1}^{n} A_k (B_k - B_{k+1});$$

(b) *the infinite version*

$$\sum_{k=1}^{\infty} a_k B_k = \lim_{n \to \infty} A_n B_{n+1} + \sum_{k=1}^{\infty} A_k (B_k - B_{k+1}),$$

if the limit is finite and the series converges.

Remark 4.15 We mention that *Abel's summation by parts formula* for sums of real or complex numbers as well as applications can be found in [11, p. 55], [22, p. 258], [57, p. 26].

4.63 The "matrix generating function" for the Fibonacci sequence.

Let $(F_n)_{n \geq 0}$ be the Fibonacci sequence defined by the recurrence formula $F_0 = 0$, $F_1 = 1$ and $F_{n+1} = F_n + F_{n-1}$, $\forall n \geq 1$. Prove that

$$\sum_{n=1}^{\infty} F_n A^{n-1} = \left(I_2 - A - A^2\right)^{-1}, \quad \forall A \in \mathcal{M}_2(\mathbb{C}) \text{ with } \rho(A) < \frac{\sqrt{5} - 1}{2}.$$

4.64 If $n \geq 1$ is an integer, the nth harmonic number H_n is defined by the formula

$$H_n = 1 + \frac{1}{2} + \frac{1}{3} + \cdots + \frac{1}{n}.$$

Let x be a real number, $\alpha \in (-1, 1)$ and let $A = \begin{pmatrix} \alpha & x \\ 0 & \alpha \end{pmatrix}$ and $B = \begin{pmatrix} 0 & x \\ 0 & 0 \end{pmatrix}$.

(a) Prove that

$$\sum_{n=1}^{\infty} H_n A^n = -\frac{\ln(1 - \alpha)}{1 - \alpha} I_2 + \frac{1 - \ln(1 - \alpha)}{(1 - \alpha)^2} B.$$

(b) Prove that

$$\sum_{n=1}^{\infty} \frac{H_n}{n+1} A^n = \frac{\ln^2(1-\alpha)}{2\alpha} I_2 - \left(\frac{\ln(1-\alpha)}{\alpha(1-\alpha)} + \frac{\ln^2(1-\alpha)}{2\alpha^2} \right) B, \quad \alpha \neq 0.$$

4.65 The generating functions of two harmonic numbers.

Let $A \in \mathcal{M}_2(\mathbb{C})$ with $\rho(A) < 1$ and let H_n denote the nth harmonic number. Prove that:

(a) $\displaystyle\sum_{n=1}^{\infty} H_n A^n = -(I_2 - A)^{-1} \ln(I_2 - A);$

(b) $\displaystyle\sum_{n=1}^{\infty} n H_n A^n = A(I_2 - \ln(I_2 - A))(I_2 - A)^{-2}.$

4.66 A power series with the tail of $\ln \frac{1}{2}$.

Let $A \in \mathcal{M}_2(\mathbb{R})$ be such that $\rho(A) < 1$ and let λ_1, λ_2 be the real eigenvalues of A. Prove that

$$\sum_{n=1}^{\infty} \left(\ln \frac{1}{2} + 1 - \frac{1}{2} + \cdots + \frac{(-1)^{n-1}}{n} \right) A^n$$

$$= \begin{cases} (I_2 - A)^{-1} (\ln(I_2 + A) - (\ln 2)A) & \text{if } 0 < |\lambda_1|, |\lambda_2| < 1, \\ \dfrac{1}{1 - \operatorname{Tr}(A)} \left(\dfrac{\ln(1 + \operatorname{Tr}(A))}{\operatorname{Tr}(A)} - \ln 2 \right) A & \text{if } \lambda_1 = 0, 0 < |\lambda_2| < 1. \end{cases}$$

4.67 A sum with the tail of the logarithmic series.

Let $A \in \mathcal{M}_2(\mathbb{R})$ be a matrix whose eigenvalues belong to $(-1, 1)$.
(a) Prove that

$$\lim_{n \to \infty} n \left(\ln(I_2 - A) + A + \frac{A^2}{2} + \cdots + \frac{A^n}{n} \right) = O_2.$$

(b) Prove that

$$\sum_{n=1}^{\infty} \left(\ln(I_2 - A) + A + \frac{A^2}{2} + \cdots + \frac{A^n}{n} \right) = -\ln(I_2 - A) - A(I_2 - A)^{-1}.$$

4.68 Logarithmic series and harmonic numbers.

Let $A \in \mathcal{M}_2(\mathbb{C})$ with $\rho(A) < 1$. Prove that:

(a) $H_1 + H_2 + \cdots + H_n = (n+1)H_n - n, \ \forall \, n \geq 1;$

(b) $\displaystyle\sum_{n=1}^{\infty} H_n \left(\ln(I_2 - A) + A + \frac{A^2}{2} + \cdots + \frac{A^n}{n} \right) = (A + \ln(I_2 - A))(I_2 - A)^{-1}.$

4.69 An arctan series.

Let $A \in \mathcal{M}_2(\mathbb{R})$ be a matrix whose eigenvalues belong to $(-1, 1)$.

(a) Prove that

$$\lim_{n \to \infty} n \left(\arctan A - A + \frac{A^3}{3} + \cdots + (-1)^n \frac{A^{2n-1}}{2n-1} \right) = O_2.$$

(b) Prove that

$$\sum_{n=1}^{\infty} \left(\arctan A - A + \frac{A^3}{3} + \cdots + (-1)^n \frac{A^{2n-1}}{2n-1} \right) = \frac{A}{2}(I_2 + A^2)^{-1} - \frac{\arctan A}{2}.$$

4.70 Summing the tail of e^A.

Let $A \in \mathcal{M}_2(\mathbb{C})$. Prove that:

(a) $\displaystyle\sum_{n=0}^{\infty} \left(e^A - I_2 - \frac{A}{1!} - \frac{A^2}{2!} - \cdots - \frac{A^n}{n!} \right) = Ae^A;$

(b) $\displaystyle\sum_{n=1}^{\infty} n \left(e^A - I_2 - \frac{A}{1!} - \frac{A^2}{2!} - \cdots - \frac{A^n}{n!} \right) = \frac{A^2}{2}e^A.$

Remark 4.16 A matrix series and Touchard's polynomials.

More generally, if $p \geq 1$ is an integer and $A \in \mathcal{M}_2(\mathbb{C})$, then

$$\sum_{n=1}^{\infty} n^p \left(e^A - I_2 - \frac{A}{1!} - \frac{A^2}{2!} - \cdots - \frac{A^n}{n!} \right) = e^A \sum_{k=1}^{p} \frac{S(p, k)}{k+1} A^{k+1},$$

where $S(p, k)$ are the *Stirling numbers of the second kind* [59, p. 58].

The polynomial Q_p defined by $Q_p(x) = \sum_{k=1}^{p} S(p,k)x^k$, where $S(p,k)$ are the Stirling numbers of the second kind is known in the mathematical literature as *Touchard's polynomial* [13].

The problem was inspired by the calculation of the series

$$\sum_{n=1}^{\infty} n^p \left(e^x - 1 - \frac{x}{1!} - \cdots - \frac{x^n}{n!} \right),$$

where $p \geq 1$ is an integer and $x \in \mathbb{R}$ (see [24]) with a solution in [44].

4.71 Let $k \geq 0$ be an integer and let $A \in \mathcal{M}_2(\mathbb{C})$. Prove that

$$\sum_{n=k}^{\infty} \binom{n}{k} \left(e^A - I_2 - \frac{A}{1!} - \frac{A^2}{2!} - \cdots - \frac{A^n}{n!} \right) = \frac{A^{k+1}}{(k+1)!} e^A.$$

4.72 Let $A \in \mathcal{M}_2(\mathbb{C})$. Prove that

(a) $\displaystyle \sum_{n=1}^{\infty} (-1)^{\lfloor \frac{n}{2} \rfloor} \left(e^A - I_2 - \frac{A}{1!} - \frac{A^2}{2!} - \cdots - \frac{A^n}{n!} \right) = I_2 - \cos A;$

(b) $\displaystyle \sum_{n=1}^{\infty} (-1)^{\lfloor \frac{n}{2} \rfloor} \left(e^A - I_2 - \frac{A}{1!} - \frac{A^2}{2!} - \cdots - \frac{A^{n-1}}{(n-1)!} \right) = \sin A.$

Here $\lfloor a \rfloor$ denotes the integer part of a.

4.73 Let f be a function which has the Taylor series expansion at 0,

$$f(z) = \sum_{n=0}^{\infty} \frac{f^{(n)}(0)}{n!} z^n, \quad |z| < R,$$

where $R \in (0, \infty]$ and let $A \in \mathcal{M}_2(\mathbb{C})$ be such that $\rho(A) < R$. Prove that:

(a) $\displaystyle \sum_{n=0}^{\infty} \left(f(A) - f(0)I_2 - \frac{f'(0)}{1!} A - \cdots - \frac{f^{(n)}(0)}{n!} A^n \right) = A f'(A);$

(b) $\displaystyle \sum_{n=1}^{\infty} n \left(f(A) - f(0)I_2 - \frac{f'(0)}{1!} A - \cdots - \frac{f^{(n)}(0)}{n!} A^n \right) = \frac{A^2}{2} f''(A).$

4.74 An exponential power series.

Let $A \in \mathscr{M}_2(\mathbb{C})$ and let $x \in \mathbb{C}$.

(a) Prove that

$$\sum_{n=1}^{\infty} x^n \left(e^A - I_2 - \frac{A}{1!} - \frac{A^2}{2!} - \cdots - \frac{A^n}{n!} \right) = \begin{cases} \dfrac{xe^A - e^{Ax}}{1-x} + I_2 & \text{if } x \neq 1, \\[2mm] Ae^A + I_2 - e^A & \text{if } x = 1. \end{cases}$$

(b) Prove that

$$\sum_{n=1}^{\infty} (-1)^{n-1} \left(e^A - I_2 - \frac{A}{1!} - \frac{A^2}{2!} - \cdots - \frac{A^n}{n!} \right) = \cosh A - I_2.$$

4.75 A variation on the same theme.

Let $A \in \mathscr{M}_2(\mathbb{C})$, let λ_1, λ_2 be the eigenvalues of A, and let

$$S(A) = \sum_{n=1}^{\infty} \left(e - 1 - \frac{1}{1!} - \frac{1}{2!} - \cdots - \frac{1}{n!} \right) A^n.$$

Prove that:

(a) $S(A) = I_2 + \dfrac{e^{\lambda_2} - e\lambda_2}{(\lambda_2 - 1)^2}(A - I_2)$ if $\lambda_1 = 1$, $\lambda_2 \neq 1$;

(b) $S(A) = I_2 + \dfrac{e^{\lambda_1} - e\lambda_1}{(\lambda_1 - \lambda_2)(\lambda_1 - 1)}(A - \lambda_2 I_2) + \dfrac{e^{\lambda_2} - e\lambda_2}{(\lambda_2 - \lambda_1)(\lambda_2 - 1)}(A - \lambda_1 I_2)$ if $\lambda_1 \neq 1$, $\lambda_2 \neq 1$, $\lambda_1 \neq \lambda_2$;

(c) $S(A) = \left(1 - \dfrac{e}{2} \right) I_2 + \dfrac{e}{2} A$ if $\lambda_1 = \lambda_2 = 1$;

(d) $S(A) = \left(\dfrac{e^{\lambda} - e\lambda}{\lambda - 1} + 1 \right) I_2 + \dfrac{\lambda e^{\lambda} - 2e^{\lambda} + e}{(\lambda - 1)^2}(A - \lambda I_2)$ if $\lambda_1 = \lambda_2 = \lambda \neq 1$.

4.76 Sine and cosine series.

Let $A \in \mathscr{M}_2(\mathbb{C})$. Prove that:

(a) $\displaystyle\sum_{n=0}^{\infty} \left(\cos A - I_2 + \frac{A^2}{2!} - \cdots - (-1)^n \frac{A^{2n}}{(2n)!} \right) = -\frac{A \sin A}{2}$;

(b) $\displaystyle\sum_{n=1}^{\infty} \left(\sin A - A + \frac{A^3}{3!} - \cdots - (-1)^{n-1} \frac{A^{2n-1}}{(2n-1)!} \right) = \frac{A \cos A - \sin A}{2}$.

4.77 (a) Let $A = \begin{pmatrix} -1 & x \\ 0 & -1 \end{pmatrix} \in \mathcal{M}_2(\mathbb{R})$. Prove that

$$\sum_{n=1}^{\infty} \left(\zeta(3) - 1 - \frac{1}{2^3} - \cdots - \frac{1}{n^3} \right) A^n = \begin{pmatrix} -\frac{\zeta(3)}{8} & \left(\frac{7\zeta(3)}{16} - \frac{\zeta(2)}{4} \right) x \\ 0 & -\frac{\zeta(3)}{8} \end{pmatrix},$$

where ζ denotes the Riemann zeta function.

(b) Let $A \in \mathcal{M}_2(\mathbb{C})$ with $\rho(A) < 1$. Prove that

$$\sum_{n=1}^{\infty} \left(\zeta(3) - 1 - \frac{1}{2^3} - \cdots - \frac{1}{n^3} \right) A^n = (\zeta(3)A - \mathrm{Li}_3(A))(I_2 - A)^{-1},$$

where Li_3 denotes the polylogarithm function.

Remark 4.17 More generally, if $A \in \mathcal{M}_2(\mathbb{C})$ with $\rho(A) < 1$, then

$$\sum_{n=1}^{\infty} \left(\zeta(k) - 1 - \frac{1}{2^k} - \cdots - \frac{1}{n^k} \right) A^n = (\zeta(k)A - \mathrm{Li}_k(A))(I_2 - A)^{-1},$$

where $k \geq 3$ is an integer and Li_k denotes the polylogarithm function.

4.78 (a) [1] If $a \in \mathbb{C}$ such that $\Re(a) > 2$, then $\displaystyle\sum_{n=1}^{\infty} \frac{n}{n^{aI_2}} = \zeta(a-1)I_2$.

(b) If $\alpha, \beta \in \mathbb{C}$ such that $\Re(\alpha) > 2$, then

$$\sum_{n=1}^{\infty} \frac{n}{n^{\begin{pmatrix} \alpha & \beta \\ 0 & \alpha \end{pmatrix}}} = \begin{pmatrix} \zeta(\alpha-1) & \beta\zeta'(\alpha-1) \\ 0 & \zeta(\alpha-1) \end{pmatrix}.$$

4.79 Computing $\zeta(A - I_2)$.

Let $A \in \mathcal{M}_2(\mathbb{C})$ and let λ_1, λ_2 be its eigenvalues.

(a) If $\lambda_1 \neq \lambda_2$, $\Re(\lambda_1) > 2$, $\Re(\lambda_2) > 2$, then

(continued)

---

[1] By an abuse of notation, if $A \in \mathcal{M}_2(\mathbb{C})$ we use the writing $\frac{1}{n^A} = \left(\frac{1}{n}\right)^A$.

4.79 (continued)

$$\sum_{n=1}^{\infty} \frac{n}{n^A} = \frac{\zeta(\lambda_1 - 1) - \zeta(\lambda_2 - 1)}{\lambda_1 - \lambda_2} A + \frac{\lambda_1 \zeta(\lambda_2 - 1) - \lambda_2 \zeta(\lambda_1 - 1)}{\lambda_1 - \lambda_2} I_2.$$

(b) If $\lambda_1 = \lambda_2 = \lambda$, $\Re(\lambda) > 2$, then

$$\sum_{n=1}^{\infty} \frac{n}{n^A} = \zeta'(\lambda - 1)A + \left(\zeta(\lambda - 1) - \lambda\zeta'(\lambda - 1)\right) I_2.$$

Remark 4.18 If $k \geq 1$ is an integer one may also express, using Theorem 4.14, the matrix zeta function

$$\zeta(A - kI_2) = \sum_{n=1}^{\infty} \frac{n^k}{n^A}$$

in terms of the eigenvalues of A, where $A \in \mathcal{M}_2(\mathbb{C})$ with $\Re(\lambda_1) > k + 1$ and $\Re(\lambda_2) > k + 1$.

4.80 Summing the tail of $\zeta(A)$.

Let $A \in \mathcal{M}_2(\mathbb{C})$ and let λ_1, λ_2 be its eigenvalues.

(a) Prove that if A is a matrix such that $\Re(\lambda_i) > 2$, $i = 1, 2$, then

$$\sum_{n=1}^{\infty} \left(\zeta(A) - \frac{1}{1^A} - \frac{1}{2^A} - \cdots - \frac{1}{n^A} \right) = \zeta(A - I_2) - \zeta(A).$$

(b) Prove that if A is a matrix such that $\Re(\lambda_i) > 3$, $i = 1, 2$, then

$$\sum_{n=1}^{\infty} n \left(\zeta(A) - \frac{1}{1^A} - \frac{1}{2^A} - \cdots - \frac{1}{n^A} \right) = \frac{\zeta(A - 2I_2) - \zeta(A - I_2)}{2}.$$

Integrals of matrices

4.81 The *Dilogarithm function* Li_2 is the special function defined, for $|z| \leq 1$, by

$$\mathrm{Li}_2(z) = \sum_{n=1}^{\infty} \frac{z^n}{n^2} = - \int_0^z \frac{\ln(1 - t)}{t} \, dt.$$

Let $a \in [-1, 1) \setminus \{0\}$, $b \in \mathbb{R}$ and let $A = \begin{pmatrix} a & b \\ 0 & a \end{pmatrix}$. Prove that

$$\int_0^1 \frac{\ln(I_2 - Ax)}{x}\,dx = \begin{pmatrix} -\mathrm{Li}_2(a) & \dfrac{b\ln(1-a)}{a} \\ 0 & -\mathrm{Li}_2(a) \end{pmatrix}.$$

4.82 Let $A \in \mathcal{M}_2(\mathbb{C})$ and let λ_1, λ_2 be the eigenvalues of A. Prove that

$$\int_0^1 \frac{\ln(I_2 - Ax)}{x}\,dx = \begin{cases} -A & \text{if } \lambda_1 = \lambda_2 = 0, \\ -\zeta(2)A & \text{if } \lambda_1 = 0,\ \lambda_2 = 1, \end{cases}$$

where ζ denotes the Riemann zeta function.

4.83 Let $A \in \mathcal{M}_2(\mathbb{C})$ and let λ_1, λ_2 be the eigenvalues of A with $\rho(A) < 1$. Prove that

$$\int_0^1 \frac{\ln(I_2 - Ax)}{x}\,dx = \begin{cases} \frac{\mathrm{Li}_2(\lambda_2) - \mathrm{Li}_2(\lambda_1)}{\lambda_1 - \lambda_2}A + \frac{\lambda_2 \mathrm{Li}_2(\lambda_1) - \lambda_1 \mathrm{Li}_2(\lambda_2)}{\lambda_1 - \lambda_2}I_2 & \text{if } \lambda_1 \neq \lambda_2, \\ -A & \text{if } \lambda_1 = \lambda_2 = 0, \\ \frac{\ln(1-\lambda)}{\lambda}A - (\mathrm{Li}_2(\lambda) + \ln(1-\lambda))I_2 & \text{if } \lambda_1 = \lambda_2 = \lambda \neq 0. \end{cases}$$

4.84 (a) Let $A \in \mathcal{M}_2(\mathbb{C})$ be a matrix having both eigenvalues equal to 0. Prove that

$$\int_0^1 \frac{\ln^2(I_2 - Ax)}{x}\,dx = O_2.$$

(b) *A matrix logarithmic integral and Apery's constant.* Let $A \in \mathcal{M}_2(\mathbb{C})$ be a matrix having the eigenvalues 0 and 1. Prove that

$$\int_0^1 \frac{\ln^2(I_2 - Ax)}{x}\,dx = 2\zeta(3)A,$$

where ζ denotes the Riemann zeta function.

Nota bene. The constant $\zeta(3) = \sum_{n=1}^\infty \frac{1}{n^3} = 1.2020569031\ldots$ is known in the mathematical literature as *Apéry's constant*. In 1979, Apéry [5] stunned the mathematical world with a miraculous proof that $\zeta(3)$ is irrational.

4.85 Let $A \in \mathcal{M}_2(\mathbb{C})$ with $\rho(A) < 1$ and let λ_1, λ_2 be the eigenvalues of A. Prove that

$$\int_0^1 \ln(I_2 - Ax)dx$$

$$= \begin{cases} -\dfrac{A}{2} & \text{if } \lambda_1 = \lambda_2 = 0, \\ -\left(\dfrac{(1 - \text{Tr}(A)) \ln(1 - \text{Tr}(A))}{\text{Tr}^2(A)} + \dfrac{1}{\text{Tr}(A)}\right) A & \text{if } \lambda_1 = 0,\, 0 < |\lambda_2| < 1, \\ (I_2 - A^{-1}) \ln(I_2 - A) - I_2 & \text{if } 0 < |\lambda_1|, |\lambda_2| < 1. \end{cases}$$

4.86 Let $A \in \mathcal{M}_2(\mathbb{R})$ and let λ_1, λ_2 be the eigenvalues of A. Prove that

$$\int_0^1 e^{Ax}dx = \begin{cases} I_2 + \dfrac{A}{2} & \text{if } \lambda_1 = \lambda_2 = 0, \\ I_2 + \dfrac{e^{\text{Tr}(A)} - 1 - \text{Tr}(A)}{\text{Tr}^2(A)}A & \text{if } \lambda_1 = 0,\, \lambda_2 \neq 0, \\ (e^A - I_2) A^{-1} & \text{if } \lambda_1, \lambda_2 \neq 0. \end{cases}$$

4.87 Let $A \in \mathcal{M}_2(\mathbb{R})$ and let λ_1, λ_2 be the eigenvalues of A.

(a) Prove that

$$\int_0^1 \cos(Ax)dx = \begin{cases} I_2 & \text{if } \lambda_1 = \lambda_2 = 0, \\ I_2 + \dfrac{\sin \text{Tr}(A) - \text{Tr}(A)}{\text{Tr}^2(A)}A & \text{if } \lambda_1 = 0,\, \lambda_2 \neq 0, \\ A^{-1} \sin A & \text{if } \lambda_1, \lambda_2 \neq 0. \end{cases}$$

(b) Prove that

$$\int_0^1 \sin(Ax)dx = \begin{cases} \dfrac{A}{2} & \text{if } \lambda_1 = \lambda_2 = 0, \\ \dfrac{1 - \cos \text{Tr}(A)}{\text{Tr}^2(A)}A & \text{if } \lambda_1 = 0,\, \lambda_2 \neq 0, \\ A^{-1}(I_2 - \cos A) & \text{if } \lambda_1, \lambda_2 \neq 0. \end{cases}$$

(c) Prove that

$$\int_0^1 \int_0^1 \sin A(x + y)dxdy = \begin{cases} A & \text{if } \lambda_1 = \lambda_2 = 0, \\ \dfrac{2 \sin \text{Tr}(A) - \sin 2\text{Tr}(A)}{\text{Tr}^3(A)}A & \text{if } \lambda_1 = 0,\, \lambda_2 \neq 0, \\ A^{-2}(2 \sin A - \sin(2A)) & \text{if } \lambda_1, \lambda_2 \neq 0. \end{cases}$$

4.88 [2] Let $A \in \mathcal{M}_2(\mathbb{R})$. Prove that

$$\int \int_{\mathbb{R}^2} v^T A v \, e^{-v^T v} dx dy = \frac{\pi}{2} \text{Tr}(A), \quad \text{where} \quad v = \begin{pmatrix} x \\ y \end{pmatrix}.$$

4.89 Let $A \in \mathcal{M}_2(\mathbb{R})$ be a symmetric matrix with positive eigenvalues.

(a) Prove that

$$\int \int_{\mathbb{R}^2} e^{-v^T A v} dx dy = \frac{\pi}{\sqrt{\det A}}.$$

(b) Let $\alpha > -1$ be a real number. Prove that

$$\int \int_{\mathbb{R}^2} (v^T A v)^\alpha e^{-v^T A v} dx dy = \frac{\pi \Gamma(\alpha + 1)}{\sqrt{\det A}},$$

where Γ denotes the Gamma function.

4.90 Let $A \in \mathcal{M}_2(\mathbb{R})$ be a symmetric matrix with positive eigenvalues and let $f : \mathbb{R}^2 \to \mathbb{R}$ be the function defined by

$$f(x, y) = \frac{e^{-\frac{1}{2} v^T A^{-1} v}}{2\pi \sqrt{\det A}}, \quad \text{where} \quad v = \begin{pmatrix} x \\ y \end{pmatrix} \in \mathbb{R}^2.$$

Prove that

$$\int \int_{\mathbb{R}^2} f(x, y) \ln f(x, y) dx dy = -\ln(2\pi e \sqrt{\det A}).$$

4.91 Let $A \in \mathcal{M}_2(\mathbb{C})$ and let α be a positive number. Prove that

$$\int_{-\infty}^{\infty} e^{Ax} e^{-\alpha x^2} dx = \sqrt{\frac{\pi}{\alpha}} e^{\frac{A^2}{4\alpha}}.$$

4.92 Let $A \in \mathcal{M}_2(\mathbb{R})$ and let α be a positive number. Prove that:

(a) $\displaystyle\int_{-\infty}^{\infty} \cos(Ax) e^{-\alpha x^2} dx = \sqrt{\frac{\pi}{\alpha}} e^{-\frac{A^2}{4\alpha}};$

(b) $\displaystyle\int_{-\infty}^{\infty} \sin(Ax) e^{-\alpha x^2} dx = O_2.$

---

[2]We mention that if $v = \begin{pmatrix} x \\ y \end{pmatrix}$ is a vector in $\mathbb{R}^2$ and $A \in \mathcal{M}_2(\mathbb{R})$, then $v^T A v$ is a matrix having one row and one column, i.e., $v^T A v = (u)$. We identify, by an abuse of notation, the matrix $v^T A v$ by the real number u. Also, we identify $v^T v$ by $x^2 + y^2$.

4.93 Let $A \in \mathcal{M}_2(\mathbb{R})$ with $\rho(A) < \alpha$ and let α be a positive number. Prove that:

(a) $\displaystyle\int_0^\infty \sin(Ax)e^{-\alpha x}dx = A\left(A^2 + \alpha^2 I_2\right)^{-1};$

(b) $\displaystyle\int_0^\infty \cos(Ax)e^{-\alpha x}dx = \alpha\left(A^2 + \alpha^2 I_2\right)^{-1}.$

4.94 Euler–Poisson matrix integrals.

(a) If $A = \begin{pmatrix} 1 & 1 \\ -1 & 3 \end{pmatrix}$ calculate $\displaystyle\int_0^\infty e^{-Ax^2}dx.$

(b) Let $A = \begin{pmatrix} a & b \\ b & a \end{pmatrix} \in \mathcal{M}_2(\mathbb{R})$ with $a > \pm b$ and let $n \in \mathbb{N}$. Prove that

$$\int_0^\infty e^{-Ax^n}dx = \frac{\Gamma\left(1 + \frac{1}{n}\right)}{2}\begin{pmatrix} \frac{1}{\sqrt[n]{a+b}} + \frac{1}{\sqrt[n]{a-b}} & \frac{1}{\sqrt[n]{a+b}} - \frac{1}{\sqrt[n]{a-b}} \\ \frac{1}{\sqrt[n]{a+b}} - \frac{1}{\sqrt[n]{a-b}} & \frac{1}{\sqrt[n]{a+b}} + \frac{1}{\sqrt[n]{a-b}} \end{pmatrix},$$

where Γ denotes the Gamma function.

4.95 When Laplace comes into play!

Let $A = \begin{pmatrix} a & b \\ b & a \end{pmatrix} \in \mathcal{M}_2(\mathbb{R})$.

(a) Prove that

$$\int_0^\infty \frac{\cos(Ax)}{x^2 + 1}dx = \frac{\pi}{4}\begin{pmatrix} e^{-|a+b|} + e^{-|a-b|} & e^{-|a+b|} - e^{-|a-b|} \\ e^{-|a+b|} - e^{-|a-b|} & e^{-|a+b|} + e^{-|a-b|} \end{pmatrix}.$$

A challenge. Calculate $\displaystyle\int_0^\infty \frac{x\sin(Ax)}{x^2 + 1}dx.$

(b) Prove that

$$\int_0^\infty e^{-x^2}\cos(Ax)dx = \frac{\sqrt{\pi}}{4}\begin{pmatrix} e^{-\frac{(a+b)^2}{4}} + e^{-\frac{(a-b)^2}{4}} & e^{-\frac{(a+b)^2}{4}} - e^{-\frac{(a-b)^2}{4}} \\ e^{-\frac{(a+b)^2}{4}} - e^{-\frac{(a-b)^2}{4}} & e^{-\frac{(a+b)^2}{4}} + e^{-\frac{(a-b)^2}{4}} \end{pmatrix}.$$

4.96 Don't forget Fresnel.

Let $A = \begin{pmatrix} a & b \\ b & a \end{pmatrix} \in \mathcal{M}_2(\mathbb{R})$ with $a \neq \pm b$.

(continued)

4.96 (continued)
(a) Prove that

$$\int_0^\infty \cos(Ax^2)dx = \frac{1}{4}\sqrt{\frac{\pi}{2}}\left(\frac{\frac{1}{\sqrt{|a+b|}} + \frac{1}{\sqrt{|a-b|}}}{\frac{1}{\sqrt{|a+b|}} - \frac{1}{\sqrt{|a-b|}}} \quad \frac{\frac{1}{\sqrt{|a+b|}} - \frac{1}{\sqrt{|a-b|}}}{\frac{1}{\sqrt{|a+b|}} + \frac{1}{\sqrt{|a-b|}}}\right).$$

(b) If $a = \pm b$ and $a \neq 0$, then $\int_0^\infty \cos(Ax^2)dx = \frac{1}{8a}\sqrt{\frac{\pi}{|a|}}A$.

A challenge. Calculate $\int_0^\infty \cos(Ax^n)dx$, where $n \geq 2$ is an integer.

4.97 Exponential matrix integrals.
(a) Let $A \in \mathcal{M}_2(\mathbb{R})$ be a matrix with positive eigenvalues. Calculate

$$\int_0^\infty e^{-Ax}dx.$$

(b) Let $\alpha > 0$ and let $A \in \mathcal{M}_2(\mathbb{R})$ be a matrix with real eigenvalues such that $\lambda_1, \lambda_2 > \alpha$. Calculate

$$\int_0^\infty e^{-Ax}e^{\alpha x}dx.$$

(c) Let $n \geq 0$ be an integer and let $A \in \mathcal{M}_2(\mathbb{R})$ be a matrix with positive eigenvalues. Calculate

$$\int_0^\infty e^{-Ax}x^n dx.$$

4.98 *Dirichlet matrix integrals.*

(a) Let $A \in \mathcal{M}_2(\mathbb{R})$ be a matrix which has two distinct real eigenvalues such that $\lambda_1\lambda_2 > 0$. Prove that

$$\int_0^\infty \frac{\sin(Ax)}{x}dx = \begin{cases} \frac{\pi}{2}I_2 & \text{if } \lambda_1, \lambda_2 > 0 \\ -\frac{\pi}{2}I_2 & \text{if } \lambda_1, \lambda_2 < 0. \end{cases}$$

(b) Let $A \in \mathcal{M}_2(\mathbb{R})$ be a matrix whose eigenvalues are real such that $\lambda_1\lambda_2 > 0$. Prove that

$$\int_0^\infty \frac{\sin^2(Ax)}{x^2}dx = \begin{cases} \frac{\pi}{2}A & \text{if } \lambda_1, \lambda_2 > 0 \\ -\frac{\pi}{2}A & \text{if } \lambda_1, \lambda_2 < 0. \end{cases}$$

4.99 Let $A \in \mathcal{M}_2(\mathbb{R})$ be a matrix which has two distinct real eigenvalues with $\rho(A) < 1$. Calculate

$$\int_0^\infty \frac{\sin(Ax)\cos x}{x}dx.$$

A beautiful result often attributed to Frullani is contained in the following formula

$$\int_0^\infty \frac{f(ax) - f(bx)}{x}dx = (f(0) - f(\infty))\ln\frac{b}{a}, \quad a, b > 0,$$

where $f : [0, \infty) \to \mathbb{R}$ is a continuous function (it may be assumed to be L-integrable over any interval of the form $0 < A \le x \le B < \infty$) and $f(\infty) = \lim_{x\to\infty} f(x)$ exists and is finite.

In the next two problems we extend this formula to square 2×2 real matrices.

4.100 *An exponential Frullani matrix integral.*

Let $A \in \mathcal{M}_2(\mathbb{R})$ be a matrix which has positive eigenvalues and let $\alpha, \beta > 0$. Prove that

$$\int_0^\infty \frac{e^{-\alpha Ax} - e^{-\beta Ax}}{x}dx = \left(\ln\frac{\beta}{\alpha}\right)I_2.$$

4.101 (a) **Frullani matrix integrals.**

Let $f : [0, \infty) \to \mathbb{R}$ be a continuous differentiable function such that $\lim_{x\to\infty} f(x) = f(\infty)$ exists and is finite. Let α, β be positive real numbers and let $A \in \mathcal{M}_2(\mathbb{R})$ be a matrix which has positive eigenvalues. Prove that

$$\int_0^\infty \frac{f(\alpha Ax) - f(\beta Ax)}{x}dx = \left[(f(0) - f(\infty))\ln\frac{\beta}{\alpha}\right]I_2.$$

(b) **Two sine Frullani integrals.**

Let $A \in \mathcal{M}_2(\mathbb{R})$ be a matrix which has positive eigenvalues. Calculate

$$\int_0^\infty \frac{\sin^4(Ax)}{x^3}dx \quad \text{and} \quad \int_0^\infty \frac{\sin^3(Ax)}{x^2}dx.$$

(c) **A quadratic Frullani integral.**

Let $A \in \mathcal{M}_2(\mathbb{R})$ be a matrix which has positive eigenvalues. Calculate

$$\int_0^\infty \left(\frac{I_2 - e^{-Ax}}{x}\right)^2 dx.$$

4.102 A spectacular double integral.

Let $A \in \mathcal{M}_2(\mathbb{R})$ be a matrix which has distinct positive eigenvalues. Prove that

$$\int_0^\infty \int_0^\infty \left(\frac{e^{-Ax} - e^{-Ay}}{x - y} \right)^2 \, dxdy = (\ln 4) \, I_2.$$

4.9 Solutions

4.2. *Solution 1.* We have, based on Theorem 3.1, that

$$A^n = \begin{cases} \lambda_1^n B + \lambda_2^n C, & \text{if } \lambda_1 \neq \lambda_2 \\ \lambda^n B + n\lambda^{n-1} C, & \text{if } \lambda_1 = \lambda_2 = \lambda \end{cases}$$

and it follows that $\lim_{n\to\infty} A^n = O_2$ if and only if $\lim_{n\to\infty} \lambda_1^n = \lim_{n\to\infty} \lambda_2^n = 0$ if $\lambda_1 \neq \lambda_2$ and $\lim_{n\to\infty} \lambda^n = \lim_{n\to\infty} n\lambda^{n-1} = 0$ if $\lambda_1 = \lambda_2 = \lambda$. In both cases the preceding limits are 0 if and only if $|\lambda_1|, |\lambda_2| < 1$.

Solution 2. We have, based on Theorem 2.9, that there exists a nonsingular matrix P such that $A = PJ_A P^{-1}$, which implies $A^n = PJ_A^n P^{-1}$. Thus,

$$A^n = \begin{cases} P \begin{pmatrix} \lambda_1^n & 0 \\ 0 & \lambda_2^n \end{pmatrix} P^{-1} & \text{if } \lambda_1 \neq \lambda_2 \\ P \begin{pmatrix} \lambda^n & n\lambda^{n-1} \\ 0 & \lambda^n \end{pmatrix} P^{-1} & \text{if } \lambda_1 = \lambda_2 = \lambda. \end{cases}$$

The problem reduces to the calculation of $\lim_{n\to\infty} \lambda^n$ and $\lim_{n\to\infty} n\lambda^{n-1}$, which are both 0 if and only if $|\lambda| < 1$.

4.3. See the solution of problem **4.2.**

4.4. We have, based on problem **4.2.**, that $\rho(A) < 1$ and $\rho(B) < 1$. An application of Theorem 2.1 shows that $\lambda_{AB} = \lambda_A \lambda_B$ and this implies that $|\lambda_{AB}| = |\lambda_A||\lambda_B| < 1$. Thus, $\rho(AB) < 1$ and we have, based on problem **4.2.**, that $\lim_{n\to\infty} (AB)^n = O_2$.

4.5. Let P be the nonsingular matrix such that $A = PJ_A P^{-1}$. We have

$$\left(I_2 + \frac{A}{n}\right)^n = P\left(I_2 + \frac{J_A}{n}\right)^n P^{-1}$$

$$= \begin{cases} P\left(\begin{matrix} \left(1 + \frac{\lambda_1}{n}\right)^n & 0 \\ 0 & \left(1 + \frac{\lambda_2}{n}\right)^n \end{matrix}\right) P^{-1} & \text{if } J_A = \begin{pmatrix} \lambda_1 & 0 \\ 0 & \lambda_2 \end{pmatrix} \\ P\left(\begin{matrix} \left(1 + \frac{\lambda}{n}\right)^n & \left(1 + \frac{\lambda}{n}\right)^{n-1} \\ 0 & \left(1 + \frac{\lambda}{n}\right)^n \end{matrix}\right) P^{-1} & \text{if } J_A = \begin{pmatrix} \lambda & 1 \\ 0 & \lambda \end{pmatrix}. \end{cases}$$

It follows that

$$\lim_{n \to \infty} \left(I_2 + \frac{A}{n}\right)^n = \begin{cases} P\begin{pmatrix} e^{\lambda_1} & 0 \\ 0 & e^{\lambda_2} \end{pmatrix} P^{-1} & \text{if } J_A = \begin{pmatrix} \lambda_1 & 0 \\ 0 & \lambda_2 \end{pmatrix} \\ P\begin{pmatrix} e^{\lambda} & e^{\lambda} \\ 0 & e^{\lambda} \end{pmatrix} P^{-1} & \text{if } J_A = \begin{pmatrix} \lambda & 1 \\ 0 & \lambda \end{pmatrix} \end{cases}$$

$$= e^A.$$

The second limit follows, from the first limit, by replacing A by $-A$.

4.6. *Solution 1.* Prove by induction that $A^n = I_2 + \frac{1-(1-a-b)^n}{a+b} B$, $n \in \mathbb{N}$.

Solution 2. Observe the eigenvalues of A are 1 and $1-a-b$ and use Theorem 3.1. On the other hand,

$$\lim_{n \to \infty} A^n = I_2 + \frac{1}{a+b} B = \frac{1}{a+b}\begin{pmatrix} b & b \\ a & a \end{pmatrix}.$$

4.7. We prove that

$$M(t) = \frac{1}{2}\begin{pmatrix} \frac{\zeta(t)}{\zeta(2t)} + \frac{1}{\zeta(t)}\frac{\zeta(t)}{\zeta(2t)} - \frac{1}{\zeta(t)} \\ \frac{\zeta(t)}{\zeta(2t)} - \frac{1}{\zeta(t)}\frac{\zeta(t)}{\zeta(2t)} + \frac{1}{\zeta(t)} \end{pmatrix},$$

where ζ denotes the Riemann zeta function. In particular, $M(2) = \frac{3}{2\pi^2}\begin{pmatrix} 7 & 3 \\ 3 & 7 \end{pmatrix}$.

A calculation shows that the eigenvalues of the matrix $B(x)$ are $1+x$ and $1-x$ with the corresponding eigenvectors $(\alpha, \alpha)^T$ and $(-\beta, \beta)^T$. Thus, $B(x) = PJ_B(x)P^{-1}$, where $J_B(x)$ denotes the Jordan canonical form and P is the matrix formed by the eigenvectors of $B(x)$, i.e.

$$J_B(x) = \begin{pmatrix} 1+x & 0 \\ 0 & 1-x \end{pmatrix} \quad \text{and} \quad P = \begin{pmatrix} 1 & -1 \\ 1 & 1 \end{pmatrix}.$$

Thus,

$$M(t) = \prod_p B(p^{-t}) = \prod_p P J_B(p^{-t}) P^{-1} = P \left(\prod_p J_B(p^{-t}) \right) P^{-1}$$

$$= P \prod_p \begin{pmatrix} 1 + p^{-t} & 0 \\ 0 & 1 - p^{-t} \end{pmatrix} P^{-1}$$

$$= P \begin{pmatrix} \prod_p (1 + p^{-t}) & 0 \\ 0 & \prod_p (1 - p^{-t}) \end{pmatrix} P^{-1}.$$

Using Euler's product formula [61, p. 272], $1/\zeta(s) = \prod_p (1 - 1/p^s)$, for $\Re(s) > 1$, we get that

$$\frac{1}{\zeta(2t)} = \prod_p \left(1 - \frac{1}{p^{2t}} \right) = \prod_p \left(1 - \frac{1}{p^t} \right) \prod_p \left(1 + \frac{1}{p^t} \right) = \frac{1}{\zeta(t)} \prod_p \left(1 + \frac{1}{p^t} \right),$$

and this implies that $\prod_p (1 + p^{-t}) = \zeta(t)/\zeta(2t)$.
Thus,

$$M(t) = P \begin{pmatrix} \zeta(t)/\zeta(2t) & 0 \\ 0 & 1/\zeta(t) \end{pmatrix} P^{-1}$$

$$= \frac{1}{2} \begin{pmatrix} 1 & -1 \\ 1 & 1 \end{pmatrix} \begin{pmatrix} \zeta(t)/\zeta(2t) & 0 \\ 0 & 1/\zeta(t) \end{pmatrix} \begin{pmatrix} 1 & 1 \\ -1 & 1 \end{pmatrix}$$

$$= \frac{1}{2} \begin{pmatrix} \frac{\zeta(t)}{\zeta(2t)} + \frac{1}{\zeta(t)} & \frac{\zeta(t)}{\zeta(2t)} - \frac{1}{\zeta(t)} \\ \frac{\zeta(t)}{\zeta(2t)} - \frac{1}{\zeta(t)} & \frac{\zeta(t)}{\zeta(2t)} + \frac{1}{\zeta(t)} \end{pmatrix}.$$

4.8. The limit equals -1. First we prove the limit does not depend on n. Let $A \in \mathcal{M}_2(\mathbb{R})$, $A = \begin{pmatrix} a & b \\ c & d \end{pmatrix}$ with $c \neq 0$ and let $A^n = \begin{pmatrix} a_n & b_n \\ c_n & d_n \end{pmatrix}$. We have, since $AA^n = A^n A = A^{n+1}$, $n \in \mathbb{N}$, that

$$\begin{cases} aa_n + bc_n = aa_n + cb_n \\ ab_n + bd_n = ba_n + db_n \end{cases} \quad \Leftrightarrow \quad \begin{cases} bc_n = cb_n \\ (a_n - d_n)b = (a - d)b_n \end{cases}$$

$\Rightarrow \frac{a_n - d_n}{c_n} = \frac{a - d}{c}$. Thus, we need to calculate $\lim_{x \to 1} \frac{x^{x^x} - x^x}{(1-x)^3}$.
More generally [20] we let $n \geq 1$ be a natural number and let

$$f_n(x) = x^{x^{\cdot^{\cdot^{x}}}},$$

where the number of x's in the definition of f_n is n. For example

$$f_1(x) = x, \quad f_2(x) = x^x, \quad f_3(x) = x^{x^x}, \ldots.$$

Then

$$L_n = \lim_{x \to 1} \frac{f_n(x) - f_{n-1}(x)}{(1 - x)^n} = (-1)^n.$$

We have, based on the Mean Value Theorem, that

$$f_n(x) - f_{n-1}(x) = \exp(\ln f_n(x)) - \exp(\ln f_{n-1}(x))$$
$$= \exp(f_{n-1}(x) \ln x) - \exp(f_{n-2}(x) \ln x)$$
$$= (f_{n-1}(x) - f_{n-2}(x)) \cdot \ln x \cdot \exp(\theta_n(x)),$$

where $\theta_n(x)$ is between $f_{n-1}(x) \ln x$ and $f_{n-2}(x) \ln x$. This implies $\lim_{x \to 1} \theta_n(x) = 0$. Thus,

$$L_n = \lim_{x \to 1} \frac{f_n(x) - f_{n-1}(x)}{(1 - x)^n} = \lim_{x \to 1} \left(\frac{f_{n-1}(x) - f_{n-2}(x)}{(1 - x)^{n-1}} \cdot \frac{\ln x}{1 - x} \cdot \exp(\theta_n(x)) \right) = -L_{n-1}.$$

It follows, since $L_2 = \lim_{x \to 1} \frac{x^x - x}{(1-x)^2} = 1$, that $L_n = (-1)^{n-2} L_2 = (-1)^n$.

4.9. Let $\alpha = \frac{1+\sqrt{5}}{2}$ and $\beta = \frac{1-\sqrt{5}}{2}$. We prove that

$$\begin{pmatrix} 1 + \frac{1}{n} & \frac{1}{n} \\ \frac{1}{n} & 1 \end{pmatrix}^n = \frac{1}{\sqrt{5}} \begin{pmatrix} \alpha \left(1 + \frac{\alpha}{n}\right)^n - \beta \left(1 + \frac{\beta}{n}\right)^n & \left(1 + \frac{\alpha}{n}\right)^n - \left(1 + \frac{\beta}{n}\right)^n \\ \left(1 + \frac{\alpha}{n}\right)^n - \left(1 + \frac{\beta}{n}\right)^n & \frac{1}{\alpha} \left(1 + \frac{\alpha}{n}\right)^n - \frac{1}{\beta} \left(1 + \frac{\beta}{n}\right)^n \end{pmatrix}.$$

Let $B = \begin{pmatrix} 1 & 1 \\ 1 & 0 \end{pmatrix}$. We have based on the binomial theorem, part (b) of problem **1.29**, and the definition of the Fibonacci sequence that

$$\begin{pmatrix} 1 + \frac{1}{n} & \frac{1}{n} \\ \frac{1}{n} & 1 \end{pmatrix}^n = \left(I_2 + \frac{1}{n} B \right)^n = \sum_{i=0}^{n} \binom{n}{i} \frac{1}{n^i} B^i = I_2 + \sum_{i=1}^{n} \binom{n}{i} \frac{1}{n^i} \begin{pmatrix} F_{i+1} & F_i \\ F_i & F_{i-1} \end{pmatrix},$$

and the result follows by straightforward calculations.

4.10. *Solution 1.* It is known that if A and B are two commutative matrices, then $e^A e^B = e^{A+B} = e^B e^A$. However, these formulae fail to hold if $AB \neq BA$. Lie's famous product formula for matrices [37] states that if $A, B \in \mathcal{M}_k(\mathbb{C})$, then

$$\lim_{n \to \infty} \left(e^{\frac{A}{n}} e^{\frac{B}{n}} \right)^n = e^{A+B}.$$

Our problem follows by taking $A = \begin{pmatrix} 0 & 0 \\ 1 & 0 \end{pmatrix}$ and $B = \begin{pmatrix} 0 & 1 \\ 0 & 0 \end{pmatrix}$.

Solution 2. Let $n \in \mathbb{N}$, let f be the polynomial function $f(x) = \left(1 + \frac{x}{n}\right)^n$, and let $A = \begin{pmatrix} 0 & 1 \\ 1 & \frac{1}{n} \end{pmatrix}$. A calculation shows that the eigenvalues of A are $\lambda_1 = \frac{1}{2n} + \sqrt{1 + \frac{1}{4n^2}}$ and $\lambda_2 = \frac{1}{2n} - \sqrt{1 + \frac{1}{4n^2}}$. We have, based on Theorem 4.7, that

$$\begin{pmatrix} 1 & \frac{1}{n} \\ \frac{1}{n} & 1 + \frac{1}{n^2} \end{pmatrix}^n$$

$$= \frac{\left(1 + \frac{1}{2n^2} + \frac{1}{n}\sqrt{1 + \frac{1}{4n^2}}\right)^n - \left(1 + \frac{1}{2n^2} - \frac{1}{n}\sqrt{1 + \frac{1}{4n^2}}\right)^n}{\sqrt{4 + \frac{1}{n^2}}} \begin{pmatrix} 0 & 1 \\ 1 & \frac{1}{n} \end{pmatrix}$$

$$+ \left[\frac{1}{2}\left(1 + \frac{1}{\sqrt{4n^2 + 1}}\right)\left(1 + \frac{1}{2n^2} - \frac{1}{n}\sqrt{1 + \frac{1}{4n^2}}\right)^n \right.$$

$$\left. + \frac{1}{2}\left(1 - \frac{1}{\sqrt{4n^2 + 1}}\right)\left(1 + \frac{1}{2n^2} + \frac{1}{n}\sqrt{1 + \frac{1}{4n^2}}\right)^n \right] I_2.$$

Passing to the limit as $n \to \infty$ in the previous equality, we get that

$$\lim_{n \to \infty} \begin{pmatrix} 1 & \frac{1}{n} \\ \frac{1}{n} & 1 + \frac{1}{n^2} \end{pmatrix}^n = \sinh 1 \begin{pmatrix} 0 & 1 \\ 1 & 0 \end{pmatrix} + \cosh 1 \begin{pmatrix} 1 & 0 \\ 0 & 1 \end{pmatrix}.$$

4.11. Let $n \in \mathbb{N}$, let f be the polynomial function $f(x) = \left(1 + \frac{x}{n}\right)^n$, and let $A = \begin{pmatrix} -\frac{1}{n} & 1 \\ 1 & \frac{1}{n} \end{pmatrix}$. The eigenvalues of A are $\lambda_1 = \sqrt{1 + \frac{1}{n^2}}$ and $\lambda_2 = -\sqrt{1 + \frac{1}{n^2}}$. We have based on Theorem 4.7 that

$$\begin{pmatrix} 1 - \frac{1}{n^2} & \frac{1}{n} \\ \frac{1}{n} & 1 + \frac{1}{n^2} \end{pmatrix}^n = \frac{\left(1 + \frac{1}{n}\sqrt{1 + \frac{1}{n^2}}\right)^n - \left(1 - \frac{1}{n}\sqrt{1 + \frac{1}{n^2}}\right)^n}{2\sqrt{1 + \frac{1}{n^2}}} \begin{pmatrix} -\frac{1}{n} & 1 \\ 1 & \frac{1}{n} \end{pmatrix}$$

$$+ \frac{\left(1 + \frac{1}{n}\sqrt{1 + \frac{1}{n^2}}\right)^n + \left(1 - \frac{1}{n}\sqrt{1 + \frac{1}{n^2}}\right)^n}{2} I_2.$$

Therefore

$$\lim_{n \to \infty} \begin{pmatrix} 1 - \frac{1}{n^2} & \frac{1}{n} \\ \frac{1}{n} & 1 + \frac{1}{n^2} \end{pmatrix}^n = \sinh 1 \begin{pmatrix} 0 & 1 \\ 1 & 0 \end{pmatrix} + \cosh 1 \begin{pmatrix} 1 & 0 \\ 0 & 1 \end{pmatrix}.$$

4.12. The first limit equals $e \begin{pmatrix} 2 & 1 \\ -4 & -2 \end{pmatrix}$.

First, we observe that both eigenvalues of A are equal to 1. Let P be the nonsingular matrix such that $P^{-1}AP = J_A = \begin{pmatrix} 1 & 1 \\ 0 & 1 \end{pmatrix}$. A calculation shows that $P = \begin{pmatrix} 1 & 0 \\ -2 & 1 \end{pmatrix}$ and $P^{-1} = \begin{pmatrix} 1 & 0 \\ 2 & 1 \end{pmatrix}$. We have

$$\frac{1}{n}\left(I_2 + \frac{A^n}{n}\right)^n = P\left[\frac{1}{n}\left(I_2 + \frac{1}{n}J_A^n\right)^n\right]P^{-1}.$$

On the other hand, since $J_A^n = \begin{pmatrix} 1 & n \\ 0 & 1 \end{pmatrix}$, we get that

$$\frac{1}{n}\left(I_2 + \frac{1}{n}J_A^n\right)^n = \frac{1}{n}\left[\begin{pmatrix} 1 & 0 \\ 0 & 1 \end{pmatrix} + \begin{pmatrix} \frac{1}{n} & 1 \\ 0 & \frac{1}{n} \end{pmatrix}\right]^n$$

$$= \frac{1}{n}\begin{pmatrix} 1+\frac{1}{n} & 1 \\ 0 & 1+\frac{1}{n} \end{pmatrix}^n$$

$$= \frac{1}{n}\begin{pmatrix} \left(1+\frac{1}{n}\right)^n & n\left(1+\frac{1}{n}\right)^{n-1} \\ 0 & \left(1+\frac{1}{n}\right)^n \end{pmatrix},$$

and this implies that

$$\lim_{n\to\infty}\frac{1}{n}\left(I_2 + \frac{A^n}{n}\right)^n = \lim_{n\to\infty}P\left[\frac{1}{n}\begin{pmatrix} \left(1+\frac{1}{n}\right)^n & n\left(1+\frac{1}{n}\right)^{n-1} \\ 0 & \left(1+\frac{1}{n}\right)^n \end{pmatrix}\right]P^{-1}$$

$$= P\begin{pmatrix} 0 & e \\ 0 & 0 \end{pmatrix}P^{-1}$$

$$= e\begin{pmatrix} 2 & 1 \\ -4 & -2 \end{pmatrix}.$$

Similarly one can prove the second limit equals $\frac{1}{e}\begin{pmatrix} -2 & -1 \\ 4 & 2 \end{pmatrix}$.

Another approach for solving the problem would be to use Theorem 4.7.

Remark 4.19 Let $f_n(x) = \frac{1}{n}\left(1 + \frac{x^n}{n}\right)^n$, $n \in \mathbb{N}$. If both eigenvalues of A are equal real numbers $\lambda_1 = \lambda_2 = \lambda$ and $A \neq \lambda I_2$, we have based on Theorem 4.7 that

$$f_n(A) = \left(1 + \frac{\lambda^n}{n}\right)^{n-1}\lambda^{n-1}A + \left[\frac{1}{n}\left(1 + \frac{\lambda^n}{n}\right)^n - \lambda^n\left(1 + \frac{\lambda^n}{n}\right)^{n-1}\right]I_2.$$

It follows that $\lim_{n\to\infty} f_n(A)$ is O_2 if $|\lambda| < 1$ and $e(A - I_2)$, when $\lambda = 1$.

If $A = \lambda I_2$, $\lambda \in \mathbb{R}$, then $f_n(A) = \frac{1}{n}\left(1 + \frac{\lambda^n}{n}\right)^n I_2$ and the limit equals ∞ if $\lambda > 1$, the limit does not exist if $\lambda \le -1$ and the limit equals O_2 if $-1 < \lambda \le 1$.

4.13. The limit equals $(\ln\sqrt{2})I_2$. Use that if $\lambda \in \mathbb{R}$, $\lambda \ne k\pi$, $k \in \mathbb{Z}$, then

$$\lim_{n\to\infty}\left(\frac{\cos(2\lambda)}{n+1} + \frac{\cos(4\lambda)}{n+2} + \cdots + \frac{\cos(2n\lambda)}{2n}\right) = 0.$$

4.14. Use Theorem 3.1 and the formula

$$\lim_{n\to\infty}\sqrt[n]{|a|^n + |b|^n + |c|^n + |d|^n} = \max\{|a|, |b|, |c|, |d|\}, \quad \text{where } a, b, c, d \in \mathbb{C}.$$

4.15. (a) $A = PJ_A P^{-1}$, where $J_A = \begin{pmatrix} 2 & 0 \\ 0 & -1 \end{pmatrix}$, $P = \begin{pmatrix} 1 & 1 \\ 1 & 4 \end{pmatrix}$ and $P^{-1} = \frac{1}{3}\begin{pmatrix} 4 & -1 \\ -1 & 1 \end{pmatrix}$.
We have

$$A(A + I_2)(A + 2I_2)\cdots(A + nI_2) = PJ_A(J_A + I_2)(J_A + 2I_2)\cdots(J_A + nI_2)P^{-1}$$

$$= P\begin{pmatrix} (n+2)! & 0 \\ 0 & 0 \end{pmatrix}P^{-1}$$

$$= \frac{(n+2)!}{3}\begin{pmatrix} 4 & -1 \\ 4 & 1 \end{pmatrix}.$$

It follows that

$$\lim_{n\to\infty}\frac{\sqrt[n]{\|A(A + I_2)(A + 2I_2)\cdots(A + nI_2)\|}}{n} = \lim_{n\to\infty}\frac{\sqrt[n]{(n+2)!}}{n}\sqrt[n]{\frac{\sqrt{34}}{3}} = \frac{1}{e}.$$

(b) The limit equals 2. Use Cauchy-d'Alembert's criteria[3] and the problem reduces to the calculation of the limit

$$\lim_{n\to\infty}\frac{\|A(A + 2I_2)(A + 4I_2)\cdots(A + (2n+2)I_2)\|}{(n+1)\|A(A + 2I_2)(A + 4I_2)\cdots(A + 2nI_2)\|} = 2.$$

(c) We have

$$(A + (n+1)I_2)(A + (n+4)I_2)\cdots(A + (4n-2)I_2)$$

$$= P(J_A + (n+1)I_2)(J_A + (n+4)I_2)\cdots(J_A + (4n-2)I_2)P^{-1}$$

$$= P\begin{pmatrix} (n+3)(n+6)\cdots(4n) & 0 \\ 0 & n(n+3)\cdots(4n-3) \end{pmatrix}P^{-1}$$

$$= \frac{1}{3}\begin{pmatrix} 4\alpha_n - \beta_n & -\alpha_n + \beta_n \\ 4\alpha_n - 4\beta_n & -\alpha_n + 4\beta_n \end{pmatrix},$$

---

[3]The Cauchy-d'Alembert criteria states that if $(a_n)_{n\ge 1}$ is a sequence of positive real numbers such that $\lim_{n\to\infty}\frac{a_{n+1}}{a_n} = l \in \overline{\mathbb{R}}$, then $\lim_{n\to\infty}\sqrt[n]{a_n} = l$.

where $\alpha_n = (n+3)(n+6)\cdots(4n)$ and $\beta_n = n(n+3)\cdots(4n-3)$. It follows that

$$||(A+(n+1)I_2)(A+(n+4)I_2)\cdots(A+(4n-2)I_2)||$$

$$= \frac{1}{3}\sqrt{34\alpha_n^2 + 34\beta_n^2 - 50\alpha_n\beta_n}.$$

Similarly,

$$(A+nI_2)(A+(n+3)I_2)\cdots(A+(4n-3)I_2) = \frac{1}{3}\begin{pmatrix} 4u_n - v_n & -u_n + v_n \\ 4u_n - 4v_n & -u_n + 4v_n \end{pmatrix},$$

where $u_n = (n+2)(n+5)\cdots(4n-1)$ and $v_n = (n-1)(n+2)\cdots(4n-4)$. This implies that

$$||(A+nI_2)(A+(n+3)I_2)\cdots(A+(4n-3)I_2)|| = \frac{1}{3}\sqrt{34u_n^2 + 34v_n^2 - 50u_nv_n}.$$

Therefore

$$\lim_{n\to\infty} \frac{||(A+(n+1)I_2)(A+(n+4)I_2)\cdots(A+(4n-2)I_2)||}{||(A+nI_2)(A+(n+3)I_2)\cdots(A+(4n-3)I_2)||}$$

$$= \lim_{n\to\infty} \frac{(n+3)(n+6)\cdots(4n)}{(n+2)(n+5)\cdots(4n-1)} \sqrt{\frac{34 + \frac{34}{4^2} - \frac{50}{4}}{34 + 34\frac{(n-1)^2}{(4n-1)^2} - 50\frac{n-1}{4n-1}}}$$

$$= \sqrt[3]{4},$$

since (prove it!)

$$\lim_{n\to\infty} \frac{(n+3)(n+6)\cdots(4n)}{(n+2)(n+5)\cdots(4n-1)} = \sqrt[3]{4}.$$

Remark 4.20 We mention that beautiful limits, like in the previous formula, involving products of sequences of integer numbers in arithmetic progression which can be solved by elementary techniques based on the Squeeze Theorem can be found in [53, problem 59, p. 19], [54, 55, 56].

4.16. $A = \Re(A) + i\Im(A)$ and $A^* = \Re(A)^T - i\Im(A)^T$. It follows that

$$||A||^2 = \text{Tr}\left[(\Re(A) + i\Im(A))(\Re(A)^T - i\Im(A)^T)\right]$$

$$= \text{Tr}\left[\Re(A)\Re(A)^T + \Im(A)\Im(A)^T + i(\Re(A)^T\Im(A) - \Im(A)^T\Re(A))\right]$$

$$= \text{Tr}\left(\Re(A)\Re(A)^T\right) + \text{Tr}\left(\Im(A)\Im(A)^T\right)$$

$$= ||\Re(A)||^2 + ||\Im(A)||^2.$$

4.17. *Solution 1.* Use that $(A - I_2)^2 = O_2$.

Solution 2. (a) Use mathematical induction.
(b) Part (b) follows from part (c) when $x = 1$.
(c) We have, based on part (a), that

$$e^{Ax} = \sum_{n=0}^{\infty} \frac{A^n x^n}{n!} = \begin{pmatrix} \sum_{n=0}^{\infty} \frac{x^n(n+1)}{n!} & \sum_{n=1}^{\infty} \frac{x^n}{(n-1)!} \\ -\sum_{n=1}^{\infty} \frac{x^n}{(n-1)!} & \sum_{n=0}^{\infty} \frac{x^n(1-n)}{n!} \end{pmatrix} = e^x \begin{pmatrix} x+1 & x \\ -x & 1-x \end{pmatrix}.$$

4.18. (a) $e^A = e^2 \begin{pmatrix} 2 & -1 \\ 1 & 0 \end{pmatrix}$. We have that $A^2 - 4A + 4I_2 = O_2 \Leftrightarrow (A - 2I_2)^2 = O_2$.

Let $B = A - 2I_2$. This implies that $B^2 = O_2$ and $A = B + 2I_2$. We have

$$e^{2I_2} = e^2 I_2 \quad \text{and} \quad e^B = I_2 + \frac{B}{1!} + \frac{B^2}{2!} + \cdots + \frac{B^n}{n!} + \cdots = I_2 + B.$$

It follows, since matrices $2I_2$ and B commute, that

$$e^A = e^{2I_2 + B} = e^{2I_2} e^B = e^2(I_2 + B) = e^2(A - I_2) = e^2 \begin{pmatrix} 2 & -1 \\ 1 & 0 \end{pmatrix}.$$

(b) $e^A = \begin{pmatrix} 4e - 3 & 2 - 2e \\ 6e - 6 & 4 - 3e \end{pmatrix}$. We have, based on Theorem 2.2, that $A^2 = A \Rightarrow A^n = A$, for all $n \geq 1$. It follows that

$$e^A = I_2 + \frac{A}{1!} + \frac{A^2}{2!} + \cdots + \frac{A^n}{n!} + \cdots$$

$$= I_2 + A\left(\frac{1}{1!} + \frac{1}{2!} + \cdots + \frac{1}{n!} + \cdots\right)$$

$$= I_2 + (e - 1)A$$

$$= \begin{pmatrix} 4e - 3 & 2 - 2e \\ 6c - 6 & 4 - 3e \end{pmatrix}.$$

4.19. Let $A = \begin{pmatrix} 0 & 1 \\ 1 & 0 \end{pmatrix}$. Observe that $A = E_p$ is the permutation matrix. A calculation shows that $A^{2n} = I_2$ and $A^{2n-1} = A$, for all $n \in \mathbb{N}$. It follows that

$$e^A = \sum_{n=0}^{\infty} \frac{A^n}{n!} = \sum_{n=0}^{\infty} \frac{A^{2n}}{(2n)!} + \sum_{n=1}^{\infty} \frac{A^{2n-1}}{(2n-1)!}$$

$$= \sum_{n=0}^{\infty} \frac{1}{(2n)!} I_2 + \sum_{n=1}^{\infty} \frac{1}{(2n-1)!} A$$

$$= (\cosh 1)I_2 + (\sinh 1)A$$

$$= \begin{pmatrix} \cosh 1 & \sinh 1 \\ \sinh 1 & \cosh 1 \end{pmatrix}.$$

4.21. *Solution 1.* We have $J_2^{2k} = (-1)^k I_2$ and $J_2^{2k-1} = (-1)^{k-1} J_2$, for all $k \geq 1$. It follows that

$$e^{-\theta J_2} = \sum_{n=0}^{\infty} \frac{(-\theta J_2)^n}{n!}$$

$$= \sum_{k=0}^{\infty} \frac{(-\theta J_2)^{2k}}{(2k)!} + \sum_{k=1}^{\infty} \frac{(-\theta J_2)^{(2k-1)}}{(2k-1)!}$$

$$= \sum_{k=0}^{\infty} (-1)^k \frac{\theta^{2k}}{(2k)!} I_2 + \sum_{n=1}^{\infty} (-1)^k \frac{\theta^{2k-1}}{(2k-1)!} J_2$$

$$= (\cos \theta)I_2 - (\sin \theta)J_2$$

$$= \begin{pmatrix} \cos \theta & -\sin \theta \\ \sin \theta & \cos \theta \end{pmatrix}.$$

Solution 2. Observe that

$$-J_2 = \begin{pmatrix} \cos \frac{\pi}{2} & -\sin \frac{\pi}{2} \\ \sin \frac{\pi}{2} & \cos \frac{\pi}{2} \end{pmatrix} \quad \text{and} \quad J_2^n = (-1)^n \begin{pmatrix} \cos \frac{n\pi}{2} & -\sin \frac{n\pi}{2} \\ \sin \frac{n\pi}{2} & \cos \frac{n\pi}{2} \end{pmatrix}.$$

It follows that

$$e^{-\theta J_2} = \sum_{n=0}^{\infty} \frac{(-\theta J_2)^n}{n!} = \sum_{n=0}^{\infty} \frac{\theta^n}{n!} \begin{pmatrix} \cos \frac{n\pi}{2} & -\sin \frac{n\pi}{2} \\ \sin \frac{n\pi}{2} & \cos \frac{n\pi}{2} \end{pmatrix}$$

$$= \begin{pmatrix} \sum_{n=0}^{\infty} \frac{\theta^n}{n!} \cos \frac{n\pi}{2} & -\sum_{n=0}^{\infty} \frac{\theta^n}{n!} \sin \frac{n\pi}{2} \\ \sum_{n=0}^{\infty} \frac{\theta^n}{n!} \sin \frac{n\pi}{2} & \sum_{n=0}^{\infty} \frac{\theta^n}{n!} \cos \frac{n\pi}{2} \end{pmatrix}.$$

Let $S_1 = \displaystyle\sum_{n=0}^{\infty} \frac{\theta^n}{n!} \cos \frac{n\pi}{2}$ and $S_2 = \displaystyle\sum_{n=0}^{\infty} \frac{\theta^n}{n!} \sin \frac{n\pi}{2}$. We have

$$S_1 + iS_2 = \sum_{n=0}^{\infty} \frac{\theta^n}{n!} \left(\cos \frac{\pi}{2} + i \sin \frac{\pi}{2}\right)^n = e^{i\theta} = \cos \theta + i \sin \theta,$$

and it follows that $S_1 = \cos \theta$ and $S_2 = \sin \theta$.

4.22. (a) Use that $e^A = \alpha A + \beta I_2$, for some $\alpha, \beta \in \mathbb{C}$.

(b) Let $A = \begin{pmatrix} 0 & -\pi \\ \pi & 0 \end{pmatrix}$. Then $e^A = -I_2$ (see problem **4.21**).

4.23. See the solution of problem **4.24**.

4.24. Observe that $A = aI_2 + bJ_2$. We have, since matrices aI_2 and bJ_2 commute, that

$$e^A = e^{aI_2 + bJ_2} = e^{aI_2} e^{bJ_2} = e^a I_2 \begin{pmatrix} \cos(-b) & -\sin(-b) \\ \sin(-b) & \cos(-b) \end{pmatrix} = e^a \begin{pmatrix} \cos b & \sin b \\ -\sin b & \cos b \end{pmatrix}.$$

We used in our calculations the result in problem **4.21** with $\theta = -b$.

4.25. First we consider the case when $t = 1$. We have, since $A(1) = I_2$, that $e^{A(1)} = e^{I_2} = eI_2 = eA(e^0)$.

Now we consider the case when $t \neq 1$. The eigenvalues of $A(t)$ are 1 and t. We have that $A(t) = PJ_{A(t)}P^{-1}$, where

$$J_{A(t)} = \begin{pmatrix} 1 & 0 \\ 0 & t \end{pmatrix}, \quad P = \begin{pmatrix} -1 & 1 \\ 1 & 0 \end{pmatrix} \quad \text{and} \quad P^{-1} = \begin{pmatrix} 0 & 1 \\ 1 & 1 \end{pmatrix}.$$

It follows that

$$e^{A(t)} = Pe^{J_{A(t)}}P^{-1} = P\begin{pmatrix} e & 0 \\ 0 & e^t \end{pmatrix}P^{-1} = \begin{pmatrix} e^t & e^t - e \\ 0 & e \end{pmatrix} = eA\left(e^{t-1}\right).$$

4.26. See Theorem 4.6. Another "solution" is based on a formal computation. If λ is an eigenvalue of A there exists a nonzero vector X such that $AX = \lambda X$. We have

$$e^A X = \left(\sum_{n=0}^{\infty} \frac{A^n}{n!}\right) X = \sum_{n=0}^{\infty} \frac{A^n X}{n!} = \sum_{n=0}^{\infty} \frac{\lambda^n X}{n!} = \left(\sum_{n=0}^{\infty} \frac{\lambda^n}{n!}\right) X = e^\lambda X,$$

which shows that e^λ is an eigenvalue of e^A and X is its corresponding eigenvector.

Recall that the determinant of a matrix equals the product of the eigenvalues and the trace equals the sum of the eigenvalues. We have $\det\left(e^A\right) = e^{\lambda_1} e^{\lambda_2} = e^{\lambda_1 + \lambda_2} = e^{\text{Tr}(A)}$.

4.27. $e^A e^B = e^B e^A$. We consider the following two cases.

Case $e^A = \alpha I_2$, $\alpha \in \mathbb{C}$. If J_A is the Jordan canonical form of A we get that $e^{J_A} = \alpha I_2$ and this implies that J_A is diagonal. If $J_A = \begin{pmatrix} \lambda_1 & 0 \\ 0 & \lambda_2 \end{pmatrix} \in \mathcal{M}_2(\mathbb{R})$, we have that $e^{\lambda_1} = e^{\lambda_2} = \alpha$ which implies, since $\lambda_1, \lambda_2 \in \mathbb{R}$, that $\alpha \in \mathbb{R}$, $\alpha > 0$, $\lambda_1 = \lambda_2 = \ln \alpha$ and $A = (\ln \alpha) I_2$. Clearly, in this case A commutes with B.

Case $e^A \neq \alpha I_2$, $\alpha \in \mathbb{C}$. We have, based on Theorem 1.1, that $e^B = a e^A + b I_2$, for some $a, b \in \mathbb{C}$. If $a = 0$ we get that $e^B = b I_2$ and, like in the previous case, we get that $b \in \mathbb{R}$, $b > 0$ and $B = (\ln b) I_2$. Clearly in this case B commutes with A.

If $a \neq 0$ we have, since B commutes with e^B, that $a \left(B e^A - e^A B \right) = O_2 \Rightarrow B e^A = e^A B$. It follows, based on Theorem 1.1, that $B = c e^A + d I_2$, for some $c, d \in \mathbb{C}$. Thus, $AB = A(c e^A + d I_2) = (c e^A + d I_2)A = BA$.

4.28. First we prove that if $A^2 = O_2$, then $e^A \in \mathcal{M}_2(\mathbb{Z})$. We have

$$e^A = I_2 + \frac{A}{1!} + \frac{A^2}{2!} + \cdots + \frac{A^n}{n!} + \cdots = I_2 + A \in \mathcal{M}_2(\mathbb{Z}).$$

Now we prove that if $e^A \in \mathcal{M}_2(\mathbb{Z})$, then $A^2 = O_2$. If λ_1, λ_2 are the eigenvalues of A, recall the eigenvalues of e^A are e^{λ_1} and e^{λ_2}. Observe that both λ_1 and λ_2 are algebraic numbers being the roots of the characteristic polynomial of A which has integer coefficients. On the other hand, e^{λ_1} and e^{λ_2} are also algebraic numbers since they are the roots of the characteristic polynomial of e^A which has integer coefficients. It follows that λ_1 and λ_2 are both 0, otherwise this would contradict the Lindemann–Weierstrass Theorem which states that if α is a nonzero algebraic number, then e^α is transcendental. Thus, $\lambda_1 = \lambda_2 = 0$ and we have based on Theorem 2.2 that $A^2 = O_2$.

4.29. We solve only the first part of the problem, the second part can be solved similarly.

We need the following lemma.

Lemma 4.7 *If $q \in \mathbb{Q}^*$, then $\cos q$ is transcendental.*

Proof We assume that $\cos q = a$ is algebraic. Then, $\sin q = \pm\sqrt{1 - a^2}$ is also algebraic. It follows, since the sum of two algebraic numbers is algebraic, that $e^{iq} = \cos q + i \sin q$ is algebraic. However, this contradicts the Lindemann–Weierstrass theorem which states that if $\alpha \neq 0$ is algebraic, then e^α is transcendental. $\square$

First we prove that if $A^2 = O_2$, then $\sin A \in \mathcal{M}_2(\mathbb{Z})$. We have

$$\sin A = \sum_{n=0}^{\infty} (-1)^n \frac{A^{2n+1}}{(2n+1)!} = A - \frac{A^3}{3!} + \cdots = A \in \mathcal{M}_2(\mathbb{Z}).$$

Now we prove the reverse implication. Let $\lambda_1 = c + id$ and $\lambda_2 = c - id$ be the eigenvalues of A and let $\sin \lambda_1 = a + ib$ and $\sin \lambda_2 = a - ib$ be the eigenvalues of $\sin A$. Observe that $c = \frac{1}{2} \text{Tr}(A) \in \mathbb{Q}$ and $d = \frac{1}{2} \sqrt{4 \det A - \text{Tr}^2(A)}$ is an algebraic number. We have, $\lambda_1 + \lambda_2 = 2c \in \mathbb{Z}$, $\lambda_1 \lambda_2 = c^2 + d^2 \in \mathbb{Z}$, $\sin \lambda_1 + \sin \lambda_2 = 2a \in \mathbb{Z}$, and $\sin \lambda_1 \sin \lambda_2 = a^2 + b^2 = v \in \mathbb{Z}$. We calculate

$$\sin \lambda_1 + \sin \lambda_2 = 2 \sin \frac{\lambda_1 + \lambda_2}{2} \cos \frac{\lambda_1 - \lambda_2}{2} = 2 \sin c \cos \frac{\lambda_1 - \lambda_2}{2} = 2a$$

which implies that $\sin c \cos \frac{\lambda_1 - \lambda_2}{2} = a$. On the other hand,

$$\sin \lambda_1 \sin \lambda_2 = \frac{\cos(\lambda_1 - \lambda_2) - \cos(\lambda_1 + \lambda_2)}{2} = v \in \mathbb{Z}.$$

This implies that $\cos^2 \frac{\lambda_1 - \lambda_2}{2} - \cos^2 c = v$.

If $\sin c \neq 0$, and this implies since $c \in \mathbb{Q}$ that $c \neq 0$, then $\cos \frac{\lambda_1 - \lambda_2}{2} = \frac{a}{\sin c}$. It follows, after simple calculations, that $\cos^4 c + (v - 1) \cos^2 c + a^2 - v = 0$ which implies that $\cos c$ is algebraic. This contradicts Lemma 4.7. Thus, $c = 0 \Rightarrow \lambda_1 = id$ and $\lambda_2 = -id$. A calculation shows that $\sin \lambda_1 \sin \lambda_2 = \sin(id) \sin(-id) = \frac{(e^d - e^{-d})^2}{4} = v \in \mathbb{Z}$. This implies $d = 0$.

If $d \neq 0$, then e^d is the solution of the equation $x^4 - (2 + 4v)x^2 + 1 = 0$, hence e^d is algebraic. However, this contradicts the Lindemann–Weierstrass theorem. Therefore, $d = 0 \Rightarrow \lambda_1 - \lambda_2 = 0 \Rightarrow A^2 = O_2$.

Remark 4.21 The problem has an equivalent formulation. If $A \in \mathcal{M}_2(\mathbb{Z})$ then:

- $\sin A \in \mathcal{M}_2(\mathbb{Z})$ if and only if $\sin A = A$;
- $\cos A \in \mathcal{M}_2(\mathbb{Z})$ if and only if $\cos A = I_2$.

4.30. We solve only the first part of the problem. If $A^2 = O_2$, then

$$\ln(I_2 - A) = -\sum_{n=1}^{\infty} \frac{A^n}{n} = -A \in \mathcal{M}_2(\mathbb{Q}).$$

Now we prove the other implication. Let λ_1, λ_2 be the eigenvalues of A. We have $\lambda_1 + \lambda_2 = k \in \mathbb{Q}$ and $\lambda_1 \lambda_2 = i \in \mathbb{Q}$. Recall the eigenvalues of $\ln(I_2 - A)$ are $\ln(1 - \lambda_1)$ and $\ln(1 - \lambda_2)$ and we get, since $\ln(I_2 - A) \in \mathcal{M}_2(\mathbb{Q})$, that $\ln(1 - \lambda_1) + \ln(1 - \lambda_2) = \ln[(1 - \lambda_1)(1 - \lambda_2)] \in \mathbb{Q}$ and $\ln(1 - \lambda_1) \ln(1 - \lambda_2) \in \mathbb{Q}$. We have $\ln[(1 - \lambda_1)(1 - \lambda_2)] = \ln(1 - \lambda_1 - \lambda_2 + \lambda_1 \lambda_2) = \ln(1 - k + i) = a \in \mathbb{Q}$.

If $a \neq 0$ we have that $e^a = 1 - k + i \in \mathbb{Q} \Rightarrow e^a$ is algebraic. However, this contradicts the Lindemann–Weierstrass Theorem which states that if α is a nonzero algebraic number, then e^α is transcendental. Therefore $a = 0 \Rightarrow 1 - k + i = 1 \Rightarrow (1 - \lambda_1)(1 - \lambda_2) = 1$. This implies $-\ln^2(1 - \lambda_1) \in \mathbb{Q}$. Let $\ln^2(1 - \lambda_1) = b \in \mathbb{Q} \Rightarrow \ln(1 - \lambda_1) = \pm \sqrt{b} \Rightarrow \lambda_1 = 1 - e^{\pm \sqrt{b}} \Rightarrow \lambda_2 = 1 - \frac{1}{1 - \lambda_1} = 1 - e^{\mp \sqrt{b}}$. Since $\lambda_1 + \lambda_2 = k \in \mathbb{Q}$ we get that $2 - k = e^{\pm \sqrt{b}} + e^{\mp \sqrt{b}}$ and this in turn implies that $e^{\pm\sqrt{b}}$ is algebraic. This implies that $b = 0$ otherwise, if $b \neq 0$ we get based on Lindemann–Weierstrass Theorem, since $\pm\sqrt{b}$ is nonzero algebraic, that $e^{\pm\sqrt{b}}$ is transcendental. Therefore $b = 0$ and this implies that $\lambda_1 = \lambda_2 = 0$. Thus, both eigenvalues of A are 0 and we get based on the Cayley–Hamilton Theorem that $A^2 = O_2$.

The second part of the problem can be solved similarly.

Remark 4.22 The problem has an equivalent formulation. If $A \in \mathcal{M}_2(\mathbb{Q})$ such that $\rho(A) < 1$ then:

- $\ln(I_2 - A) \in \mathcal{M}_2(\mathbb{Q})$ if and only if $\ln(I_2 - A) = -A$;
- $\ln(I_2 + A) \in \mathcal{M}_2(\mathbb{Q})$ if and only if $\ln(I_2 + A) = A$.

4.31. We have $e^{iA} + e^{iB} + e^{iC} = O_2$ and $e^{-iA} + e^{-iB} + e^{-iC} = O_2$. It follows that

$$O_2 = \left(e^{iA} + e^{iB} + e^{iC}\right)^2 = e^{2iA} + e^{2iB} + e^{2iC} + 2e^{i(A+B+C)}\left(e^{-iA} + e^{-iB} + e^{-iC}\right).$$

This implies that $e^{2iA} + e^{2iB} + e^{2iC} = O_2 \Rightarrow \cos(2A) + \cos(2B) + \cos(2C) = O_2$ and $\sin(2A) + \sin(2B) + \sin(2C) = O_2$. On the other hand, $O_2 = e^{3iA} + e^{3iB} + e^{3iC} - 3e^{i(A+B+C)}$ and parts (c) and (d) follow.

4.32. (a) Let $A = \begin{pmatrix} a & b \\ b & a \end{pmatrix}$ and observe that $A = aI_2 + bE_p$, where $E_p = \begin{pmatrix} 0 & 1 \\ 1 & 0 \end{pmatrix}$ is the permutation matrix. We use a technique similar to the one in the solution of problem **4.19** and we get that

$$e^A = e^{aI_2 + bE_p} = e^{aI_2}e^{bE_p} = e^a \begin{pmatrix} \cosh b & \sinh b \\ \sinh b & \cosh b \end{pmatrix}.$$

Parts (b) and (c) can be solved using Theorem 4.10.
(d) We have

$$\sum_{n=1}^{\infty} \frac{1}{(2n-1)^A} = \sum_{n=1}^{\infty} \frac{1}{n^A} - \sum_{n=1}^{\infty} \frac{1}{(2n)^A}$$

$$= \zeta(A) - 2^{-A} \sum_{n=1}^{\infty} \frac{1}{n^A}$$

$$= \zeta(A) - 2^{-A}\zeta(A)$$

$$= \left(I_2 - 2^{-A}\right)\zeta(A).$$

4.36. Let $\alpha = \rho(\cos t + i\sin t)$, $\rho > 0$, $t \in (-\pi, \pi]$. Then,

$$A = P \begin{pmatrix} \ln\rho + (t + 2k\pi)i & 0 \\ 0 & \ln\rho + (t + 2l\pi)i \end{pmatrix} P^{-1},$$

where $k, l \in \mathbb{Z}$ and $P \in GL_2(\mathbb{C})$.

4.37. Let $a = \rho_a(\cos t_a + i \sin t_a)$, $\rho_a > 0$, $t_a \in (-\pi, \pi]$ and $b = \rho_b(\cos t_b + i \sin t_b)$, $\rho_b > 0$, $t_b \in (-\pi, \pi]$. Then,

$$A = \begin{pmatrix} \ln \rho_a + (t_a + 2k\pi)i & 0 \\ 0 & \ln \rho_b + (t_b + 2l\pi)i \end{pmatrix},$$

where $k, n \in \mathbb{Z}$.

4.38. $A = \begin{pmatrix} \left(2k + \frac{1}{2}\right)\pi i & \left(2n - \frac{1}{2}\right)\pi i \\ \left(2n - \frac{1}{2}\right)\pi i & \left(2k + \frac{1}{2}\right)\pi i \end{pmatrix}$ or $A = \begin{pmatrix} \left(2k - \frac{1}{2}\right)\pi i & \left(2n + \frac{1}{2}\right)\pi i \\ \left(2n + \frac{1}{2}\right)\pi i & \left(2k - \frac{1}{2}\right)\pi i \end{pmatrix}$,

where $k, n \in \mathbb{Z}$.

Observe that A commutes with e^A and use Theorem 1.1 to deduce that A is a circulant matrix, i.e., $A = \begin{pmatrix} \beta & \alpha \\ \alpha & \beta \end{pmatrix}$, $\alpha, \beta \in \mathbb{C}$. It follows, based on part (a) of problem **4.32**, that

$$e^A = e^\beta \begin{pmatrix} \cosh\alpha & \sinh\alpha \\ \sinh\alpha & \cosh\alpha \end{pmatrix} = \begin{pmatrix} 0 & 1 \\ 1 & 0 \end{pmatrix}.$$

This implies that $e^\beta \cosh\alpha = 0$ and $e^\beta \sinh\alpha = 1$. The first equation implies that $\cosh\alpha = 0 \Rightarrow e^{2\alpha} = -1 \Rightarrow \alpha = \frac{2p+1}{2}\pi i$, $p \in \mathbb{Z}$. A calculation shows that $\sinh\alpha = (-1)^p i$ and the second equation implies that $e^\beta = (-1)^{p-1} i$. The cases when p is even or odd lead to the desired solution.

4.39. Let $a = \rho(\cos\theta + i\sin\theta)$, $\rho > 0$, and $\theta \in (-\pi, \pi]$. Then,

$$A = \begin{pmatrix} \ln\rho + (\theta + 2k\pi)i & 1 \\ 0 & \ln\rho + (\theta + 2k\pi)i \end{pmatrix}, \quad k \in \mathbb{Z}.$$

4.40. We observe that $A = \pi I_2 + B$, where $B = \begin{pmatrix} -1 & 1 \\ -1 & 1 \end{pmatrix}$ and note that $B^2 = O_2$. We have

$$\sin A = \sin(\pi I_2 + B) = \sin(\pi I_2)\cos B + \cos(\pi I_2)\sin B = \cos(\pi I_2)\sin B = -B$$

$$\cos A = \cos(\pi I_2 + B) = \cos(\pi I_2)\cos B - \sin(\pi I_2)\sin B = \cos(\pi I_2)\cos B = -I_2$$

and similarly $\sin(2A) = 2B$.

4.41. We have, since $A^2 = A$, that

$$\sin(k\pi A) = \sum_{n=0}^{\infty} \frac{(-1)^n}{(2n+1)!}(k\pi A)^{2n+1} = \sum_{n=0}^{\infty} \frac{(-1)^n}{(2n+1)!}(k\pi)^{2n+1}A = \sin(k\pi)A = O_2.$$

Similarly, if A is idempotent, then $\cos(k\pi A) = I_2 + ((-1)^k - 1)A$, for all $k \in \mathbb{Z}$.

4.42. (a) If such a matrix would exist, then $\sin^2 A = \begin{pmatrix} 1 & 4032 \\ 0 & 1 \end{pmatrix}$ and this implies,

since $\cos^2 A = I_2 - \sin^2 A$, that $\cos^2 A = \begin{pmatrix} 0 & -4032 \\ 0 & 0 \end{pmatrix}$. However, there is no $X \in$

$\mathcal{M}_2(\mathbb{C})$ such that $X^2 = \begin{pmatrix} 0 & a \\ 0 & 0 \end{pmatrix}$, with $a \neq 0$.

(b) If such a matrix would exist, then $\cosh^2 A = \begin{pmatrix} 1 & 2\alpha \\ 0 & 1 \end{pmatrix}$ and this implies,

since $\cosh^2 A - \sinh^2 A = I_2$, that $\sinh^2 A = \begin{pmatrix} 0 & 2\alpha \\ 0 & 0 \end{pmatrix}$. This is impossible since

the equation $X^2 = \begin{pmatrix} 0 & a \\ 0 & 0 \end{pmatrix}$, with $a \neq 0$, has no solutions in $\mathcal{M}_2(\mathbb{C})$.

4.43. If $\text{Tr}(A) = 0$ we have based on the Cayley–Hamilton Theorem that $A^2 = O_2$. Thus,

$$2^A = e^{(\ln 2)A} = \sum_{n=0}^{\infty} \frac{((\ln 2)A)^n}{n!} = I_2 + (\ln 2)A.$$

If $\text{Tr}(A) \neq 0$, the Cayley–Hamilton Theorem implies that $A^2 = \text{Tr}(A)A$ which in turn implies that $A^n = \text{Tr}^{n-1}(A)A$, for all $n \in \mathbb{N}$. It follows that

$$2^A = e^{(\ln 2)A} = \sum_{n=0}^{\infty} \frac{((\ln 2)A)^n}{n!} = I_2 + \frac{1}{\text{Tr}(A)} \sum_{n=1}^{\infty} \frac{\text{Tr}^n(A) \ln^n 2}{n!} A = I_2 + \frac{2^{\text{Tr}(A)} - 1}{\text{Tr}(A)} A.$$

4.45. There are no such matrices. If λ_1, λ_2 are the eigenvalues of A, then $e^{\lambda_1}, e^{\lambda_2}$ are the eigenvalues of e^A. We have

$$e^{a+d} = e^{\text{Tr}(A)} = e^{\lambda_1 + \lambda_2} = e^{\lambda_1} e^{\lambda_2} = \det\left(e^A\right) = e^a e^d - e^b e^c = e^{a+d} - e^{b+c}.$$

This implies that $e^{b+c} = 0$, which is impossible.

4.46. (a) Let $A = (a_{ij})_{i,j=1,2}$, $B = (b_{ij})_{i,j=1,2}$ and let $C = (c_{ij})_{i,j=1,2}$, where $C = AB$. We have $(c_{ij})' = (a_{i1}b_{1j} + a_{i2}b_{2j})' = a'_{i1}b_{1j} + a_{i1}b'_{1j} + a'_{i2}b_{2j} + a_{i2}b'_{2j}$, for all $i, j = 1, 2$, which implies that $(AB)' = A'B + AB'$.

(b) We have that $A^{-1}A = I_2$ and it follows, based on part (a), that $(A^{-1}A)' = O_2$. Therefore $(A^{-1})'A + A^{-1}A' = O_2 \Rightarrow (A^{-1})'A = -A^{-1}A' \Rightarrow (A^{-1})' = -A^{-1}A'A^{-1}$.

We have, based on part (a) and the first formula in part (b), that

$$(A^{-n})' = (A^{-(n-1)}A^{-1})'$$

$$= (A^{-(n-1)})'A^{-1} + A^{-(n-1)}(A^{-1})'$$

$$= (A^{-(n-1)})'A^{-1} - A^{-n}A'A^{-1}.$$

Let $Y_n = A^{-n}$. The previous formula implies that $Y_n' = Y_{n-1}'A^{-1} - A^{-n}A'A^{-1} \Rightarrow$
$Y_n'A^n = Y_{n-1}'A^{n-1} - A^{-n}A'A^{n-1}$. It follows that

$$Y_n'A^n = -\left(A^{-1}A' + A^{-2}A'A + A^{-3}A'A^2 + \cdots + A^{-n}A'A^{n-1}\right)$$

and this implies

$$\left(A^{-n}\right)' = -\left(A^{-1}A' + A^{-2}A'A + A^{-3}A'A^2 + \cdots + A^{-n}A'A^{n-1}\right)A^{-n}.$$

4.47. First we observe that $A^2(t) = A(t)$ which implies that $A^n(t) = A(t)$, for all $n \geq 1$. It follows that

$$e^{A(t)} = I_2 + \sum_{n=1}^{\infty} \frac{A^n(t)}{n!} = I_2 + \left(\sum_{n=1}^{\infty} \frac{1}{n!}\right) A(t) = I_2 + (e-1)A(t) = \begin{pmatrix} e & (e-1)t \\ 0 & 1 \end{pmatrix},$$

and this implies

$$\left(e^{A(t)}\right)' = \begin{pmatrix} 0 & e-1 \\ 0 & 0 \end{pmatrix}.$$

On the other hand,

$$e^{A(t)}A'(t) = \begin{pmatrix} e & (e-1)t \\ 0 & 1 \end{pmatrix}\begin{pmatrix} 0 & 1 \\ 0 & 0 \end{pmatrix} = \begin{pmatrix} 0 & e \\ 0 & 0 \end{pmatrix} \neq \left(e^{A(t)}\right)'$$

and

$$A'(t)e^{A(t)} = \begin{pmatrix} 0 & 1 \\ 0 & 0 \end{pmatrix}\begin{pmatrix} e & (e-1)t \\ 0 & 1 \end{pmatrix} = \begin{pmatrix} 0 & 1 \\ 0 & 0 \end{pmatrix} \neq \left(e^{A(t)}\right)'.$$

4.48. The solution of the system is

$$\begin{cases} x_1(t) = \left(2e^{-t} + e^{5t}\right)c_1 + \left(2e^{-t} - 2e^{5t}\right)c_2 \\ x_2(t) = \left(e^{-t} - e^{5t}\right)c_1 + \left(e^{-t} + 2e^{5t}\right)c_2, \end{cases}$$

with $c_1, c_2 \in \mathbb{R}$.

4.49. (a) The characteristic equation of A is $(x-1)^2 + a = 0$ and it follows that $(A - I_2)^2 + aI_2 = O_2$. Let $B = A - I_2$ and we have that $B^2 = -aI_2$. This implies that $B^{2k} = (-1)^k a^k I_2$ and $B^{2k-1} = (-1)^{k-1}a^{k-1}B$, for all $k \geq 1$. We have $e^{At} = e^{I_2 t + Bt} = e^t e^{Bt}$. On the other hand,

$$e^{Bt} = \sum_{k=0}^{\infty} \frac{(Bt)^{2k}}{(2k)!} + \sum_{k=1}^{\infty} \frac{(Bt)^{2k-1}}{(2k-1)!}$$

$$= \sum_{k=0}^{\infty} (-1)^k \frac{(t\sqrt{a})^{2k}}{(2k)!} I_2 + \frac{1}{\sqrt{a}} \sum_{k=1}^{\infty} (-1)^{k-1} \frac{(t\sqrt{a})^{2k-1}}{(2k-1)!} B$$

$$= \cos(t\sqrt{a})I_2 + \frac{\sin(t\sqrt{a})}{\sqrt{a}} B$$

$$= \begin{pmatrix} \cos(t\sqrt{a}) & \frac{\sin(t\sqrt{a})}{\sqrt{a}} \\ -\sqrt{a}\sin(t\sqrt{a}) & \cos(t\sqrt{a}) \end{pmatrix}.$$

This implies that

$$e^{At} = e^t \begin{pmatrix} \cos(t\sqrt{a}) & \frac{\sin(t\sqrt{a})}{\sqrt{a}} \\ -\sqrt{a}\sin(t\sqrt{a}) & \cos(t\sqrt{a}) \end{pmatrix}.$$

(b) The system can be written as $X' = AX$ and we have that $X(t) = e^{At}C$, where C is a constant vector. This implies that

$$\begin{cases} x(t) = c_1 e^t \cos(t\sqrt{a}) + \frac{c_2}{\sqrt{a}} e^t \sin(t\sqrt{a}) \\ y(t) = -c_1\sqrt{a}e^t \sin(t\sqrt{a}) + c_2 e^t \cos(t\sqrt{a}), \end{cases}$$

where $c_1, c_2 \in \mathbb{R}$.

4.52. (a) $tX'(t) = AX(t) \Rightarrow X'(t) - \frac{A}{t}X(t) = 0$. We multiply this equation by $t^{-A} = e^{-(\ln t)A}$ and we get that $\left(e^{-(\ln t)A}X(t)\right)' = 0 \Leftrightarrow \left(t^{-A}X(t)\right)' = 0 \Rightarrow t^{-A}X(t) = C \Rightarrow X(t) = t^A C$, where C is a constant vector.

(b) We divide by t and we multiply the system by $t^{-A} = e^{-(\ln t)A}$ and we have

$$\left(t^{-A}X(t)\right)' = \frac{1}{t}t^{-A}F(t) \quad \Leftrightarrow \quad \left(t^{-A}X(t)\right)' = t^{-(A+I_2)}F(t)$$

which implies

$$t^{-A}X(t) = \int_{t_0}^{t} u^{-(A+I_2)}F(u)du.$$

Thus

$$X(t) = t^A \int_{t_0}^{t} u^{-(A+I_2)}F(u)du.$$

4.53. We have, based on problem **4.52**, that

$$\begin{cases} x(t) = \frac{7}{4}t^2 - \frac{3}{4t^2} - \frac{1}{2} \\ y(t) = \frac{7}{4}t^2 + \frac{1}{4t^2} - \frac{1}{2}. \end{cases}$$

4.54. The eigenvalues of the matrix of the system A are the solutions of the characteristic equation $\lambda^2 - (a+d)\lambda + ad - bc = 0$ or $\lambda^2 - \text{Tr}(A)\lambda + \det A = 0$. Using remarks 4.7 and 4.8 we have the following cases:

- If $\lambda_1, \lambda_2 \in \mathbb{R}$, $\lambda_1, \lambda_2 < 0$, i.e., $\text{Tr}(A) < 0$, $\Delta \geq 0$ and $\det A > 0$, then the zero solution is asymptotically stable;
- If $\lambda_1, \lambda_2 \in \mathbb{C} \setminus \mathbb{R}$, $\lambda_{1,2} = r \pm is$, $r \in \mathbb{R}$, $s \in \mathbb{R}^*$ and $r < 0$, i.e., $\text{Tr}(A) < 0$, $\Delta < 0$ and $\det A > 0$, the zero solution is asymptotically stable. It follows, by combining this case with the previous case, that the system is asymptotically stable if and only if $\text{Tr}(A) < 0$ and $\det A > 0$;
- Using parts (c) and (d) of remark 4.8 we obtain that the zero solution is unstable if at least a solution of the characteristic equation has a positive real part or both solutions are equal to 0 (but $A \neq O_2$). We have the following possibilities: $\det A < 0 \Leftrightarrow \lambda_1, \lambda_2 \in \mathbb{R}$, $\lambda_1 < 0 < \lambda_2$ or $\det A > 0$ and $\text{Tr}(A) > 0 \Leftrightarrow \lambda_1, \lambda_2 \in \mathbb{C} \setminus \mathbb{R}$ and $\lambda_1 + \lambda_2 > 0$ or $\text{Tr}(A) = \det A = 0 \Leftrightarrow \lambda_1 = \lambda_2 = 0$. Therefore the system is unstable if and only if $\det A < 0$ or $\text{Tr}(A) > 0$ or $\text{Tr}(A) = \det A = 0$;
- In all the other cases $\text{Tr}(A) = 0$ and $\det A > 0$ or $\text{Tr}(A) < 0$ and $\det A = 0$ the system is stable but not asymptotically stable.

In conclusion we have that:

- the system is asymptotically stable if $\text{Tr}(A) < 0$ and $\det A > 0$;
- the system is stable if $\text{Tr}(A) = 0$ and $\det A > 0$ **or** $\text{Tr}(A) < 0$ and $\det A = 0$;
- the system is unstable if $\text{Tr}(A) > 0$ **or** $\det A < 0$ **or** $\text{Tr}(A) = \det A = 0$.

4.55. $\text{Tr}(A) = -(a^2 + 1)$ and $\det A = a^2 - a$. We have, based on problem **4.54**, that the system is:

- asymptotically stable for $a \in (-\infty, 0) \cup (1, \infty)$;
- stable for $a \in \{0, 1\}$;
- unstable for $a \in (0, 1)$.

4.56. $\text{Tr}(A) = -a$ and $\det A = 1 - a$. We have, based on problem **4.54**, that the system is:

- asymptotically stable for $a \in (0, 1)$;
- stable for $a \in \{0, 1\}$;
- unstable for $a \in (-\infty, 0) \cup (1, \infty)$.

4.57. $\text{Tr}(A) = 0$ and $\det A = -bc$. It follows, based on problem **4.54**, that the system is stable for $bc < 0$ and unstable if $bc \geq 0$.

4.58. $\text{Tr}(A) = -2$ and $\det A = 1 - ab$. We have, based on problem **4.54**, that

- the system is asymptotically stable if $ab < 1$;
- the system is stable if $ab = 1$;
- the system is unstable if $ab > 1$.

4.59. $\text{Tr}(A) = 2a$ and $\det A = a^2 - b$. We have, based on problem **4.54**, that

- the system is asymptotically stable if $a < 0$ and $a^2 - b > 0$;
- the system is stable if $a = 0$ and $b < 0$ **or** $a < 0$ and $a^2 = b$;
- the system is unstable if $a > 0$ **or** $a^2 - b < 0$ **or** $a = b = 0$.

4.60. Observe that $A = -I_2 + B$, where $B = \begin{pmatrix} 0 & x \\ 0 & 0 \end{pmatrix}$ and note that $B^2 = O_2$. We have, based on the Binomial Theorem, that

$$A^n = (-1)^n I_2 + n(-1)^{n-1} B = \begin{pmatrix} (-1)^n & nx(-1)^{n-1} \\ 0 & (-1)^n \end{pmatrix}.$$

It follows that

$$\sum_{n=1}^{\infty} \frac{A^n}{n^2} = \begin{pmatrix} \sum_{n=1}^{\infty} \frac{(-1)^n}{n^2} & x \sum_{n=1}^{\infty} \frac{(-1)^{n-1}}{n} \\ 0 & \sum_{n=1}^{\infty} \frac{(-1)^n}{n^2} \end{pmatrix} = \begin{pmatrix} -\frac{\pi^2}{12} & x \ln 2 \\ 0 & -\frac{\pi^2}{12} \end{pmatrix},$$

since $\sum_{n=1}^{\infty} \frac{(-1)^n}{n^2} = -\frac{\pi^2}{12}$ and $\sum_{n=1}^{\infty} \frac{(-1)^{n-1}}{n} = \ln 2$.

4.61. Let $\epsilon = \frac{-1+i\sqrt{3}}{2}$. We have

$$\sum_{n=0}^{\infty} \frac{A^{3n}}{(3n)!} = \frac{1}{3} \left(e^A + e^{\epsilon A} + e^{\epsilon^2 A} \right).$$

A calculation shows that $\epsilon^2 = \frac{-1-i\sqrt{3}}{2}$ and this implies that

$$e^{\epsilon A} + e^{\epsilon^2 A} = e^{-\frac{A}{2} + i\frac{\sqrt{3}A}{2}} + e^{-\frac{A}{2} - i\frac{\sqrt{3}A}{2}} = 2e^{-\frac{A}{2}} \cos \frac{\sqrt{3}A}{2}.$$

4.62. Part (a) of the problem can be proved by mathematical induction and part (b) follows from part (a) by passing to the limit as $n \to \infty$.

4.63. We have

$$f(A) = \sum_{n=1}^{\infty} F_n A^{n-1} = I_2 + \sum_{n=2}^{\infty} F_n A^{n-1} = I_2 + \sum_{m=1}^{\infty} F_{m+1} A^m$$

$$= I_2 + \sum_{m=1}^{\infty} (F_m + F_{m-1}) A^m = I_2 + \sum_{m=1}^{\infty} F_m A^m + \sum_{m=1}^{\infty} F_{m-1} A^m$$

$$= I_2 + Af(A) + \sum_{m=2}^{\infty} F_{m-1} A^m = I_2 + Af(A) + \sum_{k=1}^{\infty} F_k A^{k+1}$$

$$= I_2 + Af(A) + A^2 f(A),$$

and it follows that $f(A)(I_2 - A - A^2) = I_2$.

4.64. Recall the generating function for the nth harmonic numbers is given by

$$\sum_{n=1}^{\infty} H_n x^n = -\frac{\ln(1-x)}{1-x}, \quad -1 < x < 1, \tag{4.6}$$

and it follows, by differentiation and integration, that

$$\sum_{n=1}^{\infty} n H_n x^{n-1} = \frac{1 - \ln(1-x)}{(1-x)^2}, \quad -1 < x < 1 \tag{4.7}$$

and

$$\sum_{n=1}^{\infty} \frac{H_n}{n+1} x^{n+1} = \frac{\ln^2(1-x)}{2}, \quad -1 \le x < 1. \tag{4.8}$$

We have $A = \alpha I_2 + B$, $B^2 = O_2$, and it follows, by the Binomial Theorem, that $A^n = \alpha^n I_2 + n\alpha^{n-1} B$, for all $n \ge 1$.

(a) We have, based on (4.6) and (4.7), that

$$\sum_{n=1}^{\infty} H_n A^n = \sum_{n=1}^{\infty} H_n \alpha^n I_2 + \sum_{n=1}^{\infty} n H_n \alpha^{n-1} B = -\frac{\ln(1-\alpha)}{1-\alpha} I_2 + \frac{1 - \ln(1-\alpha)}{(1-\alpha)^2} B.$$

(b) If $\alpha = 0$, then $A = B$ with $B^2 = O_2$. It follows that $\sum_{n=1}^{\infty} \frac{H_n}{n+1} A^n = \frac{B}{2}$.

Let $\alpha \neq 0$. We have, based on formulae (4.6), (4.7), and (4.8), that

$$\sum_{n=1}^{\infty} \frac{H_n}{n+1} A^n = \sum_{n=1}^{\infty} \frac{H_n}{n+1} \alpha^n I_2 + \sum_{n=1}^{\infty} \frac{nH_n}{n+1} \alpha^{n-1} B$$

$$= \frac{\ln^2(1-\alpha)}{2\alpha} I_2 + \left(\sum_{n=1}^{\infty} H_n \alpha^{n-1} - \sum_{n=1}^{\infty} \frac{H_n}{n+1} \alpha^{n-1} \right) B$$

$$= \frac{\ln^2(1-\alpha)}{2\alpha} I_2 - \left(\frac{\ln(1-\alpha)}{\alpha(1-\alpha)} + \frac{\ln^2(1-\alpha)}{2\alpha^2} \right) B.$$

4.65. Use the Jordan canonical form of A combined to the power series formulae

(a) $\displaystyle \sum_{n=1}^{\infty} H_n z^n = -\frac{\ln(1-z)}{1-z}$, for $|z| < 1$;

(b) $\displaystyle \sum_{n=1}^{\infty} nH_n z^n = \frac{z(1 - \ln(1-z))}{(1-z)^2}$, for $|z| < 1$.

4.66. We use the following result [32] whose proof can be found in Appendix A.

A power series with the tail of $\ln \frac{1}{2}$.

Let $x \in \mathbb{R}$. The following equality holds:

$$\sum_{n=1}^{\infty} \left(\ln \frac{1}{2} + 1 - \frac{1}{2} + \cdots + \frac{(-1)^{n-1}}{n} \right) x^n = \begin{cases} \ln 2 - \frac{1}{2} & \text{if} \quad x = 1 \\ \frac{\ln(1+x) - x\ln 2}{1-x} & \text{if} \quad x \in (-1, 1). \end{cases}$$

First we consider the case when $\lambda_1 = 0$ and $0 < |\lambda_2| < 1$. Let $t = \lambda_2 = \text{Tr}(A)$. The Cayley–Hamilton Theorem implies that $A^2 - tA = O_2 \Rightarrow A^n = t^{n-1}A$, for all $n \geq 1$.

Let $a_n = \ln \frac{1}{2} + 1 - \frac{1}{2} + \cdots + \frac{(-1)^{n-1}}{n}$. We have

$$\sum_{n=1}^{\infty} a_n A^n = \frac{1}{t} \sum_{n=1}^{\infty} a_n t^n A = \frac{1}{1-t} \left(\frac{\ln(1+t)}{t} - \ln 2 \right) A.$$

Now we consider the case when $0 < |\lambda_1|, |\lambda_2| < 1$. Let $J_A = \begin{pmatrix} \lambda_1 & 0 \\ 0 & \lambda_2 \end{pmatrix}$. We have $A = PJ_A P^{-1} \Rightarrow A^n = PJ_A^n P^{-1}$, with $J_A^n = \begin{pmatrix} \lambda_1^n & 0 \\ 0 & \lambda_2^n \end{pmatrix}$. It follows that

$$\sum_{n=1}^{\infty} a_n A^n = P\left(\sum_{n=1}^{\infty} a_n J_A^n\right) P^{-1} = P\begin{pmatrix} \sum_{n=1}^{\infty} a_n \lambda_1^n & 0 \\ 0 & \sum_{n=1}^{\infty} a_n \lambda_2^n \end{pmatrix} P^{-1}$$

$$= P\begin{pmatrix} \frac{\ln(1+\lambda_1)-\lambda_1 \ln 2}{1-\lambda_1} & 0 \\ 0 & \frac{\ln(1+\lambda_2)-\lambda_2 \ln 2}{1-\lambda_2} \end{pmatrix} P^{-1}$$

$$= (I_2 - A)^{-1} (\ln(I_2 + A) - (\ln 2)A).$$

If $J_A = \begin{pmatrix} \lambda & 1 \\ 0 & \lambda \end{pmatrix}$, we have $A^n = PJ_A^n P^{-1}$, with $J_A^n = \begin{pmatrix} \lambda^n & n\lambda^{n-1} \\ 0 & \lambda^n \end{pmatrix}$.

Let $f(x) = \sum_{n=1}^{\infty} a_n x^n = \frac{\ln(1+x)-x\ln 2}{1-x}$, $x \in (-1,1)$. We have

$$\sum_{n=1}^{\infty} a_n A^n = P\left(\sum_{n=1}^{\infty} a_n J_A^n\right) P^{-1} = P\begin{pmatrix} f(\lambda) & f'(\lambda) \\ 0 & f(\lambda) \end{pmatrix} P^{-1} = f(A).$$

4.67. (a) Let $f_n(x) = \ln(1-x) + x + \frac{x^2}{2} + \cdots + \frac{x^n}{n}$, $x \in (-1,1)$. One can prove that

- $\lim_{n\to\infty} nf_n(x) = \lim_{n\to\infty} n\left(\ln(1-x) + x + \frac{x^2}{2} + \cdots + \frac{x^n}{n}\right) = 0$;

- $\lim_{n\to\infty} nf'_n(x) = \lim_{n\to\infty} n\left(\frac{-1}{1-x} + x + x^2 + \cdots + x^{n-1}\right) = 0$.

Let λ_1, λ_2 be the eigenvalues of A and let $A = PJ_A P^{-1}$. If

$$J_A = \begin{pmatrix} \lambda_1 & 0 \\ 0 & \lambda_2 \end{pmatrix} \quad \Rightarrow \quad nf_n(A) = P\begin{pmatrix} nf_n(\lambda_1) & 0 \\ 0 & nf_n(\lambda_2) \end{pmatrix} P^{-1},$$

and it follows, based on the first limit above, that $\lim_{n\to\infty} nf_n(A) = O_2$.

If

$$J_A = \begin{pmatrix} \lambda & 1 \\ 0 & \lambda \end{pmatrix} \quad \Rightarrow \quad nf_n(A) = P\begin{pmatrix} nf_n(\lambda) & nf'_n(\lambda) \\ 0 & nf_n(\lambda) \end{pmatrix} P^{-1},$$

and we have, based on the previous limits, that $\lim_{n\to\infty} nf_n(A) = O_2$.

(b) We calculate the series by Abel's summation formula, part (b) of problem **4.62**, with $a_n = 1$ and $B_n = \ln(I_2 - A) + A + \frac{A^2}{2} + \cdots + \frac{A^n}{n}$. We have, based on part (a), that

$$\sum_{n=1}^{\infty}\left(\ln(I_2 - A) + A + \frac{A^2}{2} + \cdots + \frac{A^n}{n}\right)$$

$$= \lim_{n\to\infty} n\left(\ln(I_2 - A) + A + \frac{A^2}{2} + \cdots + \frac{A^{n+1}}{n+1}\right) - \sum_{n=1}^{\infty}\frac{n}{n+1}A^{n+1}$$

$$= \sum_{n=1}^{\infty}\left(\frac{A^{n+1}}{n+1} - A^{n+1}\right)$$

$$= \sum_{m=1}^{\infty}\left(\frac{A^m}{m} - A^m\right)$$

$$= -\ln(I_2 - A) - A(I_2 - A)^{-1}.$$

4.68. (a) Use mathematical induction.

(b) We calculate the series using part (b) of problem **4.62**, with $a_n = H_n$ and $B_n = \ln(I_2 - A) + A + \frac{A^2}{2} + \cdots + \frac{A^n}{n}$, and we have

$$\sum_{n=1}^{\infty} H_n\left(\ln(I_2 - A) + A + \frac{A^2}{2} + \cdots + \frac{A^n}{n}\right)$$

$$= \lim_{n\to\infty}((n+1)H_n - n)\left(\ln(I_2 - A) + A + \frac{A^2}{2} + \cdots + \frac{A^{n+1}}{n+1}\right)$$

$$- \sum_{n=1}^{\infty}((n+1)H_n - n)\frac{A^{n+1}}{n+1}$$

$$= -A\sum_{n=1}^{\infty} H_n A^n + \sum_{n=1}^{\infty} A^{n+1} - \sum_{n=1}^{\infty}\frac{A^{n+1}}{n+1}$$

$$= A(I_2 - A)^{-1}\ln(I_2 - A) + A^2(I_2 - A)^{-1} + \ln(I_2 - A) + A$$

$$= (A + \ln(I_2 - A))(I_2 - A)^{-1}.$$

We used that

$$\lim_{n\to\infty}((n+1)H_n - n)\left(\ln(I_2 - A) + A + \frac{A^2}{2} + \cdots + \frac{A^{n+1}}{n+1}\right) = 0$$

combined to the first formula in problem **4.65**.

4.69. (a) See the idea in the solution of part (a) of problem **4.67**.

(b) Use Abel's summation formula, part (b) of problem **4.62**, with $a_n = 1$ and $B_n = \arctan A - A + \frac{A^3}{3} + \cdots + (-1)^n\frac{A^{2n-1}}{2n-1}$.

4.70. (a) First, one can prove that if $A \in \mathcal{M}_2(\mathbb{C})$, then

$$\lim_{n \to \infty} n \left(e^A - I_2 - \frac{A}{1!} - \frac{A^2}{2!} - \cdots - \frac{A^n}{n!} \right) = O_2.$$

We calculate the series by Abel's summation formula, part (b) of problem **4.62**, with $a_n = 1$ and $B_n = e^A - I_2 - \frac{A}{1!} - \frac{A^2}{2!} - \cdots - \frac{A^n}{n!}$, and we have

$$\sum_{n=1}^{\infty} \left(e^A - I_2 - \frac{A}{1!} - \frac{A^2}{2!} - \cdots - \frac{A^n}{n!} \right) = \lim_{n \to \infty} n \left(e^A - I_2 - \frac{A}{1!} - \frac{A^2}{2!} - \cdots - \frac{A^{n+1}}{(n+1)!} \right)$$

$$+ \sum_{n=1}^{\infty} n \frac{A^{n+1}}{(n+1)!}$$

$$= \sum_{n=1}^{\infty} \frac{A^{n+1}}{n!} - \sum_{n=1}^{\infty} \frac{A^{n+1}}{(n+1)!}$$

$$= A e^A - e^A + I_2.$$

(b) If $A \in \mathcal{M}_2(\mathbb{C})$, then (prove it!)

$$\lim_{n \to \infty} n^2 \left(e^A - I_2 - \frac{A}{1!} - \frac{A^2}{2!} - \cdots - \frac{A^n}{n!} \right) = O_2.$$

We calculate the series by Abel's summation formula, part (b) of problem **4.62**, with $a_n = n$ and $B_n = e^A - I_2 - \frac{A}{1!} - \frac{A^2}{2!} - \cdots - \frac{A^n}{n!}$, and we have

$$\sum_{n=1}^{\infty} n \left(e^A - I_2 - \frac{A}{1!} - \frac{A^2}{2!} - \cdots - \frac{A^n}{n!} \right)$$

$$= \lim_{n \to \infty} \frac{n(n+1)}{2} \left(e^A - I_2 - \frac{A}{1!} - \frac{A^2}{2!} - \cdots - \frac{A^{n+1}}{(n+1)!} \right)$$

$$+ \sum_{n=1}^{\infty} \frac{n(n+1)}{2} \cdot \frac{A^{n+1}}{(n+1)!}$$

$$= \frac{A^2}{2} \sum_{n=1}^{\infty} \frac{A^{n-1}}{(n-1)!}$$

$$= \frac{A^2}{2} e^A.$$

4.73. Apply Abel's summation formula with $a_n = 1$ and $B_n = f(A) - f(0)I_2 - \frac{f'(0)}{1!}A - \cdots - \frac{f^{(n)}(0)}{n!}A^n$, for the first series and the same formula with $a_n = n$ and $B_n = f(A) - f(0)I_2 - \frac{f'(0)}{1!}A - \cdots - \frac{f^{(n)}(0)}{n!}A^n$, for the second series.

4.74. (a) First one can show that if $x, y \in \mathbb{C}$, then

$$\lim_{n \to \infty} x^n \left(e^y - 1 - \frac{y}{1!} - \frac{y^2}{2!} - \cdots - \frac{y^n}{n!} \right) = 0.$$

When $x = 1$ this is part (a) of problem **4.70**, so we solve the problem for the case when $x \neq 1$.

We calculate the series by Abel's summation formula, part (b) of problem **4.62**, with $a_n = x^n$ and $B_n = e^A - I_2 - \frac{A}{1!} - \frac{A^2}{2!} - \cdots - \frac{A^n}{n!}$. We have

$$\sum_{n=1}^{\infty} x^n \left(e^A - I_2 - \frac{A}{1!} - \frac{A^2}{2!} - \cdots - \frac{A^n}{n!} \right)$$

$$= \lim_{n \to \infty} x \frac{1 - x^n}{1 - x} \left(e^A - I_2 - \frac{A}{1!} - \frac{A^2}{2!} - \cdots - \frac{A^{n+1}}{(n+1)!} \right)$$

$$+ \sum_{n=1}^{\infty} x \frac{1 - x^n}{1 - x} \cdot \frac{A^{n+1}}{(n+1)!}$$

$$= \frac{x}{1 - x} \sum_{n=1}^{\infty} \frac{A^{n+1}}{(n+1)!} - \frac{1}{1 - x} \sum_{n=1}^{\infty} \frac{(Ax)^{n+1}}{(n+1)!}$$

$$= \frac{x}{1 - x} \left(e^A - I_2 - A \right) - \frac{1}{1 - x} \left(e^{Ax} - I_2 - Ax \right)$$

$$= \frac{x e^A - e^{Ax}}{1 - x} + I_2.$$

(b) This part follows from part (a) when $x = -1$.

4.75. Prove, using Abel's summation formula (see Appendix A), that if $z \in \mathbb{C}$ the following equality holds

$$\sum_{n=1}^{\infty} \left(e - 1 - \frac{1}{1!} - \frac{1}{2!} - \cdots - \frac{1}{n!} \right) z^n = \begin{cases} \dfrac{e^z - ez}{z - 1} + 1 & \text{if } z \neq 1 \\ 1 & \text{if } z = 1. \end{cases}$$

It follows, by differentiation, that

$$\sum_{n=1}^{\infty} n \left(e - 1 - \frac{1}{1!} - \frac{1}{2!} - \cdots - \frac{1}{n!} \right) z^{n-1} = \begin{cases} \dfrac{z e^z - 2 e^z + e}{(z - 1)^2} & \text{if } z \neq 1 \\ \dfrac{e}{2} & \text{if } z = 1. \end{cases}$$

Calculate the matrix series using the previous two formulae combined to Theorem 3.1.

4.76. Use Abel's summation formula.

4.77. We need the following two results (see Appendix A).

Let $k \geq 3$ be an integer and let $x \in [-1, 1]$. The following formula holds

$$\sum_{n=1}^{\infty} \left(\zeta(k) - \frac{1}{1^k} - \frac{1}{2^k} - \cdots - \frac{1}{n^k} \right) x^n = \begin{cases} \dfrac{x\zeta(k) - \mathrm{Li}_k(x)}{1-x} & \text{if } x \in [-1, 1) \\ \zeta(k-1) - \zeta(k) & \text{if } x = 1, \end{cases}$$

where Li_k denotes the *polylogarithm* function.

Let $k \geq 3$ be an integer and let $x \in [-1, 1)$. The following formula holds

$$\sum_{n=1}^{\infty} n \left(\zeta(k) - \frac{1}{1^k} - \frac{1}{2^k} - \cdots - \frac{1}{n^k} \right) x^{n-1} = \frac{\zeta(k) - \frac{1-x}{x}\mathrm{Li}_{k-1}(x) - \mathrm{Li}_k(x)}{(1-x)^2},$$

where Li_k denotes the *polylogarithm* function.

(a) Use that $A^n = \begin{pmatrix} (-1)^n & n(-1)^{n-1}x \\ 0 & (-1)^n \end{pmatrix}$ and the preceding two formulae with $k = 3$ and $x = -1$

(b) Express A^n in terms of the eigenvalues of A using Theorem 4.7 and apply the preceding two formulae.

4.78. (a) Observe that $\sum\limits_{n=1}^{\infty} \frac{n}{n^{aI_2}} = \zeta((a-1)I_2)$ and use Corollary 4.8.

(b) We have

$$\sum_{n=1}^{\infty} \frac{n}{n^{\begin{pmatrix} \alpha & \beta \\ 0 & \alpha \end{pmatrix}}} = \zeta \begin{pmatrix} \alpha - 1 & \beta \\ 0 & \alpha - 1 \end{pmatrix}$$

and the problem follows based on part (b) of Corollary 4.9.

4.79. Observe that $\sum\limits_{n=1}^{\infty} \frac{n}{n^A} = \zeta(A - I_2)$, then use Theorem 4.14.

4.80. (a) Use Abel's summation formula with $a_n = 1$ and $B_n = \zeta(A) - \frac{1}{1^A} - \frac{1}{2^A} - \cdots - \frac{1}{n^A}$.

(b) Use Abel's summation formula with $a_n = n$ and $B_n = \zeta(A) - \frac{1}{1^A} - \frac{1}{2^A} - \cdots - \frac{1}{n^A}$.

4.81. $A^n = \begin{pmatrix} a^n & na^{n-1}b \\ 0 & a^n \end{pmatrix}$. It follows that

$$\int_0^1 \frac{\ln(I_2 - Ax)}{x} dx = -\int_0^1 \left(\sum_{n=1}^{\infty} \frac{x^{n-1}}{n} A^n \right) dx$$

$$= -\sum_{n=1}^{\infty} \frac{A^n}{n^2}$$

$$= -\sum_{n=1}^{\infty} \frac{1}{n^2} \begin{pmatrix} a^n & na^{n-1}b \\ 0 & a^n \end{pmatrix}$$

$$= \begin{pmatrix} -\mathrm{Li}_2(a) & \frac{b\ln(1-a)}{a} \\ 0 & -\mathrm{Li}_2(a) \end{pmatrix}.$$

4.82. If $\lambda_1 = 0$ and $\lambda_2 = 1$ we have that $A^2 = A$ and this implies that $A^n = A$, for all $n \geq 1$. Therefore

$$\int_0^1 \frac{\ln(I_2 - Ax)}{x} dx = -\sum_{n=1}^{\infty} \frac{A^n}{n^2} = -A \sum_{n=1}^{\infty} \frac{1}{n^2} = -A\zeta(2).$$

If $\lambda_1 = \lambda_2 = 0$, then $A^2 = O_2$ which implies that $A^n = O_2$, for all $n \geq 2$. Thus,

$$\int_0^1 \frac{\ln(I_2 - Ax)}{x} dx = -\sum_{n=1}^{\infty} \frac{A^n}{n^2} = -A.$$

4.83. Use Theorem 4.7 with $f(x) = \ln(1 - x)$ and A replaced by xA. Observe the eigenvalues of xA are $x\lambda_1$ and $x\lambda_2$.

4.84. We use the following series formula.

> **The quadratic logarithmic function.**
>   The following equality holds
>
> $$\ln^2(I_2 - A) = 2 \sum_{n=1}^{\infty} \frac{H_n}{n+1} A^{n+1},$$
>
> where H_n denotes the nth harmonic number and $A \in \mathcal{M}_2(\mathbb{C})$ with $\rho(A) < 1$.

We have

$$\int_0^1 \frac{\ln^2(I_2 - Ax)}{x} dx = \int_0^1 \left(2 \sum_{n=1}^{\infty} \frac{H_n}{n+1} x^n A^{n+1} \right) dx = 2 \sum_{n=1}^{\infty} \frac{H_n}{(n+1)^2} A^{n+1}.$$

(a) If $\lambda_1 = \lambda_2 = 0$, then $A^2 = O_2 \Rightarrow A^n = O_2$, for all $n \geq 2$. We have, based on the previous formula, that

$$\int_0^1 \frac{\ln^2(I_2 - Ax)}{x} dx = O_2.$$

(b) If $\lambda_1 = 0$ and $\lambda_2 = 1$, then $A^2 = A \Rightarrow A^n = A$, for all $n \geq 1$. It follows, based on the previous formula, that

$$\int_0^1 \frac{\ln^2(I_2 - Ax)}{x} dx = 2 \sum_{n=1}^{\infty} \frac{H_n}{(n+1)^2} A = 2\zeta(3)A.$$

To prove the last equality we note that

$$\sum_{n=1}^{\infty} \frac{H_n}{(n+1)^2} = \sum_{n=1}^{\infty} \frac{H_{n+1} - \frac{1}{n+1}}{(n+1)^2} = \sum_{n=1}^{\infty} \frac{H_n}{n^2} - \sum_{n=1}^{\infty} \frac{1}{n^3} = \zeta(3),$$

since $\sum_{n=1}^{\infty} \frac{H_n}{n^2} = 2\zeta(3)$ (see [22, Problem 3.55, p. 148]).

4.85. We have

$$\int_0^1 \ln(I_2 - Ax) dx = -\int_0^1 \left(\sum_{n=1}^{\infty} \frac{x^n A^n}{n} \right) dx = -\sum_{n=1}^{\infty} \frac{A^n}{n(n+1)}.$$

If $\lambda_1 = \lambda_2 = 0$, then $A^2 = O_2 \Rightarrow A^n = O_2$, for all $n \geq 2$. We have, based on the previous formula, that

$$\int_0^1 \ln(I_2 - Ax) dx = -\frac{A}{2}.$$

If $\lambda_1 = 0$ and $0 < |\lambda_2| < 1$, then $A^2 = tA$, where $t = \text{Tr}(A)$. This implies that $A^n = t^{n-1}A$, for all $n \geq 1$. It follows that

$$\int_0^1 \ln(I_2 - Ax) dx = -\sum_{n=1}^{\infty} \frac{t^{n-1}}{n(n+1)} A$$

$$= -\left(\frac{1}{t} \sum_{n=1}^{\infty} \frac{t^n}{n} - \frac{1}{t^2} \sum_{n=1}^{\infty} \frac{t^{n+1}}{n+1} \right) A$$

$$= -\left(\frac{(1-t)\ln(1-t)}{t^2} + \frac{1}{t} \right) A.$$

If $0 < |\lambda_1|, |\lambda_2| < 1$, then

$$\int_0^1 \ln(I_2 - Ax) dx = -\sum_{n=1}^{\infty} \frac{A^n}{n(n+1)}$$

$$= -\sum_{n=1}^{\infty} \left(\frac{A^n}{n} - \frac{A^n}{n+1} \right)$$

$$= \ln(I_2 - A) + A^{-1}(-\ln(I_2 - A) - A)$$

$$= (I_2 - A^{-1}) \ln(I_2 - A) - I_2.$$

4.86. We have

$$\int_0^1 e^{Ax} dx = \int_0^1 \left(\sum_{n=0}^{\infty} \frac{(xA)^n}{n!} \right) dx = \sum_{n=0}^{\infty} \frac{A^n}{(n+1)!}.$$

If $\lambda_1 = \lambda_2 = 0$, then $A^2 = O_2 \Rightarrow A^n = O_2$, for all $n \geq 2$. Thus

$$\int_0^1 e^{Ax} dx = \sum_{n=0}^{\infty} \frac{A^n}{(n+1)!} = I_2 + \frac{A}{2}.$$

If $\lambda_1 = 0$ and $\lambda_2 \neq 0$, then $A^2 = tA$, where $t = \text{Tr}(A)$. This implies that $A^n = t^{n-1}A$, for all $n \geq 1$. It follows that

$$\int_0^1 e^{xA} dx = \sum_{n=0}^{\infty} \frac{A^n}{(n+1)!} = I_2 + \sum_{n=1}^{\infty} \frac{t^{n-1}}{(n+1)!} A = I_2 + \frac{e^t - 1 - t}{t^2} A.$$

If $\lambda_1, \lambda_2 \neq 0$, then

$$\int_0^1 e^{Ax} dx = \sum_{n=0}^{\infty} \frac{A^n}{(n+1)!} = \left(e^A - I_2 \right) A^{-1}.$$

4.87. (a) We have

$$\int_0^1 \cos(Ax)dx = \int_0^1 \left(\sum_{n=0}^{\infty}(-1)^n\frac{(xA)^{2n}}{(2n)!}\right)dx = \sum_{n=0}^{\infty}(-1)^n\frac{A^{2n}}{(2n+1)!}.$$

If $\lambda_1 = \lambda_2 = 0$, then $A^2 = O_2$ which implies that $A^n = O_2$, for all $n \geq 2$. Thus

$$\int_0^1 \cos(Ax)dx = \sum_{n=0}^{\infty}(-1)^n\frac{A^{2n}}{(2n+1)!} = I_2.$$

If $\lambda_1 = 0$ and $\lambda_2 \neq 0$, then $A^2 = tA$, where $t = \mathrm{Tr}(A)$. This implies that $A^n = t^{n-1}A$, for all $n \geq 1$. It follows that

$$\int_0^1 \cos(Ax)dx = I_2 + \sum_{n=1}^{\infty}(-1)^n\frac{A^{2n}}{(2n+1)!}$$

$$= I_2 + \sum_{n=1}^{\infty}(-1)^n\frac{t^{2n-1}}{(2n+1)!}A$$

$$= I_2 + \frac{\sin t - t}{t^2}A.$$

If $\lambda_1, \lambda_2 \neq 0$, then

$$\int_0^1 \cos(Ax)dx = \sum_{n=0}^{\infty}(-1)^n\frac{A^{2n}}{(2n+1)!} = A^{-1}\sin A.$$

(b) This part of the problem is solved similarly.

(c) We use the formula $\sin A(x + y) = \sin Ax \cos Ay + \cos Ax \sin Ay$, and we get that

$$\int_0^1\int_0^1 \sin A(x+y)dxdy = \int_0^1\int_0^1 (\sin Ax \cos Ay + \cos Ax \sin Ay) \, dxdy$$

$$= 2\int_0^1 \sin Axdx \int_0^1 \cos Aydy,$$

and the result follows from parts (a) and (b).

4.89. We solve only part (b) of the problem. Let λ_1, λ_2 be the eigenvalues of A and let

$$J_A = \begin{pmatrix} \lambda_1 & 0 \\ 0 & \lambda_2 \end{pmatrix} \quad \text{and} \quad P = \begin{pmatrix} \cos\beta & -\sin\beta \\ \sin\beta & \cos\beta \end{pmatrix}.$$

Nota bene. We have that $A = PJ_AP^{-1}$ and we choose P to be a rotation matrix. We mention that the matrix P is the invertible matrix whose columns are the eigenvectors corresponding to the eigenvalues λ_1, λ_2 of A and we normalize them in order to have a rotation matrix (see Theorem 2.5).

We calculate the double integral by changing the variables according to the formula $X = PY$, i.e.

$$\begin{pmatrix} x \\ y \end{pmatrix} = \begin{pmatrix} \cos\beta & -\sin\beta \\ \sin\beta & \cos\beta \end{pmatrix} \begin{pmatrix} u \\ v \end{pmatrix}$$

and we have that

$$I(\alpha) = \int_{-\infty}^{\infty}\int_{-\infty}^{\infty} (v^TAv)^\alpha e^{-v^TAv}dxdy$$

$$= \int_{-\infty}^{\infty}\int_{-\infty}^{\infty} \left(\lambda_1u^2 + \lambda_2v^2\right)^\alpha e^{-(\lambda_1u^2+\lambda_2v^2)} \left| \frac{D(x,y)}{D(u,v)} \right| dudv$$

$$= \int_{-\infty}^{\infty}\int_{-\infty}^{\infty} \left(\lambda_1u^2 + \lambda_2v^2\right)^\alpha e^{-(\lambda_1u^2+\lambda_2v^2)} dudv,$$

where $\dfrac{D(x,y)}{D(u,v)}$ is the Jacobian of the transformation.

Using the substitutions $u = \frac{x'}{\sqrt{\lambda_1}}$ and $v = \frac{y'}{\sqrt{\lambda_2}}$ we get that

$$I(\alpha) = \frac{1}{\sqrt{\lambda_1\lambda_2}} \int_{-\infty}^{\infty}\int_{-\infty}^{\infty} \left(x'^2 + y'^2\right)^\alpha e^{-(x'^2+y'^2)}dx'dy'$$

$$= \frac{1}{\sqrt{\det A}} \int_{0}^{\infty}\int_{0}^{2\pi} \rho^{2\alpha}e^{-\rho^2}\rho d\rho d\theta$$

$$= \frac{2\pi}{\sqrt{\det A}} \int_{0}^{\infty} \rho^{2\alpha}e^{-\rho^2}\rho d\rho \quad (\rho^2 = t)$$

$$= \frac{\pi}{\sqrt{\det A}} \int_{0}^{\infty} t^\alpha e^{-t}dt$$

$$= \frac{\pi\Gamma(\alpha+1)}{\sqrt{\det A}}.$$

When $\alpha = 0$ and $A = \begin{pmatrix} 2 & -1 \\ -1 & 2 \end{pmatrix}$ we obtain [58, Problem 2.3.3, p. 40] which states that

$$\int_{-\infty}^{\infty}\int_{-\infty}^{\infty} e^{-(x^2+(x-y)^2+y^2)}dxdy = \frac{\pi}{\sqrt{3}}.$$

4.90. If λ_1, λ_2 are the eigenvalues of A, then $\frac{1}{\lambda_1}$ and $\frac{1}{\lambda_2}$ are the eigenvalues of A^{-1}. Let P be the rotation matrix which verifies the equality $A^{-1} = PJ_{A^{-1}}P^{-1}$. We calculate the integral by using the substitution $X = PY$, i.e.

$$\begin{pmatrix} x \\ y \end{pmatrix} = \begin{pmatrix} \cos\theta & -\sin\theta \\ \sin\theta & \cos\theta \end{pmatrix} \begin{pmatrix} x' \\ y' \end{pmatrix}$$

and we have that

$$\int\int_{\mathbb{R}^2} e^{-\frac{1}{2}v^T A^{-1}v} \left(-\frac{1}{2}v^T A^{-1}v - \ln(2\pi\sqrt{\det A}) \right) dxdy$$

$$= \int\int_{\mathbb{R}^2} e^{-\frac{1}{2}(\frac{x'^2}{\lambda_1}+\frac{y'^2}{\lambda_2})} \left[-\frac{1}{2}\left(\frac{x'^2}{\lambda_1} + \frac{y'^2}{\lambda_2} \right) - \ln(2\pi\sqrt{\det A}) \right] \left| \frac{D(x,y)}{D(x',y')} \right| dx'dy'$$

$$= \int\int_{\mathbb{R}^2} e^{-\frac{1}{2}(\frac{x'^2}{\lambda_1}+\frac{y'^2}{\lambda_2})} \left[-\frac{1}{2}\left(\frac{x'^2}{\lambda_1} + \frac{y'^2}{\lambda_2} \right) - \ln(2\pi\sqrt{\det A}) \right] dx'dy' \quad \begin{pmatrix} x' = \sqrt{\lambda_1}u \\ y' = \sqrt{\lambda_2}v \end{pmatrix}$$

$$= \int\int_{\mathbb{R}^2} e^{-\frac{1}{2}(u^2+v^2)} \left(-\frac{1}{2}(u^2+v^2) - \ln(2\pi\sqrt{\det A}) \right) \sqrt{\lambda_1\lambda_2}dudv$$

$$= \sqrt{\det A} \int_0^\infty \int_0^{2\pi} e^{-\frac{1}{2}\rho^2} \left(-\frac{1}{2}\rho^2 - \ln(2\pi\sqrt{\det A}) \right) \rho d\rho d\alpha$$

$$= 2\pi\sqrt{\det A} \left[-\frac{1}{2}\int_0^\infty e^{-\frac{1}{2}\rho^2}\rho^3 d\rho - \ln(2\pi\sqrt{\det A}) \int_0^\infty e^{-\frac{1}{2}\rho^2}\rho d\rho \right]$$

$$= -2\pi\sqrt{\det A} \left[1 + \ln(2\pi\sqrt{\det A}) \right].$$

4.91. Let λ_1, λ_2 be the eigenvalues of A and let P be the invertible matrix such that $A = PJ_AP^{-1}$. We have that

$$e^{Ax} = \begin{cases} P \begin{pmatrix} e^{\lambda_1 x} & 0 \\ 0 & e^{\lambda_2 x} \end{pmatrix} P^{-1} & \text{if } J_A = \begin{pmatrix} \lambda_1 & 0 \\ 0 & \lambda_2 \end{pmatrix} \\ P \begin{pmatrix} e^{\lambda x} & xe^{\lambda x} \\ 0 & e^{\lambda x} \end{pmatrix} P^{-1} & \text{if } J_A = \begin{pmatrix} \lambda & 1 \\ 0 & \lambda \end{pmatrix}. \end{cases}$$

We are going to use in our calculations the following integral formulae which can be proved by direct computations.

If $\lambda \in \mathbb{C}$ and $\alpha > 0$, then:

$$\int_{-\infty}^\infty e^{-\alpha x^2 + \lambda x}dx = \sqrt{\frac{\pi}{\alpha}}e^{\frac{\lambda^2}{4\alpha}} \quad \text{and} \quad \int_{-\infty}^\infty xe^{-\alpha x^2 + \lambda x}dx = \sqrt{\frac{\pi}{\alpha}}\frac{\lambda}{2\alpha}e^{\frac{\lambda^2}{4\alpha}}.$$

If $J_A = \begin{pmatrix} \lambda_1 & 0 \\ 0 & \lambda_2 \end{pmatrix}$ we have that

$$
\int_{-\infty}^{\infty} e^{Ax} e^{-\alpha x^2} dx = P \left(\int_{-\infty}^{\infty} e^{J_A x} e^{-\alpha x^2} dx \right) P^{-1}
$$

$$
= P \left(\int_{-\infty}^{\infty} \begin{pmatrix} e^{\lambda_1 x} & 0 \\ 0 & e^{\lambda_2 x} \end{pmatrix} e^{-\alpha x^2} dx \right) P^{-1}
$$

$$
= P \begin{pmatrix} \int_{-\infty}^{\infty} e^{-\alpha x^2 + \lambda_1 x} dx & 0 \\ 0 & \int_{-\infty}^{\infty} e^{-\alpha x^2 + \lambda_2 x} dx \end{pmatrix} P^{-1}
$$

$$
= \sqrt{\frac{\pi}{\alpha}} P \begin{pmatrix} e^{\frac{\lambda_1^2}{4\alpha}} & 0 \\ 0 & e^{\frac{\lambda_2^2}{4\alpha}} \end{pmatrix} P^{-1}
$$

$$
= \sqrt{\frac{\pi}{\alpha}} e^{\frac{A^2}{4\alpha}}.
$$

If $J_A = \begin{pmatrix} \lambda & 1 \\ 0 & \lambda \end{pmatrix}$, then

$$
\int_{-\infty}^{\infty} e^{Ax} e^{-\alpha x^2} dx = P \left(\int_{-\infty}^{\infty} e^{J_A x} e^{-\alpha x^2} dx \right) P^{-1}
$$

$$
= P \left(\int_{-\infty}^{\infty} \begin{pmatrix} e^{\lambda x} & x e^{\lambda x} \\ 0 & e^{\lambda x} \end{pmatrix} e^{-\alpha x^2} dx \right) P^{-1}
$$

$$
= P \begin{pmatrix} \int_{-\infty}^{\infty} e^{-\alpha x^2 + \lambda x} dx & \int_{-\infty}^{\infty} x e^{-\alpha x^2 + \lambda x} dx \\ 0 & \int_{-\infty}^{\infty} e^{-\alpha x^2 + \lambda x} dx \end{pmatrix} P^{-1}
$$

$$
= \sqrt{\frac{\pi}{\alpha}} P \begin{pmatrix} e^{\frac{\lambda^2}{4\alpha}} & \frac{\lambda}{2\alpha} e^{\frac{\lambda^2}{4\alpha}} \\ 0 & e^{\frac{\lambda^2}{4\alpha}} \end{pmatrix} P^{-1}
$$

$$
= \sqrt{\frac{\pi}{\alpha}} e^{\frac{A^2}{4\alpha}}.
$$

4.92. *Solution 1.* Replacing A by iA in problem **4.91** we have that

$$
\int_{-\infty}^{\infty} e^{iAx} e^{-\alpha x^2} dx = \sqrt{\frac{\pi}{\alpha}} e^{-\frac{A^2}{4\alpha}},
$$

and it follows, since $e^{iAx} = \cos(Ax) + i\sin(Ax)$, that

$$\int_{-\infty}^{\infty} (\cos(Ax) + i\sin(Ax)) \, e^{-\alpha x^2} dx = \sqrt{\frac{\pi}{\alpha}} e^{-\frac{A^2}{4\alpha}}.$$

Identifying the real and the imaginary parts in the previous formula the problem is solved.

Solution 2. Use a technique similar to the method in the solution of problem **4.91**.

4.93. We need the following formulae which can be proved by direct computation.

Three exponential integrals with sine and cosine.

If α is a positive real number and $\beta \in \mathbb{C}$ with $\alpha > |\Im(\beta)|$, then

(a) $\displaystyle\int_0^\infty e^{-\alpha x} \cos(\beta x) dx = \frac{\alpha}{\alpha^2 + \beta^2};$

(b) $\displaystyle\int_0^\infty e^{-\alpha x} \sin(\beta x) dx = \frac{\beta}{\alpha^2 + \beta^2};$

(c) $\displaystyle\int_0^\infty x e^{-\alpha x} \cos(\beta x) dx = \frac{\alpha^2 - \beta^2}{(\alpha^2 + \beta^2)^2}.$

(a) Let λ_1 and λ_2 be the eigenvalues of A. Let $J_A = \begin{pmatrix} \lambda_1 & 0 \\ 0 & \lambda_2 \end{pmatrix}$ and let P be the nonsingular matrix such that $A = PJ_AP^{-1}$. This implies that $\sin(Ax) = P\sin(J_Ax)P^{-1}$, where $\sin(J_Ax) = \begin{pmatrix} \sin(\lambda_1 x) & 0 \\ 0 & \sin(\lambda_2 x) \end{pmatrix}$. It follows that

$$\int_0^\infty \sin(Ax)e^{-\alpha x} dx = P\left(\int_0^\infty \sin(J_Ax)e^{-\alpha x} dx\right) P^{-1}$$

$$= P\begin{pmatrix} \int_0^\infty \sin(\lambda_1 x)e^{-\alpha x} dx & 0 \\ 0 & \int_0^\infty \sin(\lambda_2 x)e^{-\alpha x} dx \end{pmatrix} P^{-1}$$

$$= P\begin{pmatrix} \frac{\lambda_1}{\lambda_1^2 + \alpha^2} & 0 \\ 0 & \frac{\lambda_2}{\lambda_2^2 + \alpha^2} \end{pmatrix} P^{-1}$$

$$= A(A^2 + \alpha^2 I_2)^{-1}.$$

If $J_A = \begin{pmatrix} \lambda & 1 \\ 0 & \lambda \end{pmatrix}$, then $\sin(J_A x) = \begin{pmatrix} \sin(\lambda x) & x\cos(\lambda x) \\ 0 & \sin(\lambda x) \end{pmatrix}$. We have

$$\int_0^\infty \sin(Ax)e^{-\alpha x}dx = P\left(\int_0^\infty \sin(J_A x)e^{-\alpha x}dx\right)P^{-1}$$

$$= P\begin{pmatrix} \int_0^\infty \sin(\lambda x)e^{-\alpha x}dx & \int_0^\infty x\cos(\lambda x)e^{-\alpha x}dx \\ 0 & \int_0^\infty \sin(\lambda x)e^{-\alpha x}dx \end{pmatrix}P^{-1}$$

$$= P\begin{pmatrix} \frac{\lambda}{\lambda^2+\alpha^2} & \frac{\alpha^2-\lambda^2}{(\alpha^2+\lambda^2)^2} \\ 0 & \frac{\lambda}{\lambda^2+\alpha^2} \end{pmatrix}P^{-1}$$

$$= A(A^2 + \alpha^2 I_2)^{-1}.$$

Part (b) of the problem is solved similarly.

4.94. (a) We need the following integral formulae which can be proved by direct computation.

Euler–Poisson integrals. Let $\lambda > 0$. The following formulae hold:

(a) $\displaystyle\int_0^\infty e^{-\lambda x^2}dx = \frac{\sqrt{\pi}}{2\sqrt{\lambda}}$;

(b) $\displaystyle\int_0^\infty x^2 e^{-\lambda x^2}dx = \frac{\sqrt{\pi}}{4\lambda\sqrt{\lambda}}$;

(c) $\displaystyle\int_0^\infty e^{-\lambda x^n}dx = \frac{\Gamma\left(1+\frac{1}{n}\right)}{\sqrt[n]{\lambda}}$, where Γ denotes the Gamma function and $n \in \mathbb{N}$.

Let J_A be the Jordan canonical form of A and let P be the invertible matrix which verifies $A = PJ_AP^{-1}$. A calculation shows that $J_A = \begin{pmatrix} 2 & 1 \\ 0 & 2 \end{pmatrix}$, $P = \begin{pmatrix} 1 & 0 \\ 1 & 1 \end{pmatrix}$ and $P^{-1} = \begin{pmatrix} 1 & 0 \\ -1 & 1 \end{pmatrix}$. We have

$$e^{-Ax^2} = P\begin{pmatrix} e^{-2x^2} & -x^2 e^{-2x^2} \\ 0 & e^{-2x^2} \end{pmatrix}P^{-1}$$

and it follows that

$$\int_0^\infty e^{-Ax^2} dx = P\left(\frac{\int_0^\infty e^{-2x^2} dx - \int_0^\infty x^2 e^{-2x^2} dx}{\int_0^\infty e^{-2x^2} dx}\right) P^{-1}$$

$$= \frac{\sqrt{\pi}}{2} P \begin{pmatrix} \frac{1}{\sqrt{2}} & -\frac{1}{4\sqrt{2}} \\ 0 & \frac{1}{\sqrt{2}} \end{pmatrix} P^{-1}$$

$$= \frac{\sqrt{\pi}}{2} \begin{pmatrix} 1 & 0 \\ 1 & 1 \end{pmatrix} \begin{pmatrix} \frac{1}{\sqrt{2}} & -\frac{1}{4\sqrt{2}} \\ 0 & \frac{1}{\sqrt{2}} \end{pmatrix} \begin{pmatrix} 1 & 0 \\ -1 & 1 \end{pmatrix}$$

$$= \frac{\sqrt{\pi}}{2} \begin{pmatrix} \frac{5}{4\sqrt{2}} & -\frac{1}{4\sqrt{2}} \\ \frac{1}{4\sqrt{2}} & \frac{3}{4\sqrt{2}} \end{pmatrix}.$$

Observe that if $B = \begin{pmatrix} \frac{5}{4\sqrt{2}} & -\frac{1}{4\sqrt{2}} \\ \frac{1}{4\sqrt{2}} & \frac{3}{4\sqrt{2}} \end{pmatrix}$, then $B^2 = A^{-1}$.

(b) $A = aI_2 + bJ$, where $J = \begin{pmatrix} 0 & 1 \\ 1 & 0 \end{pmatrix}$. It follows that

$$e^{-Ax^n} = e^{(aI_2 + bJ)x^n} = e^{-ax^n I_2} e^{-bx^n J} = e^{-ax^n} e^{-bx^n J}.$$

A calculation shows that $e^{-bx^n J} = \cosh(bx^n)I_2 - \sinh(bx^n)J$ and it follows that

$$e^{-Ax^n} = \frac{e^{-(a-b)x^n} + e^{-(a+b)x^n}}{2} I_2 - \frac{e^{-(a-b)x^n} - e^{-(a+b)x^n}}{2} J.$$

Now the problem follows by integration and by using part (c) of Euler–Poisson integrals. Another method for calculating e^{-Ax^n} is to use part (a) of problem **4.32**.

4.95. We need the following integrals due to Laplace.

Three integrals of Laplace.

(a) The following formulae hold:

$$\int_0^\infty \frac{\cos ax}{1+x^2} dx = \frac{\pi}{2} e^{-|a|} \quad \text{and} \quad \int_0^\infty \frac{x \sin ax}{1+x^2} dx = \frac{\pi}{2} e^{-|a|} \operatorname{sign} a, \quad a \in \mathbb{R}.$$

(b) If $a \in \mathbb{R}$, then

$$\int_0^\infty e^{-x^2} \cos 2ax \, dx = \frac{\sqrt{\pi}}{2} e^{-a^2}.$$

(a) Observe that $A = aI_2 + bJ$, where $J = \begin{pmatrix} 0 & 1 \\ 1 & 0 \end{pmatrix}$. It follows that

$$\cos(Ax) = \cos(aI_2 + bJ)x = \cos(ax)\cos(bxJ) - \sin(ax)\sin(bxJ).$$

A calculation shows that $\cos(bxJ) = \cos(bx)I_2$ and $\sin(bxJ) = \sin(bx)J$. This implies that

$$\cos(Ax) = \cos(ax)\cos(bx)I_2 - \sin(ax)\sin(bx)J$$

$$= (\cos(a+b)x + \cos(a-b)x)\frac{I_2}{2} - (\cos(a-b)x - \cos(a+b)x)\frac{J}{2}$$

and we have

$$\int_0^\infty \frac{\cos(Ax)}{1+x^2}dx = \left(\int_0^\infty \frac{\cos(a+b)x + \cos(a-b)x}{1+x^2}dx \right) \frac{I_2}{2}$$

$$- \left(\int_0^\infty \frac{\cos(a-b)x - \cos(a+b)x}{1+x^2}dx \right) \frac{J}{2}$$

$$= \frac{\pi}{4} \begin{pmatrix} e^{-|a+b|} + e^{-|a-b|} & e^{-|a+b|} - e^{-|a-b|} \\ e^{-|a+b|} - e^{-|a-b|} & e^{-|a+b|} + e^{-|a-b|} \end{pmatrix}.$$

The challenge integral as well as part (b) of the problem can be solved similarly.

Remark 4.23 If $A \in \mathcal{M}_2(\mathbb{R})$ the reader may wish to calculate the integrals

$$\int_0^\infty \frac{\cos(Ax)}{1+x^2}dx, \quad \int_0^\infty \frac{x\sin(Ax)}{1+x^2}dx \quad \text{and} \quad \int_0^\infty e^{-x^2}\cos(Ax)dx$$

by using Theorems A.2 and A.3.

4.96. Observe that $A = aI_2 + bJ$, where $J = \begin{pmatrix} 0 & 1 \\ 1 & 0 \end{pmatrix}$, and use **Fresnel's integrals**

$$\int_0^\infty \sin x^2 dx = \int_0^\infty \cos x^2 dx = \frac{1}{2}\sqrt{\frac{\pi}{2}}.$$

The challenge integral can be calculated by using the Fresnel integral

$$\int_0^\infty \cos x^n dx = \Gamma\left(1 + \frac{1}{n}\right) \cos \frac{\pi}{2n}.$$

We leave the details to the interested reader.

4.97. (a) Let J_A be the Jordan canonical form of A and let P be the invertible matrix such that $A = PJ_AP^{-1}$. If $J_A = \begin{pmatrix} \lambda_1 & 0 \\ 0 & \lambda_2 \end{pmatrix}$, then $e^{-Ax} = P\begin{pmatrix} e^{-\lambda_1 x} & 0 \\ 0 & e^{-\lambda_2 x} \end{pmatrix}P^{-1}$. It follows that

$$\int_0^\infty e^{-Ax}dx = P\begin{pmatrix} \int_0^\infty e^{-\lambda_1 x}dx & 0 \\ 0 & \int_0^\infty e^{-\lambda_2 x}dx \end{pmatrix}P^{-1}$$

$$= P\begin{pmatrix} \frac{1}{\lambda_1} & 0 \\ 0 & \frac{1}{\lambda_2} \end{pmatrix}P^{-1}$$

$$= A^{-1}.$$

If $J_A = \begin{pmatrix} \lambda & 1 \\ 0 & \lambda \end{pmatrix}$, then $e^{-Ax} = P\begin{pmatrix} e^{-\lambda x} & -xe^{-\lambda x} \\ 0 & e^{-\lambda x} \end{pmatrix}P^{-1}$. It follows that

$$\int_0^\infty e^{-Ax}dx = P\begin{pmatrix} \int_0^\infty e^{-\lambda x}dx & -\int_0^\infty e^{-\lambda x}xdx \\ 0 & \int_0^\infty e^{-\lambda x}dx \end{pmatrix}P^{-1}$$

$$= P\begin{pmatrix} \frac{1}{\lambda} & -\frac{1}{\lambda^2} \\ 0 & \frac{1}{\lambda} \end{pmatrix}P^{-1}$$

$$= A^{-1}.$$

(b) We have, based on part (a), that

$$\int_0^\infty e^{-Ax}e^{\alpha x}dx = \int_0^\infty e^{-(A-\alpha I_2)x}dx = (A - \alpha I_2)^{-1}.$$

(c) If $J_A = \begin{pmatrix} \lambda_1 & 0 \\ 0 & \lambda_2 \end{pmatrix}$, then

$$\int_0^\infty e^{-Ax}x^n dx = P\begin{pmatrix} \int_0^\infty e^{-\lambda_1 x}x^n dx & 0 \\ 0 & \int_0^\infty e^{-\lambda_2 x}x^n dx \end{pmatrix}P^{-1}$$

$$= P\begin{pmatrix} \frac{n!}{\lambda_1^{n+1}} & 0 \\ 0 & \frac{n!}{\lambda_2^{n+1}} \end{pmatrix}P^{-1}$$

$$= n!A^{-(n+1)}.$$

If $J_A = \begin{pmatrix} \lambda & 1 \\ 0 & \lambda \end{pmatrix}$, then

$$\int_0^\infty e^{-Ax} x^n dx = P \begin{pmatrix} \int_0^\infty e^{-\lambda x} x^n dx & -\int_0^\infty e^{-\lambda x} x^{n+1} dx \\ 0 & \int_0^\infty e^{-\lambda x} x^n dx \end{pmatrix} P^{-1}$$

$$= P \begin{pmatrix} \frac{n!}{\lambda^{n+1}} & -\frac{(n+1)!}{\lambda^{n+2}} \\ 0 & \frac{n!}{\lambda^{n+1}} \end{pmatrix} P^{-1}$$

$$= n! A^{-(n+1)}.$$

4.98. (a) Let $J_A = \begin{pmatrix} \lambda_1 & 0 \\ 0 & \lambda_2 \end{pmatrix}$ be the Jordan canonical form of A and let P be the invertible matrix such that $A = PJ_A P^{-1}$. We have

$$\int_0^\infty \frac{\sin(Ax)}{x} dx = P \begin{pmatrix} \int_0^\infty \frac{\sin(\lambda_1 x)}{x} dx & 0 \\ 0 & \int_0^\infty \frac{\sin(\lambda_2 x)}{x} dx \end{pmatrix} P^{-1}$$

$$= P \begin{pmatrix} \operatorname{sign}(\lambda_1) \frac{\pi}{2} & 0 \\ 0 & \operatorname{sign}(\lambda_2) \frac{\pi}{2} \end{pmatrix} P^{-1}$$

$$= \begin{cases} \frac{\pi}{2} I_2 & \text{if } \lambda_1, \lambda_2 > 0 \\ -\frac{\pi}{2} I_2 & \text{if } \lambda_1, \lambda_2 < 0. \end{cases}$$

We used in the previous calculations Dirichlet's integral $\int_0^\infty \frac{\sin(\lambda x)}{x} dx = \operatorname{sign}(\lambda) \frac{\pi}{2}$, $\lambda \in \mathbb{R}$.

(b) If $J_A = \begin{pmatrix} \lambda_1 & 0 \\ 0 & \lambda_2 \end{pmatrix}$, then

$$\int_0^\infty \frac{\sin^2(Ax)}{x^2} dx = P \begin{pmatrix} \int_0^\infty \frac{\sin^2(\lambda_1 x)}{x^2} dx & 0 \\ 0 & \int_0^\infty \frac{\sin^2(\lambda_2 x)}{x^2} dx \end{pmatrix} P^{-1}$$

$$= P \begin{pmatrix} \operatorname{sign}(\lambda_1) \lambda_1 \frac{\pi}{2} & 0 \\ 0 & \operatorname{sign}(\lambda_2) \lambda_2 \frac{\pi}{2} \end{pmatrix} P^{-1}$$

$$= \begin{cases} \frac{\pi}{2} A & \text{if } \lambda_1, \lambda_2 > 0 \\ -\frac{\pi}{2} A & \text{if } \lambda_1, \lambda_2 < 0. \end{cases}$$

If $J_A = \begin{pmatrix} \lambda & 1 \\ 0 & \lambda \end{pmatrix}$, then

$$\int_0^\infty \frac{\sin^2(Ax)}{x^2}\,dx = P\begin{pmatrix} \int_0^\infty \frac{\sin^2(\lambda x)}{x^2}\,dx & \int_0^\infty \frac{\sin(2\lambda x)}{x}\,dx \\ 0 & \int_0^\infty \frac{\sin^2(\lambda x)}{x^2}\,dx \end{pmatrix} P^{-1}$$

$$= P\begin{pmatrix} \operatorname{sign}(\lambda)\lambda\frac{\pi}{2} & \operatorname{sign}(\lambda)\frac{\pi}{2} \\ 0 & \operatorname{sign}(\lambda)\lambda\frac{\pi}{2} \end{pmatrix} P^{-1}$$

$$= \begin{cases} \frac{\pi}{2}A & \text{if } \lambda > 0 \\ -\frac{\pi}{2}A & \text{if } \lambda < 0. \end{cases}$$

We used in our calculations the formula $\int_0^\infty \frac{\sin^2(\lambda x)}{x^2}\,dx = \operatorname{sign}(\lambda)\lambda\frac{\pi}{2}$.

4.99. Let λ_1 and λ_2 be the distinct eigenvalues of A. The eigenvalues of $A + I_2$ are the positive real numbers $\lambda_1 + 1$ and $\lambda_2 + 1$ and the eigenvalues of $A - I_2$ are the negative real numbers $\lambda_1 - 1$ and $\lambda_2 - 1$. We have, based on part (a) of problem **4.98**, that

$$\int_0^\infty \frac{\sin(Ax)\cos x}{x}\,dx = \int_0^\infty \frac{\sin(Ax)\cos(I_2 x)}{x}\,dx$$

$$= \int_0^\infty \frac{\sin(A + I_2)x + \sin(A - I_2)x}{2x}\,dx$$

$$= \frac{1}{2}\left(\frac{\pi}{2}I_2 - \left(-\frac{\pi}{2}\right)I_2\right)$$

$$= \frac{\pi}{2}I_2.$$

4.100. Let λ_1, λ_2 be the eigenvalues of A, let J_A be the Jordan canonical form of A, and let P be the invertible matrix such that $A = PJ_AP^{-1}$.

If $J_A = \begin{pmatrix} \lambda_1 & 0 \\ 0 & \lambda_2 \end{pmatrix}$, then

$$\int_0^\infty \frac{e^{-\alpha Ax} - e^{-\beta Ax}}{x}\,dx = P\begin{pmatrix} \int_0^\infty \frac{e^{-\alpha\lambda_1 x} - e^{-\beta\lambda_1 x}}{x}\,dx & 0 \\ 0 & \int_0^\infty \frac{e^{-\alpha\lambda_2 x} - e^{-\beta\lambda_2 x}}{x}\,dx \end{pmatrix} P^{-1}$$

$$= P\begin{pmatrix} \ln\frac{\beta}{\alpha} & 0 \\ 0 & \ln\frac{\beta}{\alpha} \end{pmatrix} P^{-1}$$

$$= \left(\ln\frac{\beta}{\alpha}\right)I_2.$$

If $J_A = \begin{pmatrix} \lambda & 1 \\ 0 & \lambda \end{pmatrix}$, then

$$\int_0^\infty \frac{e^{-\alpha A x} - e^{-\beta A x}}{x} dx = P \left(\begin{array}{cc} \int_0^\infty \frac{e^{-\alpha \lambda x} - e^{-\beta \lambda x}}{x} dx & \int_0^\infty \frac{-\alpha x e^{-\alpha \lambda x} + \beta x e^{-\beta \lambda x}}{x} dx \\ 0 & \int_0^\infty \frac{e^{-\alpha \lambda x} - e^{-\beta \lambda x}}{x} dx \end{array} \right) P^{-1}$$

$$= P \begin{pmatrix} \ln \frac{\beta}{\alpha} & 0 \\ 0 & \ln \frac{\beta}{\alpha} \end{pmatrix} P^{-1}$$

$$= \left(\ln \frac{\beta}{\alpha} \right) I_2.$$

4.101. (a) Let λ_1, λ_2 be the eigenvalues of A, let J_A be the Jordan canonical form of A, and let P be the invertible matrix such that $A = P J_A P^{-1}$.

If $J_A = \begin{pmatrix} \lambda_1 & 0 \\ 0 & \lambda_2 \end{pmatrix}$, then

$$\int_0^\infty \frac{f(\alpha A x) - f(\beta A x)}{x} dx = P \left(\int_0^\infty \frac{f(\alpha J_A x) - f(\beta J_A x)}{x} dx \right) P^{-1}$$

$$= P \left(\begin{array}{cc} \int_0^\infty \frac{f(\alpha \lambda_1 x) - f(\beta \lambda_1 x)}{x} dx & 0 \\ 0 & \int_0^\infty \frac{f(\alpha \lambda_2 x) - f(\beta \lambda_2 x)}{x} dx \end{array} \right) P^{-1}$$

$$= P \left(\begin{array}{cc} (f(0) - f(\infty)) \ln \frac{\beta}{\alpha} & 0 \\ 0 & (f(0) - f(\infty)) \ln \frac{\beta}{\alpha} \end{array} \right) P^{-1}$$

$$= \left[(f(0) - f(\infty)) \ln \frac{\beta}{\alpha} \right] I_2.$$

If $J_A = \begin{pmatrix} \lambda & 1 \\ 0 & \lambda \end{pmatrix}$ we have that

$$f(\alpha A x) - f(\beta A x) = \begin{pmatrix} f(\alpha \lambda x) - f(\beta \lambda x) & \alpha x f'(\alpha \lambda x) - \beta x f'(\beta \lambda x) \\ 0 & f(\alpha \lambda x) - f(\beta \lambda x) \end{pmatrix}.$$

Thus,

$$\int_0^\infty \frac{f(\alpha A x) - f(\beta A x)}{x} dx$$

$$= P \left(\begin{array}{cc} \int_0^\infty \frac{f(\alpha \lambda x) - f(\beta \lambda x)}{x} dx & \int_0^\infty (\alpha f'(\alpha \lambda x) - \beta f'(\beta \lambda x)) dx \\ 0 & \int_0^\infty \frac{f(\alpha \lambda x) - f(\beta \lambda x)}{x} dx \end{array} \right) P^{-1}$$

$$= P \left(\begin{array}{cc} (f(0) - f(\infty)) \ln \frac{\beta}{\alpha} & 0 \\ 0 & (f(0) - f(\infty)) \ln \frac{\beta}{\alpha} \end{array} \right) P^{-1}$$

$$= \left[(f(0) - f(\infty)) \ln \frac{\beta}{\alpha} \right] I_2.$$

(b) Let $f : [0, \infty) \to \mathbb{R}$ be the function defined by $f(x) = \frac{\sin^2 x}{x^2}$ if $x \neq 0$ and $f(0) = 1$. Observe that

$$\frac{\sin^4(Ax)}{x^3} = \frac{\frac{\sin^2(Ax)}{x^2} - \frac{\sin^2(2Ax)}{4x^2}}{x} = \frac{f(Ax) - f(2Ax)}{x}A^2.$$

This implies, based on part (a), that

$$\int_0^\infty \frac{\sin^4(Ax)}{x^3}\,dx = A^2 \ln 2.$$

To calculate the second integral we let $g : [0, \infty) \to \mathbb{R}$ be the function $g(x) = \frac{\sin x}{x}$ if $x \neq 0$ and $g(0) = 1$. We have

$$\frac{\sin^3(Ax)}{x^2} = \frac{g(Ax) - g(3Ax)}{x} \cdot \frac{3A}{4}.$$

It follows, based on part (a), that

$$\int_0^\infty \frac{\sin^3(Ax)}{x^2}\,dx = \frac{3\ln 3}{4}A.$$

(c) Let $f : [0, \infty) \to \mathbb{R}$ be the function $f(x) = \frac{1-e^{-x}}{x}$ if $x \neq 0$ and $f(0) = 1$. A calculation shows that

$$\left(\frac{1_2 - e^{-Ax}}{x}\right)^2 = 2A\frac{f(Ax) - f(2Ax)}{x},$$

which implies, based on part (a), that $\displaystyle\int_0^\infty \left(\frac{1_2 - e^{-Ax}}{x}\right)^2\,dx = (2\ln 2)A$.

We mention that parts (b) and (c) of this problem were inspired by the *sine Frullani integrals*

$$\int_0^\infty \frac{\sin^4 x}{x^3}\,dx = \ln 2 \quad \text{and} \quad \int_0^\infty \frac{\sin^3 x}{x^2}\,dx = \frac{3\ln 3}{4},$$

the first of which being due to Mircea Ivan, and the *quadratic Frullani integral* due to Furdui and Sîntămărian [33]

$$\int_0^\infty \left(\frac{1 - e^{-x}}{x}\right)^2\,dx = 2\ln 2.$$

4.102. Let λ_1, λ_2 be the eigenvalues of A, let $J_A = \begin{pmatrix} \lambda_1 & 0 \\ 0 & \lambda_2 \end{pmatrix}$ be the Jordan canonical form of A, and let P be the invertible matrix such that $A = PJ_AP^{-1}$. Then

$$\left(\frac{e^{-Ax} - e^{-Ay}}{x-y} \right)^2 = P \begin{pmatrix} \left(\frac{e^{-\lambda_1 x} - e^{-\lambda_1 y}}{x-y} \right)^2 & 0 \\ 0 & \left(\frac{e^{-\lambda_2 x} - e^{-\lambda_2 y}}{x-y} \right)^2 \end{pmatrix} P^{-1},$$

which implies, based on Lemma A.4, that

$$\int_0^\infty \int_0^\infty \left(\frac{e^{-Ax} - e^{-Ay}}{x-y} \right)^2 dxdy$$

$$= P \begin{pmatrix} \int_0^\infty \int_0^\infty \left(\frac{e^{-\lambda_1 x} - e^{-\lambda_1 y}}{x-y} \right)^2 dxdy & 0 \\ 0 & \int_0^\infty \int_0^\infty \left(\frac{e^{-\lambda_2 x} - e^{-\lambda_2 y}}{x-y} \right)^2 dxdy \end{pmatrix} P^{-1}$$

$$= P \begin{pmatrix} \ln 4 & 0 \\ 0 & \ln 4 \end{pmatrix} P^{-1}$$

$$= (\ln 4)I_2.$$

Chapter 5
Applications of matrices to plane geometry

Everyone wants to teach and nobody to learn.
Niels Abel (1802–1829)

5.1 Linear transformations

Definition 5.1 Let $A \in \mathscr{M}_2(\mathbb{R})$. The function $f_A : \mathbb{R}^2 \to \mathbb{R}^2$, defined by

$$f_A(x, y) = (x', y'), \quad \text{where} \quad \begin{pmatrix} x' \\ y' \end{pmatrix} = A \begin{pmatrix} x \\ y \end{pmatrix},$$

is called a *linear transformation* defined by the matrix A (or *linear map* defined by A) of $\mathbb{R}^2$.

The matrix A is called the *matrix associated* with the linear transformation f_A, in the canonical basis, and is denoted by A_f.

Proposition 5.1 *The linear transformation $f_A : \mathbb{R}^2 \to \mathbb{R}^2$, $f_A(x, y) = (x', y')$, where*

$$\begin{pmatrix} x' \\ y' \end{pmatrix} = A \begin{pmatrix} x \\ y \end{pmatrix}, \quad A \in \mathscr{M}_2(\mathbb{R}),$$

has the following properties:

(a) $f_A((x_1, y_1) + (x_2, y_2)) = f_A(x_1, y_1) + f_A(x_2, y_2), \ \forall \, (x_1, y_1), (x_2, y_2) \in \mathbb{R}^2$
(b) $f_A(\alpha(x_1, y_1)) = \alpha f_A(x_1, y_1), \ \forall \alpha \in \mathbb{R}, \ \forall \, (x_1, y_1) \in \mathbb{R}^2$,

where the operations with respect to which $\mathbb{R}^2$ is a real vector space are those defined for one line matrices or (one column matrices) as in Remark 1.3.

Proof (a) We have,

$$A \left[\begin{pmatrix} x_1 \\ y_1 \end{pmatrix} + \begin{pmatrix} x_2 \\ y_2 \end{pmatrix} \right] = A \begin{pmatrix} x_1 + x_2 \\ y_1 + y_2 \end{pmatrix} = A \begin{pmatrix} x_1 \\ y_1 \end{pmatrix} + A \begin{pmatrix} x_2 \\ y_2 \end{pmatrix},$$

and part (a) follows.

© Springer International Publishing AG 2017
V. Pop, O. Furdui, *Square Matrices of Order 2*, DOI 10.1007/978-3-319-54939-2_5

(b) On the other hand,

$$A\alpha \begin{pmatrix} x \\ y \end{pmatrix} = A \begin{pmatrix} \alpha x \\ \alpha y \end{pmatrix} = \alpha A \begin{pmatrix} x \\ y \end{pmatrix},$$

and the proposition is proved.   □

Remark 5.1 We mention that the properties in parts (a) and (b) of Proposition 5.1 are equivalent to

$$f_A(\alpha(x_1, y_1) + \beta(x_2, y_2)) = \alpha f_A(x_1, y_1) + \beta f_A(x_2, y_2),$$

$\forall \alpha, \beta \in \mathbb{R}$ and $\forall (x_1, y_1), (x_2, y_2) \in \mathbb{R}^2$.
We also have the following properties:

- *the zero vector is preserved*: $f_A(0, 0) = (0, 0)$;
- $f_A(-(x, y)) = -f_A(x, y)$;
- *the identity map*: $f_{I_2}(x, y) = (x, y)$ or $f_{I_2} = I_{\mathbb{R}^2}$.

Definition 5.2 *The kernel and the image of a linear transformation.* The set

$$\mathrm{Ker} f_A = \{(x, y) \in \mathbb{R}^2 : f_A(x, y) = (0, 0)\}$$

is called the *kernel* of f_A and the set

$$\mathrm{Im} f_A = \{f_A(x, y) : (x, y) \in \mathbb{R}^2\}$$

is called the *image* of f_A.

Thus, the kernel of a linear transformation consists of all points (vectors) in $\mathbb{R}^2$ which are mapped by f_A to zero (the zero vector in $\mathbb{R}^2$) and the image of f_A consists of all points (vectors) in $\mathbb{R}^2$ which were mapped from points (vectors) in $\mathbb{R}^2$.

It should be mentioned that the kernel and the image of a linear transformation are analogous to the zeros and the range of a function.

Lemma 5.1 *The following properties hold*:

(a) *The kernel of f_A as a vector subspace of $\mathbb{R}^2$.*
    $\forall \alpha, \beta \in \mathbb{R}$ *and* $\forall (x_1, y_1), (x_2, y_2) \in \mathrm{Ker} f_A$ *we have* $\alpha(x_1, y_1) + \beta(x_2, y_2) \in \mathrm{Ker} f_A$;
(b) *The image of f_A as a vector subspace of $\mathbb{R}^2$.*
    $\forall \alpha, \beta \in \mathbb{R}$ *and* $\forall (x', y'), (x'', y'') \in \mathrm{Im} f_A$ *we have* $\alpha(x', y') + \beta(x'', y'') \in \mathrm{Im} f_A$.

Proof The proof of the lemma is left, as an exercise, to the interested reader.   □

Theorem 5.1 *If $A, B \in \mathcal{M}_2(\mathbb{R})$ and f_A, f_B are the linear transformations determined by A and B, then the function $f_A \circ f_B : \mathbb{R}^2 \to \mathbb{R}^2$ is a linear transformation having the associated matrix AB.*

Proof If $f_B(x, y) = (x', y')$ and $f_A(x', y') = (x'', y'')$, then

$$f_A \circ f_B(x, y) = f_A(f_B(x, y)) = f_A(x', y') = (x'', y''),$$

where

$$\begin{pmatrix} x'' \\ y'' \end{pmatrix} = A \begin{pmatrix} x' \\ y' \end{pmatrix} \quad \text{and} \quad \begin{pmatrix} x' \\ y' \end{pmatrix} = B \begin{pmatrix} x \\ y \end{pmatrix}.$$

It follows that

$$\begin{pmatrix} x'' \\ y'' \end{pmatrix} = AB \begin{pmatrix} x \\ y \end{pmatrix},$$

so the matrix of the linear transformation $f_A \circ f_B$ is AB. $\qquad\square$

5.2 The matrix of special transformations

In this section and in what follows, to simplify the calculations, we identify the point $(x, y) \in \mathbb{R}^2$ by the vector $\begin{pmatrix} x \\ y \end{pmatrix}$.

Theorem 5.2 The matrix of a linear transformation.
If the linear transformation $f_A : \mathbb{R}^2 \to \mathbb{R}^2$, $f_A \begin{pmatrix} x \\ y \end{pmatrix} = A \begin{pmatrix} x \\ y \end{pmatrix}$, where $A \in \mathcal{M}_2(\mathbb{R})$, verifies

$$f_A \begin{pmatrix} 1 \\ 0 \end{pmatrix} = \begin{pmatrix} a \\ b \end{pmatrix} \quad \text{and} \quad f_A \begin{pmatrix} 0 \\ 1 \end{pmatrix} = \begin{pmatrix} c \\ d \end{pmatrix}, \quad \text{then} \quad A = \begin{pmatrix} a & c \\ b & d \end{pmatrix}.$$

Proof Let $A = \begin{pmatrix} m & n \\ p & q \end{pmatrix}$. Since

$$f_A \begin{pmatrix} 1 \\ 0 \end{pmatrix} = \begin{pmatrix} m & n \\ p & q \end{pmatrix} \begin{pmatrix} 1 \\ 0 \end{pmatrix} = \begin{pmatrix} m \\ p \end{pmatrix} = \begin{pmatrix} a \\ b \end{pmatrix}$$

and

$$f_A \begin{pmatrix} 0 \\ 1 \end{pmatrix} = \begin{pmatrix} m & n \\ p & q \end{pmatrix} \begin{pmatrix} 0 \\ 1 \end{pmatrix} = \begin{pmatrix} n \\ q \end{pmatrix} = \begin{pmatrix} c \\ d \end{pmatrix},$$

we get that $A = \begin{pmatrix} a & c \\ b & d \end{pmatrix}$ and the theorem is proved.   □

Using Theorem 5.2 we can determine the matrices associated with various linear transformations.

The matrix of the reflection through the origin

We wish to determine the matrix of the reflection of a point through the origin. This is the linear transformation which sends the point $M(x, y)$ to the point $M'(-x, -y)$, the symmetric of M about the origin.

Let $f_A : \mathbb{R}^2 \to \mathbb{R}^2, f_A \begin{pmatrix} x \\ y \end{pmatrix} = A \begin{pmatrix} x \\ y \end{pmatrix}$, where $A = \begin{pmatrix} a & c \\ b & d \end{pmatrix}$.

Since

$$f_A \begin{pmatrix} 1 \\ 0 \end{pmatrix} = \begin{pmatrix} -1 \\ 0 \end{pmatrix} \quad \text{and} \quad f_A \begin{pmatrix} 0 \\ 1 \end{pmatrix} = \begin{pmatrix} 0 \\ -1 \end{pmatrix},$$

we have, based on Theorem 5.2, that $A = \begin{pmatrix} -1 & 0 \\ 0 & -1 \end{pmatrix}$ is the *matrix of the reflection through the origin*.

The matrix of the reflection across the x-axis

Now we determine the reflection across the x-axis. This is the linear transformation which sends the point $M(x, y)$ to the point $M'(x, -y)$ which is the symmetric of M about the x-axis.

Since

$$f_A \begin{pmatrix} 1 \\ 0 \end{pmatrix} = \begin{pmatrix} 1 \\ 0 \end{pmatrix} \quad \text{and} \quad f_A \begin{pmatrix} 0 \\ 1 \end{pmatrix} = \begin{pmatrix} 0 \\ -1 \end{pmatrix},$$

we have, based on Theorem 5.2, that $A = \begin{pmatrix} 1 & 0 \\ 0 & -1 \end{pmatrix}$ is the *matrix of the reflection across the x-axis*.

The matrix of the reflection across the y-axis

As in the previous calculations we have that the *matrix of the reflection across the y-axis is given by* $A = \begin{pmatrix} -1 & 0 \\ 0 & 1 \end{pmatrix}$.

The rotation matrix of angle α and center the origin

Let $\alpha \in (0, 2\pi)$. The rotation of angle α and center the origin O is the transformation which preserves the origin (it sends O to O) and sends the point M to the point M' such that the segments $[OM]$ and $[OM']$ have the same lengths, i.e., $OM = OM'$ and $\widehat{MOM'}$ and $\widehat{\alpha}$ are equal and have the same orientation (if $\alpha > 0$ the rotation is counterclockwise, if $\alpha < 0$ the rotation is clockwise).

Let $f_A : \mathbb{R}^2 \to \mathbb{R}^2, f_A \begin{pmatrix} x \\ y \end{pmatrix} = A \begin{pmatrix} x \\ y \end{pmatrix}$, where $A = \begin{pmatrix} a & c \\ b & d \end{pmatrix}$ and $\alpha \geq 0$.

We have $x' = \cos(\theta + \alpha) = \cos\theta\cos\alpha - \sin\theta\sin\alpha$, $\cos\theta = x$, $\sin\theta = y$. It follows that $x' = x\cos\alpha - y\sin\alpha$. Similarly, $y' = \sin(\theta + \alpha) = \sin\theta\cos\alpha + \cos\theta\sin\alpha$, and we have that $y' = x\sin\alpha + y\cos\alpha$.

Since

$$f_A \begin{pmatrix} 1 \\ 0 \end{pmatrix} = \begin{pmatrix} \cos\alpha \\ \sin\alpha \end{pmatrix} \quad \text{and} \quad f_A \begin{pmatrix} 0 \\ 1 \end{pmatrix} = \begin{pmatrix} -\sin\alpha \\ \cos\alpha \end{pmatrix},$$

we have, based on Theorem 5.2, that $A = \begin{pmatrix} \cos\alpha & -\sin\alpha \\ \sin\alpha & \cos\alpha \end{pmatrix}$ is the *matrix of the rotation of angle α and center the origin.*

We have denoted this matrix (see problem **1.61**) by $R_\alpha = \begin{pmatrix} \cos\alpha & -\sin\alpha \\ \sin\alpha & \cos\alpha \end{pmatrix}$.

Any rotation of angle α is a bijective transformation and its inverse is the rotation of angle $-\alpha$.

The set of all rotations around the origin together with the composition (of transformations) is an abelian group which is called the *group of rotations*.

The rotation around an arbitrary point (x_0, y_0) has the equations

$$\begin{cases} x' = x_0 + (x - x_0)\cos\alpha - (y - y_0)\sin\alpha \\ y' = y_0 + (x - x_0)\sin\alpha + (y - y_0)\cos\alpha. \end{cases}$$

The matrix of the uniform scaling of factor k

Let $k \in \mathbb{R}^*$. The uniform scaling of factor k is the linear transformation (geometrical transformation) which associates to the point M, the point M' such that $\overrightarrow{OM'} = k\overrightarrow{OM}$, where O is the origin of the coordinate system. We immediately obtain that the image of $M(x, y)$ is the point $M'(kx, ky)$.

Let $f_A : \mathbb{R}^2 \to \mathbb{R}^2, f_A \begin{pmatrix} x \\ y \end{pmatrix} = A \begin{pmatrix} x \\ y \end{pmatrix}$, where $A = \begin{pmatrix} a & c \\ b & d \end{pmatrix}$. Since

$$f_A \begin{pmatrix} 1 \\ 0 \end{pmatrix} = \begin{pmatrix} k \\ 0 \end{pmatrix} \quad \text{and} \quad f_A \begin{pmatrix} 0 \\ 1 \end{pmatrix} = \begin{pmatrix} 0 \\ k \end{pmatrix},$$

we have, based on Theorem 5.2, that $A = \begin{pmatrix} k & 0 \\ 0 & k \end{pmatrix}$ is the *matrix of uniform scaling of factor k.*

Remark 5.2 When $k = -1$ the uniform scaling becomes the reflection through the origin.

We mention that in geometry the uniform scaling of factor $k > 0$ is also known as the homothety of center the origin and ratio k. We denote the uniform scaling of factor k by $\theta_{0,k}$ and we have $\theta_{0,k}(x, y) = (kx, ky)$.

The homothety of center (x_0, y_0) and ratio k denoted by $\theta_{(x_0,y_0),k}$ is defined by

$$\theta_{(x_0,y_0),k}(x, y) = (x_0 + k(x - x_0), y_0 + k(y - y_0)).$$

The matrix of the orthogonal projection of vectors from $\mathbb{R}^2$ onto the x-axis

Let $f_A : \mathbb{R}^2 \to \mathbb{R}^2, f_A \begin{pmatrix} x \\ y \end{pmatrix} = A \begin{pmatrix} x \\ y \end{pmatrix}$, where $A = \begin{pmatrix} a & c \\ b & d \end{pmatrix}$.

Since the projection of $M(x, y)$ onto the x-axis is the point $M'(x, 0)$, we get that

$$f_A \begin{pmatrix} 1 \\ 0 \end{pmatrix} = \begin{pmatrix} 1 \\ 0 \end{pmatrix} \quad \text{and} \quad f_A \begin{pmatrix} 0 \\ 1 \end{pmatrix} = \begin{pmatrix} 0 \\ 0 \end{pmatrix},$$

and we have, based on Theorem 5.2, that $A = \begin{pmatrix} 1 & 0 \\ 0 & 0 \end{pmatrix}$ is the *matrix of the orthogonal projection of vectors from $\mathbb{R}^2$ onto the x-axis.*

The matrix of the orthogonal projection of vectors from $\mathbb{R}^2$ onto the x-axis

Similarly one has that $A = \begin{pmatrix} 0 & 0 \\ 0 & 1 \end{pmatrix}$ is the *matrix of the orthogonal projection of vectors from $\mathbb{R}^2$ onto the y-axis.*

5.3   Projections and reflections of the plane

Definition 5.3 A linear transformation $P : \mathbb{R}^2 \to \mathbb{R}^2$ such that $P \circ P = P$ is called the *projection of the plane.*

Remark 5.3 A projection is an idempotent transformation, i.e., $\underbrace{P \circ P \circ \cdots \circ P}_{n \text{ times}} = P$,

for any integer $n \geq 2$.

Theorem 5.3 *If $P : \mathbb{R}^2 \to \mathbb{R}^2$ is a projection defined by the matrix A, then the matrix A is idempotent, i.e., $A^2 = A$.*

Proof We have, based on Theorem 5.1, that $A_{P \circ P} = A_P A_P = A_P^2$, and it follows that $A^2 = A$. $\square$

To determine all the projections of the plane we need to determine first their associated matrices, i.e., the idempotent matrices.

Theorem 5.4 Idempotent real matrices.

The matrix $A = \begin{pmatrix} a & b \\ c & d \end{pmatrix} \in \mathcal{M}_2(\mathbb{R})$ is idempotent if and only if it has one of the following forms:

(1) $A_1 = O_2$;

(2) $A_2 = I_2$;

(3) $A_3 = \begin{pmatrix} 0 & 0 \\ c & 1 \end{pmatrix}$, $c \in \mathbb{R}$;

(4) $A_4 = \begin{pmatrix} 1 & 0 \\ c & 0 \end{pmatrix}$, $c \in \mathbb{R}$;

(5) $A_5 = \begin{pmatrix} a & b \\ \frac{a-a^2}{b} & 1-a \end{pmatrix}$, $a \in \mathbb{R}$, $b \in \mathbb{R}^*$.

Proof See the solution of problem **1.14**. $\square$

The next theorem gives the geometrical interpretation of all the projections of the plane.

Theorem 5.5 The projections of the plane.

The projections of the plane are the linear maps $P_1, P_2, P_3, P_4, P_5 : \mathbb{R}^2 \to \mathbb{R}^2$ defined by:

(1) $P_1(x, y) = (0, 0)$ *(the zero projection, all points of the plane are projected to the origin)*;

(2) $P_2(x, y) = (x, y)$ *(the identity)*;

(continued)

Theorem 5.5 (continued)

(3) $P_3(x,y) = (0, cx + y)$ *(the oblique projection onto the y-axis on the direction of the line $cx + y = 0$);*

    The point $M(x,y)$ is projected onto the y-axis. The lines which connect a point (x,y) with its image $P(x,y) = (x', y')$ have the same slope

$$m = \frac{y' - y}{x' - x} = \frac{cx + y - y}{-x} = -c \quad (constant).$$

    In conclusion, P_3 is the oblique projection onto the y-axis on the direction of the line $cx + y = 0$.

(4) $P_4(x,y) = (x, cx)$ *(the vertical projection onto the line $y = cx$);*

    We have $\mathrm{Im}P_4 = \{(x, cx) : x \in \mathbb{R}\}$, *i.e., the image of this transformation is the line $y = cx$. Since the points (x,y) and (x, cx) are located on the same vertical line, we get that P_4 is the vertical projection onto the line $y = cx$.*

(5) $P_5(x,y) = \left(ax + by, \frac{a - a^2}{b}x + (1 - a)y\right)$ *(the oblique projection onto the line $y = \frac{1-a}{b}x$ on the direction of the line $ax + by = 0$);*

    We have

$$\mathrm{Im}P_5 = \left\{\left(t, \frac{1 - a}{b}t\right) : t \in \mathbb{R}\right\},$$

and it follows that the image of this transformation is the line of equation $y = \frac{1-a}{b}x$.

    The lines which connect a point $M(x,y)$ with its image $P_5(x,y) = M'(x', y')$ have the slope

$$m = \frac{y' - y}{x' - x} = -\frac{a}{b} \quad (constant).$$

In conclusion P_5 is the oblique projection onto the line $y = \frac{1-a}{b}x$ on the direction of the line $ax + by = 0$.

Proof This follows from Theorem 5.4.

Theorem 5.6 The fundamental properties of projections.

Let $A \in \mathscr{M}_2(\mathbb{R})$, $A^2 = A$, $A \neq O_2$, $A \neq I_2$ and let

$$P_A : \mathbb{R}^2 \to \mathbb{R}^2, \quad P_A\begin{pmatrix} x \\ y \end{pmatrix} = A\begin{pmatrix} x \\ y \end{pmatrix}.$$

(continued)

Theorem 5.6 (continued)

Then:

(a) $\mathrm{Ker}P_A$ *is a line;*

(b) $\mathrm{Im}P_A$ *is a line;*

(c) P_A *projects the point* (x, y) *onto the line* $\mathrm{Im}P_A$ *on the direction of the line* $\mathrm{Ker}P_A$;

(d) *any point* $(x, y) \in \mathbb{R}^2$ *can be written uniquely as the sum of a point in* $\mathrm{Ker}P_A$ *and another point in* $\mathrm{Im}P_A$ *(this means that* $\mathbb{R}^2$ *is the direct sum of the vector subspaces* $\mathrm{Ker}P_A$ *and* $\mathrm{Im}P_A$, *i.e.,* $\mathbb{R}^2 = \mathrm{Ker}P_A \oplus \mathrm{Im}P_A$);

(e) *the Jordan canonical form of an idempotent matrix A, with* $A \neq O_2$ *and* $A \neq I_2$, *is given by* $J_A = \begin{pmatrix} 1 & 0 \\ 0 & 0 \end{pmatrix}$.

Proof From the conditions of the theorem we see that the rank of A is 1, so the system $AX = 0$ has a nontrivial solution $X_0 \neq 0$ and any other solution of the system is of the following form $X = \alpha X_0$, $\alpha \in \mathbb{R}$.

(a) $\mathrm{Ker}P_A = \{(x, y) : xy_0 - yx_0\}$, where $X_0 = \begin{pmatrix} x_0 \\ y_0 \end{pmatrix}$.

(b) $Y \in \mathrm{Im}P_A \Leftrightarrow$ there exists $X \in \mathbb{R}^2$ such that $AX = Y$, i.e., the nonhomogeneous system $AX = Y$ is compatible. This is equivalent to saying that $\mathrm{rank}(A \mid Y) = \mathrm{rank}(A) = 1$ and this implies that, if A_1 is a nonzero column of A, then $Y = \alpha A_1$, $\alpha \in \mathbb{R}$. Thus, for $A_1 = \begin{pmatrix} a \\ c \end{pmatrix}$ we have that $\mathrm{Im}P_A = \{(x, y) : cx = ay\}$.

(c) Clearly P_A projects the points of the plane onto the line $\mathrm{Im}P_A$. We only need to prove that the vector which connects a point on the plane with its image is parallel to the line $\mathrm{Ker}P_A$, i.e., the vector $AX - X$ is proportional to the vector X_0, with $AX_0 = 0$. However, $A(AX - X) = A^2X - AX = (A^2 - A)X = 0$, which implies that the vector $X_1 = AX - X$ is a solution of the system $AX_1 = 0$, so $X_1 \in \mathrm{Ker}P_A$.

(d) Observe that any point $(x, y) \in \mathbb{R}^2$ can be written uniquely as $(x, y) = (x, y) - P_A(x, y) + P_A(x, y)$, where $(x, y) - P_A(x, y) \in \mathrm{Ker}P_A$ and $P_A(x, y) \in \mathrm{Im}P_A$.

The theorem is proved. □

Definition 5.4 A linear transformation $S : \mathbb{R}^2 \to \mathbb{R}^2$ such that $S \circ S = I_{\mathbb{R}^2}$ (this implies that S is bijective and $S^{-1} = S$) is called an involution or a *reflection* of the plane.

Theorem 5.7 *If $S : \mathbb{R}^2 \to \mathbb{R}^2$ is an involution defined by the matrix A, then the matrix A is involutory, i.e., $A^2 = I_2$.*

Proof We have, based on Theorem 5.1, that $A_{S \circ S} = A_S A_S = A_S^2$, and it follows that $A^2 = I_2$.   □

To determine all the reflections of the plane we need to determine first their matrices, i.e., the real involutory matrices.

Theorem 5.8 Involutory real matrices.

The matrix $B = \begin{pmatrix} a & b \\ c & d \end{pmatrix} \in \mathcal{M}_2(\mathbb{R})$ is involutory if and only if B has one of the following forms:

(1) $B_1 = -I_2$;

(2) $B_2 = I_2$;

(3) $B_3 = \begin{pmatrix} -1 & 0 \\ c & 1 \end{pmatrix}$, $c \in \mathbb{R}$;

(4) $B_4 = \begin{pmatrix} 1 & 0 \\ c & -1 \end{pmatrix}$, $c \in \mathbb{R}$;

(5) $A_5 = \begin{pmatrix} a & b \\ \frac{1-a^2}{b} & -a \end{pmatrix}$, $a \in \mathbb{R}, \ b \in \mathbb{R}^*$.

Proof See the solution of problem **1.12**.   □

Like in the case of projections, the geometrical interpretations of all the reflections of the plane are given by the next theorem.

Theorem 5.9 The reflections of the plane.

The reflections of the plane are the linear maps $S_1, S_2, S_3, S_4, S_5 : \mathbb{R}^2 \to \mathbb{R}^2$ defined by:

(1) $S_1(x, y) = (-x, -y)$ *(the reflections through the origin);*

(2) $S_2(x, y) = (x, y)$ *(the identity map);*

(3) $S_3(x, y) = (-x, cx + y)$ *(the reflection across the y-axis on the direction of the line $cx + 2y = 0$);*

(4) $S_4(x, y) = (x, cx - y)$ *(the reflection across the line $cx - 2y = 0$ on the direction of the y-axis);*

(continued)

Theorem 5.9 (continued)

(5) $S_5(x, y) = \left(ax + by, \dfrac{1-a^2}{b}x - ay\right)$ (the reflection across the line $(a-1)x + by = 0$ on the direction of the line $(a+1)x + by = 0$).

Proof This follows from Theorem 5.8.   □

Theorem 5.10 The fundamental properties of reflections.

Let $A \in \mathcal{M}_2(\mathbb{R})$, $A^2 = I_2$, $A \neq \pm I_2$, and let

$$S_A : \mathbb{R}^2 \to \mathbb{R}^2, \quad S_A\begin{pmatrix} x \\ y \end{pmatrix} = A\begin{pmatrix} x \\ y \end{pmatrix}.$$

Then:

(a) *the set $\mathrm{Inv}S_A = \{(x, y) \in \mathbb{R}^2 : S_A(x, y) = (-x, -y)\}$ is a line;*

(b) *the set of fixed points $\mathrm{Fix}S_A = \{(x, y) \in \mathbb{R}^2 : S_A(x, y) = (x, y)\}$ is a line;*

(c) *for any $X = \begin{pmatrix} x \\ y \end{pmatrix}$ there exist and are unique the vectors $X_1, X_2 \in \mathbb{R}^2$ with $AX_1 = -X_1$, $AX_2 = X_2$ and $X = X_1 + X_2$;*

(d) *S_A is the reflection across the line $\mathrm{Fix}S_A$, on the direction of the line $\mathrm{Inv}S_A$;*

(e) *any point $(x, y) \in \mathbb{R}^2$ can be written uniquely as the sum of a point in $\mathrm{Fix}S_A$ and a point in $\mathrm{Inv}S_A$ (this means that $\mathbb{R}^2$ is the direct sum of the vector subspaces $\mathrm{Fix}S_A$ and $\mathrm{Inv}S_A$, i.e., $\mathbb{R}^2 = \mathrm{Fix}S_A \oplus \mathrm{Inv}S_A$);*

(f) *the Jordan canonical form of an involutory matrix A, with $A \neq \pm I_2$, is given by $J_A = \begin{pmatrix} 1 & 0 \\ 0 & -1 \end{pmatrix}$.*

Proof (a) The matrix $A + I_2$ has rank 1, so the system $AX = -X \Leftrightarrow (A + I_2)X = O_2$ has nontrivial solutions, all of them being of the form αX_0, $X_0 \in \mathbb{R}^2$, $X_0 \neq 0$ and $\alpha \in \mathbb{R}$.

(b) The matrix $A - I_2$ has rank 1, so the system $AX = X \Leftrightarrow (A - I_2)X = O_2$ has nontrivial solutions, all of them being of the form βX_1, $X_1 \in \mathbb{R}^2$, $X_1 \neq 0$ and $\beta \in \mathbb{R}$.

(c) If X_1 and X_2 would exist, then $AX = AX_1 + AX_2 = -X_1 + X_2$ and $X = X_1 + X_2$, so $X_1 = \frac{1}{2}(X - AX)$ and $X_2 = \frac{1}{2}(X + AX)$, which verify the conditions $AX_1 = -X_1$ and $AX_2 = X_2$.

(d) We prove that the line which connects a point with its image has a fixed direction. We have $A(AX - X) = A^2X - AX = X - AX$, which implies the vector $X_1 = AX - X$ verifies the equality $AX_1 = -X_1$, so $X_1 \in \mathrm{Inv}S_A$. On the other hand,

$A\left(\frac{1}{2}(AX + X)\right) = \frac{1}{2}(A^2X + AX) = \frac{1}{2}(X + AX)$, so $X_2 = \frac{1}{2}(X + AX)$ is a fixed vector, i.e., $AX_2 = X_2$.

(e) Any vector $\mathbf{v} = (x, y) \in \mathbb{R}^2$ can be written uniquely as $\mathbf{v} = \frac{1}{2}(\mathbf{v} + S_A(\mathbf{v})) + \frac{1}{2}(\mathbf{v} - S_A(\mathbf{v}))$, where $\frac{1}{2}(\mathbf{v} + S_A(\mathbf{v})) \in \mathrm{Fix}S_A$ and $\frac{1}{2}(\mathbf{v} - S_A(\mathbf{v})) \in \mathrm{Inv}S_A$. $\square$

Now, we establish a connection between projections and involutions. Intuitively, we have the formula

$$P(x, y) = \frac{1}{2}\left((x, y) + S(x, y)\right), \quad (x, y) \in \mathbb{R}^2,$$

which can be viewed geometrically as *the point $P(x, y)$ is the midpoint of the segment determined by (x, y) and $S(x, y)$.*

Theorem 5.11   The link between projections and involutions.

If $P : \mathbb{R}^2 \to \mathbb{R}^2$ is a projection, then $S = 2P - I$ is an involution and conversely, if $S : \mathbb{R}^2 \to \mathbb{R}^2$ is an involution, then $P = \frac{1}{2}(I + S)$ is a projection, where $I : \mathbb{R}^2 \to \mathbb{R}^2$ is the identity map.

Proof Let $A \in \mathcal{M}_2(\mathbb{R})$ be the matrix of P and let $B \in \mathcal{M}_2(\mathbb{R})$ be the matrix of S. Then, $A^2 = A$ and $B^2 = I_2$. If $B = 2A - I_2$, then

$$B^2 = 4A^2 - 4A + I_2 = 4A - 4A + I_2 = I_2.$$

On the other hand, if $A = \frac{1}{2}(I_2 + B)$, then

$$A^2 = \frac{1}{4}(I_2 + 2B + B^2) = \frac{1}{4}(I_2 + 2B + I_2) = \frac{1}{2}(I_2 + B) = A,$$

and the theorem is proved. $\square$

5.4   Gems on projections and reflections

In this section we collect gems and miscellaneous results about the projections and the reflections of the plane.

Theorem 5.12 *Let $\mathcal{D}_1 : ax + by = 0$ and $\mathcal{D}_2 : cx + dy = 0$, $a^2 + b^2 \neq 0$, $c^2 + d^2 \neq 0$ and $ad - bc \neq 0$ be two lines passing through the origin.*

(continued)

Theorem 5.12 (continued)

(a) *The projection P onto the line $\mathcal{D}_1$ on the direction of the line $\mathcal{D}_2$ is the linear transformation defined by $P : \mathbb{R}^2 \to \mathbb{R}^2$*

$$P(x, y) = \left(\frac{-bcx - bdy}{ad - bc}, \frac{acx + ady}{ad - bc} \right).$$

(b) *The reflection S across the line $\mathcal{D}_1$ on the direction of the line $\mathcal{D}_2$ is the linear transformation defined by $S : \mathbb{R}^2 \to \mathbb{R}^2$*

$$S(x, y) = \left(\frac{-(ad + bc)x - 2bdy}{ad - bc}, \frac{2acx + (ad + bc)y}{ad - bc} \right).$$

Proof For any $(x, y) \in \mathbb{R}^2$ there exist and are unique $(x_1, y_1) \in \mathcal{D}_1$ and $(x_2, y_2) \in \mathcal{D}_2$ such that $(x, y) = (x_1, y_1) + (x_2, y_2)$. Since $\mathrm{Im}P = \mathrm{Fix}S = \mathcal{D}_1$ and $\mathrm{Ker}P = \mathrm{Inv}S = \mathcal{D}_2$, we get that $P(x, y) = (x_1, y_1)$ and $S(x, y) = (x_1, y_1) - (x_2, y_2)$.

Solving the system

$$\begin{cases} ax_1 + by_1 = 0 \\ cx_2 + dy_2 = 0 \\ x_1 + x_2 = x \\ y_1 + y_2 = y, \end{cases}$$

we get that

$$x_1 = \frac{-bcx - bdy}{ad - bc}, \; y_1 = \frac{acx + ady}{ad - bc}, \; x_2 = \frac{adx + bdy}{ad - bc}, \; y_2 = \frac{-acx - bcy}{ad - bc}$$

and it follows that

$$P(x, y) = \left(\frac{-bcx - bdy}{ad - bc}, \frac{acx + ady}{ad - bc} \right)$$

and

$$S(x, y) = \left(\frac{-(ad + bc)x - 2bdy}{ad - bc}, \frac{2acx + (ad + bc)y}{ad - bc} \right).$$

The matrices of P and S are

$$M_P = \frac{1}{ad - bc} \begin{pmatrix} -bc & -bd \\ ac & ad \end{pmatrix} \quad \text{and} \quad M_S = \frac{1}{ad - bc} \begin{pmatrix} -(ad + bc) & -2bd \\ 2ac & ad + bc \end{pmatrix}.$$

The theorem is proved.   □

Lemma 5.2 *Let $\mathcal{D}_1$, $\mathcal{D}_2$, $\mathcal{D}_3$, $\mathcal{D}_4$ be four lines passing through the origin such that $\mathcal{D}_1 \perp \mathcal{D}_3$ and $\mathcal{D}_2 \perp \mathcal{D}_4$. If the matrix of the projection onto the line $\mathcal{D}_1$ on the direction of line $\mathcal{D}_2$ is M, then the matrix of the projection onto the line $\mathcal{D}_4$ on the direction of line $\mathcal{D}_3$ is M^T.*

Proof Let $\mathcal{D}_1 : ax + by = 0$, $\mathcal{D}_2 : cx + dy = 0$, $\mathcal{D}_3 : -bx + ay = 0$ and $\mathcal{D}_4 : -dx + cy = 0$ be the lines through the origin. We have, based on Theorem 5.12, that the matrix of the projection onto the line $\mathcal{D}_1$ on the direction of $\mathcal{D}_2$ is

$$M = \frac{1}{ad - bc} \begin{pmatrix} -bc & -bd \\ ac & ad \end{pmatrix}$$

and the matrix of the projection onto the line $\mathcal{D}_4$ on the direction of $\mathcal{D}_3$ is obtained from matrix M via the substitutions $a \to -d$, $b \to c$, $c \to -b$, $d \to a$ and we get that

$$\frac{1}{-ad + bc} \begin{pmatrix} bc & -ac \\ bd & -ad \end{pmatrix} = \frac{1}{ad - bc} \begin{pmatrix} -bc & ac \\ -bd & ad \end{pmatrix} = M^T.$$

Similarly one can prove that if A is the matrix of the reflection across the line $\mathcal{D}_1$ on the direction of $\mathcal{D}_2$, then A^T is the matrix of the reflection across the line $\mathcal{D}_4$ on the direction of $\mathcal{D}_3$. ☐

Lemma 5.3 When is a linear map an orthogonal projection?

Let $A \in \mathcal{M}_2(\mathbb{R})$, $A \neq O_2$, $A \neq I_2$ and let $f_A : \mathbb{R}^2 \to \mathbb{R}^2$ be the linear transformation defined by the matrix A. Then, f_A is an orthogonal projection if and only if $AA^T = A$.

Proof Since $AA^T = A$ we get that $AA^T = A^T$ which implies that $A = A^T$. Thus, $A^2 = A$ which shows that A is an idempotent matrix and f_A is a projection. We have, based on Lemma 5.2, that the projection on line $\mathcal{D}_1$ on the direction of line $\mathcal{D}_2$ is orthogonal if $\mathcal{D}_1 \perp \mathcal{D}_2$, so $\mathcal{D}_3 \equiv \mathcal{D}_2$ and $\mathcal{D}_4 \equiv \mathcal{D}_1$. It follows that the projection f_A is orthogonal if and only if $A = A^T$. ☐

Nota bene. The matrix of an orthogonal projection $P : \mathbb{R}^2 \to \mathbb{R}^2$ is given by (see problem **5.8**)

(continued)

$$M_P = \begin{pmatrix} a & b \\ b & 1-a \end{pmatrix}, \quad a, b \in \mathbb{R} \quad \text{with} \quad a^2 + b^2 = a,$$

and $P(x, y) = (ax + by, \; bx + (1-a)y), \; \forall \, (x, y) \in \mathbb{R}^2$.

Observe this is not an orthogonal matrix, i.e., the matrix corresponding to an orthogonal projection is *symmetric* and *not orthogonal*!

Lemma 5.4 Projections and their matrices.

Let $A \in \mathcal{M}_2(\mathbb{R})$ be a matrix having the eigenvalues $\lambda_1 = 1$ and $\lambda_2 = 0$ and the corresponding eigenvectors $X_1 = \begin{pmatrix} a \\ b \end{pmatrix}$ and $X_2 = \begin{pmatrix} c \\ d \end{pmatrix}$. Then, A is the matrix of the projection onto the line $\mathscr{D}_1 : bx - ay = 0$ on the direction of the line $\mathscr{D}_2 : dx - cy = 0$.

Proof The Jordan canonical form of A is $J_A = \begin{pmatrix} 1 & 0 \\ 0 & 0 \end{pmatrix}$ and the invertible matrix P is given by $P = (X_1 \,|\, X_2) = \begin{pmatrix} a & c \\ b & d \end{pmatrix}$. A calculation shows that

$$A = P J_A P^{-1} = \frac{1}{ad - bc} \begin{pmatrix} ad & -ac \\ bd & -bc \end{pmatrix}.$$

We obtain, by replacing $a \to b$, $b \to -a$, $c \to d$, and $d \to -c$ in the formula of matrix M_P given at the end of the proof of Theorem 5.12, the matrix A and this proves the lemma. $\qquad\qquad\qquad\qquad\qquad\qquad\qquad\qquad\qquad\qquad\qquad\quad\square$

Lemma 5.5 Reflections and their matrices.

Let $B \in \mathcal{M}_2(\mathbb{R})$ be a matrix having the eigenvalues $\lambda_1 = 1$ and $\lambda_2 = -1$ and the corresponding eigenvectors $X_1 = \begin{pmatrix} a \\ b \end{pmatrix}$ and $X_2 = \begin{pmatrix} c \\ d \end{pmatrix}$. Then, B is the matrix of the reflection across the line $\mathscr{D}_1 : bx - ay = 0$ on the direction of the line $\mathscr{D}_2 : dx - cy = 0$.

Proof The Jordan canonical form of B is $J_B = \begin{pmatrix} 1 & 0 \\ 0 & -1 \end{pmatrix}$ and the invertible matrix Q is given by $Q = (X_1 \,|\, X_2) = \begin{pmatrix} a & c \\ b & d \end{pmatrix}$ with $Q^{-1} = \frac{1}{ad - bc} \begin{pmatrix} d & -c \\ -b & a \end{pmatrix}$.

A calculation shows that

$$B = QJ_BQ^{-1} = \frac{1}{ad - bc}\begin{pmatrix} ad + bc & -2ac \\ 2bd & -(ad + bc) \end{pmatrix}.$$

We obtain by replacing in the formula of matrix M_S given at the end of the proof of Theorem 5.12 by $a \to b, b \to -a, c \to d$, and $d \to -c$ the matrix B and this proves the lemma.   □

5.5   The isometries of the plane

Definition 5.5 A linear transformation $T : \mathbb{R}^2 \to \mathbb{R}^2$ with $T(x, y) = (x', y')$, such that $x^2 + y^2 = x'^2 + y'^2$, for all $(x, y) \in \mathbb{R}^2$, is called *a linear isometry* of the plane.

Lemma 5.6 *An isometry preserves the inner product, the distance between points and the angle between vectors in $\mathbb{R}^2$.*

Proof If $T(x_1, y_1) = (x_1', y_1')$ and $T(x_2, y_2) = (x_2', y_2')$, we need to prove that

$$x_1x_2 + y_1y_2 = x_1'x_2' + y_1'y_2'.$$

We have

$$T(x_1 + x_2, y_1 + y_2) = (x_1' + x_2', y_1' + y_2')$$

and $(x_1 + x_2)^2 + (y_1 + y_2)^2 = (x_1' + x_2')^2 + (y_1' + y_2')^2$.
This implies that

$$x_1^2 + 2x_1x_2 + x_2^2 + y_1^2 + 2y_1y_2 + y_2^2 = x_1'^2 + 2x_1'x_2' + x_2'^2 + y_1'^2 + 2y_1'y_2' + y_2'^2.$$

Since $x_1^2 + y_1^2 = x_1'^2 + y_1'^2$ and $x_2^2 + y_2^2 = x_2'^2 + y_2'^2$ we get that $x_1x_2 + y_1y_2 = x_1'x_2' + y_1'y_2'$.
If $\alpha = \angle((x_1, y_1), (x_2, y_2))$ and $\alpha' = \angle(T(x_1, y_1), T(x_2, y_2))$, then

$$\cos \alpha = \frac{x_1x_2 + y_1y_2}{\sqrt{x_1^2 + y_1^2}\sqrt{x_2^2 + y_2^2}} \quad \text{and} \quad \cos \alpha' = \frac{x_1'x_2' + y_1'y_2'}{\sqrt{x_1'^2 + y_1'^2}\sqrt{x_2'^2 + y_2'^2}}$$

which are equal based on the first part of the theorem.
    To prove that $d((x_1, y_1), (x_2, y_2)) = d(T(x_1, y_1), T(x_2, y_2))$ we need to show that $(x_1 - x_2)^2 + (y_1 - y_2)^2 = (x_1' - x_2')^2 + (y_1' - y_2')^2$, which reduces to proving that $x_1x_2 + y_1y_2 = x_1'x_2' + y_1'y_2'$.   □

Definition 5.6 A function $F : \mathbb{R}^2 \to \mathbb{R}^2$, not necessarily a linear transformation, which preserves the distance between points is called an isometry.

Theorem 5.13 *A linear transformation* $T : \mathbb{R}^2 \to \mathbb{R}^2$ *is an isometry if and only if its associated matrix M_T has one of the following forms*

$$M_T = \begin{pmatrix} \cos t & -\sin t \\ \sin t & \cos t \end{pmatrix} \quad or \quad M_T = \begin{pmatrix} \cos t & \sin t \\ \sin t & -\cos t \end{pmatrix}.$$

Proof Let $M_T = \begin{pmatrix} a & b \\ c & d \end{pmatrix}$. We have $T(x, y) = (x', y') = (ax + by, cx + dy)$ and

$$x^2 + y^2 = (ax + by)^2 + (cx + dy)^2, \quad \forall\, (x, y) \in \mathbb{R}^2.$$

This implies by identifying the coefficients of x^2, y^2, and xy that $a^2 + c^2 = 1$, $b^2 + d^2 = 1$, and $ab + cd = 0$. Since $a^2 + c^2 = 1$ and $b^2 + d^2 = 1$ we get that there exist $t, s \in \mathbb{R}$ such that $\cos t = a$, $\sin t = c$ and $\cos s = d$, $\sin s = b$. The equality $ab + cd = 0$ implies that

$$\cos t \sin s + \sin t \cos s = 0 \quad \Leftrightarrow \quad \sin(s + t) = 0$$

and it follows that $t + s \subset \{k\pi : k \in \mathbb{Z}\}$.

When $s + t = 0$ we get $s = -t$ and this implies that $M_{T_1} = \begin{pmatrix} \cos t & -\sin t \\ \sin t & \cos t \end{pmatrix}$.

When $s + t = \pi$ we get $s = \pi - t$ and this implies that $M_{T_2} = \begin{pmatrix} \cos t & \sin t \\ \sin t & -\cos t \end{pmatrix}$.

The theorem is proved.   □

Remark 5.4 We mention that the matrix

$$M_{T_1} = \begin{pmatrix} \cos t & -\sin t \\ \sin t & \cos t \end{pmatrix}$$

corresponds to a counterclockwise rotation of angle t, while the matrix

$$M_{T_2} = \begin{pmatrix} \cos t & \sin t \\ \sin t & -\cos t \end{pmatrix}$$

corresponds to the composition of a rotation and a reflection, i.e., $M_{T_2} = M_{T_1} M_S$, where $M_S = \begin{pmatrix} 1 & 0 \\ 0 & -1 \end{pmatrix}$ is the matrix of the reflection across the x-axis.

Definition 5.7 Let $(x_0, y_0) \in \mathbb{R}^2$ be fixed. A function $T_{(x_0, y_0)} : \mathbb{R}^2 \to \mathbb{R}^2$ defined by $T_{(x_0, y_0)}(x, y) = (x + x_0, y + y_0)$ is called the *translation of vector* (x_0, y_0).

Thus, the equations of the translation are

$$T_{(x_0,y_0)}(x,y) = (x',y') \quad \Leftrightarrow \quad \begin{cases} x' = x_0 + x \\ y' = y_0 + y. \end{cases}$$

The origin $(0,0)$ is translated to the point (x_0, y_0).

The composition of two translations is a translation

$$T_{(x_0,y_0)} \circ T_{(x_0',y_0')} = T_{(x_0+x_0',y_0+y_0')}$$

and the inverse of a translation is also a translation $T_{(x_0,y_0)}^{-1} = T_{(-x_0,-y_0)}$.

We mention that a translation preserves the distances between two points, the angles between lines, transforms parallel lines to parallel lines, and sends circles to circles.

The set of all translations together with the composition of applications is a group which is called the *group of translations* of the plane.

Definition 5.8 If $f : \mathbb{R}^2 \to \mathbb{R}^2$ is a linear transformation and $T : \mathbb{R}^2 \to \mathbb{R}^2$ is a translation, then the functions $g_1, g_2 : \mathbb{R}^2 \to \mathbb{R}^2$, defined by $g_1 = T \circ f$ and $g_2 = f \circ T$ are called *affine transformations*.

Thus, the affine transformations are translations composed to linear transformations.

If $A = \begin{pmatrix} a & b \\ c & d \end{pmatrix}$, is the matrix associated with f and (x_0, y_0) is the vector of the translation, then $g(x,y) = (x', y')$ where

$$\begin{pmatrix} x' \\ y' \end{pmatrix} = A \begin{pmatrix} x \\ y \end{pmatrix} + \begin{pmatrix} x_0 \\ y_0 \end{pmatrix},$$

which implies that $g(x,y) = (ax + by + x_0, cx + dy + y_0)$, $(x,y) \in \mathbb{R}^2$, and

$$\begin{cases} x' = ax + by + x_0 \\ y' = cx + dy + y_0, \end{cases}$$

are the *equations of the affine transformation*.

The set of all affine applications together with the composition of functions is a group which is called the *group of affine transformations*.

5.6   Systems of coordinates on the plane

The standard *coordinate cartesian system xOy* in the plane $\mathbb{R}^2 = \{(x, y) : x, y \in \mathbb{R}\}$ consists of two orthogonal lines, the *x*-axis, $Ox = \{(x, 0) : x \in \mathbb{R}\}$ and the *y*-axis, $Oy = \{(0, y) : y \in \mathbb{R}\}$, which intersect at the *origin* of the cartesian system $O(0, 0)$.

By rotating the cartesian system around the origin counterclockwise by an angle α we obtain a new system of coordinates which we denote by $x'Oy'$. Any point M on the plane is uniquely determined with respect to the system xOy by the pair of real numbers (x, y) and the same point considered with respect to the system $x'Oy'$ is determined by the pair of real numbers (x', y'). These two pairs are related to one another by the formulae

$$\begin{pmatrix} x \\ y \end{pmatrix} = \begin{pmatrix} \cos\alpha & -\sin\alpha \\ \sin\alpha & \cos\alpha \end{pmatrix} \begin{pmatrix} x' \\ y' \end{pmatrix}$$

or

$$\begin{pmatrix} x' \\ y' \end{pmatrix} = \begin{pmatrix} \cos\alpha & \sin\alpha \\ -\sin\alpha & \cos\alpha \end{pmatrix} \begin{pmatrix} x \\ y \end{pmatrix}$$

which allows one to pass from one coordinate system to another via the rotation matrices R_α or $R_{-\alpha}$.

By translating the coordinate system $x'Oy'$, so that the origin $O(0, 0)$ is translated to the point $O''(x_0, y_0)$, we obtain a new coordinate system $x''O''y''$ of the plane. We denote by (x'', y'') the coordinate of M with respect to the new system $x''O''y''$, then we have the formula

$$\begin{pmatrix} x'' \\ y'' \end{pmatrix} = R_{-\alpha} \begin{pmatrix} x - x_0 \\ y - y_0 \end{pmatrix}$$

or

$$\begin{pmatrix} x \\ y \end{pmatrix} = \begin{pmatrix} x_0 \\ y_0 \end{pmatrix} + R_\alpha \begin{pmatrix} x'' \\ y'' \end{pmatrix}.$$

Example 5.1 If the coordinate system $x''O''y''$ is obtained by rotating the cartesian system xOy counterclockwise by the angle $\frac{\pi}{6}$ and then by translating it to the point $O''(1, 2)$, then a point M on the plane which has coordinates (x, y) with respect to the standard coordinate system and with respect to the new system $x''O''y''$ has coordinates (x'', y'') are given by the formula

$$\begin{pmatrix} x'' \\ y'' \end{pmatrix} = \begin{pmatrix} \frac{\sqrt{3}}{2} & \frac{1}{2} \\ -\frac{1}{2} & \frac{\sqrt{3}}{2} \end{pmatrix} \begin{pmatrix} x - 1 \\ y - 2 \end{pmatrix}$$

or

$$\begin{pmatrix} x \\ y \end{pmatrix} = \begin{pmatrix} 1 \\ 2 \end{pmatrix} + \begin{pmatrix} \frac{\sqrt{3}}{2} & -\frac{1}{2} \\ \frac{1}{2} & \frac{\sqrt{3}}{2} \end{pmatrix} \begin{pmatrix} x'' \\ y'' \end{pmatrix}.$$

5.7   Problems

5.1 Let S_x be the reflection across the x-axis and S_y be the reflection across the y-axis. Find the matrix associated with $S_x \circ S_y$.

5.2 Let $ABCDEF$ be a regular hexagon with side length 2 which viewed with respect to the system xCy has the vertices B and E on Cx respectively Cy. We consider another system $x'Fy'$ positively oriented, the x'-axis being FA. Determine:

(a) the formula of passing from the system xCy to the system $x'Fy'$;

(b) the coordinates of vertices C and E with respect to the system $x'Fy'$.

5.3 What becomes the equation $x^2 - y^2 = 2$ when the system xOy is rotated counterclockwise by an angle of $\frac{\pi}{4}$ around the origin?

5.4 *Projections.* Give the geometrical interpretation of the following linear transformations $f : \mathbb{R}^2 \to \mathbb{R}^2$:

(a) $f(x, y) = (0, 2x + y)$;

(b) $f(x, y) = (x, 2x)$;

(c) $f(x, y) = (3x - y, 6x - 2y)$.

5.5 *Reflections.* Give the geometrical interpretation of the following linear transformations $f : \mathbb{R}^2 \to \mathbb{R}^2$:

(a) $f(x, y) = (-x, 2x + y)$;

(b) $f(x, y) = (x, 2x - y)$;

(c) $f(x, y) = (3x - y, 8x - 3y)$.

5.6 Find $a, b \in \mathbb{R}$ such that the following matrices are projection matrices and give the geometrical interpretation of these projections:

(a) $M_1 = \begin{pmatrix} 2 & a \\ 1 & b \end{pmatrix}$;

(b) $M_2 = \begin{pmatrix} 2 & 1 \\ a & b \end{pmatrix}$.

5.7 Prove that the function

$$f : \mathbb{R}^2 \to \mathbb{R}^2, \quad f(x, y) = (ax - by + b, \ bx + ay - a + 1),$$

is a rotation for any $(a, b) \neq (1, 0)$ with $a^2 + b^2 = 1$. Determine the center and the angle of the rotation.

5.8 Orthogonal projections and their matrices.

Prove that the linear transformation $P : \mathbb{R}^2 \to \mathbb{R}^2$ is an orthogonal projection onto a line passing through the origin if and only if there exist $a, b \in \mathbb{R}$ such that $a^2 + b^2 = a$ and $P(x, y) = (ax + by, \; bx + (1 - a)y)$, $\forall (x, y) \in \mathbb{R}^2$.

5.9 Orthogonal reflections and their matrices.

Prove that the linear transformation $S : \mathbb{R}^2 \to \mathbb{R}^2$ is an orthogonal reflection across a line passing through the origin if and only if there exist $a, b \in \mathbb{R}$ such that $a^2 + b^2 = 1$ and $S(x, y) = (ax + by, \; bx - ay)$, $\forall (x, y) \in \mathbb{R}^2$.

5.10 When is the sum of two projections a projection?

Let $P_1, P_2 : \mathbb{R}^2 \to \mathbb{R}^2$ be two nonzero projections. Prove that if $P_1 + P_2$ is a projection, then $P_1 + P_2 = I_{\mathbb{R}^2}$.

5.11 When is the sum of two projections an involution?

Let $P_1, P_2 : \mathbb{R}^2 \to \mathbb{R}^2$ be two nonzero projections. Prove that if $P_1 + P_2$ is an involution, then $P_1 + P_2 = I_{\mathbb{R}^2}$.

5.12 Prove that an isometry of the plane is uniquely determined by the images of three noncollinear points.

5.13 Prove that any isometry of the plane is of the form $F = R \circ T$ or $F = S \circ R \circ T$, where T is a translation, R is an affine rotation (around a point), and S is a reflection across a line.

5.14 Prove that the composition of two orthogonal reflections is a rotation.

5.15 Let $\mathscr{C}$ be the curve $5x^2 + 8xy + 5y^2 = 1$. Prove that there exits a rotation in the plane, $(x, y) \to (x', y') = R(x, y)$, such that with respect to the system of coordinates $x'Oy'$ the curve $\mathscr{C}$ has the equation $ax'^2 + by'^2 = 1$, for some $a, b \in \mathbb{R}$.

5.16 Write the affine application $f : \mathbb{R}^2 \to \mathbb{R}^2, f(x, y) = (2x - 3y + 1, 3x + 2y - 1)$ as a composition of elementary transformations.

5.17 Let xOy be the coordinate system of the plane and let $\mathscr{C}$ be the curve $2x^2 - y^2 - 4xy = 1$. Determine a rotation $R_\alpha : \mathbb{R}^2 \to \mathbb{R}^2$, $R_\alpha(x, y) = (x', y')$ such that in the new system of coordinates $x'Oy'$ the equation of the curve $\mathscr{C}$ becomes $ax'^2 + by'^2 = 1$.

5.18 Find the image of the square $ABCD$, where $A(1, 1)$, $B(-1, 1)$, $C(-1, -1)$, and $D(1, -1)$, under the transformation whose matrix is $\begin{pmatrix} 2 & -2 \\ 1 & 3 \end{pmatrix}$.

5.19 Let xOy be the coordinate system of the plane and let $A(2, 0)$, $B(2, 2)$, $C(0, 2)$ be the vertices of a square. We consider the transformation which sends the origin to $O'(3, -1)$ and such that the new axis of coordinate $O'C'$ makes with the x-axis an angle α with $\tan \alpha = \frac{3}{4}$. Determine the coordinates of the vertices of the square $O'A'B'C'$ with respect to the coordinate system xOy.

5.20 What is the image of the line $x - y + 1 = 0$ under the rotation R of center the origin and angle $\frac{\pi}{3}$?

5.21 Give the geometrical interpretation of the linear transformation $f : \mathbb{R}^2 \to \mathbb{R}^2$:

$$f(x, y) = (\sqrt{3}x - y, \, x + \sqrt{3}y).$$

5.22 Orthogonal reflections revisited.

Prove that for any orthogonal reflection S across a line which passes through the origin, there exists $t \in \mathbb{R}$ such that the matrix associated with S is

$$M_S = \begin{pmatrix} \cos t & \sin t \\ \sin t & -\cos t \end{pmatrix}.$$

5.23 Determine the projection onto the line $\mathscr{D}_1 : 2x + y = 0$ on the direction of the line $\mathscr{D}_2 : x - 3y = 0$.

5.24 Orthogonal projection and reflection across a line passing through the origin.

Determine the equations of the orthogonal projection onto the line $\mathscr{D}: ax + by = 0$, $a^2 + b^2 \neq 0$ and the equations of the orthogonal reflection across the line $\mathscr{D}$.

5.25 Orthogonal projection and reflection across a line not passing through the origin.

Determine the equations of the orthogonal projection and the orthogonal reflection across the line $\mathscr{D}: a(x - x_0) + b(y - y_0) = 0$.

5.26 The isometries of the square. Let $\mathscr{P}$ be the set of points in the plane located in the interior or on the square $ABCD$. Determine all isometries of the square $ABCD$, i.e., all isometries $f : \mathscr{P} \to \mathscr{P}$.

5.27 The billiard problem. Let $\mathscr{D}$ be a line and let A and B be two distinct points on the same side of $\mathscr{D}$. Determine the point M on $\mathscr{D}$ such that $AM + MB$ is minimum.

5.28 Pompeiu's Theorem. Let $\triangle ABC$ be an equilateral triangle and let M be a point on the plane of $\triangle ABC$ not on the circumscribed circle of $\triangle ABC$. Prove that the segments $[MA]$, $[MB]$ and $[MC]$ are the sides of a triangle.

5.29 Torricelli's point. Determine a point on the plane of $\triangle ABC$ such that the sum of the distances to the vertices of the triangle is minimum.

5.8 Solutions

5.1. The matrix associated with S_x is $A = \begin{pmatrix} 1 & 0 \\ 0 & -1 \end{pmatrix}$ and the matrix associated with S_y is $B = \begin{pmatrix} -1 & 0 \\ 0 & 1 \end{pmatrix}$. It follows that the matrix associated with $S_x \circ S_y$ is $AB = -I_2$.

5.2. (a) The coordinate systems xCy and $x'Fy'$ have different orientation and the matrix of the linear application is of the form $A = \begin{pmatrix} \cos \alpha & -\sin \alpha \\ -\sin \alpha & -\cos \alpha \end{pmatrix}$, where $\alpha = \frac{\pi}{3}$ is the angle between the axes Cx and Fx'. The change of coordinates is given by the formula

$$\begin{pmatrix} x \\ y \end{pmatrix} = A \begin{pmatrix} x' \\ y' \end{pmatrix} + \begin{pmatrix} 2 \\ 2\sqrt{3} \end{pmatrix} \quad \text{or} \quad \begin{cases} x = \frac{1}{2}x' - \frac{\sqrt{3}}{2}y' + 2 \\ y = -\frac{\sqrt{3}}{2}x' - \frac{1}{2}y' + 2\sqrt{3}. \end{cases}$$

(b) For the point C, which has the coordinates $x = 0$, $y = 0$, we get that $x' = 2$, $y' = 2\sqrt{2}$, so $C(2, 2\sqrt{2})$. For the point E, with coordinates $x = 0$ and $y = 2\sqrt{3}$, we have that $x' = -1$ and $y' = \sqrt{3}$, so $E(-1, \sqrt{3})$.

5.3. Let x and y be the coordinates of a point on the hyperbola $x^2 - y^2 - 2 = 0$ before the rotation and let X and Y be the coordinates of the same point on the hyperbola after the rotation. We have

$$\begin{cases} x = X\cos\dfrac{\pi}{4} - Y\sin\dfrac{\pi}{4} = \dfrac{\sqrt{2}}{2}(X - Y) \\[2mm] y = X\sin\dfrac{\pi}{4} + Y\cos\dfrac{\pi}{4} = \dfrac{\sqrt{2}}{2}(X + Y). \end{cases}$$

It follows that

$$x^2 - y^2 - 2 = \left(\frac{\sqrt{2}}{2}(X - Y)\right)^2 - \left(\frac{\sqrt{2}}{2}(X + Y)\right)^2 - 2 = -2XY - 2 = 0.$$

Thus, in the new system of coordinates the hyperbola $x^2 - y^2 - 2 = 0$ has the equation $XY + 1 = 0$ and the x-axis is the symmetry axis of the hyperbola.

5.4. (a) The matrix of the linear transformation is $A = \begin{pmatrix} 0 & 0 \\ 2 & 1 \end{pmatrix}$, so we have a projection onto the y-axis on the direction of the line $\mathscr{D} : 2x + y = 0$.

(b) The matrix of the linear transformation is $A = \begin{pmatrix} 1 & 0 \\ 2 & 0 \end{pmatrix}$ and this is a vertical projection onto the line $\mathscr{D} : y - 2x = 0$.

(c) The matrix of the linear transformation is $A = \begin{pmatrix} 3 & -1 \\ 6 & -2 \end{pmatrix}$ and we have a projection onto the line $\mathscr{D} : y - 2x = 0$ on the direction of the line $\mathscr{D}' : 3x - y = 0$.

5.5. (a) The matrix of the linear transformation is $A = \begin{pmatrix} -1 & 0 \\ 2 & 1 \end{pmatrix}$, $A^2 = I_2$, so f is a reflection across the y-axis on the direction of the line $x + y = 0$.

(b) The matrix of the linear transformation is $A = \begin{pmatrix} 1 & 0 \\ 2 & -1 \end{pmatrix}$, $A^2 = I_2$, so f is a vertical reflection across the line $x - y = 0$.

(c) The matrix of the linear transformation is $A = \begin{pmatrix} 3 & -1 \\ 8 & -3 \end{pmatrix}$, $A^2 = I_2$, so f is a reflection across the line $2x - y = 0$ on the direction of the line $4x - y = 0$.

5.6. (a) The condition $M_1^2 = M_1$ implies that $a = -2$, $b = -1$, and $M_1 = \begin{pmatrix} 2 & -2 \\ 1 & -1 \end{pmatrix}$.
We have

$$P_1\begin{pmatrix} x \\ y \end{pmatrix} = M_1\begin{pmatrix} x \\ y \end{pmatrix} = \begin{pmatrix} 2x - 2y \\ x - y \end{pmatrix}.$$

It follows that P_1 is the projection onto the line $\mathscr{D}_1 : x = 2y$ on the direction of the line $\mathscr{D}_2 : x - y = 0$.

(b) From $M_2^2 = M_2$ we get that $a = -2, b = -1$ and $M_2 = \begin{pmatrix} 2 & 1 \\ -2 & -1 \end{pmatrix}$. We have $P_2(x, y) = (2x + y, -2x - y)$, which is the projection onto the line $\mathcal{D}_3 : x + y = 0$ on the direction of the line $\mathcal{D}_4 : 2x + y = 0$.

5.7. The center of the rotation is the unique fix point of the rotation. Thus, $f(x_0, y_0) = (x_0, y_0)$ and we get that

$$\begin{cases} ax_0 - by_0 + b = x_0 \\ bx_0 + ay_0 - a + 1 = y_0 \end{cases} \Leftrightarrow \begin{cases} (a-1)x_0 - by_0 = -b \\ bx_0 + (a-1)y_0 = a-1. \end{cases}$$

The determinant of the system is $\begin{vmatrix} a-1 & -b \\ b & a-1 \end{vmatrix} = (a-1)^2 + b^2 \neq 0$, since $(a, b) \neq (1, 0)$. The system has a unique solution $x_0 = 0$ and $y_0 = 1$ and we have that the center of the rotation is $C(0, 1)$.

The equations of a rotation of center $C(x_0, y_0)$ and angle α are

$$\begin{pmatrix} x' \\ y' \end{pmatrix} = \begin{pmatrix} x_0 \\ y_0 \end{pmatrix} + \begin{pmatrix} \cos\alpha & -\sin\alpha \\ \sin\alpha & \cos\alpha \end{pmatrix} \begin{pmatrix} x - x_0 \\ y - y_0 \end{pmatrix}.$$

In our case these equations become

$$\begin{pmatrix} ax - by + b \\ bx + ay - a + 1 \end{pmatrix} = \begin{pmatrix} 0 \\ 1 \end{pmatrix} + \begin{pmatrix} \cos\alpha & -\sin\alpha \\ \sin\alpha & \cos\alpha \end{pmatrix} \begin{pmatrix} x \\ y - 1 \end{pmatrix}.$$

These imply that

$$\begin{cases} ax - by + b = x\cos\alpha - (y-1)\sin\alpha \\ bx + ay - a + 1 = 1 + x\sin\alpha + (y-1)\cos\alpha, \end{cases}$$

for all $x, y \in \mathbb{R}$. We get the necessary conditions $\cos\alpha = a$ and $\sin\alpha = b$ and we note that these conditions can be satisfied since $a^2 + b^2 = 1$. There is $\alpha \in (0, 2\pi)$ such that $\cos\alpha = a$ and $\sin\alpha = b$.

5.8. We have, based on Lemma 5.3, that P is an orthogonal projection if and only if its matrix is a symmetric matrix, i.e., $M_P = M_P^T$. However, Theorem 5.4 shows that the symmetric matrices of rank 1 are

$$A_3 = \begin{pmatrix} 0 & 0 \\ 0 & 1 \end{pmatrix}, \quad A_4 = \begin{pmatrix} 1 & 0 \\ 0 & 0 \end{pmatrix} \quad \text{and} \quad A_5 = \begin{pmatrix} a & b \\ \frac{a-a^2}{b} & 1-a \end{pmatrix}, \quad a \in \mathbb{R}, \ b \in \mathbb{R}^*.$$

Using the condition $A_5^T = A_5$ we get that $\frac{a-a^2}{b} = b \Leftrightarrow a^2 + b^2 = a$, so

$$A_5 = \begin{pmatrix} a & b \\ b & 1-a \end{pmatrix}, \quad a \in \mathbb{R}, \ b \in \mathbb{R}^*,$$

and if we allow $b = 0$ we recover the matrices A_3 and A_4.

It follows that the matrix M_P is of the following form

$$M_P = \begin{pmatrix} a & b \\ b & 1-a \end{pmatrix}, \quad a, b \in \mathbb{R} \quad \text{with} \quad a^2 + b^2 = a,$$

and $P(x, y) = (ax + by, \; bx + (1-a)y), \; \forall \, (x, y) \in \mathbb{R}^2$.

5.9. We have, based on Lemma 5.2, that the matrix of an orthogonal reflection is a symmetric matrix so that we choose from Theorem 5.8 only the symmetric matrices. These are

$$B_3 = \begin{pmatrix} -1 & 0 \\ 0 & 1 \end{pmatrix}, \quad B_4 = \begin{pmatrix} 1 & 0 \\ 0 & -1 \end{pmatrix} \quad \text{and} \quad B_5 = \begin{pmatrix} a & b \\ \frac{1-a^2}{b} & -a \end{pmatrix}, \quad a \in \mathbb{R}, \; b \in \mathbb{R}^*.$$

The symmetry condition on matrix B_5 implies that $a^2 + b^2 = 1$. Thus, matrix B_5 is of the following form

$$B_5 = \begin{pmatrix} a & b \\ b & -a \end{pmatrix},$$

and if we allow $b = 0$ we get matrices B_3 and B_4.

It follows that the matrix M_S of the orthogonal reflection S across a line passing through the origin is of the following form

$$M_S = \begin{pmatrix} a & b \\ b & -a \end{pmatrix}, \quad a, b \in \mathbb{R} \quad \text{with} \quad a^2 + b^2 = 1.$$

If we let $a = \cos t$ and $b = \sin t$ we get that $M_S = \begin{pmatrix} \cos t & \sin t \\ \sin t & -\cos t \end{pmatrix}$. This is, according to Theorem 5.13, the matrix of an isometry which appears in problem **5.22** by a different reasoning.

5.10. Since $(P_1 + P_2)^2 = P_1 + P_2$, $P_1^2 = P_1$, and $P_2^2 = P_2$ we get that

$$P_1 \circ P_2 + P_2 \circ P_1 = 0. \tag{5.1}$$

Applying P_2 to the left and to the right in (5.1) we get that $P_2 \circ P_1 \circ P_2 + P_2 \circ P_1 = 0$ and $P_1 \circ P_2 + P_2 \circ P_1 \circ P_2 = 0$ which implies that $P_1 \circ P_2 = P_2 \circ P_1$. It follows, based on (5.1), that $P_1 \circ P_2 = P_2 \circ P_1 = 0$.

We make the observation that if P is a projection, then $\text{Fix}P = \text{Im}P$.

If $x \in \text{Fix}P_1$, then $P_2 \circ P_1(x) = 0$ implies that $P_2(x) = 0$, so $\text{Fix}P_1 \subseteq \text{Ker}P_2$. Analogously $\text{Fix}P_2 \subseteq \text{Ker}P_1$. Since $P_1 \neq 0$ and $P_2 \neq 0$ we get that $\text{Fix}P_1 \neq \{(0,0)\}$, $\text{Ker}P_2 \neq \mathbb{R}^2$, so $\text{Fix}P_1 = \text{Ker}P_2$ and $\text{Fix}P_2 = \text{Ker}P_1$. However, $\text{Fix}P_1 = \mathcal{D}_1 = \text{Ker}P_2$ and $\text{Fix}P_2 = \mathcal{D}_2 = \text{Ker}P_1$ are distinct lines passing through the origin. It follows, since $\mathcal{D}_1 \oplus \mathcal{D}_2 = \mathbb{R}^2$, that if $(x, y) = (x_1, y_1) + (x_2, y_2)$, $(x_1, y_1) \in \mathcal{D}_1$,

$(x_2, y_2) \in \mathcal{D}_2$, then $P_1(x, y) = (x_1, y_1)$ and $P_2(x, y) = (x_2, y_2)$. Thus, $P_1(x, y) + P_2(x, y) = (x, y) \Rightarrow P_1 + P_2 = I$.

Remark 5.5 It is worth mentioning that if P_1 is the projection onto the line $\mathcal{D}_1$ on the direction of line $\mathcal{D}_2$, then P_2 is the projection onto the line $\mathcal{D}_2$ on the direction of line $\mathcal{D}_1$.

5.11. Since $(P_1 + P_2)^2 = I$, $P_1^2 = P_1$ and $P_2^2 = P_2$ we get that $P_1 + P_2 + P_1 \circ P_2 + P_2 \circ P_1 = I$. Applying P_2 to the left and to the right in the preceding equality we get that $2P_2 \circ P_1 + P_2 \circ P_1 \circ P_2 = 0$ and $2P_1 \circ P_2 + P_2 \circ P_1 \circ P_2 = 0$ and it follows that $P_1 \circ P_2 = P_2 \circ P_1$. However, the equality $2P_2 \circ P_1 + P_2 \circ P_1 \circ P_2 = 0$ implies that $3P_2 \circ P_1 = 0 \Rightarrow P_1 \circ P_2 = P_2 \circ P_1 = 0$. Now the solution is the same as the solution of problem **5.10**.

5.12. Let A_1, A_2, and A_3 be three noncollinear points. First we prove that the only isometry F which satisfies the conditions $F(A_1) = A_1$, $F(A_2) = A_2$, and $F(A_3) = A_3$ is the identity. For any point M on the plane let r_1, r_2, and r_3 be the distances from M to A_1, A_2, and A_3 respectively. Since $d(F(M), F(A_i)) = d(M, A_i) = r_i$, $i = 1, 2, 3$, we get that $F(M)$ is the point located at the intersection of the circles with centers A_1, A_2, and A_3 and radius r_1, r_2, and r_3. This point is unique, so $F(M) = M$.

Now, if we assume that there are two isometries which satisfy $F_1(A_i) = F_2(A_i)$, $i = 1, 2, 3$, then $(F_1^{-1} \circ F_2)(A_i) = A_i$, $i = 1, 2, 3$, so $F_1 = F_2$.

5.13. We have based on problem **5.12** that any isometry is uniquely determined by images of three noncollinear points. Let A, B, C be the vertices of a triangle with different side lengths and let $A' = F(A)$, $B' = F(B)$, and $C' = F(C)$, where F is an isometry of the plane. Observe that $\triangle ABC$ and $\triangle A'B'C'$ are congruent since $AB = A'B'$, $AC = A'C'$, and $BC = B'C'$.

If $\triangle ABC$ and $\triangle A'B'C'$ have the same orientation, they may overlap via a translation defined by $T(A) = A'$ followed by a rotation R, around point A', of angle $\widehat{AB, A'B'}$.

If $\triangle ABC$ and $\triangle A'B'C'$ have different orientation, they may overlap after a translation T followed by a rotation R, like in the previous case (the side AB overlaps onto the side $A'B'$) and a reflection across the line $A'B'$.

5.14. The matrices of such reflections are given, based on problem **5.9**, by $M_1 = \begin{pmatrix} \cos t_1 & \sin t_1 \\ \sin t_1 & -\cos t_1 \end{pmatrix}$ and $M_2 = \begin{pmatrix} \cos t_2 & \sin t_2 \\ \sin t_2 & -\cos t_2 \end{pmatrix}$. The matrix of the composition of the two reflections is $M_1 M_2 = \begin{pmatrix} \cos(t_1 - t_2) & -\sin(t_1 - t_2) \\ \sin(t_1 - t_2) & \cos(t_1 - t_2) \end{pmatrix} = R_{t_1 - t_2}$, which is the matrix of a rotation of angle $t_1 - t_2$.

5.15. The matrix of the rotation is $R_\alpha = \begin{pmatrix} \cos \alpha & -\sin \alpha \\ \sin \alpha & \cos \alpha \end{pmatrix}$ and the equations of the rotation are $\begin{pmatrix} x' \\ y' \end{pmatrix} = R_\alpha \begin{pmatrix} x \\ y \end{pmatrix}$ or $\begin{pmatrix} x \\ y \end{pmatrix} = R_{-\alpha} \begin{pmatrix} x' \\ y' \end{pmatrix}$. This implies that

$$\begin{cases} x = x' \cos \alpha + y' \sin \alpha \\ y = -x' \sin \alpha + y' \cos \alpha. \end{cases}$$

Putting these values of x and y in the equation of the curve $\mathscr{C}$ we get that

$$\mathscr{C} : 5(x'^2 + y'^2) - 8x'^2 \sin \alpha \cos \alpha + 8y'^2 \sin \alpha \cos \alpha + 8(\cos^2 \alpha - \sin^2 \alpha)x'y' = 1.$$

Since the term $x'y'$ should vanish we obtain that angle α verifies the equation $\cos^2 \alpha - \sin^2 \alpha = 0$, so we can choose $\alpha = \frac{\pi}{4}$. Thus, by rotating the system of coordinates xOy by an angle of $\frac{\pi}{4}$ the equation of the conic $\mathscr{C}$ becomes $x'^2 + 9y'^2 = 1$, so $a = 1$ and $b = 9$.

5.16. We have

$$\begin{pmatrix} x' \\ y' \end{pmatrix} = \begin{pmatrix} 2 & -3 \\ 3 & 2 \end{pmatrix} \begin{pmatrix} x \\ y \end{pmatrix} + \begin{pmatrix} 1 \\ -1 \end{pmatrix}$$

$$= \sqrt{13} \begin{pmatrix} \cos t & -\sin t \\ \sin t & \cos t \end{pmatrix} \begin{pmatrix} x \\ y \end{pmatrix} + \begin{pmatrix} 1 \\ -1 \end{pmatrix}$$

$$= \begin{pmatrix} \sqrt{13} & 0 \\ 0 & \sqrt{13} \end{pmatrix} R_t \begin{pmatrix} x \\ y \end{pmatrix} + \begin{pmatrix} 1 \\ -1 \end{pmatrix}$$

$$= O_{\sqrt{13}} R_t \begin{pmatrix} x \\ y \end{pmatrix} + \mathbf{v},$$

so $f = T_{\mathbf{v}} \circ O_{\sqrt{13}} \circ R_t$, where R_t is a rotation of angle $t = \arctan \frac{3}{2}$, $O_{\sqrt{13}}$ is the uniform scaling of factor $k = \sqrt{13}$ and $T_{\mathbf{v}}$ is the translation of vector $\mathbf{v} = \begin{pmatrix} 1 \\ -1 \end{pmatrix}$.

5.17. Since the matrix of the rotation is $R_\alpha = \begin{pmatrix} \cos \alpha & -\sin \alpha \\ \sin \alpha & \cos \alpha \end{pmatrix}$ and the equations of the rotation are $\begin{pmatrix} x' \\ y' \end{pmatrix} = R_\alpha \begin{pmatrix} x \\ y \end{pmatrix}$ or $\begin{pmatrix} x \\ y \end{pmatrix} = R_{-\alpha} \begin{pmatrix} x' \\ y' \end{pmatrix}$. This implies that

$$\begin{cases} x = x' \cos \alpha + y' \sin \alpha \\ y = -x' \sin \alpha + y' \cos \alpha. \end{cases}$$

Putting these values of x and y in the equation of the curve $\mathscr{C}$ we get that

$$(2 \cos^2 \alpha - \sin^2 \alpha + 4 \sin \alpha \cos \alpha)x'^2 + y'^2(2 \sin^2 \alpha - \cos^2 \alpha - 4 \sin \alpha \cos \alpha)$$

$$+ (6 \sin \alpha \cos \alpha - 4 \cos^2 \alpha + 4 \sin^2 \alpha)x'y' = 1.$$

Since the coefficient of $x'y'$ should vanish we have that $6\sin\alpha\cos\alpha - 4\cos^2\alpha + 4\sin^2\alpha = 0 \Rightarrow \alpha = \frac{1}{2}\arctan\frac{4}{3}$ and the equation of the conic $\mathscr{C}$ becomes $3x'^2 - 2y'^2 = 1$, so $a = 3$ and $b = -2$.

5.18. $\begin{pmatrix} 2 & -2 \\ 1 & 3 \end{pmatrix}\begin{pmatrix} 1 \\ 1 \end{pmatrix} = \begin{pmatrix} 0 \\ 4 \end{pmatrix}$, so $A'(0, 4)$. Similarly we get that $B'(-4, 2)$, $C'(0, -4)$, and $D'(4, -2)$, so the square $ABCD$ is mapped to the parallelogram $A'B'C'D'$.

5.19. Observe the rotation angle is $\theta = \frac{3\pi}{2} + \alpha$. Therefore, $\cos\theta = \cos\left(\frac{3\pi}{2} + \alpha\right) = \sin\alpha = \frac{3}{5}$ and $\sin\theta = -\cos\alpha = -\frac{4}{5}$. The transformation formulae are

$$\begin{cases} x = \dfrac{3}{5}x' + \dfrac{4}{5}y' + 3 \\ y = -\dfrac{4}{5}x' + \dfrac{3}{5}y' - 1. \end{cases}$$

With respect to the new coordinate system $x'O'y'$, the vertices A', B', C', and D' have the same coordinates like the vertices O, A, B, and C with respect to the old system of coordinates xOy. The coordinates of A', B', and C' with respect to the system xOy are obtained from the transformation formulae and we have $A'\left(\frac{21}{5}, -\frac{13}{5}\right)$, $B'\left(\frac{29}{5}, -\frac{7}{5}\right)$ and $C'\left(\frac{23}{5}, \frac{1}{5}\right)$.

5.20. The rotation of center the origin and angle $\frac{\pi}{3}$ is given by the equations

$$\begin{cases} x' = x\cos\dfrac{\pi}{3} - y\sin\dfrac{\pi}{3} \\ y' = x\sin\dfrac{\pi}{3} + y\sin\dfrac{\pi}{3} \end{cases} \Leftrightarrow \begin{cases} x = \dfrac{1}{2}x' + \dfrac{\sqrt{3}}{2}y' \\ y = -\dfrac{\sqrt{3}}{2}x' + \dfrac{1}{2}y'. \end{cases}$$

Replacing x and y in the equation $x - y + 1 = 0$ we get $\frac{\sqrt{3}+1}{2}x' + \frac{\sqrt{3}-1}{2}y' + 1 = 0$, which is the equation of a line.

5.21. The matrix of the linear transformation is

$$\begin{pmatrix} \sqrt{3} & -1 \\ 1 & \sqrt{3} \end{pmatrix} = \begin{pmatrix} 2 & 0 \\ 0 & 2 \end{pmatrix}\begin{pmatrix} \cos\frac{\pi}{6} & -\sin\frac{\pi}{6} \\ \sin\frac{\pi}{6} & \cos\frac{\pi}{6} \end{pmatrix},$$

so f is the composition of a counterclockwise rotation of angle $\frac{\pi}{6}$ and a uniform scaling of factor 2.

5.22. The matrix of a reflection is of the form $M_S = \begin{pmatrix} a & b \\ \frac{1-a^2}{b} & -a \end{pmatrix}$ and represents the reflection across the line $\mathscr{D}_1 : (a - 1)x + by = 0$ on the direction of the line $\mathscr{D}_2 : (a + 1)x + by = 0$. The condition that the lines $\mathscr{D}_1$ and $\mathscr{D}_2$ are perpendicular is $a^2 - 1 + b^2 = 0 \Leftrightarrow a^2 + b^2 = 1$. This implies there exists $t \in [0, 2\pi)$ such that $a = \cos t$, $b = \sin t$, so the matrix M_S has the required form.

5.23. We write $(x, y) = (x_1, y_1) + (x_2, y_2)$, with $(x_1, y_1) \in \mathscr{D}_1$ and $(x_2, y_2) \in \mathscr{D}_2$. This implies that $2x_1 + y_1 = 0$, $x_2 - 3y_2 = 0$, $x_1 + x_2 = x$, $y_1 + y_2 = y$. It follows that $x_1 = \frac{1}{7}(x-3y)$, $x_2 = \frac{3}{7}(2x+y)$, $y_1 = -\frac{2}{7}(x-3y)$, $y_2 = \frac{1}{7}(2x+y)$ and we have

$$P(x,y) = \left(\frac{x-3y}{7}, \frac{-2x+6y}{7}\right).$$

5.24. The direction line to which the projection and the reflection are done is the line $\mathscr{D}' : -bx + ay = 0$, the perpendicular line to $\mathscr{D}$. Any point $(x,y) \in \mathbb{R}^2$ can be written in the form $(x,y) = (x_1, y_1) + (x_2, y_2)$, with $(x_1, y_1) \in \mathscr{D}$, $(x_2, y_2) \in \mathscr{D}'$ and we have $P(x,y) = (x_1, y_1)$ and $S(x,y) = 2P(x,y) - (x,y) = (x_1, y_1) - (x_2, y_2)$. We have the system of equations

$$\begin{cases} x_1 + x_2 = x \\ y_1 + y_2 = y \\ ax_1 + by_1 = 0 \\ -bx_2 + ay_2 = 0, \end{cases}$$

from which it follows that

$$x_1 = \frac{b^2 x - aby}{a^2 + b^2}, \quad y_1 = \frac{-abx + a^2 y}{a^2 + b^2}, \quad x_2 = \frac{a^2 x + aby}{a^2 + b^2}, \quad y_2 = \frac{abx + b^2 y}{a^2 + b^2}.$$

The matrices of the two linear transformations are

$$M_P = \frac{1}{a^2 + b^2} \begin{pmatrix} b^2 & -ab \\ -ab & a^2 \end{pmatrix} \quad \text{and} \quad M_S = \frac{1}{a^2 + b^2} \begin{pmatrix} -a^2 + b^2 & -2ab \\ -2ab & a^2 - b^2 \end{pmatrix}.$$

One can check that $M_P^2 = M_P$ and $M_S^2 = I_2$.

5.25. Let P_1 and S_1 be the orthogonal projection and the orthogonal reflection across the line $\mathscr{D} : a(x-x_0) + b(y-y_0) = 0$ and let P and S be the orthogonal projection and the orthogonal reflection across the line $ax + by = 0$. Then, $P_1(x,y) = (x_0, y_0) + P(x - x_0, y - y_0)$ and $S_1(x,y) = (x_0, y_0) + S(x - x_0, y - y_0)$. We have, based on problem **5.24**, that

$$P_1(x,y) = \left(\frac{a^2 x_0 + aby_0}{a^2 + b^2}, \frac{abx_0 + b^2 y_0}{a^2 + b^2}\right) + \left(\frac{b^2 x - aby}{a^2 + b^2}, \frac{-abx + a^2 y}{a^2 + b^2}\right)$$

and

$$S_1(x,y) = 2P_1(x,y) - (x,y) = \left(\frac{2a^2 x_0 + 2aby_0}{a^2 + b^2}, \frac{2abx_0 + 2b^2 y_0}{a^2 + b^2}\right)$$
$$+ \left(\frac{(b^2 - a^2)x - 2aby}{a^2 + b^2}, \frac{-2abx + (a^2 - b^2)y}{a^2 + b^2}\right).$$

If $P_1(x, y) = (x_1, y_1)$ and $S_1(x, y) = (x_2, y_2)$, then we have the matrix equations

$$\begin{pmatrix} x_1 \\ y_1 \end{pmatrix} = \frac{1}{a^2 + b^2} \begin{pmatrix} a^2 & ab \\ ab & b^2 \end{pmatrix} \begin{pmatrix} x_0 \\ y_0 \end{pmatrix} + \frac{1}{a^2 + b^2} \begin{pmatrix} b^2 & -ab \\ -ab & a^2 \end{pmatrix} \begin{pmatrix} x \\ y \end{pmatrix}$$

and

$$\begin{pmatrix} x_2 \\ y_2 \end{pmatrix} = \frac{2}{a^2 + b^2} \begin{pmatrix} a^2 & ab \\ ab & b^2 \end{pmatrix} \begin{pmatrix} x_0 \\ y_0 \end{pmatrix} + \frac{1}{a^2 + b^2} \begin{pmatrix} b^2 - a^2 & -2ab \\ -2ab & a^2 - b^2 \end{pmatrix} \begin{pmatrix} x \\ y \end{pmatrix}.$$

5.26. It suffices to solve the problem for the square with vertices $A(1, 0)$, $B(0, 1)$, $C(-1, 0)$, and $D(0, -1)$.

First, we observe that since $d(A, C) = d(B, D) = 2$, then for any isometry f we have $d(f(A), f(C)) = d(f(B), f(D)) = 2$ and this implies that the vertices of the square $ABCD$ are sent to vertices, i.e., $f(A)$, $f(C)$ and $f(B)$, $f(D)$ are opposite vertices of the square. The value $f(A)$ is chosen from the set $\{A, B, C, D\}$ in four possible ways and $f(C)$ is the opposite vertex of $f(A)$. In each of these cases $f(B)$ is chosen in two possible ways from the other two vertices. We obtain 8 such functions which are, so far, only isometries of the set of vertices of the square $ABCD$.

Second, we note that if $M \in \mathscr{P}$, which is different from the vertices of the square, then M is uniquely determined by the distances from M to the vertices A, B, C, i.e., $a = d(M, A)$, $b = d(M, B)$ and $c = d(M, C)$. We note that M is located at the intersection of the circles $\mathscr{C}(A, a)$, $\mathscr{C}(B, b)$ and $\mathscr{C}(C, c)$. It follows that the point $f(M)$ is located at the intersection of the circles with centers $f(A), f(B), f(C)$ and radius a, b, c respectively. The 8 isometries of the vertices extend to the 8 isometries of the square. These are:

- $f(A) = A, f(C) = C, f(B) = B$ and $f(D) = D \Rightarrow f = 1_{\mathscr{P}}$;
- $f(A) = A, f(C) = C, f(B) = D$ and $f(D) = B \Rightarrow f = \sigma_x$ the symmetry across the x-axis;
- $f(A) = C, f(C) = A, f(B) = B$ and $f(D) = D \Rightarrow f = \sigma_y$ the symmetry across the y-axis;
- $f(A) = C, f(C) = A, f(B) = D$ and $f(D) = B \Rightarrow f = \sigma_O$ the symmetry through the origin;
- $f(A) = B, f(C) = D, f(B) = A$ and $f(D) = C \Rightarrow f = \sigma_{y-x=0}$ the symmetry across the line $y - x = 0$;
- $f(A) = B, f(C) = D, f(B) = C$ and $f(D) = A \Rightarrow f = \mathscr{R}_{\frac{\pi}{2}}$ the rotation of angle $\frac{\pi}{2}$ with center the origin;
- $f(A) = D, f(C) = B, f(B) = C$ and $f(D) = A \Rightarrow f = \sigma_{y+x=0}$ the symmetry across the line $y + x = 0$;
- $f(A) = D, f(C) = B, f(B) = A$ and $f(D) = C \Rightarrow f = \mathscr{R}_{-\frac{\pi}{2}}$ the rotation of angle $-\frac{\pi}{2}$ with center the origin.

These isometries form the dihedral group D_8.

5.27. Let $s_{\mathscr{D}}$ be the orthogonal reflection across the line $\mathscr{D}$ and let $A' = s_{\mathscr{D}}(A)$ and $\{M\} = \mathscr{D} \cap BA'$. We prove that M is the point for which the minimum is attained. If $P \in \mathscr{D}$, is an arbitrary point, then we apply triangle's inequality in $\triangle A'PB$ and we get that $A'P + PB \geq A'B$, with equality if and only if $P = M$. On the other hand, $A'P = AP$, $A'M = AM$ and we have $AP + PB = A'P + PB \geq A'B = AM + MB$, which implies that the minimum is attained when $P = M$ (Fig. 5.1).

5.28. Let $r = r_{B,-\frac{\pi}{3}}$ be the clockwise rotation of angle $\frac{\pi}{3}$ around B. Then, $r(A) = C$, $r(C) = C'$, $r(M) = M'$ and point B is fixed. We have that $\triangle MBM'$ is an isosceles triangle and since $\widehat{B} = \frac{\pi}{3}$ we get that $\triangle MBM'$ is equilateral. Therefore $\triangle CMM'$ has its sides congruent to the segments $[MC]$, $[MB]$ and $[MA]$. Observe that $\triangle CMM'$ degenerates if and only if M is located on the circle circumscribed to $\triangle ABC$ (Fig. 5.2).

5.29. Let $r = r_{A,\frac{\pi}{3}}$ be the counterclockwise rotation around point A of angle $\frac{\pi}{3}$ and let $C' = r(C)$, $M' = r(M)$, where M is an arbitrary point on the plane of $\triangle ABC$. We have $MA + MB + MC = BM + MM' + M'C' \geq BC'$, with equality if and

Fig. 5.1 The billiard problem

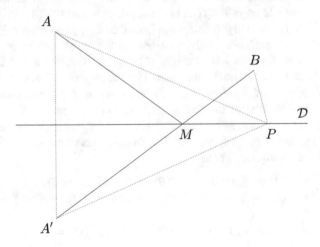

Fig. 5.2 Pompeiu's Theorem

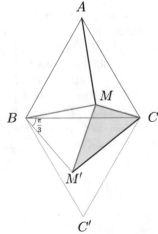

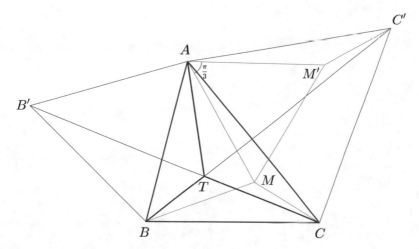

Fig. 5.3 The Torricelli point

only if the points B, M, M' and C' are collinear. Since $\triangle AMM'$ is equilateral we get that $\widehat{AMB} = 120°$ and $\widehat{AM'C'} = 120°$. It follows that $\widehat{AMB} = \widehat{AMC} = 120°$. The construction of the Torricelli's point T is as follows: we construct the equilateral triangles $\triangle AC'C$ and $\triangle AB'B$ to the exterior of $\triangle ABC$ and we have $\{T\} = BC' \cap CB'$ (Fig. 5.3).

Chapter 6
Conics

> It is easy to teach someone, but to show him
> an easy way to realize the learned things, this
> is something to admire.
>
> St. John Chrysostom (347–407)

6.1 Conics

Definition 6.1 An algebraic plane curve is a curve whose implicit equation is of the following form

$$\mathscr{C}: \quad F(x, y) = 0,$$

where F is a polynomial in variables x and y. The degree of the polynomial is called the degree of the algebraic curve.

Definition 6.2 A conic is an algebraic plane curve of degree two. The general equation of a conic is

$$\mathscr{C}: \quad a_{11}x^2 + 2a_{12}xy + a_{22}y^2 + b_1x + b_2y + c = 0,$$

where $a_{11}, a_{12}, a_{22}, b_1, b_2, c \in \mathbb{R}$ and $a_{11}^2 + a_{12}^2 + a_{22}^2 \neq 0$.

When the system of plane coordinates is specially chosen the equation of the conic has a simple form, called the canonical form. We review the nondegenerate conics.

The nondegenerate conics

- **The ellipse** is defined as the set of points $M(x, y)$ in the plane whose sum of the distances to two distinct points, $F(c, 0)$ and $F'(-c, 0)$, $c > 0$, called *foci*, is constant. Thus, the set $\mathscr{E}$ of points $M(x, y)$ with the property that $MF + MF' = 2a$, $a > c$ is called an ellipse (Fig. 6.1).
  Let $b^2 = a^2 - c^2$. The equations of the ellipse are:

$$\mathscr{E}: \quad \frac{x^2}{a^2} + \frac{y^2}{b^2} = 1 \quad \text{the implicit equation}$$

© Springer International Publishing AG 2017
V. Pop, O. Furdui, *Square Matrices of Order 2*, DOI 10.1007/978-3-319-54939-2_6

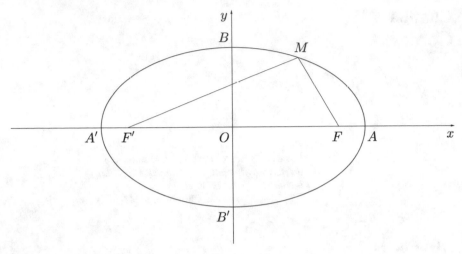

Fig. 6.1 The ellipse $\frac{x^2}{a^2} + \frac{y^2}{b^2} = 1, \quad a, b > 0$

$$\mathcal{E}: \quad \begin{cases} y = \dfrac{b}{a}\sqrt{a^2 - x^2} \\ y = -\dfrac{b}{a}\sqrt{a^2 - x^2} \end{cases} \qquad x \in [-a, a] \quad \text{the Cartesian equations}$$

$$\mathcal{E}: \quad \begin{cases} x = a\cos t \\ y = b\sin t \end{cases} \qquad t \in [0, 2\pi) \quad \text{the parametric equations}$$

When $a = b = r$ the ellipse becomes the circle

$$x^2 + y^2 = r^2 \quad \text{or} \quad \begin{cases} x = r\cos t \\ y = r\sin t \end{cases} \qquad t \in [0, 2\pi).$$

The optical property. The tangent and the normal line at a point on an ellipse are the bisectors of the angles determined by the *focal radii*.

■ **The hyperbola** is the set of points $M(x, y)$ in the plane for which the absolute value of the difference between the distances from two fixed points, $F(c, 0)$ and $F'(-c, 0)$, $c > 0$, called *foci*, is constant. Thus, the set $\mathcal{H}$ of points $M(x, y)$ with the property that $|MF - MF'| = 2a, 0 < a < c$ is called a hyperbola (Fig. 6.2). The line determined by F and F' is called the *focal axis*, the length of the segment $FF' = 2c$ is called the *focal distance* and the segments MF and MF' are called the *focal radii*. Direct calculations show that the equations of the hyperbola are:

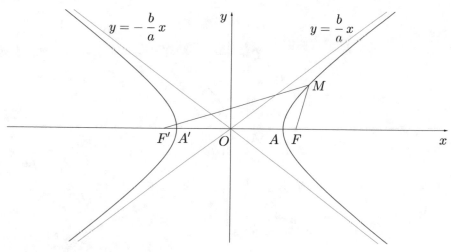

Fig. 6.2 The hyperbola $\frac{x^2}{a^2} - \frac{y^2}{b^2} = 1$, $a, b > 0$

$\mathscr{H}$: $\dfrac{x^2}{a^2} - \dfrac{y^2}{b^2} = 1$ the implicit equation

$\mathscr{H}$: $\begin{cases} y = \dfrac{b}{a}\sqrt{x^2 - a^2} \\ y = -\dfrac{b}{a}\sqrt{x^2 - a^2} \end{cases}$ $x \in (-\infty, -a] \cup [a, \infty)$ the Cartesian equations

$\mathscr{H}$: $\begin{cases} x = \pm a \cosh t \\ y = b \sinh t \end{cases}$ $t \in \mathbb{R}$ the parametric equations

where

$$\cosh t = \frac{e^t + e^{-t}}{2} \quad \text{and} \quad \sinh t = \frac{e^t - e^{-t}}{2}.$$

The hyperbola is an unbounded curve which has the inclined asymptotes $y = \frac{b}{a}x$ and $y = -\frac{b}{a}x$. A hyperbola with perpendicular asymptotes is called equilateral. *The optical property.* The tangent and the normal line at a point on a hyperbola are the bisectors of the angles determined by the focal radii.

■ **The parabola** is defined as the set of points on the plane $M(x, y)$ whose distances to a fixed line $x = -\frac{p}{2}, p > 0$, called the *directrix* and a fixed point $F\left(\frac{p}{2}, 0\right)$ called *focus* are equal (Fig. 6.3).

Thus, the equations of the parabola are:

$\mathscr{P}$: $y^2 = 2px$ the implicit equation

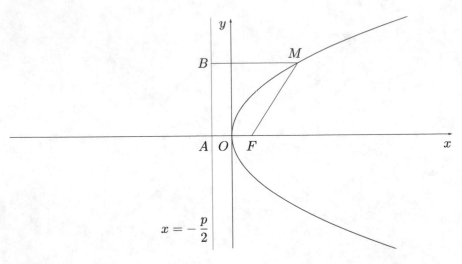

Fig. 6.3 The parabola $y^2 = 2px, \ p > 0$

$$\mathscr{P}: \quad \begin{cases} x = \dfrac{t^2}{2p} \\ y = t \end{cases} \quad t \in \mathbb{R} \quad \text{the parametric equations.}$$

A parabola, in general, is also defined as the graph of the functions of the following form

$$y = ax^2 + bx + c, \quad a \neq 0 \quad \text{or} \quad x = a'y^2 + b'y + c', \quad a' \neq 0.$$

The optical property. The tangent and the normal lines at a point on a parabola are the bisectors of the angles determined by the focal radius and the parallel line through the point to the axis of the parabola.

The degenerate conics

The algebraic curves of second degree which are degenerate conics are:

- $\mathscr{C}: (a_1x + b_1y + c_1)(a_2x + b_2y + c_2) = 0$ (the union of two lines)
- $\mathscr{C}: \alpha(x - x_0)^2 + \beta(y - y_0)^2 = 0, \ \alpha, \beta > 0$ (a point)
- $\mathscr{C}: \alpha(x - x_0)^2 + \beta(y - y_0)^2 + \delta = 0, \ \alpha, \beta, \delta > 0$ (the void set).

Elementary properties of conics that can be formulated using elementary geometry can be found in [?].

6.2   The reduction of conics to their canonical form

Let

$$\mathscr{C}: \quad a_{11}x^2 + 2a_{12}xy + a_{22}y^2 + b_1 x + b_2 y + c = 0, \qquad (6.1)$$

where $a_{11}, a_{12}, a_{22}, b_1, b_2, c \in \mathbb{R}$ and $a_{11}^2 + a_{12}^2 + a_{22}^2 \neq 0$ be a conic in the xOy plane.

To reduce a conic to its canonical form we understand to chose a system of coordinates $x'O'y'$ in the plane such that in the new coordinates the conic would have a simplified equation, the so-called reduced equation. We shall see that any such change of coordinates consists of two geometrical transformations, a translation and a rotation (and eventually a reflection across an axis). These transformations are determined based on a technique involving the Jordan canonical form of symmetric matrices of order 2.

Let $f(x, y) = a_{11}x^2 + 2a_{12}xy + a_{22}y^2$ be the quadratic part from the equation (6.1) and let A_f be the symmetric matrix associated with f

$$A_f = \begin{pmatrix} a_{11} & a_{12} \\ a_{12} & a_{22} \end{pmatrix}.$$

Nota bene. The coefficients on the second diagonal of the matrix A_f are equal to half of the coefficient of xy in the equation of the conic.

It is known (see Theorem 2.5) that the matrix A_f is diagonalizable and the matrix $P \in \mathscr{M}_2(\mathbb{R})$ can be chosen to be an orthogonal matrix, i.e., $P^T = P^{-1}$. In fact P is a rotation matrix. The eigenvalues λ_1, λ_2 of A_f are real numbers (see Theorem 2.5), at least one of them being nonzero, since $A_f \neq O_2$.

Let

$$J_{A_f} = \begin{pmatrix} \lambda_1 & 0 \\ 0 & \lambda_2 \end{pmatrix} = P^T A_f P,$$

where P is the matrix formed with the eigenvectors corresponding to the eigenvalues λ_1 and λ_2.

The rotation. We change the coordinate system by making an orthogonal transformation in the xOy plane, a rotation defined by the matrix P

$$\begin{pmatrix} x \\ y \end{pmatrix} = P \begin{pmatrix} x' \\ y' \end{pmatrix},$$

and the coordinate system xOy is changed to $x'Oy'$.

Using the formula

$$[f(x, y)] = \begin{pmatrix} x \\ y \end{pmatrix}^T A_f \begin{pmatrix} x \\ y \end{pmatrix},$$

we get that

$$[f(x, y)] = \left(P\begin{pmatrix} x' \\ y' \end{pmatrix} \right)^T A_f P\begin{pmatrix} x' \\ y' \end{pmatrix} = \begin{pmatrix} x' \\ y' \end{pmatrix}^T J_{A_f} \begin{pmatrix} x' \\ y' \end{pmatrix} = \lambda_1 x'^2 + \lambda_2 y'^2.$$

Thus, in the new system of coordinates $x'Oy'$ the equation of the conic becomes

$$\mathscr{C} : \quad \lambda_1 x'^2 + \lambda_2 y'^2 + b'_1 x' + b'_2 y' + c = 0, \tag{6.2}$$

where the coefficients b'_1, b'_2 are determined by the formulae

$$b_1 x + b_2 y = b'_1 x' + b'_2 y' \quad \Leftrightarrow \quad [b_1 \ b_2]\begin{pmatrix} x \\ y \end{pmatrix} = [b'_1 \ b'_2]\begin{pmatrix} x' \\ y' \end{pmatrix}$$

which is equivalent to

$$[b_1 \ b_2]P\begin{pmatrix} x' \\ y' \end{pmatrix} = [b'_1 \ b'_2]\begin{pmatrix} x' \\ y' \end{pmatrix} \quad \text{so} \quad [b'_1 \ b'_2] = [b_1 \ b_2]P.$$

Nota bene. The purpose of the rotation is to make the term xy disappear.

The translation. We distinguish between the cases when both eigenvalues of A_f are nonzero and one is zero.

Case 1. If $\lambda_1 \neq 0$ and $\lambda_2 \neq 0$ we write equation (6.2) in the following form

$$\mathscr{C} : \quad \lambda_1 \left(x' + \frac{b'_1}{2\lambda_1} \right)^2 + \lambda_2 \left(y' + \frac{b'_2}{2\lambda_2} \right)^2 + c' = 0,$$

where

$$c' = c - \frac{b'^2_1}{4\lambda_1} - \frac{b'^2_2}{4\lambda_2}.$$

Now we translate the coordinate system $x'Oy'$ to the coordinate system $x''O''y''$ and the equations of the translation are

$$T : \begin{cases} x'' = x' + \dfrac{b'_1}{2\lambda_1} \\ y'' = y' + \dfrac{b'_2}{2\lambda_2}. \end{cases}$$

The center of the new system of coordinates is the point O'' whose coordinates are determined by

$$x'' = y'' = 0 \quad \Leftrightarrow \quad x' = -\frac{b'_1}{2\lambda_1}, \ y' = -\frac{b'_2}{2\lambda_2} \quad \Leftrightarrow \quad \begin{pmatrix} x \\ y \end{pmatrix} = P\begin{pmatrix} x' \\ y' \end{pmatrix}.$$

The equation of the conic becomes

$$\mathscr{C}: \quad \lambda_1 x''^2 + \lambda_2 y''^2 + c' = 0.$$

If $c' = 0$ we get a degenerate conic which could be a point or a union of two lines.

If $c' \neq 0$ the conic is either an ellipse or a hyperbola according to whether the eigenvalues λ_1, λ_2 have the same sign or different signs.

Case 2. If one of the eigenvalues is 0, say $\lambda_2 = 0$ and $\lambda_1 \neq 0$ we have

$$\mathscr{C}: \quad \lambda_1 \left(x' + \frac{b_1'}{2\lambda_1} \right)^2 + b_2' y' + c' = 0, \quad c' = c - \frac{b_1'^2}{4\lambda_1},$$

and in this case the equations of the translation are

$$T: \begin{cases} x'' = x' + \dfrac{b_1'}{2\lambda_1} \\ y'' = y' + \dfrac{c'}{b_2'}. \end{cases}$$

The conic is a parabola of equation

$$\mathscr{C}: \quad \lambda_1 x''^2 + b_2' y'' = 0.$$

Nota bene. The equations of the translation are determined by completing the squares (square) in x' and/or y'.

Remark 6.1 If the conic is nondegenerate, then its nature can be determined only by analyzing the sign of the eigenvalues of A_f. More precisely, if

$\lambda_1 \lambda_2 > 0$ the conic is an ellipse;

$\lambda_1 \lambda_2 < 0$ the conic is a hyperbola;

$\lambda_1 \lambda_2 = 0$ the conic is a parabola.

Now we summarize the technique used above and we give an algorithm for reducing a conic to its canonical form.

Algorithm for reducing a conic to its canonical form

- **Step 1.** Write the matrix A_f and find its eigenvalues.

- **Step 2.** Determine the Jordan canonical form J_{A_f} and the orthogonal matrix $P, P^T = P^{-1}$, which verifies the identity

(continued)

$$J_{A_f} = \begin{pmatrix} \lambda_1 & 0 \\ 0 & \lambda_2 \end{pmatrix} = P^T A_f P.$$

■ **Step 3.** Write the equations of the rotation in the form

$$X = PY, \quad \text{where} \quad X = \begin{pmatrix} x \\ y \end{pmatrix} \quad \text{and} \quad Y = \begin{pmatrix} x' \\ y' \end{pmatrix}.$$

Determine the angle of the rotation by writing $P = R_\alpha$. Thus, the coordinate system xOy is rotated counterclockwise by an angle α and becomes the system $x'Oy'$.

■ **Step 4.** Write the equations of the translation by completing both, or possible one, squares in x' and y'.

■ **Step 5.** Determine the nature of the conic by inspecting the equation in x'' and y''.

Example 6.1 **A hyperbola.** We consider the conic

$$\mathscr{C}: \quad 3x^2 + 10xy + 3y^2 - 2x - 14y - 13 = 0,$$

which we reduce to its canonical form and determine its nature (Fig. 6.4).

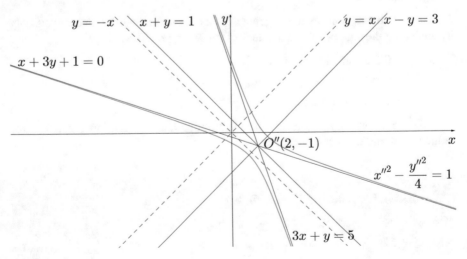

Fig. 6.4 The hyperbola $3x^2 + 10xy + 3y^2 - 2x - 14y - 13 = 0$

Step 1. The quadratic form associated with our conic is $f(x, y) = 3x^2 + 10xy + 3y^2$ and its corresponding symmetric matrix is

$$A_f = \begin{pmatrix} 3 & 5 \\ 5 & 3 \end{pmatrix}.$$

The eigenvalues of A_f are obtained by solving the equation $\det(A_f - \lambda I_2) = 0$ which implies $(3 - \lambda)^2 - 25 = 0 \Rightarrow \lambda_1 = 8$ and $\lambda_2 = -2$.

Step 2. We determine the eigenvectors corresponding to the eigenvalues $\lambda_1 = 8$ and $\lambda_2 = -2$. The eigenvector corresponding to $\lambda_1 = 8$ is determined by solving the system $(A_f - 8I_2)X = 0$ and we have

$$\begin{cases} -5x_1 + 5x_2 = 0 \\ 5x_1 - 5x_2 = 0, \end{cases}$$

which implies that $x_1 = x_2$. The solution of the system is $\begin{pmatrix} \alpha \\ \alpha \end{pmatrix}$, $\alpha \in \mathbb{R}^*$. We let $\alpha = 1$ and we divide our vector by its norm[1] (length) and we get the eigenvector

$$X_1 = \begin{pmatrix} \frac{1}{\sqrt{2}} \\ \frac{1}{\sqrt{2}} \end{pmatrix}.$$

Similarly, the eigenvectors corresponding to $\lambda_2 = -2$ are determined by solving the system $(A_f + 2I_2)X = 0$ and we obtain the eigenvector

$$X_2 = \begin{pmatrix} -\frac{1}{\sqrt{2}} \\ \frac{1}{\sqrt{2}} \end{pmatrix}.$$

Thus,

$$P = \begin{pmatrix} \frac{1}{\sqrt{2}} & -\frac{1}{\sqrt{2}} \\ \frac{1}{\sqrt{2}} & \frac{1}{\sqrt{2}} \end{pmatrix} = \begin{pmatrix} \cos\frac{\pi}{4} & -\sin\frac{\pi}{4} \\ \sin\frac{\pi}{4} & \cos\frac{\pi}{4} \end{pmatrix} = R_{\frac{\pi}{4}},$$

which is a rotation matrix of angle $\frac{\pi}{4}$.

Step 3. The equations of the rotation are

$$\begin{pmatrix} x \\ y \end{pmatrix} = P \begin{pmatrix} x' \\ y' \end{pmatrix} \quad \text{or} \quad \begin{cases} x = \frac{1}{\sqrt{2}}x' - \frac{1}{\sqrt{2}}y' \\ y = \frac{1}{\sqrt{2}}x' + \frac{1}{\sqrt{2}}y'. \end{cases} \tag{6.3}$$

---

[1] Recall the norm or the length of a vector $v = \begin{pmatrix} a \\ b \end{pmatrix} \in \mathbb{R}^2$ is defined by $||v|| = \sqrt{|a|^2 + |b|^2}$.

We rotate the system of coordinates xOy by an angle of $\frac{\pi}{4}$ counterclockwise and the equation of the conic becomes

$$\mathscr{C}: \quad 8x'^2 - 2y'^2 - \frac{2}{\sqrt{2}}(x' - y') - \frac{14}{\sqrt{2}}(x' + y') - 13 = 0,$$

which can be written, after completing the squares in x' and y', as

$$\mathscr{C}: \quad 8\left(x' - \frac{1}{\sqrt{2}}\right)^2 - 2\left(y' + \frac{3}{\sqrt{2}}\right)^2 - 8 = 0.$$

Step 4. The equations of the translation are

$$T: \begin{cases} x'' = x' - \dfrac{1}{\sqrt{2}} \\ y'' = y' + \dfrac{3}{\sqrt{2}}. \end{cases}$$

Thus, we translate the system $x'Oy'$ to $x''O''y''$ and the equation of the conic becomes

$$\mathscr{C}: \quad 8x''^2 - 2y''^2 - 8 = 0 \quad \Leftrightarrow \quad x''^2 - \frac{y''^2}{4} = 1.$$

Thus, our conic is an ellipse of semi axes $a = 1$ and $b = 2$.

Next, we determine the equations of the axes of symmetry and the coordinates of the center O''.

We have, based on (6.3), that $\begin{pmatrix} x' \\ y' \end{pmatrix} = P^{-1}\begin{pmatrix} x \\ y \end{pmatrix} = P^T\begin{pmatrix} x \\ y \end{pmatrix}$ and this implies that

$$\begin{cases} x' = \frac{1}{\sqrt{2}}(x + y) \\ y' = \frac{1}{\sqrt{2}}(-x + y). \end{cases} \tag{6.4}$$

The equation of $O''x''$. To determine the equation of the $O''x''$ we set $y'' = 0$ which in turn implies that $y' = -\frac{3}{\sqrt{2}}$. The second equation in (6.4) implies that $-x + y = -3$.

The equation of $O''y''$. To determine the equation of the $O''y''$ we set $x'' = 0$ and this implies that $x' = \frac{1}{\sqrt{2}}$. However, the first equation in (6.4) implies that $x + y = 1$.

The coordinates of the center of symmetry O''. The coordinates of O'' are the solutions of the system

$$\begin{cases} x+y = 1 \\ -x+y = -3, \end{cases}$$

which implies that $x = 2$ and $y = -1$. Thus, $O''(2, -1)$, i.e., the coordinates of O'' with respect to the coordinate system xOy are $(2, -1)$.

6.3 Problems

6.1 Find the canonical form of the following conics, determine their nature, the symmetry axes, and the center:

(1) $3x^2 - 4xy + 3y^2 - 2x - 2y + 1 = 0$;

(2) $2x^2 - 4xy - y^2 + \sqrt{5}x + \sqrt{5}y - 1 = 0$;

(3) $x^2 - 2xy + y^2 + 2x - 4y + 5 = 0$;

(4) $4x^2 + 12xy + 9y^2 - 64 = 0$;

(5) $9x^2 + 24xy + 16y^2 - 40x + 30y = 0$;

(6) $2x^2 - 6xy + 10y^2 - 8x + 12y + 2 = 0$;

(7) $4x^2 - 4xy + y^2 - 2x - 14y + 7 = 0$;

(8) $4xy - 3y^2 + 4x - 14y - 7 - 0$,

(9) $3x^2 - 4xy - 2x + 4y - 3 = 0$;

(10) $x^2 + 2xy + y^2 + 2x + 2y - 3 = 0$;

(11) $x^2 - 8xy + 7y^2 + 6x - 6y + 9 = 0$;

(12) $5x^2 + 12xy - 22x - 12y - 19 = 0$;

(13) $6x^2 - 4xy + 9y^2 - 4x - 32y - 6 = 0$;

(14) $5x^2 + 4xy + 8y^2 - 32x - 56y + 80 = 0$;

(15) $xy - k = 0, \ k \in \mathbb{R}^*$.

6.2 Discuss, according to the values of the parameter a, the nature of the following conics:

(1) $5x^2 + 2axy + 5y^2 + 2x + 2y + 2 = 0$;

(2) $ax^2 + 2xy + ay^2 - 2x + 2y + 9 = 0$;

(3) $x^2 + 4xy + 4y^2 + ax = 0$;

(4) $7x^2 - 8xy + y^2 + a = 0$.

6.3 For what value (or values) of c will the graph of the Cartesian equation $2xy - 4x + 6y + c = 0$ be a pair of lines?

6.4 *An equilateral hyperbola.* Prove that the conic $(x+2y+1)(2x-y+1)+\alpha = 0$, where $\alpha \neq 0$, is a hyperbola having the asymptotes $x+2y+1 = 0$ and $2x-y+1 = 0$.

Remark 6.2 More generally, one can prove that the conic

$$(a_1x + b_1y + c_1)(a_2x + b_2y + c_2) + \alpha = 0,$$

with $a_1b_2 - a_2b_1 \neq 0$ and $\alpha \neq 0$, is a hyperbola having the asymptotes $a_1x + b_1y + c_1 = 0$ and $a_2x + b_2y + c_2 = 0$.

6.5 (a) Find the equation of a hyperbola passing through the points $(1, 1)$, $(2, 1)$ and $(-1, -2)$ which has the asymptote $x + y - 1 = 0$.

(b) Find the equation of an equilateral hyperbola passing through the points $(1, 1)$ and $(2, 1)$ which has the asymptote $x - y + 1 = 0$.

6.6 A Lamé's curve and a parabola in disguise.

Let $k > 0$ be a real number. Prove that the curves $\mathscr{C}_k^1 : \sqrt{y} - \sqrt{x} = \sqrt{k}$, $\mathscr{C}_k^2 : \sqrt{x} + \sqrt{y} = \sqrt{k}$ and $\mathscr{C}_k^3 : \sqrt{x} - \sqrt{y} = \sqrt{k}$ form together a parabola, determine its canonical form, the vertex, and the symmetry axis.

Remark 6.3 A *Lamé's curve* has the Cartesian equation $\frac{x^\alpha}{a^\alpha} + \frac{y^\alpha}{b^\alpha} = 1$, where a, b are positive real numbers and α is a real number. These curves have been studied by G. Lamé [39] in the 19th century and now they are called *Lame's curves* or *super ellipses*.

When $k = 1$ the curve $\mathscr{C}_1^2$ is discussed in [12] where it is shown, in spite the graph of $\mathscr{C}_1^2$ looks like the arc of a circle, that $\mathscr{C}_1^2$ is part of a parabola and not of a circle.

6.7 Determine the nature of the conic $\mathscr{C} : x^2 + y^2 - 4xy - 1 = 0$ and find the lattice points on $\mathscr{C}$.

6.8 Prove the hyperbola $\mathscr{H} : x^2 - 5y^2 = 4$ contains infinitely many lattice points.

6.9 (a) Determine the equation of an ellipse which, on the xOy plane, has the foci $F_1(\sqrt{3}, 2)$, $F_2(3\sqrt{3}, 4)$ and the large semi axes $a = 3$.

(b) Find the canonical form of the conic from part (a).

6.10 (a) Find the equation of a hyperbola which has the foci $F_1(-1, 2)$, $F_2(3, 6)$ and the semi axes $a = 2$.

(b) Find the canonical form of the conic from part (a).

6.11 (a) Find the locus of points $M(x, y)$ on the xOy plane with the property that the distance from M to the line $\mathscr{D} : \sqrt{3}x - 3y + 2\sqrt{3} = 0$ and the distance from M to the point $F(2, 0)$ are equal.

(b) Find the canonical form of the conic from part (a).

6.12 *An abelian group determined by un ellipse.*

Let $\mathscr{E}$ be the ellipse $\frac{x^2}{a^2} + \frac{y^2}{b^2} = 1$ and let $*$ be the binary operation on $\mathscr{E}$ defined by $* : \mathscr{E} \times \mathscr{E} \to \mathscr{E}$, $(M_1, M_2) \in \mathscr{E} \times \mathscr{E} \to M_1 * M_2 \in \mathscr{E}$, where $M_1 * M_2$ is the point on

$\mathscr{E}$ where the line passing through $A(a, 0)$ parallel to the segment $[M_1M_2]$ intersects the ellipse $\mathscr{E}$. Prove that $(\mathscr{E}, *)$ is an abelian group.

6.13 Find the extreme values of the function $f(x, y) = x^2 + xy + y^2 + x - y - 1$ and the points where these values are attained.

6.14 *Extremum problems with constraints.*

(a) Find the minimum and the maximum values of the function $f(x, y) = x + y$ subject to the constraint $3x^2 - 2xy + 3y^2 + 4x + 4y - 4 = 0$.

(b) Find the extreme values of the function $f(x, y) = 2x + y$ subject to the constraint $3x^2 + 10xy + 3y^2 - 16x - 16y - 16 = 0$.

(c) Find the extreme values of the function $f(x, y) = 2x - y$ subject to the constraint $9x^2 + 24xy + 16y^2 - 40x + 30y = 0$.

6.15 Constrained extrema of a quadratic form.

Let $f(x, y) = ax^2 + 2bxy + cy^2$, $a, b, c \in \mathbb{R}$ with $a^2 + b^2 + c^2 \neq 0$. Prove that the minimum and the maximum values of f, subject to the constraint $x^2 + y^2 = 1$, are the smallest respectively the largest of the eigenvalues of

$$A_f = \begin{pmatrix} a & b \\ b & c \end{pmatrix}.$$

6.16 Find the area of the domain bounded by the curve

$$5x^2 + 6xy + 5y^2 - 16x - 16y - 16 = 0.$$

Double integrals over elliptical domains.

6.17 Calculate:

(a) $\displaystyle\iint_{D_1} e^{x^2 - xy + y^2} dxdy$, where $D_1 = \{(x, y) \in \mathbb{R}^2 : x^2 - xy + y^2 \leq 1\}$;

(b) $\displaystyle\iint_{D_2} e^{x^2 + xy + y^2} dxdy$, where $D_2 = \{(x, y) \in \mathbb{R}^2 : x^2 + xy + y^2 \leq 1\}$;

(c) $\displaystyle\iint_{D_3} e^{-x^2 + xy - y^2} dxdy$, where $D_3 = \{(x, y) \in \mathbb{R}^2 : x^2 - xy + y^2 \geq 1\}$;

(d) $\displaystyle\iint_{D_4} e^{-x^2 - xy - y^2} dxdy$, where $D_4 = \{(x, y) \in \mathbb{R}^2 : x^2 + xy + y^2 \geq 1\}$.

6.18 Let $a, b \in \mathbb{R}$ such that $0 < b < 2a$ and let $\alpha > 0$. Calculate:

(a) $\displaystyle\iint_{D_1} e^{ax^2 - bxy + ay^2} dxdy$, where $D_1 = \{(x, y) \in \mathbb{R}^2 : ax^2 - bxy + ay^2 \leq \alpha\}$;

(b) $\displaystyle\iint_{D_2} e^{ax^2 + bxy + ay^2} dxdy$, where $D_2 = \{(x, y) \in \mathbb{R}^2 : ax^2 + bxy + ay^2 \leq \alpha\}$;

(c) $\iint_{D_3} e^{-ax^2+bxy-ay^2}\,dxdy$, where $D_3 = \{(x,y) \in \mathbb{R}^2 : ax^2 - bxy + ay^2 \geq \alpha\}$;

(d) $\iint_{D_4} e^{-ax^2-bxy-ay^2}\,dxdy$, where $D_4 = \{(x,y) \in \mathbb{R}^2 : ax^2 + bxy + ay^2 \geq \alpha\}$.

6.19 Let $a, b \in \mathbb{R}$ such that $0 < b < 2a$ and let $\alpha > 0$. Calculate

(a) $\iint_{D_\alpha} xe^{-ax^2-bxy-ay^2}\,dxdy$

(b) $\iint_{D_\alpha} xye^{-ax^2-bxy-ay^2}\,dxdy$,

where $D_\alpha = \{(x,y) \in \mathbb{R}^2 : ax^2 + bxy + ay^2 \geq \alpha\}$.

6.20 Quadratic forms and special integrals.

(a) Calculate

$$\iint_{\mathbb{R}^2} \frac{dxdy}{(1 + 3x^2 - 4xy + 3y^2)^3}.$$

(b) Let $A \in \mathcal{M}_2(\mathbb{R})$ be a symmetric matrix with positive eigenvalues and let $\alpha > 1$ be a real number. Prove that

$$\iint_{\mathbb{R}^2} \frac{dxdy}{(1 + v^T A v)^\alpha} = \frac{\pi}{(\alpha - 1)\sqrt{\det A}}, \quad \text{where} \quad v = \begin{pmatrix} x \\ y \end{pmatrix}.$$

(c) Let f be an integrable function over $[0, \infty)$ and let $\int_0^\infty f(x)dx = I$. Prove that if $A \in \mathcal{M}_2(\mathbb{R})$ is a symmetric matrix with positive eigenvalues, then

$$\iint_{\mathbb{R}^2} f\left(v^T A v\right) dxdy = \frac{\pi I}{\sqrt{\det A}}, \quad \text{where} \quad v = \begin{pmatrix} x \\ y \end{pmatrix}.$$

6.21 A particular case and a formula.

(a) Calculate

$$\iint_{\mathbb{R}^2} e^{-(3x^2-2xy+3y^2+2x+2y-1)}\,dxdy.$$

(continued)

6.21 (continued)

(b) Let $A \in \mathcal{M}_2(\mathbb{R})$ be a symmetric matrix with positive eigenvalues, let $b = \begin{pmatrix} b_1 \\ b_2 \end{pmatrix}$ be a vector in $\mathbb{R}^2$, and let c be a real number. Prove that

$$\int \int_{\mathbb{R}^2} e^{-(v^T A v + 2b^T v + c)} \, dx dy = \frac{\pi}{\sqrt{\det A}} e^{b^T A^{-1} b - c}, \quad \text{where} \quad v = \begin{pmatrix} x \\ y \end{pmatrix}.$$

6.4 Solutions

6.1. (1) The ellipse $x''^2 + 5y''^2 = 1$, the equation of the axes $O''x'' : x - y = 0$, $O''y'' : x + y = 2$, and the center $O''(1, 1)$;

(2) the hyperbola $48x''^2 - 72y''^2 = 1$, the equation of the axes $O''x'' : -2x + y = \frac{\sqrt{5}}{6}$, $O''y'' : x + 2y = \frac{3\sqrt{5}}{4}$, and the center $O'' \left(\frac{\sqrt{5}}{12}, \frac{\sqrt{5}}{3} \right)$;

(3) the parabola $x'' - \sqrt{2}y''^2 = 0$, the equation of the axes $O''x'' : -x + y = \frac{3}{2}$, $O''y'' : x + y = \frac{11}{4}$, and the vertex $O'' \left(\frac{5}{8}, \frac{17}{8} \right)$;

(4) the conic degenerates to a union of two parallel lines $(2x + 3y - 8)(2x + 3y + 8) = 0$;

(5) the parabola $x'^2 + 2y' = 0$, the equation of the axes $Ox' : -4x + 3y = 0$, $Oy' : 3x + 4y = 0$, and the vertex $O(0, 0)$;

(6) the ellipse $x''^2 + 11y''^2 - 6 = 0$, the equation of the axes $O''x'' : x - 3y = 2$, $O''y'' : 3x + y = 6$, and the center $O''(2, 0)$;

(7) the parabola $y''^2 - \frac{6}{\sqrt{5}} x'' = 0$, the equation of the axes $O''x'' : -2x + y = 1$, $O''y'' : x + 2y = 1$, and the vertex $O'' \left(-\frac{1}{5}, \frac{3}{5} \right)$;

(8) the hyperbola $-x''^2 + 4y''^2 - 4 = 0$, the equation of the axes $O''x'' : -x + 2y = -4$, $O''y'' : 2x + y = 3$, and the center $O''(2, -1)$;

(9) the hyperbola $-x''^2 + 4y''^2 = 2$, the equation of the axes $O''x'' : 2x - y = 1$, $O''y'' : x + 2y = 3$, and the center $O''(1, 1)$;

(10) the conic degenerates to a union of two parallel lines $(x + y - 1)(x + y + 3) = 0$;

(11) the hyperbola $\frac{x''^2}{9} - y''^2 = 1$, the equation of axes $O''x'' : -x + 2y = 1$, $O''y'' : 2x + y = 3$, and the center $O''(1, 1)$;

(12) the hyperbola $\frac{x''^2}{4} - \frac{y''^2}{9} = 1$, the equation of axes $O''x'' : -2x + 3y = 1$, $O''y'' : 3x + 2y = 5$, and the center $O''(1, 1)$;

(13) the ellipse $\frac{x''^2}{8} + \frac{y''^2}{4} = 1$, the equation of the axes $O''x'' : -x + 2y = 3$, $O''y'' : 2x + y = 4$, and the center $O''(1, 2)$;

(14) the ellipse $\frac{x''^2}{4} + \frac{y''^2}{9} = 1$, the equation of the axes $O''x'' : 2x - y = 1$, $O''y'' : x + 2y = 8$, and the center $O''(2, 3)$;

(15) the equilateral hyperbola $x'^2 - y'^2 - 2k = 0$, which is tangent to the x-axis and the y-axis. If $k > 0$ the branches of the hyperbola are located in the first and the third quadrant and if $k < 0$ the branches of the hyperbola are located in the second and the fourth quadrant.

6.2. (1) The matrix associated with our conic is $A = \begin{pmatrix} 5 & a \\ a & 5 \end{pmatrix}$, which has the eigenvalues $\lambda_1 = 5 - a$ and $\lambda_2 = 5 + a$. If $a \neq 0$ the eigenvectors corresponding to the eigenvalues $\lambda_1 = 5 - a$ and $\lambda_2 = 5 + a$ are $X_1 = \frac{1}{\sqrt{2}} \begin{pmatrix} 1 \\ -1 \end{pmatrix}$ and $X_2 = \frac{1}{\sqrt{2}} \begin{pmatrix} 1 \\ 1 \end{pmatrix}$

and $P = \frac{1}{\sqrt{2}} \begin{pmatrix} 1 & 1 \\ -1 & 1 \end{pmatrix} = R_{-\frac{\pi}{4}}$, which is a rotation matrix of angle $-\frac{\pi}{4}$.

- If $a = 0$ the equation of the conic becomes $\mathscr{C} : 5x^2 + 5y^2 + 2x + 2y + 2 = 0$ $\Leftrightarrow \mathscr{C} : 4x^2 + 4y^2 + (x + 1)^2 + (y + 1)^2 = 0 \Rightarrow \mathscr{C} = \emptyset$.
- If $a = 5$ we have $\mathscr{C} : 5x^2 + 10xy + 5y^2 + 2x + 2y + 2 = 0 \Leftrightarrow \mathscr{C} : 4(x + y)^2 + (x + y + 1)^2 + 1 = 0 \Rightarrow \mathscr{C} = \emptyset$.
- If $a = -5$ the equation of the conic becomes $\mathscr{C} : 5x^2 - 10xy + 5y^2 + 2x + 2y + 2 = 0 \Leftrightarrow \mathscr{C} : 5(x-y)^2 + 2(x-y) + 4y + 2 = 0 \Leftrightarrow \mathscr{C} : 5\left(x - y + \frac{1}{5}\right)^2 = -4\left(y + \frac{9}{20}\right)$, which is a parabola.

We make the rotation $\begin{pmatrix} x \\ y \end{pmatrix} = P \begin{pmatrix} x' \\ y' \end{pmatrix}$ and the equation of the conic becomes $(5 - a)x'^2 + (5 + a)y'^2 + 2\sqrt{2}y' + 2 = 0$ or

$$\mathscr{C} : \quad (5 - a)x'^2 + (5 + a)\left(y'^2 + \frac{2\sqrt{2}}{5 + a}y' + \frac{2}{(5 + a)^2}\right) + 2\frac{4 + a}{5 + a} = 0.$$

- If $a = -4$ the equation of the conic becomes $\mathscr{C} : \frac{9}{2}(x - y)^2 + \frac{1}{2}(x + y + 2)^2 = 0$, and the conic reduces to a point $\mathscr{C} = \{(-1, -1)\}$.
- If $a \in (-5, -4)$, we have $5 - a > 0$, $5 + a > 0$, $\frac{4+a}{5+a} < 0$, which implies that $\mathscr{C}$ is an ellipse.
- If $a \in (-4, 5)$, we have $5 - a > 0$, $5 + a > 0$, $\frac{4+a}{5+a} > 0 \Rightarrow \mathscr{C} = \emptyset$.
- If $a \in (-\infty, -5) \cup (5, \infty)$, since $5 + a$ and $5 - a$ have opposite signs, we get that $\mathscr{C}$ is a hyperbola.

In conclusion:

- If $a \in (-\infty, -5) \cup (5, \infty) \Rightarrow \mathscr{C}$ is a hyperbola.
- If $a = -5 \Rightarrow \mathscr{C}$ is a parabola.
- If $a \in (-5, -4) \Rightarrow \mathscr{C}$ is an ellipse.
- If $a = -4 \Rightarrow \mathscr{C}$ reduces to a point.
- If $a \in (-4, 5] \Rightarrow \mathscr{C}$ is the empty set.

(2) The matrix associated with the conic is $A = \begin{pmatrix} a & 1 \\ 1 & a \end{pmatrix}$ with eigenvalues $\lambda_1 = a + 1$, $\lambda_2 = a - 1$ and eigenvectors $X_1 = \frac{1}{\sqrt{2}} \begin{pmatrix} 1 \\ 1 \end{pmatrix}$, $X_2 = \frac{1}{\sqrt{2}} \begin{pmatrix} -1 \\ 1 \end{pmatrix}$ and $P = \frac{1}{\sqrt{2}} \begin{pmatrix} 1 & -1 \\ 1 & 1 \end{pmatrix} = R_{\frac{\pi}{4}}$, which is a rotation matrix of angle $\frac{\pi}{4}$.

- If $a = 1$ we have $(x+y)^2 - 2(x+y) + 1 = -4y - 8 \Leftrightarrow (x+y-1)^2 = -4(y+2)$, which is a parabola.
- If $a = -1$ we have $\mathscr{C} : (x - y + 1)^2 = 10 \Leftrightarrow \mathscr{C} : x - y + 1 = \pm\sqrt{10}$ and the conic reduces to a union of two lines.
- If $a \in \mathbb{R} \setminus \{\pm 1\}$ we make the rotation $\begin{pmatrix} x \\ y \end{pmatrix} = P \begin{pmatrix} x' \\ y' \end{pmatrix}$ and the equation of the conic becomes $(a+1)x'^2 + (a-1)y'^2 + 2\sqrt{2}y' + 9 = 0$. We complete the square in y' and we get that

$$(a+1)x'^2 + (a-1)\left(y' + \frac{\sqrt{2}}{a-1}\right)^2 + \frac{9a-11}{a-1} = 0.$$

We distinguish between the following cases:

If $a = \frac{11}{9}$, since $a+1 > 0$, $a-1 > 0$, the conic reduces to the point $\left(\frac{9}{2}, -\frac{9}{2}\right)$.

If $a > \frac{11}{9}$, since $a+1 > 0$, $a-1 > 0$, $\frac{9a-11}{a-1} > 0$, the conic reduces to the empty set.

If $a \in \left(1, \frac{11}{9}\right)$, then $a+1 > 0$, $a-1 > 0$, $\frac{9a-11}{a-1} < 0$ and the conic is an ellipse.

If $a \in (-1, 1)$, then $a + 1 > 0$, $a - 1 < 0$, $\frac{9a-11}{a-1} > 0$ and the conic is a hyperbola.

If $a \in (-\infty, -1)$, then $a + 1 < 0$, $a - 1 < 0$, $\frac{9a-11}{a-1} > 0$ and the conic is an ellipse.

In conclusion:

- If $a \in (-\infty, -1) \cup \left(1, \frac{11}{9}\right)$, the conic is an ellipse.
- If $a = -1$, the conic is a union of two lines.
- If $a \in (-1, 1)$, the conic is a hyperbola.
- If $a = 1$, the conic is a parabola.
- If $a = \frac{11}{9}$, the conic reduces to a point.
- If $a \in \left(\frac{11}{9}, \infty\right)$, the conic is the empty set.

(3) We have $x^2 + 4xy + 4y^2 + ax = 0 \Leftrightarrow (x + 2y)^2 = -ax$. If $a = 0$ the conic degenerates to the line $x + 2y = 0$ and if $a \neq 0$ the conic is a parabola.

(4) The matrix associated with the conic is $A = \begin{pmatrix} 7 & -4 \\ -4 & 1 \end{pmatrix}$ with eigenvalues $\lambda_1 = 9$, $\lambda_2 = -1$. Making the rotation $X = PY$ the equation of the conic becomes $9x'^2 - y'^2 + a = 0$. If $a \neq 0$ the conic is a hyperbola and if $a = 0$ the conic

degenerates to a union of two lines. This follows since $7x^2 - 8xy + y^2 = 0 \Leftrightarrow$
$(x-y)(7x-y) = 0$, which implies that $x-y = 0$ or $7x-y = 0$. Thus, when $a = 0$
we have that $\mathscr{C} = \mathscr{D}_1 \cup \mathscr{D}_2$, where $\mathscr{D}_1 : x - y = 0$ and $\mathscr{D}_2 : 7x - y = 0$.

6.3. $c = -12$.

6.5. (a) We have, based on Remark 6.2, that the equation of the hyperbola which has
$x + y - 1 = 0$ as an asymptote is $(ax + by + c)(x + y - 1) + \alpha = 0$, $\alpha \neq 0$. Since
the points $(1, 1)$, $(2, 1)$, and $(-1, -2)$ are on the graph of the hyperbola we have the
system of linear equations

$$\begin{cases} a + b + c + \alpha = 0 \\ 4a + 2b + 2c + \alpha = 0 \\ 4a + 8b - 4c + \alpha = 0. \end{cases}$$

This implies that the equation of the hyperbola is $(2x - 3y - 3)(x + y - 1) + 4 = 0$.

(b) Since an equilateral hyperbola has perpendicular asymptotes we obtain that
the equation of the hyperbola is $(x + y + c)(x - y + 1) + \alpha = 0$, $\alpha \neq 0$. It follows
that $(x + y - 4)(x - y + 1) + 2 = 0$.

6.6. We consider the curve $\mathscr{C}_k^2 : \sqrt{x} + \sqrt{y} = \sqrt{k}$ and observe that $x, y \in [0, k]$. We
square both sides of this equation and we get $x + y + 2\sqrt{xy} = k \Leftrightarrow 2\sqrt{xy} = k - x - y$.
This implies $k \geq x + y$ and since $x, y \in [0, k]$ we get that our curve is contained in
the triangle having the vertices $A(0, k)$, $B(k, 0)$, and $O(0, 0)$. We square both sides
of the preceding equation and we have $x^2 - 2xy + y^2 - 2kx - 2ky + k^2 = 0$. Thus,
our curve is a conic, more precisely the arc of a conic contained in $\triangle OAB$, i.e.

$$\mathscr{C}_k^2 : \quad x^2 - 2xy + y^2 - 2kx - 2ky + k^2 = 0, \quad 0 \leq x \leq k, \quad 0 \leq y \leq k.$$

The matrix associated with the quadratic form $x^2 - 2xy + y^2$ is $A = \begin{pmatrix} 1 & -1 \\ -1 & 1 \end{pmatrix}$, with

eigenvalues $\lambda_1 = 0$, $\lambda_2 = 2$ and the corresponding eigenvectors

$$X_1 = \begin{pmatrix} \frac{1}{\sqrt{2}} \\ \frac{1}{\sqrt{2}} \end{pmatrix} \quad \text{and} \quad X_2 = \begin{pmatrix} -\frac{1}{\sqrt{2}} \\ \frac{1}{\sqrt{2}} \end{pmatrix}.$$

The rotation matrix P is given by $P = \begin{pmatrix} \frac{\sqrt{2}}{2} & -\frac{\sqrt{2}}{2} \\ \frac{\sqrt{2}}{2} & \frac{\sqrt{2}}{2} \end{pmatrix} = R_{\frac{\pi}{4}}$. We make a rotation

which has the equations $X = PY$,

$$\begin{cases} x = \frac{1}{\sqrt{2}}x' - \frac{1}{\sqrt{2}}y' \\ y = \frac{1}{\sqrt{2}}x' + \frac{1}{\sqrt{2}}y' \end{cases} \quad \text{or} \quad \begin{cases} x' = \frac{1}{\sqrt{2}}x + \frac{1}{\sqrt{2}}y \\ y' = -\frac{1}{\sqrt{2}}x + \frac{1}{\sqrt{2}}y \end{cases}$$

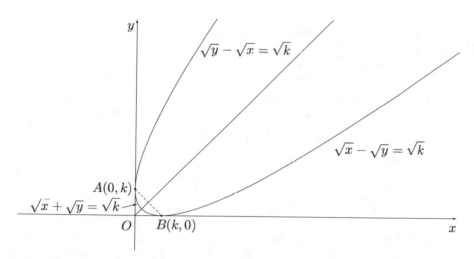

Fig. 6.5 Lamé's parabola

and the equation of the conic with respect to the coordinate system $Ox'y'$ becomes $y'^2 - \sqrt{2}kx' + \frac{k^2}{2} = 0 \Leftrightarrow y'^2 - \sqrt{2}k\left(x' - \frac{k}{2\sqrt{2}}\right) = 0$. The translation of equations $x'' = x' - \frac{k}{2\sqrt{2}}$, $y'' = y'$, reduces the conic to its canonical form which has, with respect to the system of coordinates $O''x''y''$, the equation $y''^2 - \sqrt{2}kx'' = 0$. This is a parabola, so our curve $\mathscr{C}_k^2$ is the arc of this parabola, tangent to the x and the y axes, which is contained in $\triangle OAB$.

The axes of the new system of coordinates are $O''x'' : y'' = 0 \Leftrightarrow y' = 0 \Leftrightarrow x - y = 0$ and $O''y'' : x'' = 0 \Leftrightarrow x' = \frac{k}{2\sqrt{2}} \Leftrightarrow x + y = \frac{k}{2}$. The vertex of the parabola, the point where the two axes intersect, has coordinates $\left(\frac{k}{4}, \frac{k}{4}\right)$.

Similarly, one can prove that the other two curves $\mathscr{C}_k^1$ and $\mathscr{C}_k^3$ are parts of the same parabola, i.e., $\mathscr{C}_k^1$ is the unbounded arc of the parabola, above the line $y - x = 0$, which is tangent to the y axis at $A(0, k)$ and $\mathscr{C}_k^3$ is the unbounded arc, below the line $y - x = 0$, which is tangent to the x axis at $B(k, 0)$ (Fig. 6.5).

6.7. The matrix corresponding to the quadratic form of $\mathscr{C}$ is $A = \begin{pmatrix} 1 & -2 \\ -2 & 1 \end{pmatrix}$, which has the eigenvalues $\lambda_1 = 3$ and $\lambda_2 = -1$. Since $\lambda_1\lambda_2 < 0$ the conic is a hyperbola. We write the equation of the conic as $(x - 2y)^2 - 3y^2 = 1$ and we note this is a Pell's equation which has the minimal solution $x_0 = 4$, $y_0 = 1$. The general solution of this equation is given by $x_n - 2y_n + y_n\sqrt{3} = (2 + \sqrt{3})^n$ and this implies that

$$x_n - 2y_n = \frac{(2 + \sqrt{3})^n + (2 - \sqrt{3})^n}{2} \quad \text{and} \quad y_n = \frac{(2 + \sqrt{3})^n - (2 - \sqrt{3})^n}{2\sqrt{3}}, \quad n \geq 1.$$

The lattice points on the hyperbola are $(\pm 1, 0)$ and $(\pm x_n, \pm y_n)$, where

$$x_n = \left(\frac{1}{2} + \frac{1}{\sqrt{3}}\right)(2 + \sqrt{3})^n + \left(\frac{1}{2} - \frac{1}{\sqrt{3}}\right)(2 - \sqrt{3})^n,$$

$$y_n = \frac{1}{2\sqrt{3}}(2 + \sqrt{3})^n - \frac{1}{2\sqrt{3}}(2 - \sqrt{3})^n, \quad n \geq 1.$$

6.8. A direct computation shows that $(2,0)$, $(3,1)$ and $(7,3)$ are on the hyperbola. We determine $A \in \mathcal{M}_2(\mathbb{Q})$ which generates infinitely many lattice points $(x_n, y_n) \in \mathcal{H}$ via the recurrence relation $\begin{pmatrix} x_{n+1} \\ y_{n+1} \end{pmatrix} = A \begin{pmatrix} x_n \\ y_n \end{pmatrix}$, $n \geq 0$. Let $\begin{pmatrix} x_0 \\ y_0 \end{pmatrix} = \begin{pmatrix} 2 \\ 0 \end{pmatrix}$, $\begin{pmatrix} x_1 \\ y_1 \end{pmatrix} = \begin{pmatrix} 3 \\ 1 \end{pmatrix}$ and $\begin{pmatrix} x_2 \\ y_2 \end{pmatrix} = \begin{pmatrix} 7 \\ 3 \end{pmatrix}$. We have $\begin{pmatrix} 3 \\ 1 \end{pmatrix} = A \begin{pmatrix} 2 \\ 0 \end{pmatrix}$, $\begin{pmatrix} 7 \\ 3 \end{pmatrix} = A \begin{pmatrix} 3 \\ 1 \end{pmatrix}$ and it follows that $\begin{pmatrix} 3 & 7 \\ 1 & 3 \end{pmatrix} = A \begin{pmatrix} 2 & 3 \\ 0 & 1 \end{pmatrix}$. This implies that $A = \frac{1}{2}\begin{pmatrix} 3 & 5 \\ 1 & 3 \end{pmatrix}$. The matrix recurrence relation implies that $x_{n+1} = \frac{1}{2}(3x_n + 5y_n)$ and $y_{n+1} = \frac{1}{2}(x_n + 3y_n)$, $n \geq 0$. One can check that $x_{n+1}^2 - 5y_{n+1}^2 = x_n^2 - 5y_n^2 = 4$, for all $n \geq 0$, so $(x_n, y_n) \in \mathcal{H}$, for all $n \geq 0$.

6.9. (a) The center of the ellipse is the midpoint of the segment $[F_1 F_2]$ which is $O''(2\sqrt{3}, 3)$. The large semi axes of the ellipse is on the line $F_1 F_2$ which has the equation $O''x'' : y = \frac{1}{\sqrt{3}}x + 1$. The small semi axes, which passes through O'' and is perpendicular to $O''x''$, has equation $O''y'' : y = -\sqrt{3}x + 9$. The angle between Ox and $O''x''$ is given by the slope of the line $O''x''$, i.e., $m = \frac{1}{\sqrt{3}} = \tan\frac{\pi}{6}$, so the coordinate system $O''x''y''$ is obtained from the canonical system Oxy by a rotation of angle $\frac{\pi}{6}$ followed by a translation which sends the origin $O(0,0)$ to $O''(2\sqrt{3}, 3)$. The coordinates x'' and y'', with respect to the coordinate system $O''x''y''$, of a point $M(x, y)$ are given by

$$\begin{pmatrix} x'' \\ y'' \end{pmatrix} = \begin{pmatrix} \frac{\sqrt{3}}{2} & \frac{1}{2} \\ -\frac{1}{2} & \frac{\sqrt{3}}{2} \end{pmatrix} \begin{pmatrix} x - 2\sqrt{3} \\ y - 3 \end{pmatrix}. \tag{6.5}$$

The ellipse has the focal distance $F_1 F_2 = 4$, so $c = 2$, $a = 3$, and $b^2 = a^2 - c^2 = 5$. The equation of the ellipse with respect to the system $O''x''y''$ is $\mathcal{C} : \frac{x''^2}{9} + \frac{y''^2}{5} - 1 = 0$. Using equation (6.5), we get that $x'' = \frac{1}{2}(\sqrt{3}x + y - 9)$ and $y'' = \frac{1}{2}(-x + \sqrt{3}y - \sqrt{3})$ and replacing them in the equation of the ellipse we obtain that $\mathcal{C} : 6x^2 + 8y^2 - 2\sqrt{3}xy - 18\sqrt{3}x - 36y + 63 = 0$.

(b) The matrix of the quadratic form of the ellipse is $A = \begin{pmatrix} 6 & -\sqrt{3} \\ -\sqrt{3} & 8 \end{pmatrix}$ which has the eigenvalues $\lambda_1 = 5$ and $\lambda_2 = 9$. The eigenvectors corresponding to the two eigenvalues are $X_1 = \begin{pmatrix} \frac{\sqrt{3}}{2} \\ \frac{1}{2} \end{pmatrix}$ and $X_2 = \begin{pmatrix} -\frac{\sqrt{3}}{2} \\ \frac{1}{2} \end{pmatrix}$. The matrix P, of passing to the Jordan canonical form of A, is given by $P = \begin{pmatrix} \frac{\sqrt{3}}{2} & -\frac{1}{2} \\ \frac{1}{2} & \frac{\sqrt{3}}{2} \end{pmatrix} = R_{\frac{\pi}{6}}$. We make a rotation of angle $\frac{\pi}{6}$ around the origin and we obtain the coordinate system $Ox'y'$. With respect to this system a point on the ellipse $\mathscr{C}$ would have the coordinates x' and y' which are defined by the system of equations

$$\begin{pmatrix} x \\ y \end{pmatrix} = \begin{pmatrix} \frac{\sqrt{3}}{2} & -\frac{1}{2} \\ \frac{1}{2} & \frac{\sqrt{3}}{2} \end{pmatrix} \begin{pmatrix} x' \\ y' \end{pmatrix}.$$

Without making the calculations the quadratic form $6x^2 + 88y^2 - 2\sqrt{3}xy$, of the conic, becomes $5x'^2 + 9y'^2$ and the linear terms from the equation of the conic are calculated, based on the previous equations and we get that the equation of the conic, with respect to the system $Ox'y'$ becomes $\mathscr{C} : 5x'^2 + 9y'^2 - 45x' - 9\sqrt{3}y' + 63 = 0$ or $\mathscr{C} : 5\left(x' - \frac{9}{2}\right)^2 + 9\left(y' - \frac{\sqrt{3}}{2}\right)^2 - 45 = 0$. We make a translation of equations $x'' = x' - \frac{9}{2}$ and $y'' = y' - \frac{\sqrt{3}}{2}$, we obtain a new system of coordinates $O''x''y''$, and the equation of the conic with respect to this system becomes $\frac{x''^2}{9} + \frac{y''^2}{5} - 1 = 0$, which is the same equation we obtained in part (a) of the problem.

The axes of the new system of coordinates are $O''x'' : y'' = 0 \Leftrightarrow y' = \frac{\sqrt{3}}{2} \Leftrightarrow x - \sqrt{3}y - \sqrt{3} = 0$ and $O''y'' : x'' = 0 \Leftrightarrow x' = \frac{9}{2} \Leftrightarrow \sqrt{3}x + y - 9 = 0$. The center of the ellipse, which is the intersection of these two lines, is $O''(2\sqrt{3}, 3)$. Therefore the conic is an ellipse of semi axes $a = 3$ and $b = \sqrt{5}$.

6.10. (a) The center of the hyperbola is the midpoint of the segment $[F_1F_2]$, i.e., $O''(1, 4)$. The focal distance is $F_1F_2 = 4\sqrt{2}$, so $c = 4\sqrt{2}$, $a = 2$ and $b^2 = c^2 - a^2 = 4$, which implies that the hyperbola is equilateral. We choose the system of coordinates $O''x''y''$ such that the x'' axis is the line F_1F_2 which has the equation $O''x'' : y = x + 3$ and the y'' axis, perpendicular to the x'' axis and passing through O'', has the equation $O''y'' : y = -x + 5$. With respect to the coordinate system $O''x''y''$ the hyperbola has the equation $\frac{x''^2}{4} - \frac{y''^2}{4} - 1 = 0 \Leftrightarrow x''^2 - y''^2 - 4 = 0$. The coordinate system $O''x''y''$ is obtained by a rotation of angle $\frac{\pi}{4}$, the angle between the lines $O''x''$ and Ox, and a translation which sends O to O''. The equations involving the coordinates (x, y) and (x'', y'') of a point M with respect to the coordinate systems Oxy and $O''x''y''$ are given by

$$\begin{pmatrix} x'' \\ y'' \end{pmatrix} = R_{-\frac{\pi}{4}} \begin{pmatrix} x - 1 \\ y - 4 \end{pmatrix} = \begin{pmatrix} \frac{\sqrt{2}}{2} & \frac{\sqrt{2}}{2} \\ -\frac{\sqrt{2}}{2} & \frac{\sqrt{2}}{2} \end{pmatrix} \begin{pmatrix} x - 1 \\ y - 4 \end{pmatrix}.$$

It follows that $x'' = \frac{\sqrt{2}}{2}(x + y - 5)$ and $y'' = \frac{\sqrt{2}}{2}(-x + y - 3)$ and replacing them in $x''^2 - y''^2 - 4 = 0$ we get the equation of the hyperbola with respect to the coordinate system Oxy, $xy - 4x - y - 2 = 0$.

(b) We write the equation of the conic as $\mathscr{C} : 2xy - 8x - 2y + 4 = 0$ and we note the matrix corresponding to the quadratic form $2xy$ is $A = \begin{pmatrix} 0 & 1 \\ 1 & 0 \end{pmatrix}$ which has the eigenvalues $\lambda_1 = 1$ and $\lambda_2 = -1$ and the eigenvectors $X_1 = \begin{pmatrix} \frac{\sqrt{2}}{2} \\ \frac{\sqrt{2}}{2} \end{pmatrix}$ and

$X_2 = \begin{pmatrix} -\frac{\sqrt{2}}{2} \\ \frac{\sqrt{2}}{2} \end{pmatrix}$. The matrix P, of passing to the Jordan canonical form of A, is given

by $P = \begin{pmatrix} \frac{\sqrt{2}}{2} & -\frac{\sqrt{2}}{2} \\ \frac{\sqrt{2}}{2} & \frac{\sqrt{2}}{2} \end{pmatrix} = R_{\frac{\pi}{4}}$. We change the coordinates by making the rotation

$\begin{pmatrix} x \\ y \end{pmatrix} = R_{\frac{\pi}{4}} \begin{pmatrix} x' \\ y' \end{pmatrix}$ and we get the equation of the conic with respect to the coordinate

system $Ox''y'$: $x'^2 - y'^2 - 5\sqrt{2}x' - 3\sqrt{2}y' + 4 = 0 \Leftrightarrow \left(x' - \frac{5\sqrt{2}}{2} \right)^2 - \left(y' - \frac{3\sqrt{2}}{2} \right)^2 - 4 = 0$. This implies that $\frac{x''^2}{4} - \frac{y''^2}{4} - 1 = 0$, where $x'' = x' - \frac{5\sqrt{2}}{2}$ and $y'' = y' - \frac{3\sqrt{2}}{2}$.

The axes of the new system of coordinates are $O''x'' : y'' = 0 \Leftrightarrow y = x + 3$ and $O''y'' : x'' = 0 \Leftrightarrow y = -x + 5$. The center of the hyperbola, which is the intersection of these two lines, is $O''(1, 4)$.

6.11. (a) The locus is a parabola with directrix $\mathcal{D}$ and focus F. The symmetry axis of the parabola is the line which passes through F and is perpendicular to $\mathcal{D}$. The slope of $\mathcal{D}$ is $m = \frac{\sqrt{3}}{3}$ and the slope of the symmetry axis is $m' = -\frac{1}{m} = -\sqrt{3}$. It follows that the symmetry axis has the equation $\mathcal{D}' : y = -\sqrt{3}(x - 2)$. The projection of F onto the directrix is the point $F'(1, \sqrt{3})$, the intersection point of lines $\mathcal{D}$ and $\mathcal{D}'$, $FF' = 2$, so $p = 4$.

The center of the coordinate system with respect to which the parabola has the equation $\mathcal{P} : y''^2 = 2px''$ is the midpoint of the segment $[FF']$, i.e., $O'' \left(\frac{3}{2}, \frac{\sqrt{3}}{2} \right)$ and the coordinate axes $O''x'' = \mathcal{D}' : y = -\sqrt{3}(x + 2)$ and $O''y'' : x - \sqrt{3}y = 0$, a line parallel to $\mathcal{D}$. With respect to the coordinate system $O''x''y''$ the equation of parabola is $\mathcal{P} : y''^2 = 8x''$. The angle between the axes $O''x''$ and Ox is given by the slope of $\mathcal{D}'$, i.e., $m = \tan \alpha = -\sqrt{3}$, so $\alpha = -\frac{\pi}{3}$. Thus, the system $O''x''y''$ is obtained by a rotation of angle $-\frac{\pi}{3}$ followed by a translation of vector $\mathbf{OO''}$. The equations that relate the coordinates of a point M with respect to the coordinate systems $O''x''y''$ and Oxy are given by

$$\begin{pmatrix} x'' \\ y'' \end{pmatrix} = R_{\frac{\pi}{3}} \begin{pmatrix} x - \frac{3}{2} \\ y - \frac{\sqrt{3}}{2} \end{pmatrix} = \begin{pmatrix} \frac{1}{2} & -\frac{\sqrt{3}}{2} \\ -\frac{\sqrt{3}}{2} & \frac{1}{2} \end{pmatrix} \begin{pmatrix} x - \frac{3}{2} \\ y - \frac{\sqrt{3}}{2} \end{pmatrix}.$$

We obtain the equation of the parabola

$$\mathcal{P} : \ 3x^2 + y^2 + 2\sqrt{3}xy - 28x + 12\sqrt{3}y + 12 = 0.$$

(b) The matrix corresponding to the quadratic form $3x^2 + y^2 + 2\sqrt{3}xy$ is $A = \begin{pmatrix} 3 & \sqrt{3} \\ \sqrt{3} & 1 \end{pmatrix}$ which has the eigenvalues $\lambda_1 = 0$ and $\lambda_2 = 4$ and the corresponding eigenvectors $X_1 = \begin{pmatrix} \frac{1}{2} \\ -\frac{\sqrt{3}}{2} \end{pmatrix}$ and $X_2 = \begin{pmatrix} \frac{\sqrt{3}}{2} \\ \frac{1}{2} \end{pmatrix}$. The matrix P, of passing to the Jordan

canonical form of A, is given by $P = \begin{pmatrix} \frac{1}{2} & \frac{\sqrt{3}}{2} \\ -\frac{\sqrt{3}}{2} & \frac{1}{2} \end{pmatrix} = R_{-\frac{\pi}{3}}$. We rotate the coordinate

system Oxy by an angle $-\frac{\pi}{3}$ and we obtain the coordinate system $Ox'y'$ and we have the equations

$$\begin{pmatrix} x \\ y \end{pmatrix} = R_{-\frac{\pi}{3}} \begin{pmatrix} x' \\ y' \end{pmatrix} \quad \text{or} \quad \begin{pmatrix} x' \\ y' \end{pmatrix} = R_{\frac{\pi}{3}} \begin{pmatrix} x \\ y \end{pmatrix}.$$

The equation of the conic with respect to the coordinate system $Ox'y'$ is $y'^2 - 2\sqrt{3}y' + 3 = 8x' \Leftrightarrow (y' - \sqrt{3})^2 = 8x' \Leftrightarrow y''^2 = 8x''$, where $x'' = x'$ and $y'' = y' - \sqrt{3}$.

The axes of the system of coordinates are $O''x'' : y'' = 0 \Leftrightarrow y' = \sqrt{3} \Leftrightarrow \sqrt{3}x + y - 2\sqrt{3} = 0$ which is line $\mathscr{D}'$, and $O''y'' : x'' = 0 \Leftrightarrow x' = 0 \Leftrightarrow x - \sqrt{3}y = 0$. The center of the system, the intersection of the coordinate axes, is the point $O'' \left(\frac{3}{2}, \frac{\sqrt{3}}{2} \right)$.

6.12. Let $M_1(a\cos t_1, b\sin t_1)$ and $M_2(a\cos t_2, b\sin t_2)$ and $M_1 * M_2 = M(a\cos t, b\sin t)$. The slope of the line M_1M_2 is given by $m = \frac{b\sin t_2 - \sin t_1}{a\cos t_2 - \cos t_1} = -\frac{b}{a}\cot\frac{t_1+t_2}{2}$. The equation of the line which passes through A parallel to M_1M_2 is given by $\mathscr{D} : y = -\frac{b}{a}\cot\frac{t_1+t_2}{2}(x - a)$. Intersecting the line with the ellipse we get the equation $b\sin t = \frac{b}{a}\cot\frac{t_1+t_2}{2}a(\cos t - 1) \Leftrightarrow \sin\frac{t}{2}\cos\frac{t}{2} = \sin^2\frac{t}{2}\cot\frac{t_1+t_2}{2}$. When $\sin\frac{t}{2} = 0$ we get the point A and when $\sin\frac{t}{2} \neq 0$ we have that $\tan\frac{t}{2} = \tan\frac{t_1+t_2}{2}$, so $t = t_1 + t_2$.

The binary operation $*$ is associative since $M(t_1) * (M(t_2) * M(t_3)) = (M(t_1) * M(t_2)) * M(t_3) = M(t_1 + t_2 + t_3)$ and commutative $M(t_1) * M(t_2) = M(t_2) * M(t_1) = M(t_1 + t_2)$. The identity element of $*$ is $A = M(0)$ and the inverse element of $M(t)$ with respect to $*$ is $M'(-t)$, which is the symmetric point of M with respect to the x axis.

6.13. The matrix associated with the quadratic term $2x^2 + 2xy + 2y^2$ of the function $2f(x, y)$ is $A = \begin{pmatrix} 2 & 1 \\ 1 & 2 \end{pmatrix}$ which has the eigenvalues $\lambda_1 = 3$ and $\lambda_2 = 1$ and the

corresponding eigenvectors $v_1 = \begin{pmatrix} \frac{1}{\sqrt{2}} \\ \frac{1}{\sqrt{2}} \end{pmatrix}$ and $v_2 = \begin{pmatrix} -\frac{1}{\sqrt{2}} \\ \frac{1}{\sqrt{2}} \end{pmatrix}$. It follows that

$$J_A = \begin{pmatrix} 3 & 0 \\ 0 & 1 \end{pmatrix} \quad \text{and} \quad P = \begin{pmatrix} \frac{1}{\sqrt{2}} & -\frac{1}{\sqrt{2}} \\ \frac{1}{\sqrt{2}} & \frac{1}{\sqrt{2}} \end{pmatrix}.$$

We make a rotation of equations

$$\begin{cases} x = \frac{1}{\sqrt{2}}x' - \frac{1}{\sqrt{2}}y' \\ y = \frac{1}{\sqrt{2}}x' + \frac{1}{\sqrt{2}}y' \end{cases}$$

and we obtain that the function $2f$ has, with respect to the coordinate system $x'Oy'$, the expression $g(x', y') = 3x'^2 + y'^2 - 2\sqrt{2}y' - 2 = 3x'^2 + (y' - \sqrt{2})^2 - 4$. The function g has the minimum value -4, when $(x', y') = (0, \sqrt{2})$. The minimum value of f is obtained when $(x, y) = (-1, 1)$ and is equal to $f(-1, 1) = -2$.

6.14. (a) The matrix associated with the quadratic form $3x^2 - 2xy + 3y^2$ is $A = \begin{pmatrix} 3 & -1 \\ -1 & 3 \end{pmatrix}$ which has the eigenvalues $\lambda_1 = 2$, $\lambda_2 = 4$ and the corresponding

eigenvectors are $v_1 = \begin{pmatrix} \frac{1}{\sqrt{2}} \\ \frac{1}{\sqrt{2}} \end{pmatrix}$ and $v_2 = \begin{pmatrix} -\frac{1}{\sqrt{2}} \\ \frac{1}{\sqrt{2}} \end{pmatrix}$. It follows that

$$J_A = \begin{pmatrix} 2 & 0 \\ 0 & 4 \end{pmatrix} \quad \text{and} \quad P = \begin{pmatrix} \frac{1}{\sqrt{2}} & -\frac{1}{\sqrt{2}} \\ \frac{1}{\sqrt{2}} & \frac{1}{\sqrt{2}} \end{pmatrix}.$$

We make the rotation

$$\begin{cases} x = \frac{1}{\sqrt{2}}x' - \frac{1}{\sqrt{2}}y' \\ y = \frac{1}{\sqrt{2}}x' + \frac{1}{\sqrt{2}}y' \end{cases}$$

and the equation of the conic becomes $x'^2 + 2y'^2 + 2\sqrt{2}x' - 2 = 0 \Leftrightarrow (x' + \sqrt{2})^2 + 2y'^2 = 4$. The translation of equations $x' + \sqrt{2} = x''$, $y' = y''$ shows that our conic is the ellipse $x''^2 + 2y''^2 = 4$. The parametric equations of this ellipse are $x'' = 2 \cos t$, $y'' = \sqrt{2} \sin t$, $t \in [0, 2\pi)$. Using the equations of the rotation and the translation we get that

$$\begin{cases} x = \sqrt{2} \cos t - \sin t - 1 \\ y = \sqrt{2} \cos t + \sin t - 1, \end{cases}$$

where $t \in [0, 2\pi)$. It follows that $f(x, y) = g(t) = 2\sqrt{2} \cos t - 2$, $t \in [0, 2\pi)$. This function has the minimum value $-2\sqrt{2} - 2$, when $t = \pi$ and the maximum value $2\sqrt{2} - 2$, when $t = 0$. Thus, the global minimum value of f, subject to the constraint, is $-2\sqrt{2} - 2$ obtained at $(x, y) = (-\sqrt{2} - 1, -\sqrt{2} - 1)$ and the global maximum value of f is $2\sqrt{2} - 2$ obtained at $(x, y) = (\sqrt{2} - 1, \sqrt{2} - 1)$.

Nota bene. Let $\mathscr{E}$ be the ellipse $3x^2 - 2xy + 3y^2 + 4x + 4y - 4 = 0$. It is worth mentioning that since f is a continuous function and $\mathscr{E}$ is a compact set we know, based on the Weierstrass Theorem, that $f|_\mathscr{E}$ has a global minimum and a global maximum. Using the Lagrange multipliers method one can only need to determine the constraint critical points of f, in our case there are two such points, and then to observe that one is a point of global minimum and the other is a point of global maximum.

(b) The matrix associated with the quadratic form $3x^2+10xy+3y^2$ is $A = \begin{pmatrix} 3 & 5 \\ 5 & 3 \end{pmatrix}$ which has the eigenvalues $\lambda_1 = 8$ and $\lambda_2 = -2$ and the corresponding eigenvectors $v_1 = \begin{pmatrix} \frac{1}{\sqrt{2}} \\ \frac{1}{\sqrt{2}} \end{pmatrix}$ and $v_2 = \begin{pmatrix} -\frac{1}{\sqrt{2}} \\ \frac{1}{\sqrt{2}} \end{pmatrix}$. It follows that

$$J_A = \begin{pmatrix} 8 & 0 \\ 0 & -2 \end{pmatrix} \quad \text{and} \quad P = \begin{pmatrix} \frac{1}{\sqrt{2}} & -\frac{1}{\sqrt{2}} \\ \frac{1}{\sqrt{2}} & \frac{1}{\sqrt{2}} \end{pmatrix}.$$

The rotation

$$\begin{cases} x = \frac{1}{\sqrt{2}}x' - \frac{1}{\sqrt{2}}y' \\ y = \frac{1}{\sqrt{2}}x' + \frac{1}{\sqrt{2}}y' \end{cases}$$

and the translation

$$\begin{cases} x' - \sqrt{2} = x'' \\ y' = y'' \end{cases}$$

show that our conic is the hyperbola $4x''^2 - y''^2 = 16$.

Let $\mathscr{H}$ be this hyperbola and let $\mathscr{H}_1$ and $\mathscr{H}_2$ be the two branches of $\mathscr{H}$, i.e., $\mathscr{H} = \mathscr{H}_1 \cup \mathscr{H}_2$.

The parametric equation of $\mathscr{H}_1$ are $x'' = 2\cosh t$, $y'' = 4\sinh t$, $t \in \mathbb{R}$. We get from the equations of the rotation and the translation that

$$\begin{cases} x = 1 + \sqrt{2}\cosh t - 2\sqrt{2}\sinh t \\ y = 1 + \sqrt{2}\cosh t + 2\sqrt{2}\sinh t, \end{cases}$$

$t \in \mathbb{R}$. Thus, we study the extreme values of the function $g(t) = f(x,y) = 2x+y = 3 + \frac{1}{2}\left(\sqrt{2}e^t + 5\sqrt{2}e^{-t}\right)$, $t \in \mathbb{R}$. A calculation shows that g has the global minimum value $3 + \sqrt{10}$ when $t = \ln\sqrt{5}$. Thus, the global minimum value of f is $3 + \sqrt{10}$ obtained when $x = 1 - \sqrt{\frac{2}{5}}$ and $y = 1 + 7\sqrt{\frac{2}{5}}$.

On the other hand, the parametric equations of $\mathscr{H}_2$ are $x'' = -2\cosh t$, $y'' = 4\sinh t$, $t \in \mathbb{R}$. This implies that

$$\begin{cases} x = 1 - \sqrt{2}\cosh t - 2\sqrt{2}\sinh t \\ y = 1 - \sqrt{2}\cosh t + 2\sqrt{2}\sinh t, \end{cases}$$

$t \in \mathbb{R}$. Now we study the extreme values of the function $h(t) = f(x, y) = 2x + y = 3 - \frac{1}{2}\left(5\sqrt{2}e^t + \sqrt{2}e^{-t}\right)$, $t \in \mathbb{R}$. This function has the absolute maximum value $3 - \sqrt{10}$ at $t = -\ln\sqrt{5}$. Thus, the absolute maximum value of f is $3 - \sqrt{10}$ obtained when $x = 1 + \sqrt{\frac{2}{5}}$ and $y = 1 - 7\sqrt{\frac{2}{5}}$.

(c) The matrix associated with the quadratic form $9x^2 + 24xy + 16y^2$ is $A = \begin{pmatrix} 9 & 12 \\ 12 & 16 \end{pmatrix}$ which has the eigenvalues $\lambda_1 = 25$ and $\lambda_2 = 0$ and the corresponding eigenvectors $v_1 = \begin{pmatrix} \frac{3}{5} \\ \frac{4}{5} \end{pmatrix}$ and $v_2 = \begin{pmatrix} -\frac{4}{5} \\ \frac{3}{5} \end{pmatrix}$. It follows that

$$J_A = \begin{pmatrix} 25 & 0 \\ 0 & 0 \end{pmatrix} \quad \text{and} \quad P = \begin{pmatrix} \frac{3}{5} & -\frac{4}{5} \\ \frac{4}{5} & \frac{3}{5} \end{pmatrix}.$$

The rotation

$$\begin{cases} x = \frac{3}{5}x' - \frac{4}{5}y' \\ y = \frac{4}{5}x' + \frac{3}{5}y' \end{cases}$$

shows that the equation of the conic becomes $x'^2 + 2y' = 0$. The parametric equations of this parabola are $x' = t$, $y' = -\frac{t^2}{2}$, $t \in \mathbb{R}$. The equations of the rotation imply that $x = \frac{3}{5}t + \frac{2}{5}t^2$ and $y = \frac{4}{5}t - \frac{3}{10}t^2$. Now we study the extreme values of the function $g(t) = f(x, y) = 2x - y = \frac{2}{5}t + \frac{11}{10}t^2$, which has a global minimum at $t = -\frac{2}{11}$. The global minimum value of f is $-\frac{2}{55}$ which is obtained when $x = -\frac{58}{605}$ and $y = -\frac{94}{605}$.

6.15. Let $f(x, y) = ax^2 + 2bxy + cy^2$ and let L be the Lagrangian $L(x, y) = ax^2 + 2bxy + cy^2 - \lambda(x^2 + y^2 - 1)$. Then

$$\begin{cases} \frac{\partial L}{\partial x} = 2ax + 2by - 2\lambda x = 0 \\ \frac{\partial L}{\partial y} = 2bx + 2cy - 2\lambda y = 0 \end{cases} \Leftrightarrow \begin{cases} (a - \lambda)x + by = 0 \\ bx + (c - \lambda)y = 0. \end{cases}$$

Thus, $X_0 = \begin{pmatrix} x_0 \\ y_0 \end{pmatrix}$ is a constrained critical point of f subject to $x^2 + y^2 = 1$ if and only if $A_f X_0 = \lambda X_0$, for some λ. That is, if and only if λ is an eigenvalue of A_f and X_0 is its corresponding unit eigenvector. If $X_0 = \begin{pmatrix} x_0 \\ y_0 \end{pmatrix}$, with $x_0^2 + y_0^2 = 1$, then $f(x_0, y_0) = (ax_0 + by_0)x_0 + (bx_0 + cy_0)y_0 = \lambda x_0^2 + \lambda y_0^2 = \lambda$. Therefore, the largest and the smallest eigenvalues of A are the maximum and the minimum of f subject to $x^2 + y^2 = 1$.

6.16. The curve is an ellipse whose canonical form is $4x''^2 + y''^2 = 16$. Since the area of the domain bounded by the ellipse $\frac{x^2}{a^2} + \frac{y^2}{b^2} = 1$ is πab and, by making a rotation and a translation the area of a domain is preserved, we get that the area of the domain bounded by the curve $5x^2 + 6xy + 5y^2 - 16x - 16y - 16 = 0$ is 8π.

6.17. (a) $\dfrac{2\pi(e-1)}{\sqrt{3}}$; (b) $\dfrac{2\pi(e-1)}{\sqrt{3}}$; (c) $\dfrac{2\pi}{e\sqrt{3}}$; (d) $\dfrac{2\pi}{e\sqrt{3}}$.

6.18. (a) The matrix associated with the quadratic form $ax^2 - bxy + ay^2$ is $A = \begin{pmatrix} a & -\frac{b}{2} \\ -\frac{b}{2} & a \end{pmatrix}$ which has eigenvalues $\lambda_1 = a - \frac{b}{2}$ and $\lambda_2 = a + \frac{b}{2}$. We have

$$J_A = \begin{pmatrix} a - \frac{b}{2} & 0 \\ 0 & a + \frac{b}{2} \end{pmatrix} \quad \text{and} \quad P = \frac{1}{\sqrt{2}} \begin{pmatrix} 1 & -1 \\ 1 & 1 \end{pmatrix}.$$

We change variables according to the equation $X = PY$, i.e.

$$\begin{pmatrix} x \\ y \end{pmatrix} = \frac{1}{\sqrt{2}} \begin{pmatrix} 1 & -1 \\ 1 & 1 \end{pmatrix} \begin{pmatrix} x' \\ y' \end{pmatrix} \quad \Rightarrow \quad \begin{cases} x = \frac{1}{\sqrt{2}}(x' - y') \\ y = \frac{1}{\sqrt{2}}(x' + y') \end{cases}$$

and we get that

$$I_1 = \iint_{D_1} e^{ax^2 - bxy + ay^2} dx\,dy = \iint_{D_1'} e^{(a-\frac{b}{2})x'^2 + (a+\frac{b}{2})y'^2} \left| \frac{D(x, y)}{D(x', y')} \right| dx'\,dy',$$

where $\dfrac{D(x, y)}{D(x', y')}$ is the Jacobian of the transformation and D_1' is the elliptical disk

$$D_1' = \left\{ (x', y') \in \mathbb{R}^2 : \left(a - \frac{b}{2} \right) x'^2 + \left(a + \frac{b}{2} \right) y'^2 \leq \alpha \right\}.$$

Passing to polar coordinates $x' = \frac{\rho \cos\theta}{\sqrt{a - \frac{b}{2}}}$ and $y' = \frac{\rho \sin\theta}{\sqrt{a + \frac{b}{2}}}$, where $\theta \in [0, 2\pi)$ and $\rho \in [0, \sqrt{\alpha})$ we get that

$$I = \int_0^{\sqrt{\alpha}} \int_0^{2\pi} e^{\rho^2} \frac{\rho}{\sqrt{a^2 - \frac{b^2}{4}}} d\rho\,d\theta = \frac{4\pi}{\sqrt{4a^2 - b^2}} \int_0^{\sqrt{\alpha}} \rho e^{\rho^2} d\rho = \frac{2\pi(e^\alpha - 1)}{\sqrt{4a^2 - b^2}}.$$

(b) The integral equals $\dfrac{2\pi(e^\alpha - 1)}{\sqrt{4a^2 - b^2}}$.

(c) The integral equals $\dfrac{2\pi}{e^\alpha \sqrt{4a^2 - b^2}}$.

(d) The matrix associated with the quadratic form $ax^2 + bxy + ay^2$ is $A = \begin{pmatrix} a & \frac{b}{2} \\ \frac{b}{2} & a \end{pmatrix}$

which has eigenvalues $\lambda_1 = a + \frac{b}{2}$ and $\lambda_2 = a - \frac{b}{2}$. We have

$$J_A = \begin{pmatrix} a + \frac{b}{2} & 0 \\ 0 & a - \frac{b}{2} \end{pmatrix} \quad \text{and} \quad P = \frac{1}{\sqrt{2}} \begin{pmatrix} 1 & -1 \\ 1 & 1 \end{pmatrix}.$$

We change variables according to the equation $X = PY$, i.e.

$$\begin{pmatrix} x \\ y \end{pmatrix} = \frac{1}{\sqrt{2}} \begin{pmatrix} 1 & -1 \\ 1 & 1 \end{pmatrix} \begin{pmatrix} x' \\ y' \end{pmatrix} \quad \Rightarrow \quad \begin{cases} x = \frac{1}{\sqrt{2}} (x' - y') \\ y = \frac{1}{\sqrt{2}} (x' + y') \end{cases}$$

and we get that

$$I_4 = \iint_{D_4} e^{-ax^2 - bxy - ay^2} dxdy = \iint_{D_4'} e^{-\left(a + \frac{b}{2}\right)x'^2 - \left(a - \frac{b}{2}\right)y'^2} \left| \frac{D(x, y)}{D(x', y')} \right| dx'dy',$$

where $\dfrac{D(x, y)}{D(x', y')}$ is the Jacobian of the transformation and D_4' is the exterior, including the boundary, of the elliptical disk

$$D_4' = \left\{ (x', y') \in \mathbb{R}^2 : \left(a + \frac{b}{2} \right) x'^2 + \left(a - \frac{b}{2} \right) y'^2 \geq \alpha \right\}.$$

Passing to polar coordinates $x' = \frac{\rho \cos \theta}{\sqrt{a + \frac{b}{2}}}$ and $y' = \frac{\rho \sin \theta}{\sqrt{a - \frac{b}{2}}}$, where $\theta \in [0, 2\pi)$ and $\rho \in [\sqrt{\alpha}, \infty)$ we get that

$$I_4 = \int_{\sqrt{\alpha}}^{\infty} \int_0^{2\pi} e^{-\rho^2} \frac{\rho}{\sqrt{a^2 - \frac{b^2}{4}}} d\rho d\theta = \frac{4\pi}{\sqrt{4a^2 - b^2}} \int_{\sqrt{\alpha}}^{\infty} \rho e^{-\rho^2} d\rho = \frac{2\pi}{e^\alpha \sqrt{4a^2 - b^2}}.$$

6.19. (a) The integral equals 0. See the solution of part (b).

(b) The matrix associated with the quadratic form $ax^2 + bxy + ay^2$ is $A = \begin{pmatrix} a & \frac{b}{2} \\ \frac{b}{2} & a \end{pmatrix}$

which has the eigenvalues $\lambda_1 = a + \frac{b}{2}$ and $\lambda_2 = a - \frac{b}{2}$. A calculation shows that the Jordan canonical form of A and the invertible matrix P which verifies the equality $A = PJ_A P^{-1}$ are given by

$$J_A = \begin{pmatrix} a + \frac{b}{2} & 0 \\ 0 & a - \frac{b}{2} \end{pmatrix} \quad \text{and} \quad P = \frac{1}{\sqrt{2}} \begin{pmatrix} 1 & -1 \\ 1 & 1 \end{pmatrix}.$$

We change variables according to the equation $X = PY$, i.e.

$$\begin{pmatrix} x \\ y \end{pmatrix} = \frac{1}{\sqrt{2}} \begin{pmatrix} 1 & -1 \\ 1 & 1 \end{pmatrix} \begin{pmatrix} x' \\ y' \end{pmatrix} \quad \Rightarrow \quad \begin{cases} x = \frac{1}{\sqrt{2}}(x' - y') \\ y = \frac{1}{\sqrt{2}}(x' + y') \end{cases}$$

and we get that

$$I = \iint_{D_\alpha} xy e^{-ax^2 - bxy - ay^2} \, dxdy$$

$$= \frac{1}{2} \iint_{D'_\alpha} \left(x'^2 - y'^2 \right) e^{-\left(a + \frac{b}{2}\right)x'^2 - \left(a - \frac{b}{2}\right)y'^2} \left| \frac{D(x, y)}{D(x', y')} \right| \, dx'dy',$$

where $\dfrac{D(x, y)}{D(x', y')}$ is the Jacobian of the transformation and D'_α is the exterior, including the boundary, of the elliptical disk

$$D'_\alpha = \left\{ (x', y') \in \mathbb{R}^2 : \left(a + \frac{b}{2} \right) x'^2 + \left(a - \frac{b}{2} \right) y'^2 \geq \alpha \right\}.$$

Passing to polar coordinates $x' = \dfrac{\rho \cos \theta}{\sqrt{a + \frac{b}{2}}}$, $y' = \dfrac{\rho \sin \theta}{\sqrt{a - \frac{b}{2}}}$, where $\theta \in [0, 2\pi)$ and $\rho \in [\sqrt{\alpha}, \infty)$, we get that

$$I = \frac{1}{\sqrt{4a^2 - b^2}} \int_{\sqrt{\alpha}}^{\infty} \int_0^{2\pi} \left(\frac{\rho^2 \cos^2 \theta}{a + \frac{b}{2}} - \frac{\rho^2 \sin^2 \theta}{a - \frac{b}{2}} \right) e^{-\rho^2} \rho \, d\rho d\theta$$

$$= \frac{1}{\sqrt{4a^2 - b^2}} \left[\frac{1}{a + \frac{b}{2}} \int_{\sqrt{\alpha}}^{\infty} \rho^3 e^{-\rho^2} \, d\rho \int_0^{2\pi} \cos^2 \theta d\theta \right.$$

$$\left. - \frac{1}{a - \frac{b}{2}} \int_{\sqrt{\alpha}}^{\infty} \rho^3 e^{-\rho^2} \, d\rho \int_0^{2\pi} \sin^2 \theta d\theta \right]$$

$$= \frac{1}{\sqrt{4a^2 - b^2}} \int_{\sqrt{\alpha}}^{\infty} \rho^3 e^{-\rho^2} \, d\rho \left(\frac{\pi}{a + \frac{b}{2}} - \frac{\pi}{a - \frac{b}{2}} \right)$$

$$= -\frac{4b\pi}{\sqrt{4a^2 - b^2}(4a^2 - b^2)} \left(-\frac{1}{2}(1 + \rho^2) e^{-\rho^2} \right) \Big|_{\sqrt{\alpha}}^{\infty}$$

$$= -\frac{2b\pi(1 + \alpha)}{\sqrt{4a^2 - b^2}(4a^2 - b^2)e^\alpha}.$$

6.21. (a) We have, based on part (b), that the integral equals $\dfrac{\pi e^2}{\sqrt{8}}$.

(b) Let λ_1, λ_2 be the eigenvalues of A, let $J_A = \begin{pmatrix} \lambda_1 & 0 \\ 0 & \lambda_2 \end{pmatrix}$ be the Jordan canonical

form of A and let P be the orthogonal (rotation) matrix which satisfies $A = PJ_AP^{-1}$.

We have that $P = (X_1 \mid X_2)$, where $X_1 = \begin{pmatrix} x_1 \\ y_1 \end{pmatrix}$ and $X_2 = \begin{pmatrix} x_2 \\ y_2 \end{pmatrix}$ are the eigenvectors

corresponding to the eigenvalues λ_1 and λ_2. Using the substitution $v = \begin{pmatrix} x \\ y \end{pmatrix} =$

$P\begin{pmatrix} x' \\ y' \end{pmatrix}$ we get that

$$
I = \int\int_{\mathbb{R}^2} e^{-(v^TAv+2b^Tv+c)}dxdy
$$

$$
= \int\int_{\mathbb{R}^2} e^{-(\lambda_1 x'^2+\lambda_2 y'^2+b_1'x'+b_2'y'+c)}\left|\frac{D(x,y)}{D(x',y')}\right|dx'dy'
$$

$$
= \int\int_{\mathbb{R}^2} e^{-(\lambda_1 x'^2+\lambda_2 y'^2+b_1'x'+b_2'y'+c)}dx'dy',
$$

where $b_1' = 2b^TX_1$ and $b_2' = 2b^TX_2$. We make the substitutions $x'\sqrt{\lambda_1} = u$, $y'\sqrt{\lambda_2} = v$ and we have that

$$
I = \frac{1}{\sqrt{\det A}}\int\int_{\mathbb{R}^2} e^{-\left(u^2+v^2+\frac{b_1'}{\sqrt{\lambda_1}}u+\frac{b_2'}{\sqrt{\lambda_2}}v+c\right)}dudv
$$

$$
= \frac{1}{\sqrt{\det A}}\int\int_{\mathbb{R}^2} e^{-\left(t^2+w^2+c-\frac{b_1'^2}{4\lambda_1}-\frac{b_2'^2}{4\lambda_2}\right)}dtdw,
$$

where the last equality follows based on the substitutions $u + \dfrac{b_1'}{2\sqrt{\lambda_1}} = t$,

$v + \dfrac{b_2'}{2\sqrt{\lambda_2}} = w$.

We have, since $X_1 = \lambda_1 A^{-1}X_1$ and $X_2 = \lambda_2 A^{-1}X_2$, that

$$
\lambda_2 b_1'^2 + \lambda_1 b_2'^2 = \lambda_2\left(2b^TX_1\right)\left(2X_1^Tb\right) + \lambda_1\left(2b^TX_2\right)\left(2X_2^Tb\right)
$$

$$
= 4\lambda_1\lambda_2 b^TA^{-1}\left(X_1X_1^T + X_2X_2^T\right)b
$$

$$
= 4\det A b^TA^{-1}I_2b
$$

$$
= 4\det A b^TA^{-1}b,
$$

and this implies that

$$
\begin{aligned}
I &= \frac{e^{b^T A^{-1} b - c}}{\sqrt{\det A}} \int \int_{\mathbb{R}^2} e^{-t^2 - w^2} \, dt \, dw \\
&= \frac{e^{b^T A^{-1} b - c}}{\sqrt{\det A}} \int_0^\infty \int_0^{2\pi} e^{-\rho^2} \rho \, d\rho \, d\theta \\
&= \frac{\pi}{\sqrt{\det A}} e^{b^T A^{-1} b - c}.
\end{aligned}
$$

Appendix A
Gems of classical analysis and linear algebra

A.1 Series mirabilis

Lemma A.1 [32] **A power series with the tail of** $\ln \frac{1}{2}$.

The convergence set of the power series

$$\sum_{n=1}^{\infty} \left(\ln \frac{1}{2} + 1 - \frac{1}{2} + \cdots + \frac{(-1)^{n-1}}{n} \right) x^n$$

is $(-1, 1]$ *and the following equality holds*

$$\sum_{n=1}^{\infty} \left(\ln \frac{1}{2} + 1 - \frac{1}{2} + \cdots + \frac{(-1)^{n-1}}{n} \right) x^n = \begin{cases} \ln 2 - \frac{1}{2} & \text{if } x = 1 \\ \dfrac{\ln(1+x) - x \ln 2}{1-x} & \text{if } x \in (-1, 1). \end{cases}$$

Proof First we show that if $n \geq 1$ is an integer, then

$$\ln \frac{1}{2} + 1 - \frac{1}{2} + \cdots + \frac{(-1)^{n-1}}{n} = (-1)^{n-1} \int_0^1 \frac{x^n}{1+x} dx.$$

We have

$$\ln\frac{1}{2} + 1 - \frac{1}{2} + \cdots + \frac{(-1)^{n-1}}{n} = \ln\frac{1}{2} - \sum_{k=1}^{n}\frac{(-1)^k}{k}$$

$$= \ln\frac{1}{2} - \sum_{k=1}^{n}(-1)^k\int_0^1 x^{k-1}dx$$

$$= \ln\frac{1}{2} + \int_0^1 \sum_{k=1}^{n}(-x)^{k-1}dx$$

$$= \ln\frac{1}{2} + \int_0^1 \frac{1-(-x)^n}{1+x}dx$$

$$= (-1)^{n-1}\int_0^1 \frac{x^n}{1+x}dx.$$

Let $a_n = \ln\frac{1}{2} + 1 - \frac{1}{2} + \cdots + \frac{(-1)^{n-1}}{n}$. The radius of convergence of the power series $\sum_{n=1}^{\infty} a_n x^n$ is given by $R = \lim_{n\to\infty}\frac{|a_n|}{|a_{n+1}|}$. A calculation shows that

$$\frac{a_{n+1}}{a_n} = 1 - \frac{1}{(n+1)\int_0^1 \frac{x^n}{1+x}dx}.$$

On the other hand,

$$(n+1)\int_0^1 \frac{x^n}{1+x}dx = \frac{x^{n+1}}{1+x}\Big|_0^1 + \int_0^1 \frac{x^{n+1}}{(1+x)^2}dx = \frac{1}{2} + \int_0^1 \frac{x^{n+1}}{(1+x)^2}dx,$$

and this implies, since $\int_0^1 \frac{x^{n+1}}{(1+x)^2}dx < \int_0^1 x^{n+1}dx = \frac{1}{n+2}$, that $\lim_{n\to\infty}(n+1)\int_0^1 \frac{x^n}{1+x}dx = \frac{1}{2}$. Thus, $R = 1$ and the series converges on $(-1, 1)$.

Now we show that the series converges when $x = 1$ and diverges when $x = -1$.

Let $x = 1$. We prove that the sum of the series equals $\ln 2 - \dfrac{1}{2}$. Let $N \in \mathbb{N}$. We calculate the Nth partial sum of the series and we get that

$$s_N = \sum_{n=1}^{N} \left(\ln \frac{1}{2} + 1 - \frac{1}{2} + \cdots + \frac{(-1)^{n-1}}{n} \right) = \sum_{n=1}^{N} (-1)^{n-1} \int_0^1 \frac{t^n}{1+t} dt$$

$$= -\int_0^1 \frac{1}{1+t} \sum_{n=1}^{N} (-t)^n dt = \int_0^1 \frac{t(1 - (-t)^N)}{(1+t)^2} dt$$

$$= \int_0^1 \frac{t}{(1+t)^2} dt - (-1)^N \int_0^1 \frac{t^{N+1}}{(1+t)^2} dt$$

$$= \ln 2 - \frac{1}{2} - (-1)^N \int_0^1 \frac{t^{N+1}}{(1+t)^2} dt.$$

This implies, since $0 < \displaystyle\int_0^1 \frac{t^{N+1}}{(1+t)^2} dt < \int_0^1 t^{N+1} dt = \frac{1}{N+2}$, that $\displaystyle\lim_{N\to\infty} s_N = \ln 2 - \frac{1}{2}$.

When $x = -1$ we get that

$$\sum_{n=1}^{\infty} (-1)^n \left(\ln \frac{1}{2} + 1 - \frac{1}{2} + \cdots + \frac{(-1)^{n-1}}{n} \right) = -\sum_{n=1}^{\infty} \int_0^1 \frac{t^n}{1+t} dt$$

$$\overset{\dagger}{=} -\int_0^1 \frac{1}{1+t} \sum_{n=1}^{\infty} t^n dt$$

$$= -\int_0^1 \frac{t}{1-t^2} dt$$

$$= \frac{1}{2} \ln(1 - t^2) \Big|_0^{1^-}$$

$$= -\infty.$$

We used at step ($\dagger$) Tonelli's Theorem for nonnegative functions which allow us to interchange the summation and the integration signs.

Let $x \in (-1, 1)$ and let $N \in \mathbb{N}$. We have

$$s_N(x) = \sum_{n=1}^{N} \left(\ln \frac{1}{2} + 1 - \frac{1}{2} + \cdots + \frac{(-1)^{n-1}}{n} \right) x^n$$

$$= \sum_{n=1}^{N} (-1)^{n-1} \int_0^1 \frac{x^n t^n}{1+t} dt = - \int_0^1 \frac{1}{1+t} \sum_{n=1}^{N} (-xt)^n dt$$

$$= \int_0^1 \frac{tx(1 - (-tx)^N)}{(1+t)(1+xt)} dt$$

$$= \int_0^1 \frac{tx}{(1+t)(1+tx)} dt - (-1)^N \int_0^1 \frac{(tx)^{N+1}}{(1+t)(1+xt)} dt.$$

On the other hand,

$$\left| \int_0^1 \frac{(tx)^{N+1}}{(1+t)(1+xt)} dt \right| \leq \int_0^1 \frac{t^{N+1}}{|1+tx|} dt \leq \frac{1}{1-|x|} \int_0^1 t^{N+1} dt = \frac{1}{(N+2)(1-|x|)},$$

and this implies that

$$\lim_{N \to \infty} s_N(x) = \int_0^1 \frac{tx}{(1+t)(1+tx)} dt.$$

A calculation shows that

$$\int_0^1 \frac{tx}{(1+t)(1+tx)} dt = \int_0^1 \frac{x}{1-x} \left(-\frac{1}{1+t} + \frac{1}{1+tx} \right) dt = \frac{\ln(1+x) - x \ln 2}{1-x}.$$

Lemma A.1 is proved.   □

The *Polylogarithm function* $\mathrm{Li}_n(z)$ is defined, for $|z| \leq 1$ and $n \neq 1, 2$, by

$$\mathrm{Li}_n(z) = \sum_{k=1}^{\infty} \frac{z^k}{k^n} = \int_0^z \frac{\mathrm{Li}_{n-1}(t)}{t} dt.$$

When $n = 1$, we define $\mathrm{Li}_1(z) = -\ln(1-z)$ and when $n = 2$, we have that $\mathrm{Li}_2(z)$, also known as the *Dilogarithm function*, is defined by

$$\mathrm{Li}_2(z) = \sum_{n=1}^{\infty} \frac{z^2}{n^2} = -\int_0^z \frac{\ln(1-t)}{t} dt.$$

Before we give the proofs of the next two lemmas we collect a result from the theory of series. Recall that Abel's summation formula [11, p. 55], [22, p. 258] states that if $(a_n)_{n \geq 1}$ and $(b_n)_{n \geq 1}$ are two sequences of real or complex numbers and $A_n = \sum_{k=1}^{n} a_k$, then

$$\sum_{k=1}^{n} a_k b_k = A_n b_{n+1} + \sum_{k=1}^{n} A_k (b_k - b_{k+1}), \quad n \in \mathbb{N}.$$

We also use, in our calculations, the infinite version of the preceding formula

$$\sum_{k=1}^{\infty} a_k b_k = \lim_{n \to \infty} (A_n b_{n+1}) + \sum_{k=1}^{\infty} A_k (b_k - b_{k+1}),$$

provided the infinite series converges and the limit is finite.

Lemma A.2 The generating function of the tail of $\zeta(k)$.

Let $k \geq 3$ be an integer and let $x \in [-1, 1]$. The following formula holds

$$\sum_{n=1}^{\infty} \left(\zeta(k) - \frac{1}{1^k} - \frac{1}{2^k} - \cdots - \frac{1}{n^k} \right) x^n = \begin{cases} \dfrac{x\zeta(k) - Li_k(x)}{1-x} & \text{if } x \in [-1, 1) \\[2mm] \zeta(k-1) - \zeta(k) & \text{if } x = 1, \end{cases}$$

where Li_k denotes the polylogarithm function.

Proof If $x = 0$ we have nothing to prove, so we consider the case when $x \in [-1, 1)$ and $x \neq 0$. We use Abel's summation formula, with $a_n = x^n$ and $b_n = \zeta(k) - \frac{1}{1^k} - \frac{1}{2^k} - \cdots - \frac{1}{n^k}$. We have

$$\sum_{n=1}^{\infty} \left(\zeta(k) - \frac{1}{1^k} - \frac{1}{2^k} - \cdots - \frac{1}{n^k} \right) x^n$$

$$= \lim_{n \to \infty} (x + x^2 + \cdots + x^n) \left(\zeta(k) - \frac{1}{1^k} - \frac{1}{2^k} - \cdots - \frac{1}{(n+1)^k} \right)$$

$$+ \sum_{n=1}^{\infty} (x + x^2 + \cdots + x^n) \frac{1}{(n+1)^k}$$

$$= \frac{x}{1-x} \sum_{n=1}^{\infty} \left(\frac{1}{(n+1)^k} - \frac{x^n}{(n+1)^k} \right)$$

$$= \frac{x}{1-x} \left[\zeta(k) - 1 - \frac{1}{x} (Li_k(x) - x) \right]$$

$$= \frac{x\zeta(k) - Li_k(x)}{1-x}.$$

Now we consider the case when $x = 1$. We use Abel's summation formula, with $a_n = 1$ and $b_n = \zeta(k) - \frac{1}{1^k} - \frac{1}{2^k} - \cdots - \frac{1}{n^k}$. We have

$$\sum_{n=1}^{\infty} \left(\zeta(k) - \frac{1}{1^k} - \frac{1}{2^k} - \cdots - \frac{1}{n^k} \right) = \lim_{n \to \infty} n \left(\zeta(k) - \frac{1}{1^k} - \frac{1}{2^k} - \cdots - \frac{1}{(n+1)^k} \right)$$

$$+ \sum_{n=1}^{\infty} \frac{n}{(n+1)^k}$$

$$= \sum_{n=1}^{\infty} \left(\frac{1}{(n+1)^{k-1}} - \frac{1}{(n+1)^k} \right)$$

$$= \zeta(k-1) - \zeta(k),$$

and the lemma is proved. $\qquad\qquad\qquad\qquad\qquad\qquad\qquad\qquad\qquad\qquad\qquad$ $\square$

Lemma A.3 The generating function of n times the tail of $\zeta(k)$.

(a) *Let $k \geq 3$ be an integer and let $x \in [-1, 1)$. The following formula holds*

$$\sum_{n=1}^{\infty} n \left(\zeta(k) - \frac{1}{1^k} - \frac{1}{2^k} - \cdots - \frac{1}{n^k} \right) x^{n-1} = \frac{\zeta(k) - \frac{1-x}{x} Li_{k-1}(x) - Li_k(x)}{(1-x)^2},$$

where Li_k denotes the polylogarithm function.

(b) *Let $k > 3$ be a real number. Then*

$$\sum_{n=1}^{\infty} n \left(\zeta(k) - \frac{1}{1^k} - \frac{1}{2^k} - \cdots - \frac{1}{n^k} \right) = \frac{1}{2} \left(\zeta(k-2) - \zeta(k-1) \right).$$

Proof (a) Differentiate the series in Lemma A.2.

(b) This part of the lemma can be proved by applying Abel's summation formula with $a_n = n$ and $b_n = \zeta(k) - \frac{1}{1^k} - \frac{1}{2^k} - \cdots - \frac{1}{n^k}$. $\qquad\qquad\qquad\qquad\qquad$ $\square$

A.2   Two quadratic Frullani integrals

In this section we prove a lemma which is used in the solution of problem **4.102**.

Lemma A.4 Frullani in disguise.

Let α be a positive real number. The following equality holds

$$\int_0^\infty \int_0^\infty \left(\frac{e^{-\alpha x} - e^{-\alpha y}}{x - y} \right)^2 dxdy = \int_0^\infty \left(\frac{1 - e^{-x}}{x} \right)^2 dx = 2 \ln 2.$$

Proof First we calculate the single integral by observing that it is a Frullani integral [33]. Let $f : [0, \infty) \to \mathbb{R}$ be the function $f(x) = \frac{1-e^{-x}}{x}$, if $x \neq 0$ and $f(0) = 1$. A calculation shows that

$$\left(\frac{1 - e^{-x}}{x} \right)^2 = 2 \frac{f(x) - f(2x)}{x}.$$

It follows, based on Frullani's formula, that

$$\int_0^\infty \left(\frac{1 - e^{-x}}{x} \right)^2 dx = 2 \int_0^\infty \frac{f(x) - f(2x)}{x} dx = 2 \left(f(0) - f(\infty) \right) \ln 2 = 2 \ln 2,$$

and the second equality of the lemma is proved.

Now we calculate the double integral by using the substitutions $\alpha x = u, \alpha y = v$ and we get that

$$I = \int_0^\infty \int_0^\infty \left(\frac{e^{-\alpha x} - e^{-\alpha y}}{x - y} \right)^2 dxdy$$

$$= \int_0^\infty \int_0^\infty \left(\frac{e^{-u} - e^{-v}}{u - v} \right)^2 dudv$$

$$= 2 \int_0^\infty \left(\int_0^u \left(\frac{e^{-u} - e^{-v}}{u - v} \right)^2 dv \right) du.$$

The substitution $u - v = t$ shows that the inner integral becomes

$$\int_0^u \left(\frac{e^{-u} - e^{-v}}{u - v} \right)^2 dv = e^{-2u} \int_0^u \left(\frac{1 - e^t}{t} \right)^2 dt.$$

This implies that $I = 2 \int_0^\infty e^{-2u} \left(\int_0^u \left(\frac{1 - e^t}{t} \right)^2 dt \right) du$. We calculate this integral by parts with

$$f(u) = \int_0^u \left(\frac{1 - e^t}{t} \right)^2 dt, \quad f'(u) = \left(\frac{1 - e^u}{u} \right)^2, \quad g'(u) = e^{-2u}, \quad g(u) = -\frac{e^{-2u}}{2},$$

and we get that

$$I = -e^{-2u} \int_0^u \left(\frac{1-e^t}{t}\right)^2 dt \Big|_0^\infty + \int_0^\infty \left(\frac{1-e^{-u}}{u}\right)^2 du$$

$$= \int_0^\infty \left(\frac{1-e^{-u}}{u}\right)^2 du$$

$$= 2\ln 2.$$

The lemma is proved.   □

More generally [21] one can prove that if $n \geq 2$ is an integer, then

$$\int_0^\infty \int_0^\infty \left(\frac{e^{-x}-e^{-y}}{x-y}\right)^n dxdy = \frac{2(-1)^n}{n} \int_0^\infty \left(\frac{1-e^{-x}}{x}\right)^n dx$$

$$= \frac{2}{n!} \sum_{j=2}^n \binom{n}{j} j^{n-1}(-1)^j \ln j.$$

However, the case when $n = 2$ reduces to the calculation of a Frullani integral.

A.3   Computing e^{Ax}

In this section we give a general method for calculating e^{Ax}, where $A \in \mathcal{M}_2(\mathbb{R})$ and $x \in \mathbb{R}$. This method is based on a combination of the Cayley–Hamilton Theorem and the power series expansion of the exponential function.

Theorem A.1   The exponential matrix e^{Ax}.

Let $A \in \mathcal{M}_2(\mathbb{R})$, $x \in \mathbb{R}$, $Tr(A) = t$, and $\det A = d$. Then:

$$e^{Ax} = \begin{cases} e^{\frac{tx}{2}} \left[\cosh \frac{\sqrt{\Delta}x}{2} I_2 + \frac{2}{\sqrt{\Delta}} \sinh \frac{\sqrt{\Delta}x}{2} \left(A - \frac{t}{2}I_2\right) \right] & \text{if } \Delta > 0 \\ e^{\frac{tx}{2}} \left[I_2 + \left(A - \frac{t}{2}I_2\right)x \right] & \text{if } \Delta = 0 \\ e^{\frac{tx}{2}} \left[\cos \frac{\sqrt{-\Delta}x}{2} I_2 + \frac{2}{\sqrt{-\Delta}} \sin \frac{\sqrt{-\Delta}x}{2} \left(A - \frac{t}{2}I_2\right) \right] & \text{if } \Delta < 0, \end{cases}$$

where $\Delta = t^2 - 4d$

Proof We have, based on the Cayley–Hamilton Theorem, that $A^2 - tA + dI_2 = O_2$ and it follows that $\left(A - \frac{t}{2}I_2\right)^2 = \frac{\Delta}{4}I_2$.

The case $\Delta > 0$. Let $b = \frac{\sqrt{\Delta}}{2}$ and let $B = A - \frac{t}{2}I_2$. We have that $B^2 = b^2 I_2$ and this implies that $B^{2k} = b^{2k}I_2$, for all $k \geq 0$ and $B^{2k-1} = b^{2k-2}B$, for all $k \geq 1$. A calculation shows that

$$
\begin{aligned}
e^{Bx} &= \sum_{k=0}^{\infty} \frac{(Bx)^{2k}}{(2k)!} + \sum_{k=1}^{\infty} \frac{(Bx)^{2k-1}}{(2k-1)!} \\
&= \sum_{k=0}^{\infty} \frac{(bx)^{2k}}{(2k)!} I_2 + \sum_{k=1}^{\infty} \frac{(bx)^{2k-1}}{(2k-1)!} \cdot \frac{B}{b} \\
&= \cosh(bx)I_2 + \frac{\sinh(bx)}{b} B \\
&= \cosh \frac{\sqrt{\Delta}x}{2} I_2 + \frac{2}{\sqrt{\Delta}} \sinh \frac{\sqrt{\Delta}x}{2} B.
\end{aligned}
$$

This implies that

$$
e^{Ax} - e^{\frac{tx}{2}I_2}e^{Bx} = e^{\frac{tx}{2}}\left[\cosh \frac{\sqrt{\Delta}x}{2} I_2 + \frac{2}{\sqrt{\Delta}} \sinh \frac{\sqrt{\Delta}x}{2}\left(A - \frac{t}{2}I_2\right)\right].
$$

The case $\Delta = 0$. We have that $B^2 = O_2$ and this implies that $B^k = O_2$, for all $k \geq 2$. This implies that $e^{Bx} = I_2 + Bx$ and

$$
e^{Ax} = e^{\frac{tx}{2}I_2}e^{Bx} = e^{\frac{tx}{2}}\left[I_2 + \left(A - \frac{t}{2}I_2\right)x\right].
$$

The case $\Delta < 0$. We have that $B^2 = \frac{\Delta}{4}I_2$ or $B^2 = -b^2 I_2$, where $b = \frac{\sqrt{-\Delta}}{2}$. This implies that $B^{2k} = (-1)^k b^{2k}I_2$, for all $k \geq 0$ and $B^{2k-1} = (-1)^{k-1}b^{2k-2}B$, for all $k \geq 1$. A calculation shows that

$$
\begin{aligned}
e^{Bx} &= \sum_{k=0}^{\infty} \frac{(Bx)^{2k}}{(2k)!} + \sum_{k=1}^{\infty} \frac{(Bx)^{2k-1}}{(2k-1)!} \\
&= \sum_{k=0}^{\infty} (-1)^k \frac{(bx)^{2k}}{(2k)!} I_2 + \sum_{k=1}^{\infty} (-1)^{k-1} \frac{(bx)^{2k-1}}{(2k-1)!} \cdot \frac{B}{b} \\
&= \cos(bx)I_2 + \frac{\sin(bx)}{b} B \\
&= \cos \frac{\sqrt{-\Delta}x}{2} I_2 + \frac{2}{\sqrt{-\Delta}} \sin \frac{\sqrt{-\Delta}x}{2}\left(A - \frac{t}{2}I_2\right).
\end{aligned}
$$

This implies that

$$
e^{Ax} = e^{\frac{tx}{2}I_2} e^{Bx} = e^{\frac{tx}{2}} \left[\cos \frac{\sqrt{-\Delta}x}{2} I_2 + \frac{2}{\sqrt{-\Delta}} \sin \frac{\sqrt{-\Delta}x}{2} \left(A - \frac{t}{2}I_2 \right) \right].
$$

The theorem is proved.   □

Remark A.1 We mention that the hyperbolic functions $\sinh(Ax)$ and $\cosh(Ax)$ can also be calculated as a consequence of Theorem A.1. We leave these calculations to the interested reader.

A.4   Computing $\sin Ax$ and $\cos Ax$

In this section we give a technique, other than the one involving the Jordan canonical form of a matrix, for calculating the trigonometric functions $\sin(Ax)$ and $\cos(Ax)$, where $A \in \mathcal{M}_2(\mathbb{R})$ and $x \in \mathbb{R}$.

Theorem A.2   The trigonometric function $\sin(Ax)$.

Let $A \in \mathcal{M}_2(\mathbb{R})$, $x \in \mathbb{R}$, $\mathrm{Tr}(A) = t$ and $\det A = d$. Then:

$$
\sin(Ax) = \begin{cases} \cos \dfrac{\sqrt{\Delta}x}{2} \sin \dfrac{tx}{2} I_2 + \dfrac{2}{\sqrt{\Delta}} \cos \dfrac{tx}{2} \sin \dfrac{\sqrt{\Delta}x}{2} \left(A - \dfrac{t}{2}I_2 \right) & \text{if } \Delta > 0 \\[3mm] \sin \dfrac{tx}{2} I_2 + x \cos \dfrac{tx}{2} \left(A - \dfrac{t}{2}I_2 \right) & \text{if } \Delta = 0 \\[3mm] \cosh \dfrac{\sqrt{-\Delta}x}{2} \sin \dfrac{tx}{2} I_2 + \dfrac{2}{\sqrt{-\Delta}} \sinh \dfrac{\sqrt{-\Delta}x}{2} \cos \dfrac{tx}{2} \left(A - \dfrac{t}{2}I_2 \right) & \text{if } \Delta < 0, \end{cases}
$$

where $\Delta = t^2 - 4d$.

Proof The Cayley–Hamilton Theorem implies that $A^2 - tA + dI_2 = O_2$ and it follows that $\left(A - \frac{t}{2}I_2 \right)^2 = \frac{\Delta}{4}I_2$.

The case $\Delta > 0$. Let $b = \frac{\sqrt{\Delta}}{2}$ and let $B = A - \frac{t}{2}I_2$. We have that $B^2 = b^2 I_2$ and this implies that $B^{2k} = b^{2k}I_2$, for all $k \geq 0$ and $B^{2k-1} = b^{2k-2}B$, for all $k \geq 1$. We have

$$\sin(Bx) = \sum_{n=1}^{\infty}(-1)^{n-1}\frac{(Bx)^{2n-1}}{(2n-1)!}$$

$$= \sum_{n=1}^{\infty}(-1)^{n-1}\frac{(bx)^{2n-1}}{(2n-1)!}\cdot\frac{B}{b}$$

$$= \frac{\sin(bx)}{b}B$$

$$= \frac{2}{\sqrt{\Delta}}\sin\frac{\sqrt{\Delta}x}{2}\left(A - \frac{t}{2}I_2\right)$$

and

$$\cos(Bx) = \sum_{n=0}^{\infty}(-1)^{n}\frac{(Bx)^{2n}}{(2n)!} = \sum_{n=0}^{\infty}(-1)^{n}\frac{(bx)^{2n}}{(2n)!}I_2 = \cos(bx)I_2 = \cos\frac{\sqrt{\Delta}x}{2}I_2.$$

It follows that

$$\sin(Ax) = \sin\left(Bx + \frac{tx}{2}I_2\right)$$

$$= \sin(Bx)\cos\left(\frac{tx}{2}I_2\right) + \cos(Bx)\sin\left(\frac{tx}{2}I_2\right)$$

$$= \sin(Bx)\cos\left(\frac{tx}{2}\right) + \cos(Bx)\sin\left(\frac{tx}{2}\right)$$

$$= \cos\left(\frac{\sqrt{\Delta}x}{2}\right)\sin\frac{tx}{2}I_2 + \frac{2}{\sqrt{\Delta}}\cos\frac{tx}{2}\sin\frac{\sqrt{\Delta}x}{2}\left(A - \frac{t}{2}I_2\right).$$

The case $\Delta = 0$. We have that $B^2 = O_2$ and this implies that $B^k = O_2$, for all $k \geq 2$. A calculation shows that

$$\sin(Bx) = Bx = \left(A - \frac{t}{2}I_2\right)x \quad \text{and} \quad \cos(Bx) = I_2.$$

Thus

$$\sin(Ax) = \sin\left(Bx + \frac{tx}{2}I_2\right)$$

$$= \sin(Bx)\cos\left(\frac{tx}{2}I_2\right) + \cos(Bx)\sin\left(\frac{tx}{2}I_2\right)$$

$$= \sin\frac{tx}{2}I_2 + x\cos\frac{tx}{2}\left(A - \frac{t}{2}I_2\right).$$

The case $\Delta < 0$. We have that $B^2 = \frac{\Delta}{4}I_2$ or $B^2 = -b^2 I_2$, where $b = \frac{\sqrt{-\Delta}}{2}$. This implies that $B^{2k} = (-1)^k b^{2k} I_2$, for all $k \geq 0$ and $B^{2k-1} = (-1)^{k-1} b^{2k-2} B$, for all $k \geq 1$. A calculation shows that

$$\sin(Bx) = \sum_{n=1}^{\infty} (-1)^{n-1} \frac{(Bx)^{2n-1}}{(2n-1)!}$$

$$= \sum_{n=1}^{\infty} \frac{(bx)^{2n-1}}{(2n-1)!} \cdot \frac{B}{b}$$

$$= \frac{\sinh(bx)}{b} B$$

$$= \frac{2}{\sqrt{-\Delta}} \sinh \frac{\sqrt{-\Delta}x}{2} \left(A - \frac{t}{2} I_2 \right)$$

and

$$\cos(Bx) = \sum_{n=0}^{\infty} (-1)^n \frac{(Bx)^{2n}}{(2n)!} = \sum_{n=0}^{\infty} \frac{(bx)^{2n}}{(2n)!} I_2 = \cosh(bx) I_2 = \cosh \frac{\sqrt{-\Delta}x}{2} I_2.$$

It follows that

$$\sin(Ax) = \sin \left(Bx + \frac{tx}{2} I_2 \right)$$

$$= \sin(Bx) \cos \left(\frac{tx}{2} I_2 \right) + \cos(Bx) \sin \left(\frac{tx}{2} I_2 \right)$$

$$= \cosh \frac{\sqrt{-\Delta}x}{2} \sin \frac{tx}{2} I_2 + \frac{2}{\sqrt{-\Delta}} \sinh \frac{\sqrt{-\Delta}x}{2} \cos \frac{tx}{2} \left(A - \frac{t}{2} I_2 \right).$$

The theorem is proved.   □

Theorem A.3 The trigonometric function $\cos(Ax)$.

Let $A \in \mathcal{M}_2(\mathbb{R})$, $x \in \mathbb{R}$, $Tr(A) = t$, and $\det A = d$. Then:

$$\cos(Ax) = \begin{cases} \cos \dfrac{\sqrt{\Delta}x}{2} \cos \dfrac{tx}{2} I_2 - \dfrac{2}{\sqrt{\Delta}} \sin \dfrac{tx}{2} \sin \dfrac{\sqrt{\Delta}x}{2} \left(A - \dfrac{t}{2} I_2 \right) & \text{if } \Delta > 0 \\[2ex] \cos \dfrac{tx}{2} I_2 - x \sin \dfrac{tx}{2} \left(A - \dfrac{t}{2} I_2 \right) & \text{if } \Delta = 0 \\[2ex] \cosh \dfrac{\sqrt{-\Delta}x}{2} \cos \dfrac{tx}{2} I_2 - \dfrac{2}{\sqrt{-\Delta}} \sinh \dfrac{\sqrt{-\Delta}x}{2} \sin \dfrac{tx}{2} \left(A - \dfrac{t}{2} I_2 \right) & \text{if } \Delta < 0, \end{cases}$$

where $\Delta = t^2 - 4d$.

Proof The proof of the theorem is similar to the proof of Theorem A.2.   □

Appendix B
Trigonometric matrix equations

B.1 Four trigonometric equations

In this appendix we solve the fundamental trigonometric matrix equations. First we record a lemma which will be used in the proofs of Lemmas B.2 and B.4.

Lemma B.1 *Let f be a function which has the Taylor series expansion at 0,*
$f(z) = \sum_{n=0}^{\infty} \frac{f^{(n)}(0)}{n!} z^n$, $|z| < R$, *where $R \in (0, \infty]$ and let $A \in \mathcal{M}_2(\mathbb{C})$ be such that $\rho(A) < R$. Let $\alpha \in \mathbb{C}$ and let $B \in \mathcal{M}_2(\mathbb{C})$ such that A and B are similar. Then, $f(A) = \alpha I_2$ if and only if $f(B) = \alpha I_2$.*

Proof The proof is left as an exercise to the interested reader. $\qquad\square$

Lemma B.2 *Let $A \in \mathcal{M}_2(\mathbb{R})$. The solutions of the equation $\sin A = O_2$ are given by*

$$A = Q \begin{pmatrix} k\pi & 0 \\ 0 & l\pi \end{pmatrix} Q^{-1},$$

where $l, k \in \mathbb{Z}$ and $Q \in \mathcal{M}_2(\mathbb{R})$ is any invertible matrix.

Proof Let J_A be the Jordan canonical form of A. Since $A \sim J_A$ we have, based on Lemma B.1, that it suffices to study the equation $\sin J_A = O_2$. Let λ_1, λ_2 be the eigenvalues of A. We distinguish here the following two cases.

© Springer International Publishing AG 2017
V. Pop, O. Furdui, *Square Matrices of Order 2*, DOI 10.1007/978-3-319-54939-2

A has real eigenvalues. If $J_A = \begin{pmatrix} \lambda_1 & 0 \\ 0 & \lambda_2 \end{pmatrix}$, then $\sin J_A = \begin{pmatrix} \sin \lambda_1 & 0 \\ 0 & \sin \lambda_2 \end{pmatrix} = O_2$

which implies that $\sin \lambda_1 = 0$ and $\sin \lambda_2 = 0$. Thus, $\lambda_1 = k\pi$ and $\lambda_2 = l\pi$, where $k, l \in \mathbb{Z}$.

If $J_A = \begin{pmatrix} \lambda & 1 \\ 0 & \lambda \end{pmatrix}$, then $\sin J_A = \begin{pmatrix} \sin \lambda & \cos \lambda \\ 0 & \sin \lambda \end{pmatrix} = O_2$ and this implies $\sin \lambda = 0$

and $\cos \lambda = 0$ which is impossible since $\sin^2 \lambda + \cos^2 \lambda = 1$.

A has complex eigenvalues. Let $\beta \in \mathbb{R}^*$ and $\lambda_1 = \alpha + i\beta$ and $\lambda_2 = \alpha - i\beta$ be

the eigenvalues of A. We have, based on Theorem 2.10, that $J_A = \begin{pmatrix} \alpha & \beta \\ -\beta & \alpha \end{pmatrix}$. The

equation $\sin A = O_2$ implies that $\sin A = P \sin J_A P^{-1} = O_2$ which in turn implies that $\sin J_A = O_2$. A calculation, based on Theorem A.2, shows that

$$\sin J_A = \frac{\sinh |\beta| \cos \alpha}{|\beta|} J_A + \left[\cosh |\beta| \sin \alpha - \frac{\alpha}{|\beta|} \sinh |\beta| \cos \alpha \right] I_2$$

$$= \begin{pmatrix} \cosh |\beta| \sin \alpha & \frac{\beta}{|\beta|} \sinh |\beta| \cos \alpha \\ -\frac{\beta}{|\beta|} \sinh |\beta| \cos \alpha & \cosh |\beta| \sin \alpha \end{pmatrix}.$$

This implies that

$$\begin{cases} \cosh |\beta| \sin \alpha = 0 \\ \frac{\beta}{|\beta|} \sinh |\beta| \cos \alpha = 0. \end{cases}$$

The first equation implies that $\sin \alpha = 0$ and, since $\beta \neq 0$, the second equation shows that $\cos \alpha = 0$ which contradicts $\sin^2 \alpha + \cos^2 \alpha = 1$. $\square$

Lemma B.3 *Let $A \in \mathcal{M}_2(\mathbb{R})$. The solutions of the equation $\cos A = O_2$ are given by*

$$A = Q \begin{pmatrix} \frac{\pi}{2} + k\pi & 0 \\ 0 & \frac{\pi}{2} + l\pi \end{pmatrix} Q^{-1},$$

where $l, k \in \mathbb{Z}$ and $Q \in \mathcal{M}_2(\mathbb{R})$ is any invertible matrix.

Proof Since $\cos A = \sin \left(\frac{\pi}{2} I_2 - A \right)$ we have that the equation to solve reads

$$\sin \left(\frac{\pi}{2} I_2 - A \right) = O_2,$$

and the result follows based on Lemma B.2. $\square$

Lemma B.4 *Let $A \in \mathcal{M}_2(\mathbb{R})$. The solutions of the equation $\sin A = I_2$ are given by*

$$A = Q \begin{pmatrix} \frac{\pi}{2} + 2k\pi & 0 \\ 0 & \frac{\pi}{2} + 2l\pi \end{pmatrix} Q^{-1},$$

where $l, k \in \mathbb{Z}$ and $Q \in \mathcal{M}_2(\mathbb{R})$ is any invertible matrix or

$$A = Q \begin{pmatrix} \frac{\pi}{2} + 2m\pi & 1 \\ 0 & \frac{\pi}{2} + 2m\pi \end{pmatrix} Q^{-1},$$

where $m \in \mathbb{Z}$ and $Q \in \mathcal{M}_2(\mathbb{R})$ is any invertible matrix.

Proof Let J_A be the Jordan canonical form of A. Since $A \sim J_A$ we have, based on Lemma B.1, that it suffices to study the equation $\sin J_A = I_2$. Let λ_1, λ_2 be the eigenvalues of A. We distinguish here the following two cases.

A has real eigenvalues. If $J_A = \begin{pmatrix} \lambda_1 & 0 \\ 0 & \lambda_2 \end{pmatrix}$, then $\sin J_A = \begin{pmatrix} \sin \lambda_1 & 0 \\ 0 & \sin \lambda_2 \end{pmatrix} = I_2$ which implies that $\sin \lambda_1 = 1$ and $\sin \lambda_2 = 1$. Thus, $\lambda_1 = \frac{\pi}{2} + 2k\pi$ and $\lambda_2 = \frac{\pi}{2} + 2l\pi$, where $k, l \in \mathbb{Z}$.

If $J_A = \begin{pmatrix} \lambda & 1 \\ 0 & \lambda \end{pmatrix}$, then $\sin J_A = \begin{pmatrix} \sin \lambda & \cos \lambda \\ 0 & \sin \lambda \end{pmatrix} = I_2$ and this implies $\sin \lambda = 1$ and $\cos \lambda = 0$ which implies that $\lambda = \frac{\pi}{2} + 2m\pi$, where $m \in \mathbb{Z}$.

A has complex eigenvalues. Let $\beta \in \mathbb{R}^*$ and $\lambda_1 = \alpha + i\beta$ and $\lambda_2 = \alpha - i\beta$ be the eigenvalues of A. We have, based on Theorem 2.10, that $J_A = \begin{pmatrix} \alpha & \beta \\ -\beta & \alpha \end{pmatrix}$. The equation $\sin A = I_2$ implies that $\sin A = P \sin J_A P^{-1} = I_2$ which in turn implies that $\sin J_A = I_2$. A calculation, based on Theorem A.2, shows that

$$\sin J_A = \frac{\sinh |\beta| \cos \alpha}{|\beta|} J_A + \left[\cosh |\beta| \sin \alpha - \frac{\alpha}{|\beta|} \sinh |\beta| \cos \alpha \right] I_2$$

$$= \begin{pmatrix} \cosh |\beta| \sin \alpha & \frac{\beta}{|\beta|} \sinh |\beta| \cos \alpha \\ -\frac{\beta}{|\beta|} \sinh |\beta| \cos \alpha & \cosh |\beta| \sin \alpha \end{pmatrix}.$$

This implies that

$$\begin{cases} \cosh |\beta| \sin \alpha = 1 \\ \frac{\beta}{|\beta|} \sinh |\beta| \cos \alpha = 0. \end{cases}$$

The second equation implies, since $\beta \neq 0$, that $\cos \alpha = 0$. This shows that $\sin \alpha = \pm 1$ and we get from the first equation that $\cosh |\beta| = 1$. The solution of this equation is $\beta = 0$ which is impossible. $\square$

Lemma B.5 *Let $A \in \mathcal{M}_2(\mathbb{R})$. The solutions of the equation $\cos A = I_2$ are given by*

$$A = Q \begin{pmatrix} 2k\pi & 0 \\ 0 & 2l\pi \end{pmatrix} Q^{-1},$$

where $l, k \in \mathbb{Z}$ and $Q \in \mathcal{M}_2(\mathbb{R})$ is any invertible matrix or

$$A = Q \begin{pmatrix} 2m\pi & -1 \\ 0 & 2m\pi \end{pmatrix} Q^{-1},$$

where $m \in \mathbb{Z}$ and $Q \in \mathcal{M}_2(\mathbb{R})$ is any invertible matrix.

Proof Observe that $\sin \left(\frac{\pi}{2} I_2 - A \right) = \cos A$ and the proof follows based on Lemma B.4.

Other equations involving trigonometric functions can be solved by reducing them to these four fundamental matrix equations. We stop our line of investigation here and invite the reader to study further other matrix equations involving trigonometric or inverse of trigonometric functions.

Epilogue

The authors wish you success in solving this collection of problems involving *only matrices of order* 2. Why only matrices of order 2? Simply because there are spectacular results involving 2×2 matrices, see for example the determinant formulae from Chapter 1 that do not hold for matrices other than those of order 2 and even if, in some cases, these results can be extended to matrices of order greater than 2, these formulae lose splendor and beauty, not to mention the finesse of their proofs.

Whether the problems turn out to be splendid or not that is for you, the reader, to decide. We hope that you will enjoy both the problems and the theory. For questions, generalizations, remarks, observations regarding the improvement of this material and why not criticism, please do not hesitate to contact us at:

Vasile Pop

Technical University of Cluj-Napoca
Department of Mathematics
Str. Memorandumului Nr. 28, 400114
Cluj-Napoca, Romania
E-mail: Vasile.Pop@math.utcluj.ro

and

Ovidiu Furdui

Technical University of Cluj-Napoca
Department of Mathematics
Str. Memorandumului Nr. 28, 400114
Cluj-Napoca, Romania
E-mail: Ovidiu.Furdui@math.utcluj.ro
E-mail: ofurdui@yahoo.com

References

1. Adams, F., Bloch, A., Lagarias, J.: Problem 11739. Am. Math. Mon. **120**(9), 855 (2013)
2. Akopyan, A.V., Zaslavsky, A.A.: Geometry of Conics, Mathematical World, vol. 26. American Mathematical Society, Providence (2007)
3. Amdeberhan, T., Medina, L.A., Moll, V.H.: Arithmetical properties of a sequence arising from an arctangent sum. J. Number Theory **128**, 1807–1846 (2008)
4. Anghel, N.: Square roots of real 2 × 2 matrices. Gazeta Matematică Ser. B **118**(11), 489–491 (2013)
5. Apéry, R.: Irrationalité de $\zeta(2)$ et $\zeta(3)$. Astérisque **61**, 11–13 (1979)
6. Apostol, T.M.: Calculus, Multi Variable Calculus and Linear Algebra, with Applications to Differential Equations and Probability, Vol. II, 2nd edn. Wiley, New York (1969)
7. Bannon, T.: The origin of quaternions. Coll. Math. J. **46**(1), 43–50 (2015)
8. Barbeau, E.J.: Pell's Equation. Springer, New York (2003)
9. Bernstein, S.D.: Matrix Mathematics. Theory, Facts, and Formulas, 2nd edn. Princeton University Press, Princeton (2009)
10. Bernstein, S.D., So, W.: Some explicit formulas for the matrix exponential. IEEE Trans. Autom. Control **38**(8), 1228–1232 (1993)
11. Bonar, D.D., Koury, M.J.: Real Infinite Series. The Mathematical Association of America, Washington, DC (2006)
12. Boualem, H., Brouzet, R.: To be (a circle) or not to be? Coll. Math. J. **46**(3), 197–206 (2015)
13. Boyadzhiev, K.N.: Exponential polynomials, Stirling numbers and evaluations of some gamma integrals. Abstr. Appl. Anal. **2009**, Article ID 168672 (2009). Available at http://www.hindawi.com/journals/aaa/2009/168672/
14. Cayley, A.: A memoir on the theory of matrices. Philos. Trans. R. Soc. Lond. **148**, 17–37 (1858)
15. Dickson, L.E.: History of the Theory of Numbers, Vol. II, Diophantine Analysis. AMS Chelsea, Providence (1999)
16. Djukić, D.: Pell's Equation, The IMO Compendium Group, Olympiad Training Materials. http://www.imomath.com/index.php?options=257&lmm=0 (2007)
17. Dummit, D.S., Foote, R.M.: Abstract Algebra, 2nd edn. Prentice Hall, Upper Saddle River, NJ (1999)
18. Faddeev, D., Sominsky, I.: Problems in Higher Algebra. Mir, Moscow. Revised from the 1968 Russian edition
19. Frobenius, G.: Ueber lineare Substitutionen und bilineare Formen. J. Reine Angew Math. (Crelle's J.) **84**, 1–63 (1878)
20. Furdui, O.: Problem 5211, problems. Sch. Sci. Math. **112**(4) (2012)
21. Furdui, O.: Problem 3800. Crux Mathematicorum **38**(10), 422 (2012)

22. Furdui, O.: Limits, Series and Fractional Part Integrals. Problems in Mathematical Analysis. Springer, New York (2013)
23. Furdui, O.: Problem 5330, problems. Sch. Sci. Math. **114**(8) (2014). Available online at http:// ssma.play-cello.com/wp-content/uploads/2016/03/Dec-2014.pdf
24. Furdui, O.: Problem 96, problems. Mathproblems **4**(2), 263 (2014)
25. Furdui, O.: Problem 446. Gazeta Matematică Ser. A **33**(3–4), 43 (2015). Available online at http://ssmr.ro/gazeta/gma/2015/gma3-4-2015-continut.pdf
26. Furdui, O.: Problem 1091, problems. Coll. Math. J. **48**(1), 58 (2017)
27. Furdui, O.: Problem 5390, problems. Sch. Sci. Math. **116**(2), 2 (2016)
28. Furdui, O.: Problem 5414, problems. Sch. Sci. Math. **116**(6), 2 (2016)
29. Furdui, O.: Problem 5420, problems. Sch. Sci. Math. **116**(7), 2 (2016)
30. Furdui, O.: Problem 11917, problems and solutions. Am. Math. Mon. **123**(6) (2016)
31. Furdui, O.: Problem 1066, quickies. Math. Mag. **89**(5), 379 (2016)
32. Furdui, O., Sîntămărian, A.: Problem 3965, problems. Crux Mathematicorum **40**(7), 300 (2014)
33. Furdui, O., Sîntămărian, A.: Problem 1997, problems. Math. Mag. **89**(3), 223 (2016)
34. Gelbaum, B.R., Olmsted, J.M.H.: Theorems and Counterexamples in Mathematics. Springer, New York (1990)
35. Gowers, T., Barrow-Green, J., Leader, I. (eds.): The Princeton Companion to Mathematics. Princeton University Press, Princeton (2008)
36. Hamilton, W.R.: Lectures on Quaternions. Hodges and Smith, Dublin (1853)
37. Herzog, G.: A proof of Lie's product formula. Am. Math. Mon. **121**(3), 254–257 (2014)
38. Horn, R.A., Johnson, C.R.: Matrix Analysis, 2nd edn. Cambridge University Press, Cambridge (2013)
39. Lamé, G.: Examen des différentes méthodes employées resoundre les problémes de geom etrie. Courcier, Paris (1818)
40. Levy, M.: Problem 5330, solutions. Sch. Sci. Math. **115**(3), 13–15 (2015). Available online at http://ssma.play-cello.com/wp-content/uploads/2016/03/March-2015.pdf
41. Lupu, C., Rădulescu, V.: Problem 11532, problems and solutions. Am. Math. Mon. **117**(9), 834 (2010)
42. Markushevich, A.I.: Theory of Functions of a Complex Variable, Part I, Rev. English edn. Translated and edited by Richard A. Silverman. AMS Chelsea, Providence (2005)
43. Martin, A.: Markov's Theorem and 100 Years of the Uniqueness Theorem. A Mathematical Journey from Irrational Numbers to Perfect Matchings. Springer, New York (2013)
44. Meghaichi, H., Levy, M., Kouba, O.: Problem 96, solutions. Mathproblems **4**(3), 304–308 (2014)
45. Opincariu, M., Stroe, M.: Despre matrice şi determinanţi de ordinul doi. Gazeta Matematică Ser. B **116**(12), 559–567 (2011)
46. Pierce, S.: Solution of problem 11532. Am. Math. Mon. **119**(9), 806 (2012)
47. Pop, V.: Matematică pentru grupele de performanţă, Clasa a XI-a, Colecţia Dacia Educaţional. Editura Dacia, Cluj-Napoca (2004)
48. Pop, V.: Algebră liniară. Matrice şi determinanţi. Pentru elevi, studenţi şi concursuri, Ediţia a doua. Editura Mediamira, Cluj-Napoca (2007)
49. Pop, V., Furdui, O.: Problem 133, problems. Mathproblems **5**(3), 442 (2015)
50. Pop, V., Heuberger, D.: Matematică de excelenţă pentru concursuri, olimpiade şi centre de excelenţă, Clasa a XI-a, Vol. I, Algebră. Editura Paralela 45, Piteşti (2014)
51. Pop, V., Raşa, I.: Linear Algebra. Editura Mediamira, Cluj-Napoca (2005)
52. Ricardo, H.J.: A Modern Introduction to Differential Equations, 2nd edn. Elsevier, Amsterdam (2009)
53. Sîntămărian, A.: Probleme selectate cu şiruri de numere reale (Selected problems with sequences of real numbers) U.T. Press, Cluj-Napoca (2008)
54. Sîntămărian, A.: Some convergent sequences and series. Int. J. Pure Appl. Math. **57**(6), 885–902 (2009)
55. Sîntămărian, A.: Problem 11528, problems and solutions. Am. Math. Mon. **117**(8), 742 (2010)

56. Sîntămărian, A.: Some applications of a limit problem. Gazeta Matematică Ser. B **115**(11), 517–520 (2010)
57. Sîntămărian, A., Furdui, O.: Teme de analiză matematică. Exerciţii şi probleme, Ediţia a II-a, Revăzută şi adăugită (Topics of mathematical analysis. Exercises and problems, 2nd edn. Revised and enlarged). Presa Universitară Clujeană, Cluj-Napoca (2015)
58. Souza, P.N., Silva, J.-N.: Berkeley Problems in Mathematics, 3rd edn. Springer, New York (2004)
59. Srivastava, H.M., Choi, J.: Series Associated with the Zeta and Related Functions. Kluwer, Dordrecht (2001)
60. Wagner, S.: Problem 1, IMC 2014, Day 1, 31 July 2014, Blagoevgrad, Bulgaria. Available online at http://www.imc-math.org.uk/imc2014/IMC2014-day1-solutions.pdf
61. Whittaker, E.T., Watson, G.N.: A Course of Modern Analysis, 4th edn. Cambridge University Press, London (1927)
62. Zhang, F.: Quaternions and matrices of quaternions. Linear Algebra Appl. **251**, 21–57 (1997)
63. Zhang, F.: Matrix Theory. Basic Results and Techniques. Universitext, 2nd edn. Springer, New York (2011)

Index

A
adjugate of a matrix, 12
affine transformation, 298
alternating bilinear application, 10
Apery's constant, 231
arithmetic progression, 114
asymptotes, 317

C
Cauchy-d'Alembert's criteria, 243
characteristic equation, 63
Chebyshev polynomials, 180
circulant matrix, 110
conjugate Pell equation, 178
cyclic group, 18

D
degenerate conics, 318
diagonal matrix, 4
dihedral group, 32, 311
direct sum, 289
distinct pairwise commuting matrices, 31
double stochastic matrix, 113, 211

E
eigenvalues, 63
elementary matrix, 7
elementary transformations, 7
Euclidean space, 112
Euler's totient function, 31

F
Frobenius norm, 213, 214

G
Gamma function, 208
Gaussian integers, 32
generalized eigenvector, 78
generating function, 257

H
Hermitian adjoint, 9
homographic function, 129

J
Jordan basis, 78
Jordan canonical form, 78
Jordan cell of order 2, 23

K
Klein group, 30

L
left stochastic matrix, 212
Lie's product formula, 240
Lindemann–Weierstrass Theorem, 248
linear functional, 10
linear groups, 11
linear isometry of the plane, 296
Lucas numbers, 24

M
matrices with blocks, 12
Matrix Hamilton Quaternions, 34
modular group $SL_2(\mathbb{Z})$, 30

© Springer International Publishing AG 2017
V. Pop, O. Furdui, *Square Matrices of Order 2*, DOI 10.1007/978-3-319-54939-2

N
nilradical, 59

O
orthogonal basis, 112

P
Pauli matrices, 35
Pell resolvent, 177
Pell's equation, 174
periodic sequence, 123
permutation matrix, 7
projection of the plane, 286

Q
quadratic Frullani integral, 279

R
rational canonical form, 81
real canonical form, 80
reciprocal matrix, 12

reflection matrix, 14
rotation matrix, 18, 31

S
sine Frullani integral, 279
special linear group, 11
special orthogonal group, 31
spectral radius, 186
spectrum of A, 63, 185
stability, 204
Stirling numbers of the second kind, 227

T
the nth harmonic number, 224
the image of a linear transformation, 282
the kernel of a linear transformation, 282
transition matrix, 211
translation, 297
triangular matrix, 4

V
vector space, 4
vector subspaces, 289

Printed in the United States
By Bookmasters